Basic Structural Design

Basic Structural Design

KURT H. GERSTLE
PROFESSOR OF CIVIL ENGINEERING
UNIVERSITY OF COLORADO

McGRAW-HILL BOOK COMPANY
NEW YORK □ ST. LOUIS
SAN FRANCISCO □ TORONTO
LONDON □ SYDNEY

Basic Structural Design

Library of Congress Catalog Card Number 66-22294
23120
1234567890 MP 7321069876

Preface

The analytical portion of the structural engineer's work may be divided into two somewhat distinct parts.

First is the determination of the external and internal forces and moments (sometimes summarized as *stress resultants*). In the academic curriculum this part of the analysis is usually broached in a course in statics, carried forward through a first course in mechanics of materials, and then pursued in subsequent courses in statically determinate and indeterminate structural analysis. Thus, the determination of stress resultants is usually well covered. A number of excellent texts present the analysis of elastic structures, and some also give an introduction to the plastic analysis of structures.

The second portion of the structural engineer's analytical work may be summarized by the term *stress analysis*, as distinct from *structural analysis*. This consists of the determination of the stresses in a structural member after the stress resultants are known. In most structural curricula this aspect of the work is covered in a much less coherent and thorough manner than the first. While an introduction to the principles of stress analysis is presented in the elementary mechanics of materials course (sometimes followed up by a second course in this field), the application of these principles is largely left to courses in structural design.

Because conventional courses in structural design usually focus the student's attention on a specific structural material and on the use of building codes applicable to this one material (the usual division of both courses and textbooks being steel design, reinforced concrete design, and timber design), the rational and consistent use of basic principles of stress analysis falls largely by the wayside. A student may successfully go through several design courses without realizing that identical approaches and solutions are applicable in each case. Because of the emphasis on familiarity with code provisions and professional practice, a student often finds himself at ease only with standard formulas and graphical aids, without having much confidence either to verify existing methods or to establish new ones.

Because of the use of new materials and their combinations, the increasing attention to inelastic behavior of structures, and the growing number of structural engineers engaged in aircraft and missile work where no standardized methods exist, the student's ability to work with basic principles becomes of increasing importance. Furthermore, both the steel and concrete specifications allow the use of two different sets of rules, one based on elastic, the other on ultimate-strength, considerations. Thus an understanding of the underlying rational foundations is necessary so that the engineer may make a sound choice among these alternatives.

In order to instill in the student this basic understanding of the rational principles of stress analysis as applied to structural design, the curriculum in civil engineering at the University of Colorado offers in the last semester of the junior year, along with the first course of structural analysis, a course in basic structural design. This course begins with an outline of the basic approach common to the stress analysis of structural members, no matter what the material, loading condition, or range of stress. Using this approach (which may be called the "strength of materials" approach since it is usually associated with courses of this name), the methods of stress analysis and design, based on both elastic and inelastic considerations, are established in a unified fashion. This is accomplished by organizing the sequence of topics in terms of the type of loading on the member and the appropriate stress-strain relation, rather than speaking in terms of the different structural materials. In particular, the relation between elastic and inelastic response is established as early as possible. Finally, the course leads to rational derivations of the more important provisions of the relevant building codes.

The emphasis here, then, is on a thorough understanding of the rational bases of the rules used in structural design, rather than on a thorough familiarity with code provisions.

The approach to design followed in this text, which is intended to accompany the outlined course, is essentially the conventional one followed by structural engineers: straightforward methods of stress analysis are applied, and, if possible, the resulting formulas are inverted to solve for the required properties of the structural member being designed. In more complicated cases, trial-and-error methods are followed. It is realized that at this time more efficient methods of design are being developed. The entire field of "systems engineering" represents a highly scientific approach to engineering design, and the "optimum design methods" of structural theory which are primarily associated with limit design fall into this category. However, their development at the present time is not sufficiently far advanced to warrant their inclusion in a basic course. The same remark can be made about design methods involving the use of digital computers. These methods must be founded first of all on a sound understanding of the basic mechanics of the structural systems considered; it is therefore considered that such sophisticated methods might properly be covered in courses following the first one which is treated in this text.

The emphasis in this volume is placed on the analysis of members subjected to uniaxial states of stress; this simple theory can account for the majority of cases which occur in professional practice, and, more importantly at this stage, can convey a better structural feeling to the student than more elaborate theories. The thought here is that the remarkable power of uniaxial theory, which is able to account for the main features of the behavior of structural members—elastic and inelastic stress and deformations and ultimate strength—should be fully exploited before proceeding to higher theories.

On the other hand, for the instructor who prefers to include some multiaxial theory (here called the "continuum mechanics approach"), Sec. 2.6 introduces the principles of this theory, and Chap. 12 applies them to the analysis of torsion members. In addition, some other advanced topics are covered: plate buckling in Sec. 3.9, lateral buckling in Sec. 4.8, unsymmetric bending in Sec. 4.6; however, all these sections (which are marked with an asterisk in the Table of Contents) are so arranged that they can be deleted without interrupting the continuity.

Chapter 2, dealing with basic principles, is the heart of the presentation, and should under no conditions be deleted. This chapter contains outlines for three types of analysis: the strength of materials approach in Secs. 2.2 to 2.5, the continuum mechanics approach in Sec. 2.6, and stability analysis in Sec. 2.7. Of these, the continuum mechanics approach is used further only in Chap. 12, dealing with torsion, and if this is to be left out then Sec. 2.6 might also be deleted. The strength of materials approach is used throughout and should be covered in a more or less thorough fashion, depending on the students' previous exposure to this material. Stability analysis is necessary for an understanding of column buckling and several other buckling phenomena.

It is suggested that for a professionally oriented course the asterisked sections dealing with advanced topics be deleted; on the other hand, for an engineering science–oriented course these topics might be covered, presumably at the expense of some of the other material. It is believed that sufficient material for judicious selection is provided.

It is fully recognized that this course covers only a part of the training of the structural designer. The *art* of the profession, that is the basic conception, the decisions leading to the choice of material, configuration, and proportioning of the structure to fulfill a specific purpose, the interaction between structural and stress analysis, and other matters largely depending on the designer's judgment, are left to further courses in structural design in the student's senior year. It is in these courses that the student obtains practice in considering entire structures (rather than individual members), fluency in the use of the various building codes, and confidence in his ability to express his design on paper. The organization of the design sequence which is advocated here may also leave time for future developments associated with structural applications of systems engineering to be accommodated in these subsequent courses.

The aim of this presentation has been to outline the basic theory as clearly as possible, and to effect a smooth transition from fundamental approach to practical applications. It is hoped that this will provide a firm base for the further professional development of the structural engineer.

Appreciation is expressed to the University of Colorado and to SEATO Graduate School of Engineering, Bangkok, Thailand, for secretarial assistance, and to Mrs. Jane Blackburn, Mrs. Ann Clark, and Mrs. Florence Petersen for capable typing. Thanks are also due several of the author's colleagues and students for fruitful discussion, and above all to his wife for unfailing patience and encouragement.

Kurt H. Gerstle

Contents

part one

Principles of Structural Design

1

Design of Structures

1.1 Design Criteria

From the engineer's viewpoint, there are two main criteria for the adequacy of a structure which must be satisfied in the most economical fashion: safety, that is, sufficient strength to resist applied loads without danger of collapse, and serviceability, that is, the ability to carry these loads without excessive deformations or local distress.

Safety of structures The strength requirement may be satisfied in several ways: In the so-called "conventional" or "elastic" design, the strength is checked by ensuring that no point of the structure is stressed above a value called the *allowable* or *working stress*. The choice of an appropriate allowable stress will be discussed in a later section.

There are several questionable points about the use of allowable stress as criterion of strength. In the first place, in any structure there are innumerable regions of stress concentration in which it becomes impractical, or even impossible, to analyze the stresses; in such cases, we often tend just to overlook the stress peak and let the material fend for itself; this is particularly true in the vicinity of connections.

In the second place, in many structures there is no obvious reason why the attainment of a critical stress at some point of the structure should necessarily coincide with the exhaustion of the safe carrying capacity of the structure. If the material is a ductile one, a phenomenon called *redistribution of stresses* occurs which will tend to relieve overstressed portions of the structure, and throw a fair share of the load to the understressed portions.

An example of this type of action may be illustrated by considering the action of a riveted lap joint under load, as shown in Fig. 1.1. If owing to faulty fabrication two out of the three rivets are loose, one rivet is called on to transmit all of the load; because of this overstress (which according to the elastic theory will cause failure of the joint), yielding will occur in the rivet with consequent deformation which enables the two other rivets to participate in the transmission of the load. Before the joint can fail, all three rivets are forced to share the load.

The preceding example indicates the fallacies underlying elastic design of ductile structures; because of manufacturing errors (some of them unavoidable) and other irregularities, the actual stress distribution is far different from that obtained by elastic analysis, and prior to failure of the joint, plastic action has shifted part of the load to the understressed parts.

In spite of the above critique it should be appreciated that the allowable-stress criterion of strength has served in the past, and will continue to serve, a useful purpose. If well done, it will always yield a safe structure, though not always the most economical; and in the case of structures of brittle material, or those exposed to dynamic or fatigue loads, it is the only safe method of design.

In the so-called method of "limit" design, the total load is investigated under which the structure will fail. The safety of the structure is in this case controlled by the choice of a *load factor*, defined as the ratio of the collapse load to the working load.

The method of limit analysis is the classical approach to the investigation of structures subject to buckling; in such cases, a *critical load* is calculated under which collapse will take place. More recently, an analogous approach has been taken to calculate the total applied load under which a structure of ductile material will collapse as a so-called "mechanism." This "plastic" analysis has been widely used in the design of steel structures. Because of its simplicity (the previously covered example of a riveted joint was an example of this approach), and because it enables the analyst to predict conditions of failure quite closely, it will probably continue to gain ground. In this approach, attention is shifted from stress to the *collapse load*, and, as the example of the riveted joint indicated, some local sections of the structure may even be exposed to plastic stresses under working loads.

Here, we shall focus our attention on both elastic and plastic methods of design. Only by acquaintance with both approaches will the analyst be enabled to make a sound decision as to the suitability of one method or the other for any particular case.

Serviceability of structures The investigation of deformations is usually considered a secondary problem, to be checked after the structure is designed.

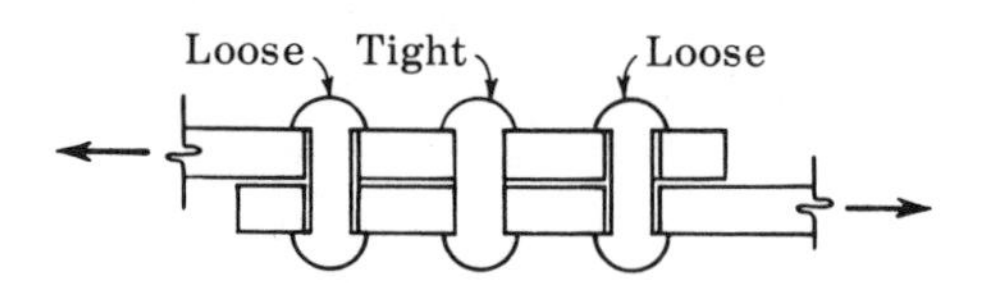

(*a*) Elastic assumption: One rivet carries entire load

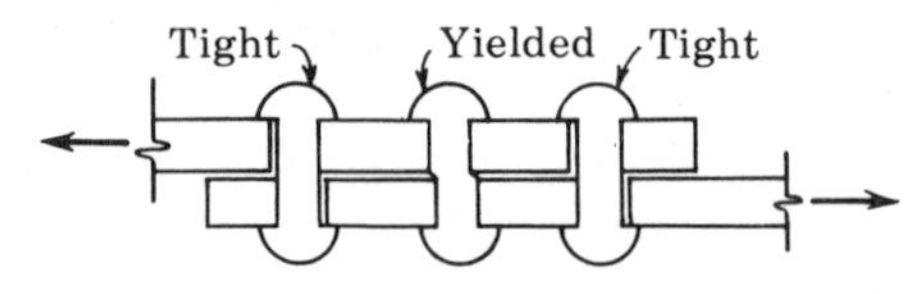

(*b*) Actual behavior of riveted joint: Overloaded rivet yields, thereby enabling other rivets to participate

Fig. 1.1 Behavior of riveted joint

Nevertheless, the designer ought to keep in mind the necessity of avoiding excessive deflections, or vibration, in selecting members and connections. While methods for calculating inelastic deformations are available (see Sec. 7.7), even structures designed according to plastic methods will generally act elastically under working loads, and the elastic deflection analyses in texts on structural theory are usually appropriate to check structural stiffness.

Another aspect of serviceability is the necessity to avoid local concentrations of deformations which may impair the appearance or usefulness of the structure; an example of this is excessive cracking of reinforced concrete structures; even though such cracking may not be fatal, it is undesirable and can be avoided by proper arrangement of details.

1.2 Safety Factors

The question of an appropriate value for the allowable stress or load factor to be used in the design of members should depend on several factors:

1. The strength of the material to be used. As a rule, the allowable stress in elastic design is taken at some fraction of the proportional limit or the yield stress. Thus, for structural steel with a minimum specification yield point f_{YP} (according to ASTM Code A 7-61[1]) the AISC Specifications[2] (sec. 1.5.1) permit an allowable stress of 0.6 f_{YP} in tension members, thus providing a safety factor of 1.67 against yielding of the section. For other materials allowable stresses are provided by corresponding codes, or in the absence of such the choice may be at the judgment of the engineer.
2. The methods of analysis and design used. Generally, the more exacting the engineering determination of actual conditions under possible loadings, the higher the allowable stresses may be taken. In this connection, the safety factor may be considered a "factor of ignorance"; if the safety factor to be used is at the discretion of the designer, it may well be that a thorough analysis considering all factors, though lengthy, is a wise investment, resulting in a light and elegant yet economical structure.
3. The predictability of loads. In some structures, such as fluid containers, it may be a physical impossibility to apply more than the design load; in others, unexpected loads or overloads may be routine, and provision must be made for their occurrence. In bridges it may be wise to anticipate future increase in traffic loads; the fact that many

[1] American Society for Testing and Materials, "Selected A.S.T.M. Standards for Civil Engineering Students," ASTM, 1916 Race St., Philadelphia, Pa., 1965.

[2] American Institute of Steel Construction, "Specification for the Design, Fabrication, and Erection of Structural Steel for Buildings," 1963 edition.

railway bridges of venerable age are still in use may be ascribed largely to generous safety factors provided by designers.

4. The seriousness of failure. The type of failure and its consequences should be well considered.
5. Possibility of deterioration of the structure with time. Corrosion of metals, spalling and weakening of concrete, rotting of timber may seriously impair the carrying capacity of a structure; if such conditions are anticipated, the structure may be designed for a definite service life.

In most cases faced by the professional engineer, applicable codes prescribe safety factors or allowable stresses; however, in the case of unusual structures or those in locations where no codes are provided, the designer should consider the above factors in selecting allowable stresses; recently, some elaborate studies have used statistical and probabilistic methods to arrive rationally at suitable values for the safety factor, considering a number of the above factors as variables.[1,2]

1.3 Statically Determinate and Indeterminate Structures

Statically determinate structures are those which may be analyzed for internal forces and moments by use of conditions of statics alone; accordingly, they are convenient from the designer's viewpoint. However, this convenience is paid for with a high price in the final economy and elegance of the structure as compared to a statically indeterminate one. Generally, statically indeterminate structures have vastly superior strength and stiffness properties, and a more even distribution of internal forces results in reduction of member size compared to the statically determinate form. Besides, the nature of modern building methods in reinforced concrete and welded steel favors the construction of statically indeterminate structures with continuous beams and rigid joints.

The dependence of the distribution of the internal forces of an indeterminate structure on the member stiffness means that the design of such a structure must be a trial-and-error process. Once the general configuration of the structure is laid out, most likely member sizes are assumed (possibly on basis of approximate methods of analysis) and the structure analyzed accordingly. Once the member forces and moments are obtained, members are designed to resist these forces; if their stiffness and weights turn out to coincide with those of the members previously assumed, the design is satisfactory. If not, then the structure must be reanalyzed on basis of the new member stiffness and this process continued to convergence. Design of

[1] A. M. Freudenthal, Safety and the Probability of Structural Failure, *Trans. ASCE*, vol. 121, p. 1337, 1956.

[2] S. O. Asplund, Probabilities of Traffic Loads on Bridges, *Proc. Struct. Div. ASCE*, vol. 81, January, 1955.

statically indeterminate structures is thus a process of successive approximation, in which the rapidity of convergence depends to a large extent on the experience of the designer. Recently, the use of high-speed computers has served to increase the speed with which such a design can be accomplished and to eliminate much of the human factor in this process.

In recent years, so-called "optimum design methods"[1] are being introduced which aim directly at the creation of a structure of the required properties which at the same time satisfies a certain efficiency criterion, such as minimum weight. New mathematical methods, such as the techniques of "mathematical programming" and computers, are called upon to achieve this. This approach shows great promise, but presently is considered insufficiently far advanced to be covered in an elementary course.

1.4 Codes and Design Aids

In professional practice it is customary for the designer to follow codes of practice; the need for such codes arises for several reasons. Local building authorities generally insist on certain minimum requirements to ensure the structural safety of buildings and bridges. These minimum requirements are usually laid down in building codes. The designer who works in a variety of different structures and materials generally must refer to an authoritative source to determine reasonable allowable stresses for structural materials as well as load factors for different types of loads. Codes are sources of such information.

Typical of such codes are the "AISC Specifications,"[2] the most commonly followed set of design rules for structural steel buildings, the "AASHO Code"[3] for design of highway bridges of different materials, the "ACI Building Code"[4] for reinforced concrete structures, and the "National Design Specifications for Stress Grade Lumber"[5] for timber structures. Copies of relevant sections of these rules are frequently referred to in the text. Most of the locally used building codes in this country base their requirements on the appropriate codes enumerated above, with changes or additions to suit local conditions. Generally, familiarity with one of these codes will enable the designer to readily follow any other.

[1] F. Moses, Optimum Structural Design Using Linear Programming, *Jnl. Struct. Div., Proc. ASCE*, vol. 90, p. 89, December, 1964.

[2] American Institute of Steel Construction, "Specification for the Design, Fabrication, and Erection of Structural Steel for Buildings," AISC, 101 Park Avenue, New York, 1963.

[3] American Association of State Highway Officials, "Standard Specifications for Highway Bridges," 8th ed., AASHO, 917 National Press Building, Washington 4, D.C., 1961.

[4] American Concrete Institute, "Building Code Requirements for Reinforced Concrete (ACI 318-63)," ACI, P.O. Box 4754, Redford Station, Detroit 19, Mich., 1963.

[5] National Lumber Manufacturers Association, "National Design Specifications for Stress Grade Lumber," NLMA, Washington 6, D.C., 1962.

Code provisions generally originate from three different sources of knowledge:

1. Rational theory: A logical approach resulting in conclusions which can be verified experimentally is considered the most satisfactory method, since it can not only predict behavior under certain conditions quantitatively, but also give reasons for this behavior. This understanding enables predictions to be made for a wide variety of conditions, and lets the analyst apply the theory to new uses as the need for them arises.

   Sometimes unrealistic assumptions, or a break in the chain of logic, result in conclusions which cannot be backed up experimentally; if, nevertheless, theoretical results are consistently on the safe side, they may find their way into professional practice. Meanwhile, it is hoped that steady progress is being made by researchers to reconcile theory and experiment.
2. Experimental evidence: In cases too complicated to analyze theoretically, experiments or field measurements must be used. Such empirical data can be plotted to isolate the effect of different variables, and analytical representations of such plots find their way into codes. Because a thorough understanding of the behavior represented by such data is lacking, they should be used with caution, making sure not to extrapolate them beyond experimental conditions in an unwarranted fashion.
3. Professional practice: A tremendous portion of our total fund of engineering knowledge is based on our professional heritage; that part of our engineering practice which through the years has been shown to produce good results tends to be carried over into current practice. This portion represents the "art" of our profession as compared to the "science" incorporated in the preceding two points, and as such is to be highly valued. In modern structures, conditions of such complexity arise that even a relatively sophisticated theory, which must nevertheless be based on a series of simplifying assumptions, could hardly do justice to the situation; in such cases, practices which have proved their worth through the years are often adhered to. Nevertheless, with new types of structures under loading conditions never envisioned in traditional practice, the trend is steadily in the direction of rational explanation of structural behavior.

Design codes are intended to be of help to the engineer. It is not intended that he blindly follow the provisions; as a matter of fact, unless the designer has a good understanding of the whys and wherefores of the various rules, it is practically impossible for him to apply them correctly and with discretion. For this reason, an understanding of their underlying background is essential, and it is one of the purposes of this treatise to point out the origins of code provisions; in particular, attention will be directed at their correlation with rational theory.

There is a real danger that too strict reliance on building codes may result in stagnation of engineering progress, and no code is intended to stand in the way of innovation. It should be understood that no engineer worthy of the name has to be a slave to the code—every specification makes provision for designs not covered by its provisions. So, for instance, sec. 104 of the ACI Code states in part: "The sponsors of any system of design or construction . . . shall have the right to present the data on which their design is based to a Board . . . appointed by the Building Official. . . ." The introduction of prestressed concrete, for instance, or the magnificent thin-shell structures which are rising all over the country bear testimony that competent hands and minds need not be hobbled by building regulations. It goes without saying that such ventures can be successfully accomplished only by men in full command of the theory of structures.

There are also many areas of structural engineering for which no codes exist; among these the aerospace industry is predominant. To perform an efficient design in these fields, intimate knowledge of the material behavior under various environmental conditions and excellent understanding of the basic theory are necessary.

Design aids A considerable portion of the technical literature concerns itself with design aids—graphic or tabular compilations of numerical results which are of such routine nature that, once computed, they need not be repeated each time they are needed. A thorough knowledge of such available data is indispensable for the efficient designer, and their use will enable him to perform in a fraction of the time otherwise required. Among such design aids are the tables giving cross-sectional properties of rolled steel sections, interaction curves showing the strength of members under combined loads, moment coefficients as given in the AISC and CRSI[1] Handbooks, the extensive tables and graphs giving strength properties of reinforced concrete members in the CRSI and RCDH[2] Manuals. Again, the conscientious engineer will avoid the danger of using such information without a firm understanding of its theoretical basis.

While such design aids will be referred to frequently in this book, and while the reader is encouraged to acquaint himself with them and to correlate them with both theory and specifications, they will not be unduly emphasized here, beyond a short discussion here and there. A thorough understanding of structural mechanics will enable the engineer both to draw up and to use design aids to help him with his structural design problems.

[1] Concrete Reinforcing Steel Institute, "Design Handbook," revised ed., CRSI, 228 N. LaSalle St., Chicago, Ill., 1966.

[2] American Concrete Institute, "Reinforced Concrete Design Handbook," 3d ed., 1965.

1.5 Structural Materials

To build economical structures use is made of commercially available materials; in this section we shall acquaint ourselves with the usual components of structures of steel, aluminum, reinforced concrete, and timber. To draw up plans which are economically feasible it is necessary for the designer to be familiar with the nomenclature, dimensioning, and mechanical characteristics of commonly used elements. We shall shortly discuss those in connection with the different types of construction.

Steel structures The basic components of steel structures are rolled shapes of structural steel; the standard cross sections are listed in the tables of the handbook of the AISC, together with their cross-sectional properties. For design purposes, it is generally desired to find the lightest-weight member which satisfies the design requirements on strength and stiffness, and these tables allow one to do this efficiently. The most widely used shapes are the so-called wide-flange beams of I cross section, angles, channel sections, rods, and flat plates which are used in connections and welded together to form built-up shapes such as plate girders.

The most commonly used structural steels conform to ASTM Specifications A-7 and A-373, which specify a minimum yield point of 33 ksi on which allowable stress should be based. ASTM A-36 steel of minimum yield point 36 ksi is also frequently used, and steels of higher strength are available for special conditions where savings in dead weight are important, such as long-span bridges. Minimum ductility and ultimate tensile strength are also specified. Allowable stresses for these steels are specified in AISC sec. 1.5 if used for buildings, and in AASHO sec. 1.4 if used for highway bridges.

If the plastic-design method is used for steel structures, part 2 of the 1963 AISC Code should be consulted for load factors and other design criteria.

Steel members are joined by riveting, bolting, or welding, and rules for the design of such connections are also laid down in the above sources.

Aluminum structures Structural elements of aluminum are somewhat similar to those of structural steel, but because aluminum can be shaped by extrusion through dies, it can be obtained in more complicated and diversified shapes. Commercially available standard shapes and their section properties are listed in the "Aluminum Construction Manual."[1] Many different aluminum alloys with widely differing mechanical properties are available, and considerable knowledge on the part of the designer is necessary to specify the most suitable material for a particular purpose.

[1] The Aluminum Association, "Aluminum Construction Manual," 1st ed., The Aluminum Association, 420 Lexington Avenue, New York 17, 1959.

Reinforced concrete structures Reinforced concrete structures are combinations of two structural materials, plain concrete and steel reinforcing bars.

Most characteristics of plain concrete—strength in axial and shear stress, stiffness, durability—are functions of its compressive strength f'_c, and only this property is generally specified. The strength of newly placed concrete increases with time, reaching about 90 percent of its final strength at an age of 28 days, and it is this 28-day strength which is stipulated. Standard test cylinders are cast along with the structure, cured under similar conditions, and expected to conform to the specified strength requirement at an age of 28 days. Because reinforced concrete is generally constructed under field conditions, it is wise to expect considerable variance from optimum placement and curing conditions, and allowable stresses, as for instance specified in the ACI Code, chap. 10, reflect a considerable safety factor. Concrete strength is controlled by the water-cement ratio, and commonly used 28-day strengths are from 3,000 to 5,000 psi.

Reinforcing bars are usually rolled of intermediate-grade, hard-grade, or high-strength steels, of minimum yield stress 40, 50, and 60 ksi respectively. It is wise to remember that the higher strengths are obtained at the expense of ductility, and therefore in structures which may be called on to absorb large amounts of energy—due to earthquakes or blast, for instance—it may be prudent to specify a lower-grade steel.

Allowable stresses in reinforcing bars are also specified in the ACI Code, sec. 1003.

Reinforcing bars are rolled in standard sizes designated by the number of eighths of an inch contained in their diameter: a No. 4 bar, for instance, has a diameter of $\frac{4}{8}$ in. $= \frac{1}{2}$ in. Above a No. 8 bar (diameter $= 1$ in.), this numbering system is invalid, and it is therefore better to refer to the table showing cross-sectional properties.

Table of reinforcing bars

No.	*Diameter, in.*	*Area, in.*^{2}	*Perimeter, in.*
2	0.250	0.05	0.79
3	0.375	0.11	1.18
4	0.500	0.20	1.57
5	0.625	0.31	1.96
6	0.750	0.44	2.36
7	0.875	0.60	2.75
8	1.000	0.79	3.14
9	1.128	1.00	3.54
10	1.270	1.27	3.99
11	1.410	1.56	4.43
14S	1.693	2.25	5.32
18S	2.257	4.00	7.09

Bars above No. 2 come with a ribbed surface to ensure bond between the steel and the surrounding concrete.

For ultimate-strength design of concrete, in which the failure load on the member rather than the attainment of allowable stresses forms the criterion of safety, part IVB of the ACI Code gives load factors and other design criteria.

Timber structures Allowable stresses and stiffness of members of different kinds of wood are specified in "National Design Specifications for Stress Grade Lumber," along with the allowed capacities of different type of connections.

Structural timber comes in actual sizes which are less than the nominal dimensions by which it is specified, due to the amount of lumber removed in planing the piece to a finished surface. Thus, a 3 by 12 in. nominal size member is actually $2\frac{5}{8}$ by $11\frac{1}{2}$ in., and these latter values should be used to compute its engineering properties. Tables are available which give cross-sectional properties based on the actual dimension.

It is suggested the reader peruse the publications mentioned above and acquaint himself with their organization and contents, since we shall frequently refer to them in our further discussions.

Other materials Modern technology demands and produces ever new materials to serve different purposes. Plastics are finding their way into building construction; exotic metals for light weight and ceramic materials for heat resistance are finding use in the aerospace industry; and composite constructions such as fiberglas-wound pressure vessels have certain advantages over more conventional types. Generally no rules or codes are available for such materials, and the engineer may therefore have to use his own knowledge and judgment; indeed, the formulation of specifications to be followed by others in such cases may be in his hands.

2

Principles of Structural Mechanics

2.1 Basic Approaches in Structural Mechanics

The extensive literature of structural mechanics reveals a large number of problems and methods for their solution. To obtain a coherent understanding of this field, as well as to extend our knowledge to still unexplored or only partly solved problem areas, it is important to have a good grasp of the basic principles involved. Fortunately these fundamentals are surprisingly simple and few in number.

To grasp the extent of the field of structural mechanics, reference is made to the table on page 14, which breaks this field down into subtopics. Three major headings are evident: structural analysis, stress analysis, and ultimate-strength analysis.

The field of *structural* analysis is concerned with the determination of stress resultants (that is, total forces and moments). These notes do not concern themselves with this class of problem, though a knowledge of statically determinate analysis is necessary, and of statically indeterminate analysis is desirable, for the study of these notes.

The field of *stress* analysis seeks to find the internal stresses, strains, and deformations in solid bodies under given conditions of loading, constraint, or temperature.

When uniaxial theory is applicable, that is, any fiber is stressed in one direction only, the "strength of materials" approach may be used, and the main emphasis here will be on the application of this method to the stress analysis and design of structural members. Whereas the classical treatment of this topic is restricted to elastic conditions, here an attempt is made at a unified treatment of elastic and inelastic stress analysis. Secs. 2.2 to 2.5 will be concerned with a detailed outline of this approach.

In more complicated cases of stress analysis, a two- or three-dimensional state of stress must be considered, and then the "continuum mechanics" method is called for. The classical theories of elasticity and inelasticity, of plates, and of shells fall into this category; this approach is outlined in Sec. 2.6, and some structural applications are indicated in Chap. 12. Lastly, the "ultimate strength" analysis is used to determine conditions of collapse of structures, and such failures can be attributed to two main causes: buckling or instability of the structure (the analysis for such a condition is discussed in Sec. 2.7 and applied in Secs. 3.6 and 4.6) and collapse of perfectly plastic

Outline of structural mechanics

Structural analysis *Determination of stress resultants*			*Stress analysis* *Determination of internal stresses*		*Ultimate-strength analysis* *Determination of collapse loads*	
Statically determinate structures	Statically indeterminate structures (More unknowns than equations of statics)		Uniaxial stress states (Most structural members)	Multiaxial stress states (Theory of elasticity and plasticity; theory of plates and shells)	Collapse load of ductile structures	Buckling of slender members
Statics only	Force methods (Forces are unknowns)	Deformation methods (Displacements are unknowns)	Strength of materials method	Continuum mechanics method	Limit or plastic analysis	Buckling or stability analysis
	Consistent deformation (Maxwell-Mohr) Min. compl. energy Matrix flexibility method	Slope deflection Moment distribution Min. potent. energy Matrix stiffness method	1. Finite free body 2. Equilibrium (in terms of stresses) 3. *Assume* deformations (in terms of strains) 4. Stress-strain relations (elastic or inelastic)	1. Infinitesimally small free body 2. Diff. eq. of equilibrium 3. Compatibility (strain-displacement relations) 4. Stress-strain relations (elastic or plastic) 5. Boundary conditions	Upper and lower bound theorems; Mechanism method	1. Equilibrium of buckled shape 2. Force-deformation relation 3. Eigenvalue solution of homogeneous equations

structures under a load large enough so that sufficient portions of the structure yield to enable it to deform excessively (this aspect is covered in Secs. 7.3 to 7.6 under the heading of "limit analysis").

One important aspect which is not covered here is the dynamic response of structures. This is important if effects of blast, earthquake, aerodynamic loading, or vibrations must be considered. All of the different headings which have been mentioned under the above outline have their counterparts in the field of dynamics.

2.2 The Strength of Materials Method—Equilibrium

The strength of materials method makes use of three basic tools of mechanics: equilibrium, geometry of the deformed structure, and the stress-strain relations. The first of these will be considered in this section, the others in the following sections.

For any body in equilibrium,

$$\bar{F} = 0 \tag{2.1}$$

and $$\bar{M} = 0 \tag{2.2}$$

where $\bar{F}$ and $\bar{M}$ are, respectively, the resultant force vector and the resultant moment vector on the body. If a structure or structural member is at rest, then any portion of it, isolated as a free body, must satisfy Eqs. (2.1) and (2.2).

A commonly occurring force system will be the parallel noncoplanar force system shown in Fig. 2.1; for such a force system, the study of statics tells us that we have three independent equilibrium conditions:

$$\Sigma F_z = 0 \qquad \Sigma M_x = 0 \qquad \Sigma M_y = 0$$

For the important special case in which the forces are distributed symmetrically with respect to one axis (say the Y axis), the moment-equilibrium condition about this axis is always satisfied by symmetry, and we are left

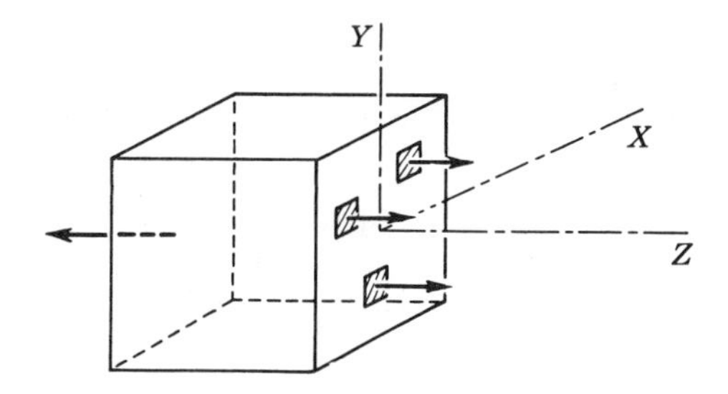

Fig. 2.1 Parallel force system

with the equilibrium conditions of a coplanar parallel force system in the YZ plane:

$$\Sigma F_z = 0 \tag{2.3}$$
$$\Sigma M_x = 0 \tag{2.4}$$

We shall frequently consider a cut plane of a free body with stresses acting on it; consider the cut section of the body shown in Fig. 2.2 under the action of the normal stresses of intensity σ lb force per unit area.

A typical small area element of size dA will have acting on it a small force

$$dF = \sigma \cdot dA$$

The small moment of this small force about the X axis is obtained by multiplying force by appropriate lever arm y:

$$dM = dF \cdot y = \sigma y \, dA$$

The total force of the stresses acting on the cut area will be obtained by adding all the small forces acting on the area:

$$F_z = \int_A dF = \int_A \sigma \, dA \tag{2.5}$$

and its total moment about the X axis, similarly, by

$$M_x = \int_A dM = \int_A \sigma y \, dA \tag{2.6}$$

We now consider a surface superimposed on the cut plane, the distance of which from the plane at every point represents the intensity of stress σ. We recognize that the volume of this so-called "stress block," with due regard to sign, is

$$\text{Volume} = \int_A \sigma \, dA = F$$

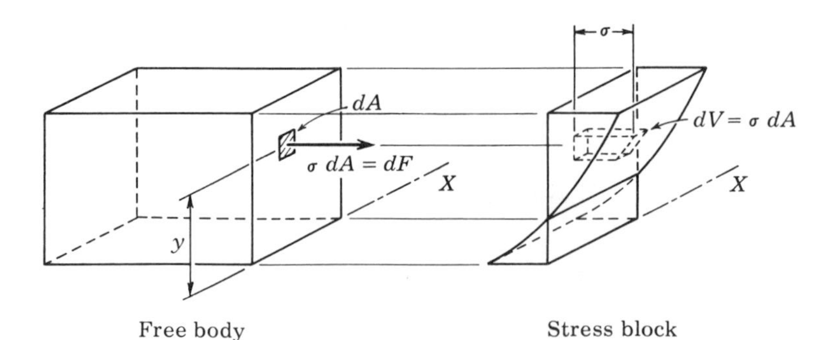

Fig. 2.2

according to Eq. (2.5); likewise, the moment of the volume about the X axis is given by

$$M = \int_A \sigma y \, dA$$

equal to the static moment of the stresses. We can thus represent the resultant effects of a set of stresses by means of the stress block; in some cases, it is convenient to express conditions of statics in terms of the resultant of the stress block.

Example Problem 2.1 A rectangular area as shown is subjected to normal stresses varying according to the expression $\sigma = Cy$.

(*a*) Find the total force acting on the area, and its moments about the X and Y axes, by integration of stresses.
(*b*) Draw the stress block, and, by making use of its geometric properties, find the total force and moments.

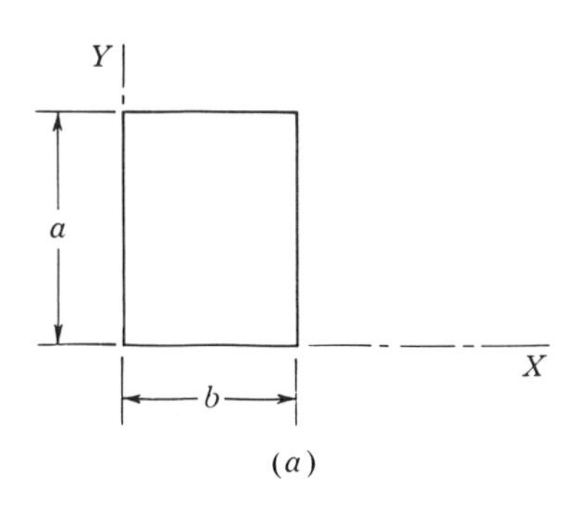

(*a*)

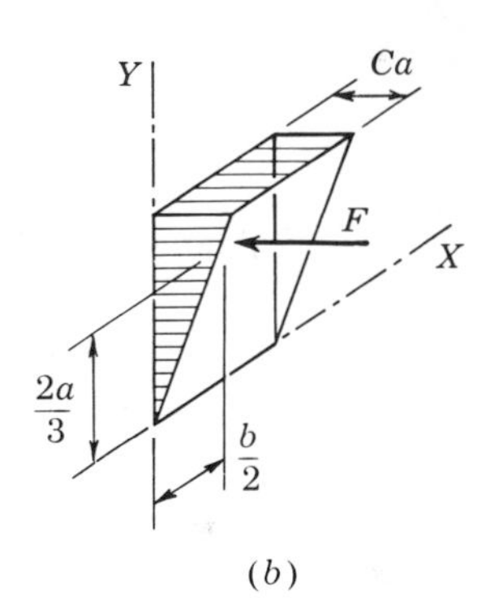

(*b*)

Solution:

$$(a)\quad F = \int_A \sigma \, dA = \int_{y=0}^{a} \int_{x=0}^{b} Cy \, dx \, dy = C \frac{y^2}{2}\Big|_0^a x \Big|_0^b = \underline{\frac{C}{2} a^2 b}$$

$$M_x = \int_A \sigma y \, dA = \int_{y=0}^{a} \int_{x=0}^{b} Cy^2 \, dx \, dy = \underline{\frac{C}{3} a^3 b}$$

$$M_y = \int_A \sigma x \, dA = \int_{y=0}^{a} \int_{x=0}^{b} Cyx \, dx \, dy \, \underline{\frac{C}{4} a^2 b^2}$$

(*b*) See figure above

$$F = \tfrac{1}{2}(ab)(Ca) = \underline{\frac{C}{2} a^2 b}$$

$$M_x = F \frac{2a}{3} = \underline{\frac{C}{3} a^3 b}$$

$$M_y = F \frac{b}{2} = \underline{\frac{C}{4} a^2 b^2}$$

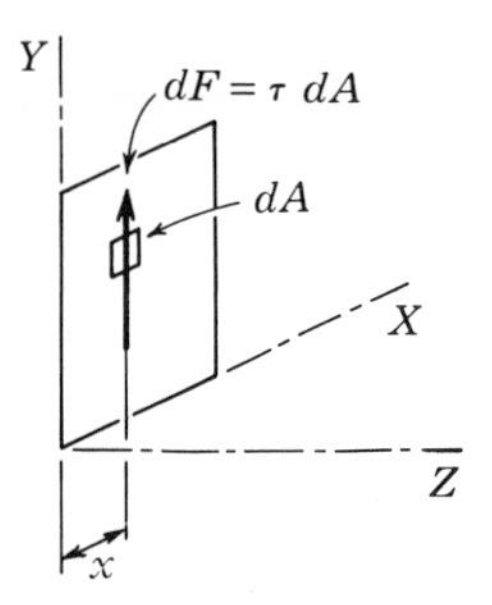

Fig. 2.3 Shear stresses on section

The identical procedure can be used to evaluate the resultant of shearing stresses; as before, the lever arm used to find the moment of the stresses is to be taken normal to the force; twisting in this case occurs about an axis normal to the plane on which the stresses are acting. By integration, and referring to Fig. 2.3,

$$F_y = \int_A \tau \, dA \tag{2.7}$$

and
$$M_z = \int_A \tau x \, dA \tag{2.8}$$

Having found the resultant of the stresses as indicated, it may be substituted into Eqs. (2.3) and (2.4) to ensure equilibrium of the free body.

Example Problem 2.2 A rectangular area as shown is subjected to shear stresses parallel to the Y axis, varying according to

$$\tau = \tau_{max}\left[1 - 4\left(\frac{y}{d}\right)^2\right]$$

(*a*) By integration, find the resultant shear force on the area.
(*b*) Draw the stress block and verify result of part (*a*).
(*c*) If the total applied shear force is V, find the value of τ_{max}.

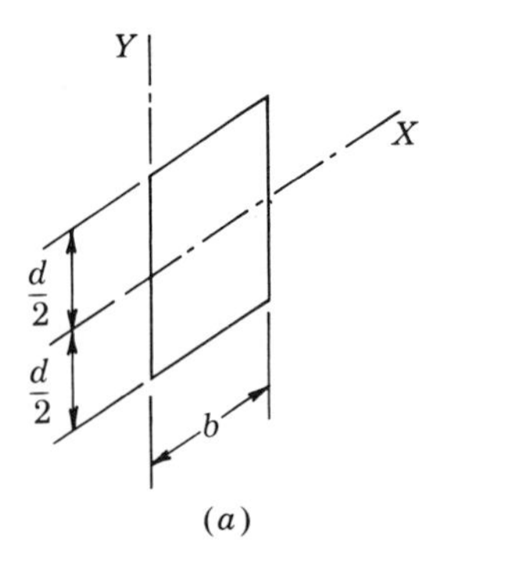

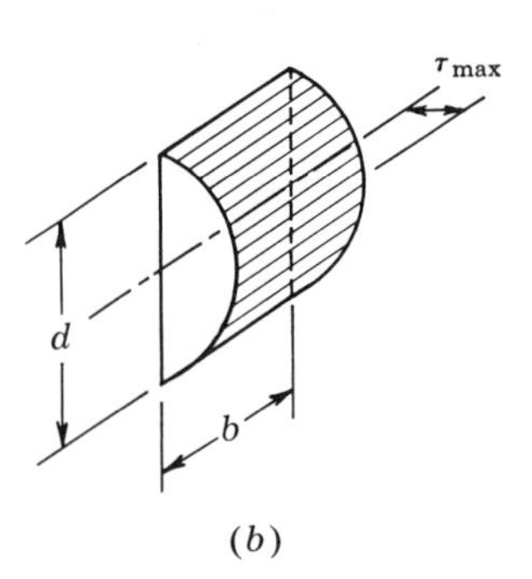

Solution:

$$(a)\quad \text{Shear force} = \int_A \tau\, dA = \int_{x=0}^{b} \int_{y=-\frac{d}{2}}^{\frac{d}{2}} \tau_{max} \left[1 - 4\left(\frac{y}{d}\right)^2\right] dx\, dy$$

$$= \tau_{max} b \left[y - \frac{4}{d^2}\frac{y^3}{3} \right] \Bigg|_{-\frac{d}{2}}^{\frac{d}{2}}$$

$$= \tau_{max} b[d - \tfrac{1}{3}d] = \underline{\tfrac{2}{3} bd\tau_{max}}$$

(*b*) Variation of τ is parabolic. Stress block is as shown in figure above.

$$\text{Volume of stress block} = \text{shear force}$$
$$= b\tfrac{2}{3}\tau_{max} d = \underline{\tfrac{2}{3} bd\tau_{max}}$$

$$(c)\quad \Sigma F_y = 0: \qquad \int_A \tau\, dA - V = 0$$

$$\text{or}\quad \tfrac{2}{3} bd\tau_{max} - V = 0 \qquad \tau_{max} = \frac{3}{2}\frac{V}{bd} = \underline{\frac{3}{2}\frac{V}{A}}$$

2.3 The Strength of Materials Method—Geometry

The second major tool in structural mechanics is the geometry of the deformed structure. Quantitatively the deformations may be expressed by the normal, or axial, strains ϵ and shearing strains γ, or else, by the displacement components at any point in the body in three orthogonal directions.

For our purposes, considerations of normal strains will in most cases be sufficient; in order to obtain a complete stress analysis, it is necessary to make some statement about the variation of strains.

For more advanced analyses, it is desirable to express the deformations in terms of displacements of any point; then, relationships between displacements and strains have to be obtained, resulting in so-called "compatibility equations." This approach will be used in Sec. 2.6.

Normal strain is defined as the change in length between two points originally a small distance unity apart; *shear strain*, as the change of angle, in radians, between two axes originally perpendicular to each other. We shall restrict our considerations to small strains, that is, of such magnitude that the deformed shape of the strained body varies from the original shape by very little. For most structural applications this is sufficient, since large deformations are usually inadmissible during the useful life of a structure.

The total change of length of a line between two points A and B is given by the sum of the changes of length of the elements of the original line; referring to Fig. 2.4, the short element ds subjected to strain ϵ will

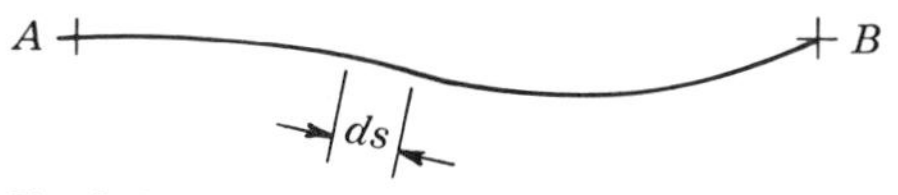

Fig. 2.4

deform a distance $\epsilon\, ds$; the total deformation between A and B will be the sum of the individual small deformations:

$$\Delta = \int_A^B \epsilon\, ds \tag{2.9}$$

In the important case where the strain is constant between A and B,

$$\Delta = \epsilon \int_A^B ds = \epsilon L \tag{2.10}$$

where L is the gage length between A and B. If the gage length is chosen very short, the strain may be assumed at a constant average value ϵ, and Eq. (2.10) applies.

Example Problem 2.3 Two parallel reference lines, a short distance dx apart, are scribed on the surface of a structural member before deformation. The displacement u parallel to the X axis during deformation follows the variation

$$u = Cxy$$

(*a*) Sketch the deformed position of the reference lines, and give the distance between lines and the strain ϵ_x.
(*b*) Calculate the angle between the reference lines after deformation.
(*c*) Find the shear strain between the X axis and the line $x = dx$ corresponding to this strain.

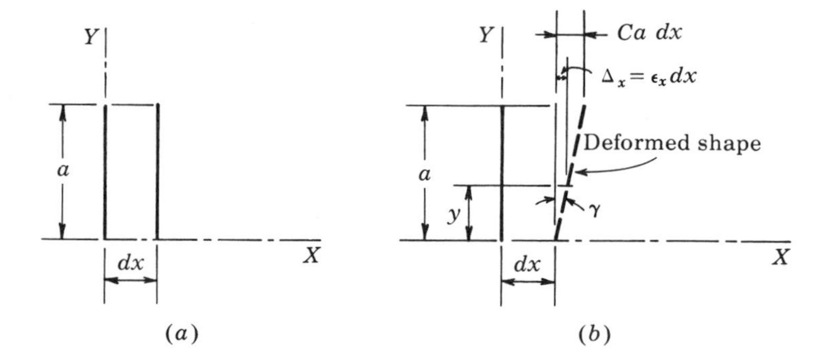

Solution:
(*a*) The displacement parallel to the X axis of any point on the line $x = dx$ is

$$u = \underline{Cy\, dx}$$

C and dx are constant; therefore the line will remain straight but become inclined as shown in the figure above. The distance between the lines at $y = a$ will be

$$dx + Ca\, dx = \underline{(1 + Ca)\, dx}$$

the strain

$$\epsilon_x = \frac{ds}{S} = \frac{(dx + Cy\, dx) - dx}{dx} = \underline{Cy}$$

(*b*) The angle γ between the two lines will be given by

$$\gamma = \tan^{-1}\frac{Cy\,dx}{y} = \tan^{-1} C\,dx$$

For small displacements the tangent can be set equal to the angle in radians, and

$$\underline{\gamma = C\,dx}$$

(*c*) The shear strain is defined as the change of angle between two lines originally at right angles. This change of angle is given by γ, and accordingly the shear strain is given by

$$\underline{\text{Shear strain} = \gamma = C\,dx}$$

Example Problem 2.4 Two reference lines are scribed on the surface of a structural member as shown, at an infinitesimal distance dx apart, and inclined with respect to each other by a small angle $d\theta$.

Under load, one reference line inclines with respect to the other by an additional small angle ϕ, as shown in the figure.

Find the variation of the strain ϵ_x with y.

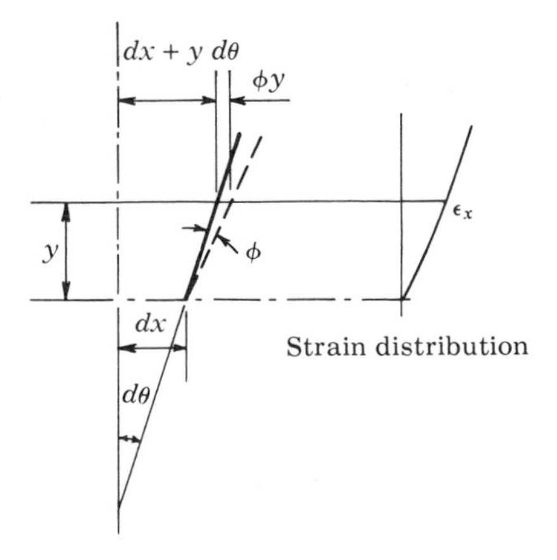

Strain distribution

Solution: The original distance (parallel to the X axis) of two points is, according to the geometry of the figure, given by

$$dx + y\,d\theta$$

The change of this distance due to the given load is ϕy. The strain is then, by definition, the change of distance divided by the original distance, or

$$\epsilon_x = \frac{\phi y}{dx + y d\theta} = \frac{1}{(dx/y) + d\theta}\,\phi$$

This indicates a hyperbolic variation with y.

2.4 The Strength of Materials Method—Stress-strain Relations

The deformation of a material under uniaxial load is characterized by the stress-strain relation and depicted by the stress-strain curve.

Generally, the stress can be expressed as a function of strain:

$$\sigma = f(\epsilon) \tag{2.11}$$

The character of the function differs for various materials; however, practically all engineering materials have a proportional range at relatively low stress, during which Hooke's law holds:

$$\sigma = E\epsilon \tag{2.12}$$

During this range, stress and strain are proportional, the proportionality constant being the modulus of elasticity E. The conventional methods of stress analysis and design make use only of the elastic proportional range, and Eq. (2.12) is the basis for most of our elementary procedures and formulas, which are therefore generally restricted to stresses below the proportional limit of the material.

During recent decades considerable thought and research have resulted in our getting a clearer understanding of the effects of the inelastic portion of the stress-strain curve on the behavior and strength of structures. The "plastic" or "limit design" method which is coming into use for steel structures and the "ultimate strength" method for design of reinforced concrete structures, both of which are based on the inelastic properties of the stress-strain curve of the materials and have recently been incorporated into building codes, are the result of this research. One reason for the approach followed in these notes is to point out clearly the connections, similarities, and differences in reasoning and application between the conventional elastic and newer inelastic methods of analysis and design.

One property which merits special discussion is the stiffness of a material, that is to say its resistance to deformation. Mathematically, this is specified by the change of stress necessary to produce a small increment of strain, or

$$\text{Stiffness} = \frac{d\sigma}{d\epsilon} = \text{slope of stress-strain curve}$$

In the elastic range we recognize this as given by the modulus of elasticity. In the inelastic range, we can either perform the indicated differentiation of the mathematical stress-strain relation or else measure the slope of the stress-strain curve by drawing its tangent. For this reason, the inelastic stiffness is called the *tangent modulus.*

We shall now consider the complete stress-strain curves of some important engineering materials:

Structural steel (Fig. 2.5) The material is elastic nearly up to the yield point, when continuing deformation becomes possible under a constant value of stress σ_{YP}. This "yield range" continues up to a value of strain about 20 times the elastics train ϵ_{YP}, when additional strength and stiffness become available at the onset of "strain hardening." Additional deformation is possible up to a value of strain about 200 times the strain value at first yielding prior to rupture. It is seen that structural steel has a large degree of ductility or plasticity.

For analytical purposes this rather complicated response is simplified into two regions, shown by the dashed line in Fig. 2.5:

1. An elastic region, up to a stress σ_{YP}, during which Eq. (2.12) prevails:

$$\sigma = E\epsilon \qquad 0 < \epsilon < \epsilon_{YP}$$

2. A plastic region which is assumed to extend to an infinitely large value of strain:

$$\sigma = \sigma_{YP} \qquad \epsilon_{YP} < \epsilon < \infty \tag{2.13}$$

The assumption of indefinitely continuing deformation is called *perfectly plastic.* The strain hardening is usually neglected; this simplification leads to error on the side of safety, since a portion of the material's strength is neglected.

Tensile and compressive material properties of steel are assumed identical.

Aluminum A typical stress-strain curve for a structural aluminum alloy is shown in Fig. 2.6. It has a distinct elastic range, but no abrupt yield

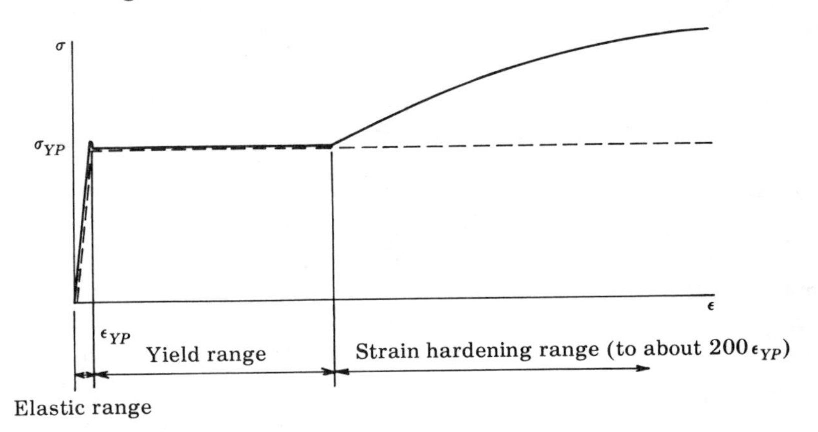

Fig. 2.5 Stress-strain curve for structural steel

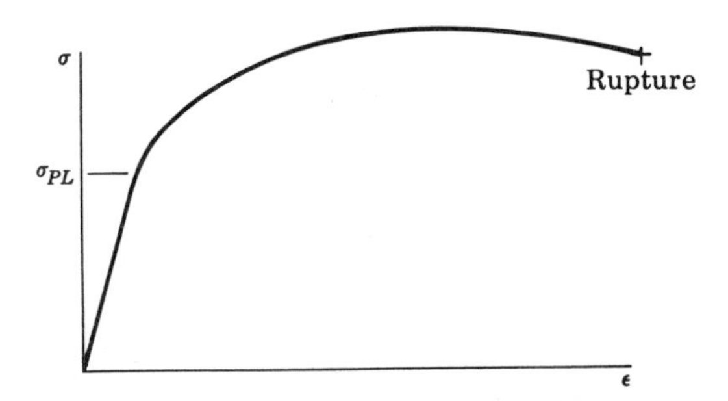

Fig. 2.6 Stress-strain curve for aluminum

point, nor the tremendous ductility of structural steel. For this reason an analytical treatment of its inelastic behavior is more difficult than that of steel; most design methods in aluminum restrict themselves to the elastic range. Again, compressive and tensile properties are very similar.

An analytical formulation for the stress-strain relation of some aluminum alloys used in the aircraft industry, cited by Shanley,[1] is

$$\epsilon = \frac{\sigma}{E} + \left(\frac{\sigma}{B}\right)^n \tag{2.14}$$

where E, B, and n are experimental constants of the material, with values as shown in the table below:

	E, psi	*B, psi*	*n*
Aluminum alloy 2024-T3: Ultimate stress = 65 ksi	10×10^6	72,300	10
Aluminum alloy 7075-T6: Ultimate stress = 83 ksi	10×10^6	101,200	20

The first term of Eq. (2.14) represents the elastic strain, while the second term represents the additional strain due to plastic deformation. Equation (2.14) represents the behavior of these materials with reasonable accuracy up to the point of ultimate strength.

Timber Figure 2.7 shows that the transition from the elastic to the inelastic behavior of timber is rather gradual. Because of the lack of uniformity in the strength of wood, due to its nonhomogeneous and irregular structure, only low stresses are allowed in timber design, and analysis is usually based on purely elastic behavior. As compared to steel and aluminum, wood is also a rather brittle material, with little ability to deform plastically prior to rupture.

[1] F. R. Shanley, "Strength of Materials," p. 159, McGraw-Hill, 1957.

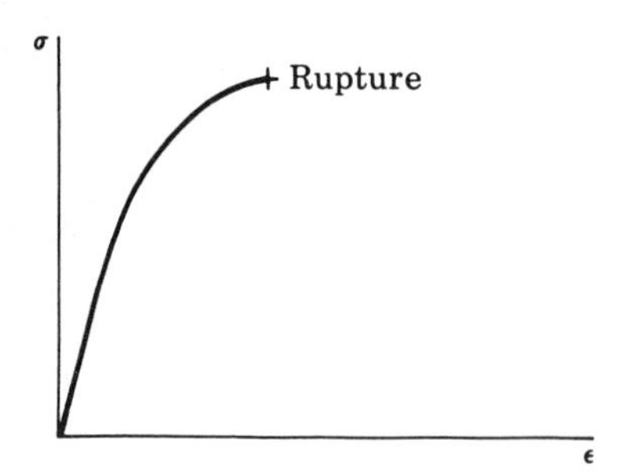

Fig. 2.7 Stress-strain curve for timber

Plain concrete The mechanical properties of concrete exhibit some remarkable features. There is hardly a trace of proportionality between stress and strain; tensile and compressive properties differ widely from each other. These properties are exhibited by the solid line in Fig. 2.8; in practice, the slight tensile strength of about 10 percent of f'_c of concrete is neglected, and it is assumed that cracking will occur as soon as tension sets in; with very few exceptions, therefore, no tensile stresses are admitted in reinforced concrete theory. In plain concrete, tensile stresses of the order of 3 percent of the compressive strength are allowed. Compressive crushing occurs at strains of the order of 0.003, but all values are greatly affected by time effects.

To express the compressive stress-strain relation, several possibilities are open:

1. The actual curve could be represented with fair accuracy by a polynomial, such as

$$\sigma = f(\epsilon) = a + b\epsilon + c\epsilon^2 \tag{2.15}$$

where a, b, and c are experimental constants, or by an exponential function such as

$$\sigma = f(\epsilon) = k\epsilon^n \tag{2.16}$$

where k and n are constants. While useful for research purposes, expressions of this type are usually too complicated for practice.

2. For small values of stress, the dashed line shows that fair results might be obtained by assuming a linear Hookean relationship:

$$\sigma = E\epsilon \tag{2.12}$$

This approximation forms the basis of the widely used "elastic," "working load," or "straight line" theory of reinforced concrete design.

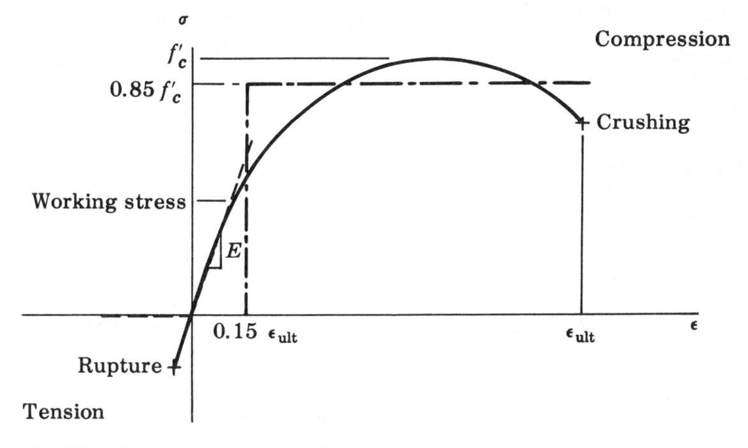

Fig. 2.8 Stress-strain curve for concrete

As long as stresses are held low by suitable choice of safety factor, the response to working loads (but not the safety against collapse) is easily obtained using Eq. (2.12).

3. To obtain information on the ultimate strength of concrete structures, a rough approximation to the actual stress-strain curve is taken in the form of a rectangle shown by the dash-dotted line. When used in beam theory, it has been shown by Whitney[1] that good results can be obtained when the values of stress and strain for this idealized function are assumed as shown in Fig. 2.8, in which case the stress-strain relation becomes

$$\begin{aligned} \sigma &= 0 & 0 &< \epsilon < 0.15\epsilon_{ult} \\ \epsilon &= 0.15\epsilon_{ult} & 0 &< \sigma < 0.85f'_c \\ \sigma &= 0.85f'_c & 0.15\epsilon_{ult} &< \epsilon < \epsilon_{ult} \end{aligned} \qquad (2.17)$$

More attention will be devoted to the Whitney stress block later on; the ultimate-strength method of reinforced concrete design is based on this assumption.

In the following, theories will be developed which are valid for any material, no matter what its stress-strain relation; to do this, this relation will be expressed by the general Eq. (2.11); it should be understood that before definite results can be obtained, Eq. (2.11) must be replaced by a more specific relation of the type of Eqs. (2.12) to (2.17) suitable for the particular material under consideration.

It should be noted that the entire preceding discussion was concerned with the deformations of material due to uniaxial tension or compression stress. In the more complicated case of a biaxial or triaxial state of stress acting at a point of a structural member under load, deformations may be obtained by means of the "generalized Hooke's law" in the elastic range, and by means of "yield condition" and "flow rule" in the plastic range. Section 2.6 will indicate the use of some of these concepts in continuum mechanics.

Example Problem 2.5

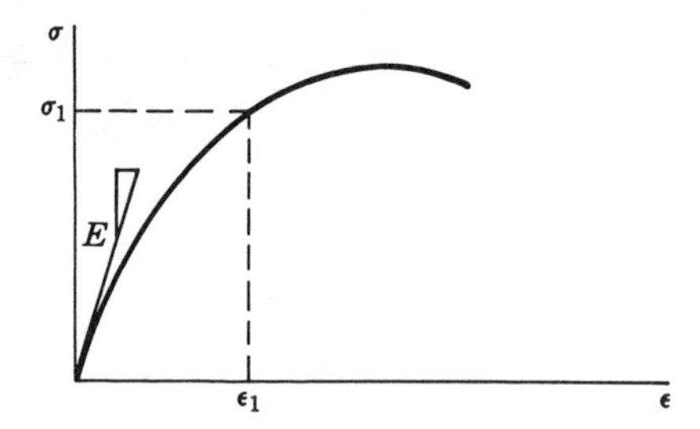

The tension test of a certain material results in a stress-strain curve with values as shown; it is to be expressed analytically by an expression of the form

[1] C. S. Whitney, Plastic Theory of Reinforced Concrete Design, *Trans. ASCE*, vol. 68, p. 251, 1942.

$\sigma = f(\epsilon) = a + b\epsilon + c\epsilon^2$

Determine the values of the experimental constants a, b, c, and find an analytical expression for the tangent modulus.

Solution: To determine the three constants, we need three equations, obtained from known conditions:

(1) $\sigma \Big|_{\epsilon=0} = 0: \qquad 0 = a + 0 + 0; \qquad a = 0$

(2) Slope $\Big|_{\epsilon=0} = \dfrac{d\sigma}{d\epsilon}\Big|_{\epsilon=0} = E:$

$$\frac{d\sigma}{d\epsilon}\Big|_{\epsilon=0} = (b + 2c\epsilon)\Big|_{\epsilon=0} = b + 0 = E; \qquad b = E$$

(3) $\sigma \Big|_{\epsilon=\epsilon_1} = \sigma_1: \qquad \sigma_1 = E\epsilon_1 + c\epsilon_1^2$

or $\quad c = \dfrac{1}{\epsilon_1^2}(\sigma_1 - E\epsilon_1)$

Note that since $E\epsilon_1$ is larger than σ_1, c is negative. Therefore, the stress-strain relation is

$$\sigma = E\epsilon + \frac{\sigma_1 - E\epsilon_1}{\epsilon_1^2}\epsilon^2$$

The tangent modulus is given by

$$E_T = \frac{d\sigma}{d\epsilon} = b + 2c\epsilon = E + \frac{2}{\epsilon_1^2}(\sigma_1 - E\epsilon_1)\epsilon$$

Note that for $\epsilon = 0$, the tangent modulus becomes the initial modulus E. For larger strains, the stiffness decreases since c is negative.

Example Problem 2.6

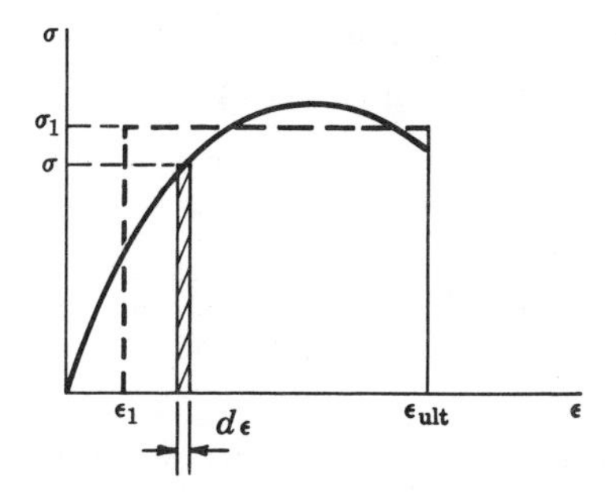

The experimental stress-strain curve can be expressed by a parabola:

$\sigma = f(\epsilon) = b\epsilon + c\epsilon^2$

It is required to idealize this relation by a rectangular stress block as shown. This stress block is to satisfy two conditions:

(*a*) Area under stress block = area under parabola.
(*b*) First moment with respect to the σ axis of the rectangular area to be equal to that of the area under parabolic curve.

On the basis of these conditions, find the values of σ_1 and ϵ_1.

Solution: *Equating areas*

$$(\epsilon_{ult} - \epsilon_1)\sigma_1 = \int_{\epsilon=0}^{\epsilon_{ult}} \sigma\, d\epsilon = \int_{\epsilon=0}^{\epsilon_{ult}} (b\epsilon + c\epsilon^2)\, d\epsilon$$

$$= b\frac{\epsilon_{ult}^2}{2} + c\frac{\epsilon_{ult}^3}{3}$$

Equating first moments

$$[\sigma_1(\epsilon_{ult} - \epsilon_1)][\tfrac{1}{2}(\epsilon_{ult} + \epsilon_1)] = \int_{\epsilon=0}^{\epsilon_{ult}} \sigma\epsilon\, d\epsilon$$

$$= b\frac{\epsilon_{ult}^3}{3} + c\frac{\epsilon_{ult}^4}{4}$$

We can solve these equations simultaneously for the two unknowns σ_1 and ϵ_1 by substituting the right-hand side of the first equation for the first bracketed quantity on the left side of the second equation:

$$\left[b\frac{\epsilon_{ult}^2}{2} + c\frac{\epsilon_{ult}^3}{3}\right][\tfrac{1}{2}(\epsilon_{ult} + \epsilon_1)] = b\frac{\epsilon_{ult}^3}{3} + c\frac{\epsilon_{ult}^4}{4}$$

and solving for ϵ_1 we get

$$\epsilon_1 = -\epsilon_{ult} + 2\frac{(b/3)\epsilon_{ult}^3 + (c/4)\epsilon_{ult}^4}{(b/2)\epsilon_{ult}^2 + (c/3)\epsilon_{ult}^3} = \left[2\frac{(b/3) + (c/4)\epsilon_{ult}}{(b/2) + (c/3)\epsilon_{ult}} - 1\right]\epsilon_{ult}$$

Substituting this value back into the first equation, we solve for σ_1:

$$\sigma_1 = 3\frac{[(b/2) + (c/3)\epsilon_{ult}]^2}{b + (c/2)\epsilon_{ult}}\epsilon_{ult}$$

2.5 The Strength of Materials Method—Applications

To apply the tools which have been presented to the analysis of structural members, it is desirable to outline a generally applicable coherent approach to the calculation of stresses and deformation. Such an approach is given by the "strength of materials" method.

The individual steps of this method may be summarized as:

Step 1:

Cut an appropriate free body of finite size; by finite size we mean that while some of its dimensions may be infinitely small, at least one length is to be of finite dimension. The free body is selected in such a fashion that the desired stresses act as external forces on one of the cut faces of the free body.

Step 2:

Establish the conditions of equilibrium for the free body selected in step 1; this will result in a set of equations in which the stresses are unknowns.

Step 3:

Make a reasonable assumption on the distribution of strains. This step will involve a visualization of the deformation of the structure under load; it represents also the severe limitation of the strength of materials approach, and restricts this method to the analysis of relatively simple situations (which however will suffice for the analysis of most structural members). In complicated cases of irregular loads or difficult outlines it may be impossible to arrive at a reasonable assumption of the deformation, and in such cases recourse must be had either to experimental methods, or to the analytical "continuum mechanics" approach, which is introduced in Sec. 2.6, and which can be studied in detail in books on elasticity or plasticity.

To illustrate the point, we consider a flat strap under tensile load; at one section, a small hole is punched in the strap, as shown in Fig. 2.9*a*. At a typical section, A, away from any irregularity, we may reasonably assume that two lines drawn on the surface normal to the axis of the strap, and a unit distance apart, will, under load, move a certain distance apart but will remain parallel to each other; thus, we can deduce that the change in distance, according to its definition the tensile strain ϵ, will be constant over the cross section of the member.

Considering the deformation at section B, drawn through the hole, we could visualize (and verify experimentally by deforming a rubber strap, as shown in Fig. 2.9*b*) that two transverse parallel lines, originally unity apart, will upon load application deform in an irregular fashion, with those portions nearest the hole subjected to the largest strains. An assumption of uniform strain would be entirely unrealistic for the vicinity of the hole, and accordingly an analysis with strength of materials methods would be of questionable value. The problem here is one of a "stress concentration" and should be analyzed by other methods.

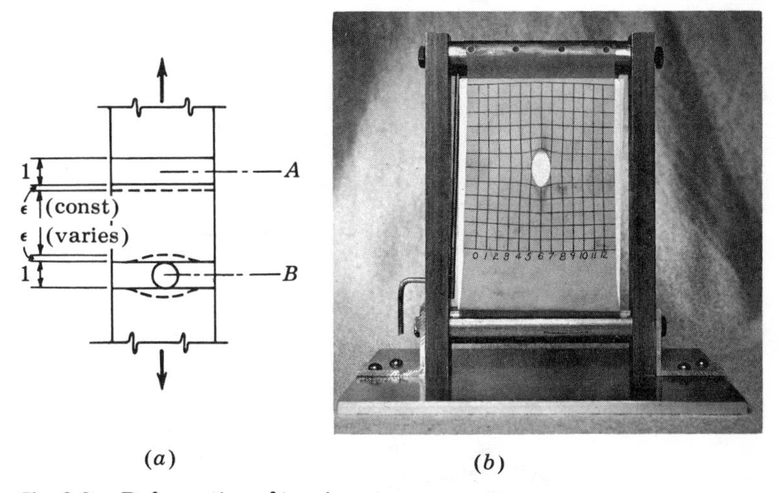

Fig. 2.9 Deformation of tension strap
Courtesy L.G. Tulin, University of Colorado

It is, of course, possible to base any analysis on simplified strain distributions, and this is often done in engineering practice; but then it must be realized that the resulting strains (and stresses) are average values; to obtain the maximum values, it is necessary to apply a "stress concentration factor" based on previous analysis; otherwise, the allowable stresses should be suitably reduced, and this is often done. It will also be seen later that irregular stress and strain distributions can be ironed out by the plastic behavior of ductile materials.

An assumption on the strains will result in equations containing strain as an unknown.

Step 4:

Apply stress-strain relations for the material under consideration. This will serve to express the unknown stresses in the equations of step 2 as functions of the cross-sectional dimensions of the free body; the equations of step 2 can then be integrated and solved for stresses and deflections.

To illustrate the basic steps outlined, we shall apply them in reviewing a problem in stress analysis encountered in strength of materials:

Example Problem 2.7 Beam in bending, elastic material.

Step 1: We select a finite free body which contains as an exterior face the surface on which the desired stresses are acting, as shown in Fig. 2.10*a*.

Step 2: To put this free body into equilibrium, we write

$$\Sigma F_z = 0: \quad \int_A \sigma \, dA = 0 \tag{2.18}$$

$$\Sigma M_x = 0: \quad \int_A \sigma y \, dA = M \tag{2.19}$$

In order to integrate these expressions, we must find the variation of the stresses over the cross-sectional area; to do this, we consider the strains.

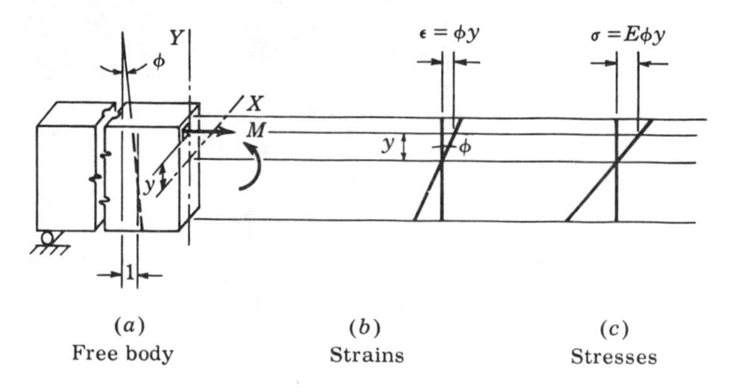

(*a*) Free body (*b*) Strains (*c*) Stresses

Fig. 2.10

Step 3: We may reasonably assume that two transverse lines drawn on the side of the beam, parallel and a unit distance apart, may during bending incline with respect to each other, but will remain straight lines; thus, the change of distance is linear across the depth of the beam; but the change of distance per unit distance is defined as the strain, and we may thus conclude that the strain is linearly distributed over the depth of the beam, with the point of zero strain defined as the *neutral axis*, as shown in Fig. 2.10*b*. Analytically, this strain can be expressed as

$$\epsilon = y\phi \tag{2.20}$$

where ϕ is the small angle between the two transverse lines after bending has occurred; since these lines were originally unity apart, the angle ϕ represents the change of slope between two sections unity apart; this quantity is defined as the curvature of the deformed beam; we thus identify ϕ as the curvature of the beam caused by bending.

Step 4: We now relate the strains found in step 3 to the corresponding stresses; this will involve the mechanical properties of the material indicated by the stress-strain curve, according to the statement of the problem, elastic:

$$\sigma = \epsilon E = E\phi y \tag{2.21}$$

Equation (2.21) gives us the desired variation of stresses which is illustrated in Fig. 2.10*c*, which we now substitute into Eqs. (2.18) and (2.19):

$$\int_A E\phi y \, dA = E\phi \int_A y \, dA = 0 \tag{2.22}$$

and

$$\int_A E\phi y^2 \, dA = E\phi \int_A y^2 \, dA = M \tag{2.23}$$

E and ϕ can be taken outside the integral since they are constants for any cross section.

Equation (2.22) indicates that, since $E\phi$ is a constant other than zero,

$$\int_A y \, dA = 0 \tag{2.24}$$

This defines the neutral axis (from which y is measured) as the centroidal axis of the cross section.

Turning now to Eq. (2.23), we identify the integral $\int_A y^2 \, dA$ as the moment of inertia of the cross section with respect to the neutral axis, and can therefore write, solving for ϕ:

$$\phi = \frac{M}{EI} \tag{2.25}$$

This is the starting point for the calculation of the deformation of elastic beams; to find stresses, we substitute Eq. (2.25) into Eq. (2.21) and obtain

$$\sigma = E\left(\frac{M}{EI}\right)y = \frac{My}{I} \tag{2.26}$$

the well-known flexure formula.

For beams of nonhomogeneous cross section, or for those made of inelastic material, the only change necessary is to substitute appropriate stress-strain relations to replace Eq. (2.21). The former of these cases is covered in Sec. 5.1, the latter in Sec. 7.1.

To indicate the versatility of the strength of materials method, we shall apply the same basic steps to an entirely different problem.

Example Problem 2.8

Given: Riveted lap joint as shown; the three rivets are identical, and the rivet shear distortion Δ_R is related linearly to the rivet shear force P_R by the relation $P_R = K_R\Delta_R$, where K_R is the given stiffness of the rivet. The plate is elastic, and its elongation per rivet pitch Δ_P is related linearly to the tension force P_P in the plate by the relation $P_P = K_P\Delta_P$, where K_P is the stiffness of the plate (if stress concentrations are neglected, then $K_P = AE/L$). These force-deformation relations are plotted in Fig. 2.11*c*.

Required: The distribution of the total transmitted force P among the rivets. Neglect bending of joint.

Solution: *Equilibrium:* The only applicable equilibrium condition is, referring to Fig. 2.11*a* and using symmetry,

$\Sigma F = 0: \qquad 2P_1 + P_2 = P$

where P_1 and P_2 are the shear forces transmitted through rivets 1 and 2 respectively.

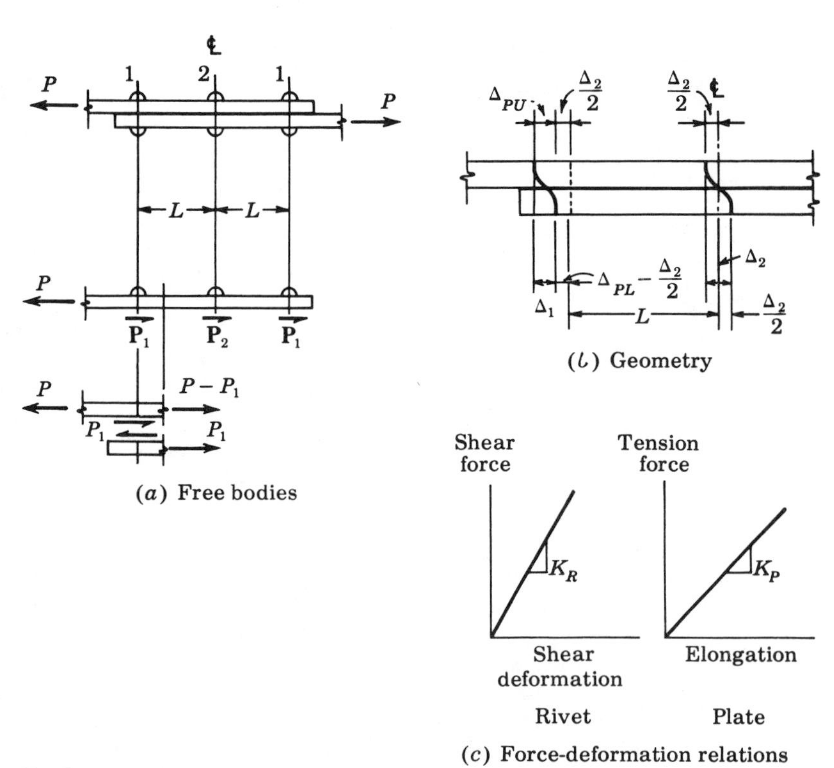

Fig. 2.11

In addition, by considering partial free bodies, we find the tension forces in the upper and lower plates as $P - P_1$ and P_1, respectively.

Geometry: With one equation and two unknowns, we resort to geometry, and carefully draw the deformed joint, as in Fig. 2.11*b*. The deformations are due to the rivet distortions Δ_1 and Δ_2 and the plate elongations between rivet rows Δ_{PU} and Δ_{PL} in the upper and lower plates respectively.

Matching the deformations of the upper and lower plates, as in Fig. 2.11*b*, we get the geometric relation

$$\Delta_{PU} + \frac{\Delta_2}{2} = \left(\Delta_{PL} - \frac{\Delta_2}{2}\right) + \Delta_1$$

or

$$\Delta_{PU} - \Delta_{PL} = \Delta_1 - \Delta_2$$

Force-deformation relation: It remains to relate the static and geometric relations by the force-deformation relations

$$\Delta_{PU} = \frac{1}{K_P}(P - P_1)$$

$$\Delta_{PL} = \frac{1}{K_P}P_1$$

$$\Delta_1 = \frac{1}{K_R}P_1$$

$$\Delta_2 = \frac{1}{K_R}P_2$$

Substituting these values into the geometric relations, we get

$$\left(2 + \frac{K_P}{K_R}\right)P_1 - \frac{K_P}{K_R}P_2 = P$$

which, when solved simultaneously with the force equation, yields

$$\frac{P_1}{P_2} = \frac{K_R}{K_P} + 1$$

We can now evaluate results for special cases:

When the rivets are perfectly rigid compared to the plate, then $K_R/K_P = \infty$, and $P_2 = 0$, $P_1 = P/2$; in this case, the outer rivets take all the load.

When the rivets are perfectly flexible, then $K_R/K_P = 0$, and $P_1 = P_2 = P/3$; this would apply when the rivets are perfectly plastic, and verifies the reasoning of Sec. 1.1 regarding the ultimate strength of such a joint.

For an intermediate case, say, $K_R/K_P = 1$, $P_1 = \frac{2}{5}P$, $P_2 = \frac{1}{5}P$, and in this manner we are able to make predictions about the response of different joints.

2.6 Continuum Mechanics*

The severe limitations of the strength of materials approach were already discussed in Sec. 2.5. It was indicated that in cases where geometric distortions could not be predicted, a more exact method, the continuum mechanics approach, is appropriate. In this section, such an approach will be outlined and carried forward for the special case of elastic bodies.

The continuum mechanics method incorporates the same basic tools of mechanics as the strength of materials method, that is, equilibrium, geometry, and stress-strain (or, generally speaking, constitutive) relations, but these are applied in a somewhat different manner. A generally valid outline is as follows:

1. Select a typical infinitesimal element in an appropriate coordinate system, and put in equilibrium. Because the element is infinitesimal, a set of differential equations results in which the stresses are unknowns. Usually there will be more unknown stresses than equilibrium equations, indicating that additional equations of geometry are necessary.
2. The geometrical conditions must ensure that the deformations of adjacent elements are compatible, that is, no discontinuities such as gaps or jags arise in the body (in some cases of plasticity, discontinuities may be allowed, but in elastic bodies, to which the application of this method will be restricted here, continuity is required). This is ensured by imposing so-called "compatibility equations" on the strains, or by expressing the strains in terms of continuous displacement functions, resulting in so-called "strain-displacement" equations. This will result in a set of differential equations in terms of strains.
3. To relate the stresses of the equilibrium equations of step 1 to the strains of the compatibility equations of step 2, we invoke the stress-strain or constitutive equations. In the case of elastic material, this will be the generalized Hooke's law; in the case of plasticity the "yield condition" and "flow rule" will serve this purpose. Other constitutive relations will obtain for time-dependent or fluid materials.

   The equations of steps 1, 2, and 3 are combined to yield the governing differential equations of the body. These equations may be formulated in terms of stresses, strains, or deformations. The solution of these differential equations leads to constants of integration.
4. To solve for the constants of integration, we must write an appropriate number of boundary conditions. At this time, the support conditions and external loads will be introduced.

It should be understood that the execution of this outline is by no means as simple as it sounds. The direct solution of a boundary-value problem of the type which has been outlined is possible only in special cases. In the vast majority of cases, the use of series or numerical methods is indicated to obtain approximate solutions. At this time, solutions to a large number of problems in plane or two-dimensional elasticity are available; solutions to three-dimensional elasticity problems are restricted to symmetrical cases, and only a few plasticity solutions have so far been obtained.

An approach similar to the one which has been outlined can be developed for the analysis of plate and shell structures; a command over these methods is necessary for the modern structural engineer.

We shall elaborate on the several steps of the theory of elasticity in order to establish a mathematical framework for solving some problems in elasticity.

Equilibrium: We consider an infinitesimal element, shown in Fig. 2.12 within a set of cartesian X, Y, Z axes, of dimensions dx, dy, dz, and subject it to a state of stress. The normal components are designated by like subscripts, the shearing stresses by unlike subscripts, the first of which denotes the plane on which the stress is acting, the second the axis to which it is parallel. Thus, the complete state of stress on the element can be specified by an array of nine quantities.

Moment equilibrium demands that conjugate shear stresses are equal, so that subscripts may be exchanged, that is, $\sigma_{XY} = \sigma_{YX}$, and this reduces the number of independent stress components to six, but on the other hand this leaves only three force-equilibrium conditions to be written. To do this, we express the stress intensity on opposing sides of the element in terms of its rate of change, $\partial\sigma_{XY}/\partial x$, and so on, multiply these intensities by the area of the face on which they act, and write, referring to Fig. 2.12:

$$\Sigma F_X = 0: \quad \left[\left(\sigma_{XX} + \frac{\partial\sigma_{XX}}{\partial x}\,dx\right) - \sigma_{XX}\right] dy\,dz + \left[\left(\sigma_{YX} + \frac{\partial\sigma_{YX}}{\partial y}\,dy\right) - \sigma_{YX}\right] dz\,dx + \left[\left(\sigma_{ZX} + \frac{\partial\sigma_{ZX}}{\partial z}\,dz\right) - \sigma_{ZX}\right] dx\,dy = 0$$

or, canceling, simplifying, and invoking the equality of conjugate shears,

$$\frac{\partial\sigma_{XX}}{\partial x} + \frac{\partial\sigma_{XY}}{\partial y} + \frac{\partial\sigma_{XZ}}{\partial z} = 0 \tag{2.27a}$$

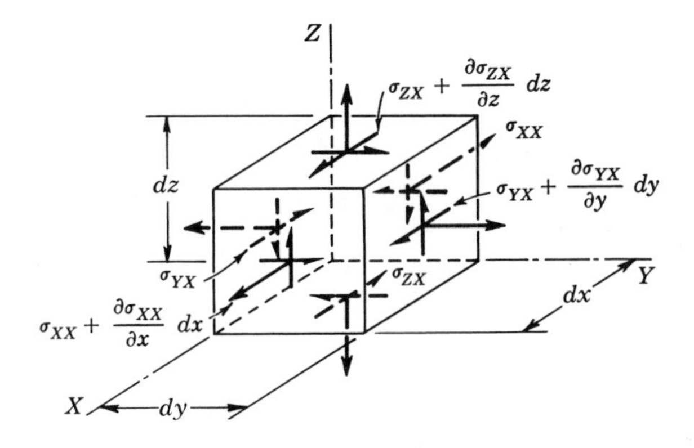

Fig. 2.12 State of stress

Similarly, equating forces along the Y and Z axes to zero:

$$\frac{\partial \sigma_{YX}}{\partial x} + \frac{\partial \sigma_{YY}}{\partial y} + \frac{\partial \sigma_{YZ}}{\partial z} = 0 \tag{2.27b}$$

$$\frac{\partial \sigma_{ZX}}{\partial x} + \frac{\partial \sigma_{ZY}}{\partial y} + \frac{\partial \sigma_{ZZ}}{\partial z} = 0 \tag{2.27c}$$

Since, by the conjugate shear-stress theory, $\sigma_{XY} = \sigma_{YX}$, and so on, there are six unknown stress components and only three equations, so we now turn to geometry for more equations.

Geometry: Because consideration of deformations of a strained body is easier in two dimensions, we shall consider a projection of the deformed cartesian element in the XY plane and then extrapolate in the other dimensions.

Figure 2.13a considers the effect of normal strains only, because of which the element suffers displacements u, v along the X, Y axes. The strain ϵ_X is defined as the change of length per unit of original length, or

$$\epsilon_X = \frac{u + (\partial u/\partial x)\,dx - u}{dx} = \frac{\partial u}{\partial x} \tag{2.28a}$$

and likewise

$$\epsilon_Y = \frac{\partial v}{\partial y} \tag{2.28b}$$

$$\epsilon_Z = \frac{\partial w}{\partial z} \tag{2.28c}$$

The distortions due to shear strains are shown in Fig. 2.13b, and, remembering that the shear strain is defined as the change of angle between the sides of the element, we write them as

$$\gamma_{XY} = \frac{\partial u}{\partial y} + \frac{\partial v}{\partial x} \tag{2.29a}$$

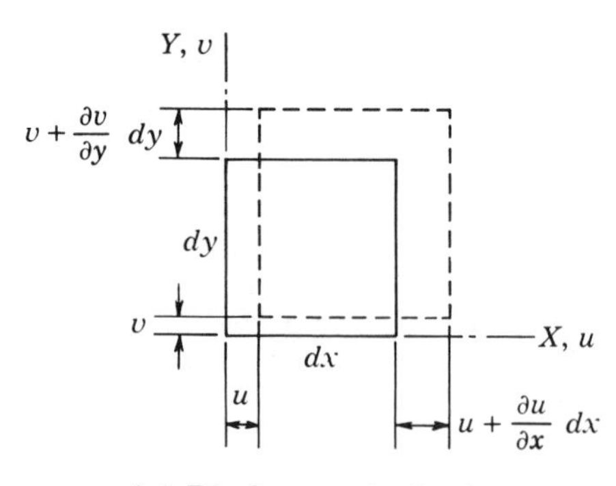

(a) Displacements due to normal strains

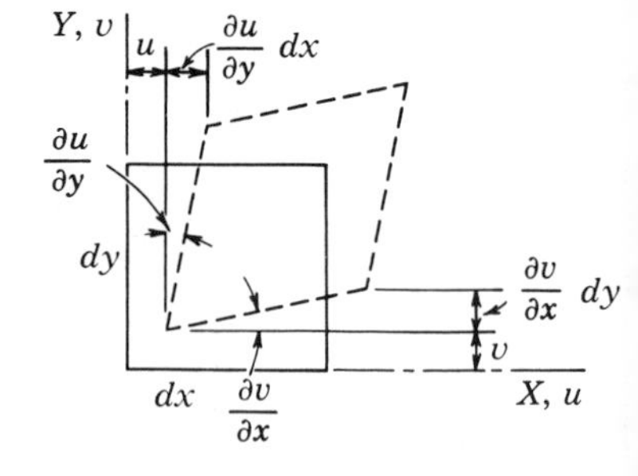

(b) Displacements due to shearing strains

Fig. 2.13

and similarly

$$\gamma_{YZ} = \frac{\partial v}{\partial z} + \frac{\partial w}{\partial y} \tag{2.29b}$$

$$\gamma_{ZX} = \frac{\partial w}{\partial x} + \frac{\partial u}{\partial z} \tag{2.29c}$$

Equations (2.28*a*) to (2.29*c*) introduce nine more unknowns in the system, six strain components and three displacement components.

All relations presented so far are independent of the material properties and are therefore quite general. It is only at the next stage that distinction arises between elastic and inelastic theories.

Stress-strain relations: To relate the strains of Eqs. (2.28) to (2.29) to the stresses of Eqs. (2.27), the material properties are invoked; from here on, we shall restrict ourselves to elastic, isotropic bodies, for which the stress-strain relations are given by the generalized Hooke's law:

$$\epsilon_X = \frac{1}{E}[\sigma_{XX} - \mu(\sigma_{YY} + \sigma_{ZZ})] \tag{2.30a}$$

$$\epsilon_Y = \frac{1}{E}[\sigma_{YY} - \mu(\sigma_{ZZ} + \sigma_{XX})] \tag{2.30b}$$

$$\epsilon_Z = \frac{1}{E}[\sigma_{ZZ} - \mu(\sigma_{XX} + \sigma_{YY})] \tag{2.30c}$$

$$\gamma_{XY} = \frac{1}{G}\sigma_{XY} = \frac{2(1+\mu)}{E}\sigma_{XY} \tag{2.30d}$$

$$\gamma_{YZ} = \frac{1}{G}\sigma_{YZ} = \frac{2(1+\mu)}{E}\sigma_{YZ} \tag{2.30e}$$

$$\gamma_{ZX} = \frac{1}{G}\sigma_{ZX} = \frac{2(1+\mu)}{E}\sigma_{ZX} \tag{2.30f}$$

We have now assembled enough equations to match the number of unknowns: a total of fifteen equations [three equilibrium equations, (2.27*a*) to (2.27*c*), six strain-displacement equations, (2.28*a*) to (2.29*c*), and six stress-strain equations, (2.30*a*) to (2.30*f*)] to solve for fifteen unknowns (six stress components, six strain components, and three displacement components). The formulation and solution of these equations, subject to given boundary conditions, constitute the classical problem of the theory of elasticity.

The solution of the boundary-value problem expressed by these fifteen equations and the appropriate boundary conditions is by no means an easy one in most cases, and often roundabout or approximate methods of solution are indicated. One way of approaching the problem is to assume likely displacements or stresses, say those given by a strength of materials solution, and check to what extent the continuum mechanics equations are satisfied; if by chance they are, we can conclude that the assumed solution is the correct one.

Let us follow this approach in investigating the pure bending of an elastic beam of rectangular cross section (see Fig. 2.14) subjected to a moment M. To simplify the matter, we consider only stresses and displacements in the XY plane; for such a "plane stress" problem, all terms involving Z components can be deleted.

We start with the stresses given by technical beam theory:

$$\sigma_{XX} = -\frac{M}{I}y \qquad \sigma_{YY} = 0 \qquad \sigma_{XY} = 0$$

We know that these stresses satisfy external equilibrium; to check equilibrium of each internal element, we substitute these stresses into Eqs. (2.27), suitably shortened for plane stress:

$$\frac{\partial \sigma_{XX}}{\partial x} + \frac{\partial \sigma_{XY}}{\partial y} = 0 + 0 = 0$$

$$\frac{\partial \sigma_{XY}}{\partial x} + \frac{\partial \sigma_{YY}}{\partial y} = 0 + 0 = 0$$

We see that equilibrium is satisfied.

To obtain strains, we use the generalized Hooke's law, Eqs. (2.30):

$$\epsilon_X = \frac{1}{E}(\sigma_{XX} - \mu\sigma_{YY}) = -\frac{M}{EI}y$$

$$\epsilon_Y = \frac{1}{E}(-\mu\sigma_{XX} + \sigma_{YY}) = +\mu\frac{M}{EI}y$$

$$\gamma_{XY} = \frac{\sigma_{XY}}{G} = 0$$

The strain-displacement equations, (2.28) and (2.29), are used to find the displacements:

$$\epsilon_X = \frac{\partial u}{\partial x}$$

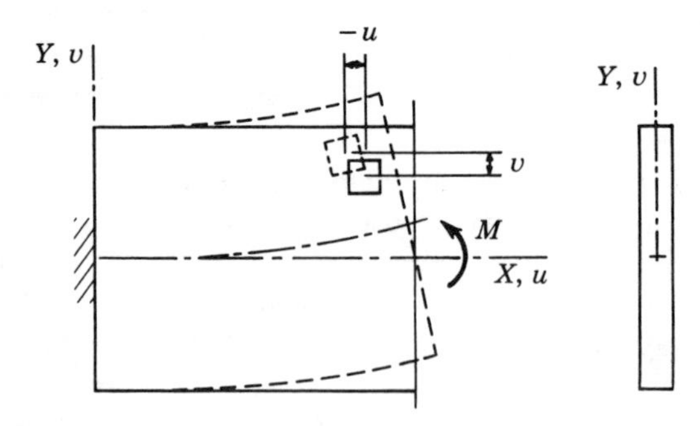

Fig. 2.14

therefore

$$u = \int_X \epsilon_X \, dx$$
$$= -\frac{M}{EI} yx + f_1(y)$$
$$\epsilon_Y = \frac{\partial v}{\partial y}$$

therefore

$$v = \int_Y \epsilon_Y \, dy$$
$$= +\mu \frac{M}{EI} \frac{y^2}{2} + f_2(x)$$

Also, $\gamma_{xy} = \dfrac{\partial u}{\partial y} + \dfrac{\partial v}{\partial x} = 0$

$$\left[-\frac{M}{EI} x + \frac{\partial f_1(y)}{\partial y} \right] + \left[0 + \frac{\partial f_2(x)}{\partial x} \right] = 0$$

or $\left[\dfrac{\partial f_1(y)}{\partial y} \right] + \left[\dfrac{\partial f_2(x)}{\partial x} - \dfrac{M}{EI} x \right] = 0$

Because the integrations are with respect to x and y respectively, the "constants" of integration may be functions of y and x respectively. The third equation will be used to determine these functions in the following manner: The bracketed terms are functions of the independent variables x and y, and the equation can be satisfied for arbitrary values of x and y only if each expression is equal to a constant, thus

$$\frac{\partial f_1(y)}{\partial y} = C_1 \qquad \frac{\partial f_2(x)}{\partial x} - \frac{M}{EI} x = -C_1$$

Solving now for f_1 and f_2, we get

$$f_1(y) = \int C_1 \, dy = C_1 y + C_2$$
$$f_2(x) = \int \left(\frac{M}{EI} x - C_1 \right) dx = \frac{M}{EI} \frac{x^2}{2} - C_1 x + C_3$$

where C_1, C_2, and C_3 are constants of integration which can be found by use of the boundary conditions; assume that the bar is held against translation and rotation at the origin, then

$$u(0,0) = 0: \qquad -\frac{M}{EI}(0) + C_1(0) + C_2 = 0$$
$$v(0,0) = 0: \qquad \mu \frac{M}{EI}(0) + \frac{M}{EI}(0) - C_1(0) + C_3 = 0$$
$$\frac{dv}{dx}(0,0) = 0: \qquad \mu \frac{M}{EI}(0) + \frac{M}{EI}(0) - C_1 = 0$$

therefore

$$C_1 = C_2 = C_3 = 0$$

and the displacements of the beam are

$$u = \frac{M}{EI}(-xy)$$

$$v = \frac{M}{EI}\left(\frac{x^2}{2} + \mu\frac{y^2}{2}\right)$$

At the neutral axis $y = 0$, and the u and v displacements show the values predicted by the usual beam theory. For other points of the beam section, the u component of displacement is that associated with the plane section assumption, and the v component shows the effect of Poisson's ratio.

It is concluded that for the case of pure bending of a rectangular-section beam, the technical beam theory is exact. For other shapes of beam, and for the case in which moment and shear prevail at a section, similar studies would show slight discrepancies between exact and strength of materials results.

2.7 Instability

In a sufficiently slender structure under load, failure may occur by buckling at a stress level lower than the fracture or yield stress of the material. In such cases, the buckling strength of the structure must be investigated, and this section will present an approach to the problem. We shall see that this analysis will be one of failure or ultimate load, rather than one of stress analysis. It is interesting to note that even at a time when design practice was based almost exclusively on the concept of allowable stresses, design of slender compression members always had as its foundation a concept of ultimate or limit design.

In contrast with the gradual failure which would result from exceeding the safe stresses of ductile materials and which is announced by excessive deformation or sag, buckling failure occurs suddenly without giving advance warning. Most of the major catastrophic structural failures which involved loss of life resulted from buckling failures.

To understand the phenomenon of buckling, and to devise ways to analyze it, let us consider a rigid bar which is pinned at its ends, hinged elastically at its center, and loaded axially, as shown in Fig. 2.15*a*.

On this column under axial compression P let us apply some lateral force F which causes some lateral displacement of the type shown in Fig. 2.15*b*. For a sufficiently low value of P, this deflection will disappear when F is removed. If, however, we repeat this process under progressively higher values of axial load P, we shall eventually find that the lateral deflection will remain even after the lateral load F is removed. In other words, this axial load, called the *critical load* P_{cr}, is sufficient to keep the column in equilibrium in its deflected shape, and it is this fact upon which further analysis is based. It may be visualized that under an axial load larger than P_{cr} any lateral deflections would keep increasing till the column might be considered as having buckled.

In actual members, the place of the lateral force F is taken by unavoidable imperfections, eccentricities, or other disturbances, so that the above-described behavior will take place even in the absence of lateral forces.

Because the problem is not one of stress analysis, the typical steps of Sec. 2.5 cannot be followed here. Rather, we base our further development on the fact that only under the critical or buckling load can the deformed shape be an equilibrium configuration. Accordingly, the first step is to write an equilibrium condition in terms of the geometry of the buckled shape (assumed small); for the upper free body of Fig. 2.16,

$$\Sigma M = 0: \qquad M - P_{cr}\Delta = 0 \tag{2.31}$$

The second step is to relate the internal moment M to the deflected shape;

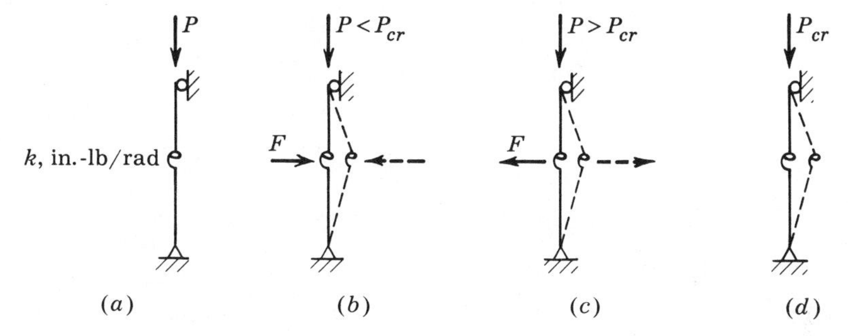

Fig. 2.15

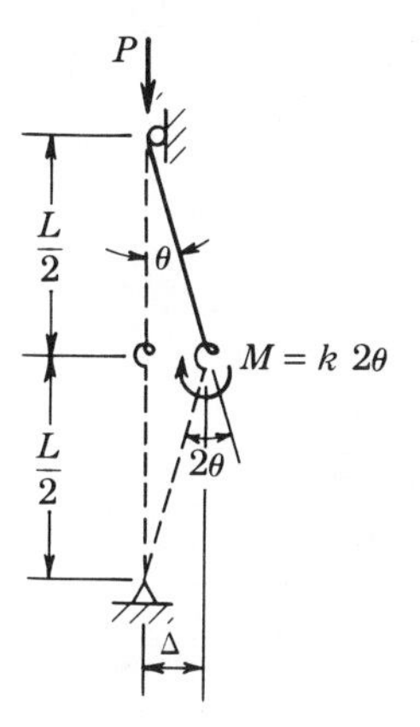

Fig. 2.16

the spring constant at the hinge being k in.-lb/rad, the moment is

$$M = k2\theta = k2\frac{\Delta}{L/2} = k4\frac{\Delta}{L}$$

Substituting this into Eq. (2.31), we find that equilibrium demands that

$$\left(4\frac{k}{L} - P_{cr}\right)\Delta = 0$$

The third step consists of finding a solution for this homogeneous, so-called "characteristic" equation. Here it is obvious that, for $\Delta \neq 0$, the only possible solution is

$$P_{cr} = \frac{4k}{L}$$

Note that in this solution, the deflection Δ does not have a unique value, and therefore the moment at the elastic hinge, which is proportional to the deflection, is also nonunique. This characteristic sets it apart from the problem of stress analysis; in technical language, the latter is a "response" problem, while the buckling problem is a "stability" problem.

With more than one hinge, a larger number of degrees of freedom would be possible, and instead of one, we could have written a set of simultaneous, homogeneous equations. Such a set can have a solution only if the determinant of the coefficients of the unknown deformations vanishes, and step 3 then consists of the solution of this determinant. Mathematically, this will be recognized as constituting an *eigenvalue problem*. The example problem of this section outlines the procedure in such a case.

In the case of a continuous flexural structure, the set of simultaneous homogeneous algebraic equations is replaced by a homogeneous differential equation, whose boundary conditions then lead to a set of homogeneous equations, the determinant of whose coefficients then is equated to zero to determine the characteristic equation. The simplest case of this type is the buckling of a Euler column, and this case will be covered in Sec. 3.6.

Example Problem 2.9

Given: Rigid bar, pinned at its ends, with two elastic hinges, each of rotational stiffness k in.-lb/rad, at its third points, as shown in Fig. 2.17*a*.

Required: To find the critical axial load P_{cr}.

Solution: With two unknown degrees of freedom Δ_B and Δ_C corresponding to the displacements of the two intermediate hinges, we can write two independent equilibrium conditions. Taking, in sequence, appropriate free bodies of the buckled shape, as shown in Fig. 2.17*c*, we write, following step 1:

$$\Sigma M_B = 0: \qquad M_B - P_{cr}\Delta_B = 0$$
$$\Sigma M_C = 0: \qquad M_C - P_{cr}\Delta_C = 0$$

In step 2, we express the moments in terms of the unknown deformations, following the geometry of Fig. 2.17*b*, as

$$\Sigma M_B = 0: \qquad \frac{k}{L}(6\Delta_B - 3\Delta_C) - P_{cr}\Delta_B = 0$$

$$\Sigma M_C = 0: \qquad \frac{k}{L}(-3\Delta_B + 6\Delta_C) - P_{cr}\Delta_C = 0$$

or

$$\left(6 - P_{cr}\frac{L}{k}\right)\Delta_B \qquad -3 \qquad \Delta_C = 0$$

$$-3 \qquad \Delta_B + \left(6 - P_{cr}\frac{L}{k}\right)\Delta_C = 0$$

To solve for a nontrivial solution to this set of homogeneous equations constitutes the eigenvalue problem. Step 3 involves equating the determinant of the coefficients of the unknowns to zero.

$$\begin{vmatrix} \left(6 - P_{cr}\dfrac{L}{k}\right) & -3 \\ -3 & \left(6 - P_{cr}\dfrac{L}{k}\right) \end{vmatrix} = 0$$

or

$$\left(P_{cr}\frac{L}{k}\right)^2 - 12\left(P_{cr}\frac{L}{k}\right) + 27 = 0$$

from which, solving for both roots of the quadratic equation,

$$P_{cr} = 3\frac{k}{L} \quad \text{or} \quad 9\frac{k}{L}$$

Note that of these two possible solutions, the lower one is the critical value.

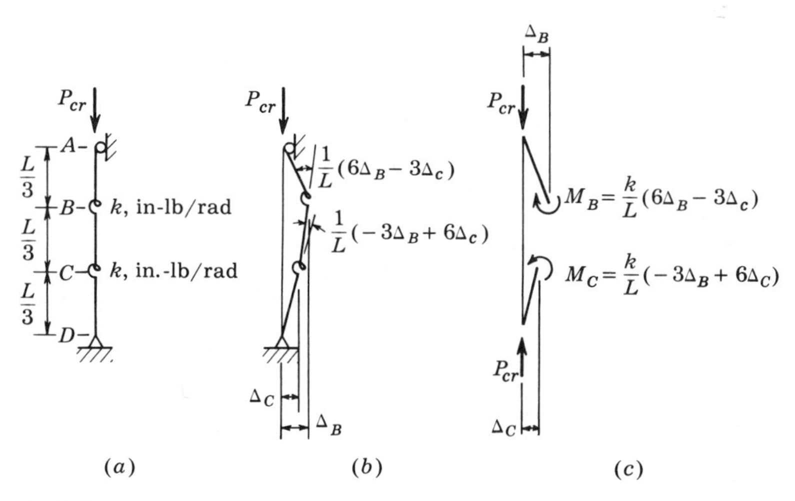

Fig. 2.17

General Readings

Statics and structural analysis

Higdon, Archie, and William B. Stiles: "Engineering Mechanics," Prentice-Hall, 1962. A sound, readable presentation of elementary mechanics in the conventional style. There is also an edition which makes somewhat more extensive use of vector methods.

Michalos, James, and Edward N. Wilson: "Structural Mechanics and Analysis," Macmillan, 1965. A presentation of statically determinate and indeterminate structural analysis. The advantage of this over many other similar texts is the fact that in chap. 7 a rational outline of the major methods of indeterminate analysis is provided.

Shames, Irving H.: "Engineering Mechanics," Prentice-Hall, 1959. A more modern treatment of mechanics which makes better use of mathematical concepts such as vector analysis and tensors than do older mechanics texts.

Strength of materials and continuum mechanics

Borg, S. F.: "Matrix-Tensor Methods in Continuum Mechanics," Van Nostrand, 1963. A unified presentation of continuum mechanics using matrix and tensor notation, including its application to problems in elasticity, plasticity, fluid mechanics, and structures.

Crandall, Stephen H., and Norman C. Dahl (eds.): "An Introduction to the Mechanics of Solids," McGraw-Hill, 1959. A somewhat more modern approach to mechanics of materials which tries to integrate the traditional with a continuum mechanics attack. This, like other similar texts, discards some of the traditional engineering topics in favor of a more rigorous engineering science approach that is intended to provide a firm foundation for a more theoretical curriculum than the conventional professionally oriented engineering program.

Popov, Egor P.: "Mechanics of Materials," Prentice-Hall, 1952. An excellent mechanics of materials text in the classic vein. Sound and clear derivations and engineering applications.

Shames, Irving H.: "Mechanics of Deformable Bodies," Prentice-Hall, 1964. A very good continuum mechanics approach to solid mechanics for undergraduates, with appropriate use of vector and tensor mathematics. As in several other recent books of this type, the transition from rigorous theory to engineering applications is a little rough.

Timoshenko, Stephen P.: "History of Strength of Materials," McGraw-Hill, 1953. Fascinating reading about the development of structural mechanics and structural engineering, with simple theoretical discussion. Should be read by every structural engineer.

Timoshenko, Stephen P., and J. N. Goodier: "Theory of Elasticity," 2d ed., McGraw-Hill, 1951. A classical introduction to the theory of elasticity, with emphasis on engineering problems.

Elastic stability

Gerard, George: "Introduction to Structural Stability Theory," McGraw-Hill, 1962. A concise summary of elastic and inelastic buckling theory and its applications to buckling of columns, plates, and shells.

Timoshenko, Stephen P., and James M. Gere: "Theory of Elastic Stability," 2d ed., McGraw-Hill, 1961. A classic treatise of structural buckling problems, with emphasis on elastic problems.

Problems

2.1 A square area of sides a, as shown, is subjected to normal stresses varying according to the expression

$$\sigma = C(x + y)$$

(*a*) By integration of stresses find the total force acting on the area and its moments about the X and Y axes.
(*b*) Draw the stress block, and, by making use of its geometrical properties, find total force and moments.

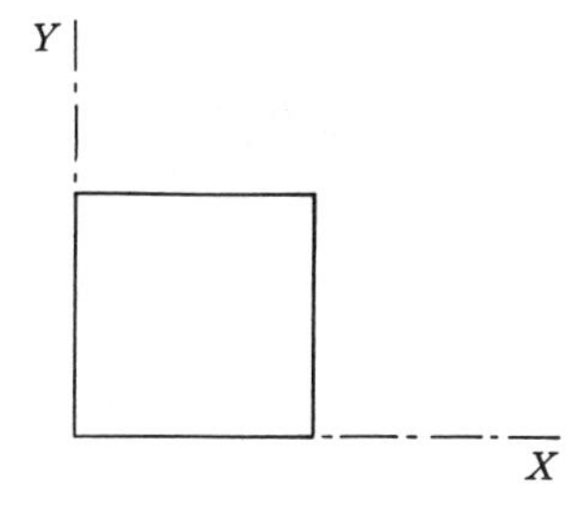

2.2 Repeat Prob. 2.1, but now assume normal-stress distribution varying as

$$\sigma = Cxy$$

2.3 The triangular area is subjected to normal stresses which vary according to

$$\sigma = Cy$$

(*a*) By integration of stresses, find the total force acting on the area, and its moment about the X axis.
(*b*) Draw the stress block, and by making use of its geometric properties, verify results of part (*a*).

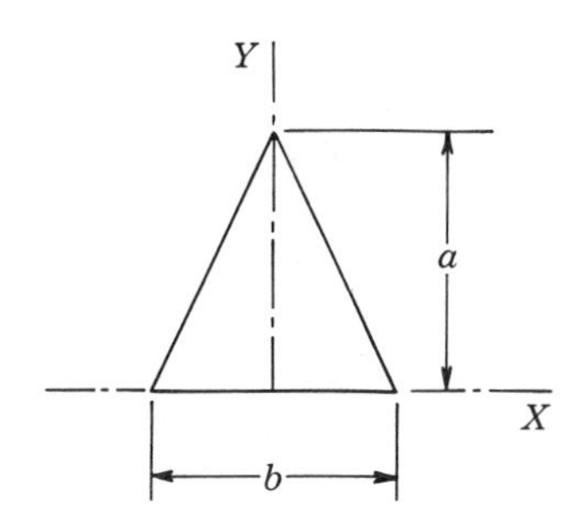

2.4 The triangular area of Prob. 2.3 is subjected to uniform shear stress, parallel to the Y axis. Find an expression for the total shear force, and draw its stress block.

2.5 A circular area of radius R is subjected to tangential shearing stresses τ varying as

$$\tau = Cr$$

where r is the radial distance to any point. Find the resultant moment about an axis normal to the area and passing through its center.

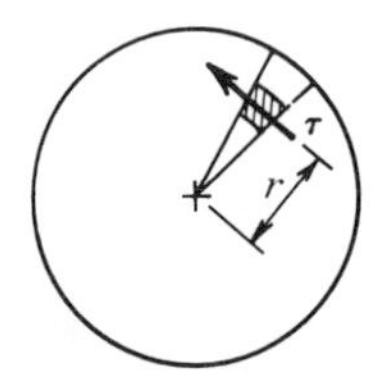

2.6 Repeat Prob. 2.5, but now consider tangential shear stresses τ which are constant in intensity.

2.7 Every point of the square section shown has acting on it a shear stress whose components parallel to the X and Y axes are τ_x and τ_y, of intensity

$$\tau_x = -Cy\left[1 - \left(\frac{x}{a}\right)^2\right]$$

and $$\tau_y = +Cx\left[1 - \left(\frac{y}{a}\right)^2\right]$$

Calculate the resultant moment about an axis normal to the area.

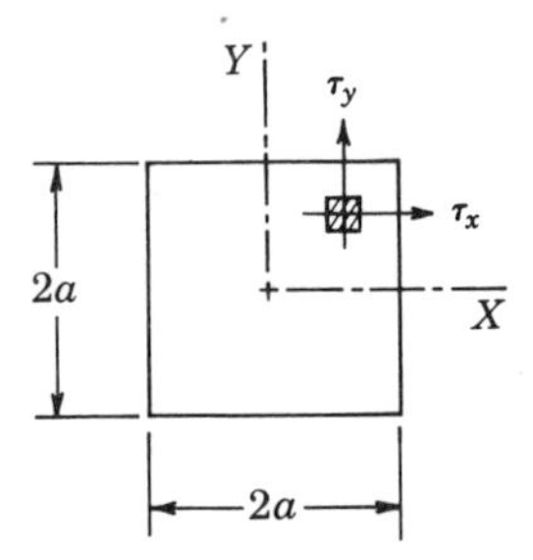

2.8 Two parallel reference lines, a finite distance b apart, are scribed on the surface of a structural member before deformation. The strain variation parallel to the X axis, during deformation, is

$$\epsilon_x = Cxy^2$$

Assuming the reference line $x = 0$ is fixed, find the equation for the deformed shape of the reference line originally at $x = b$.

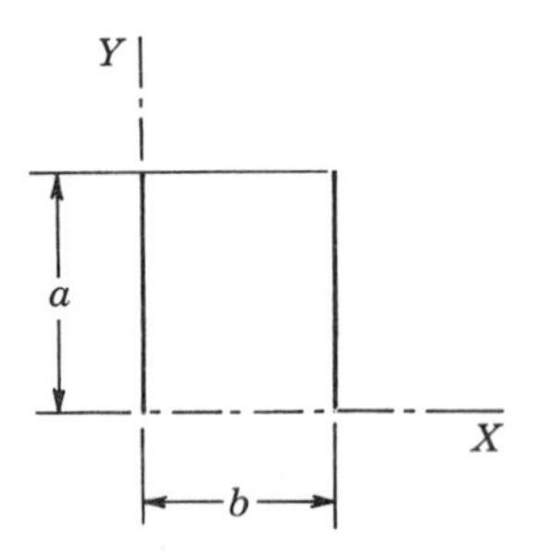

2.9 The straight line AB along the X axis is subjected to strain varying as

$$\epsilon = Ax$$

Find the deformed length of the line.

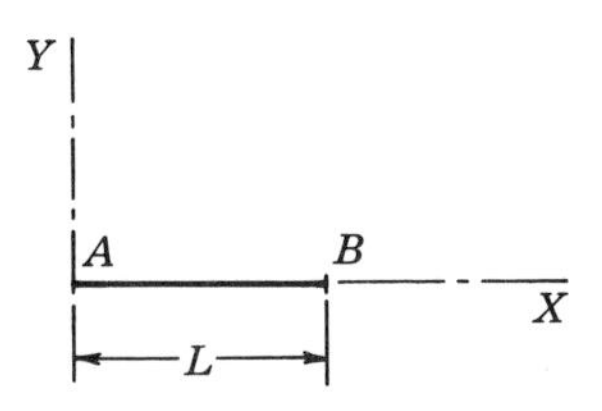

2.10 Point A is subjected to a displacement u parallel to the X axis and v parallel to the Y axis. If $u = 0.1\ L$ and $v = 0.2\ L$, find the average strain along the diagonal line OA as it deforms into the line OA'.

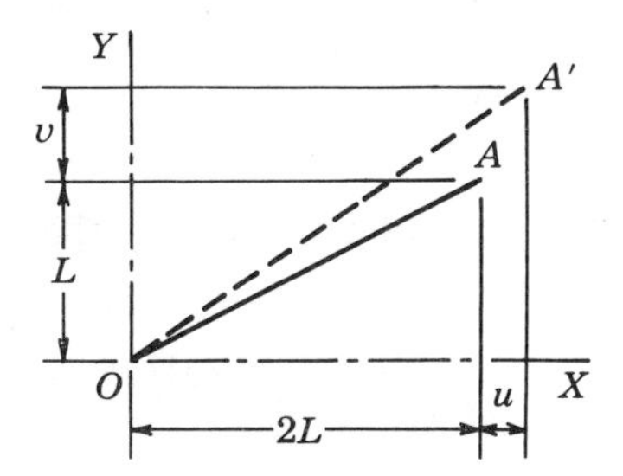

2.11 The circle of initial radius r is stretched such that its radius in the deformed state is $r + u$. Find the strain along the circle (the so-called *tangential strain*).

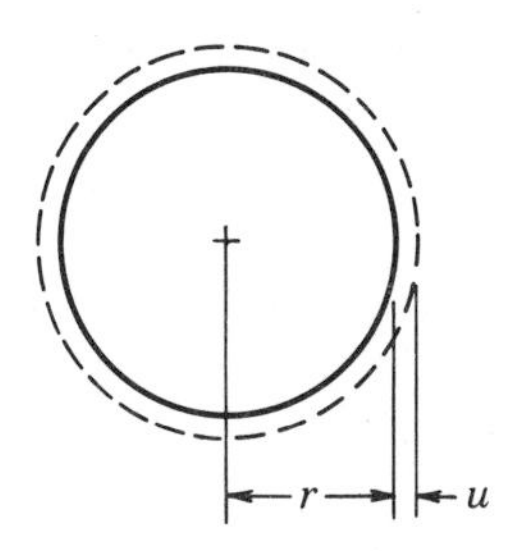

2.12 A circular shaft of radius r and length L is subjected to an angle of twist θ. Calculate the shear strain of an element on the outer surface due to this twist.

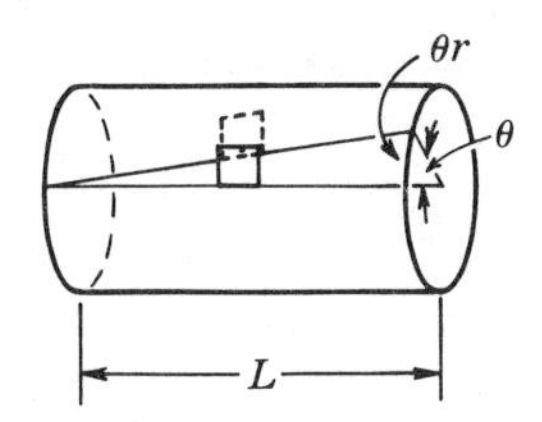

2.13 The stress-strain curve shown can be expressed by the parabola

$$\sigma = a + b\epsilon + c\epsilon^2$$

The maximum stress has a value σ_M and coincides with a strain ϵ_M. Find the values of the coefficients a, b, and c, and an expression for the tangent modulus.

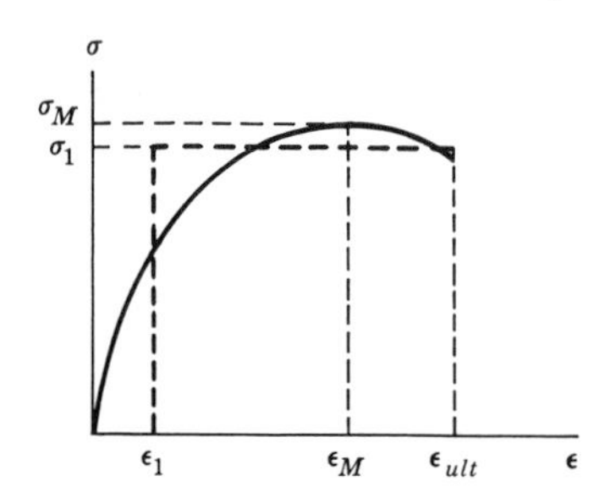

2.14 Find a rectangular stress block (of height σ_1 and base from ϵ_1 to ϵ_{ult}, as shown in dashed line in Prob. 2.13) which has the same area and the same first moment about the stress axis as the parabolic area of the stress-strain curve of Prob. 2.13.

(*a*) Specify the values of σ_1 and ϵ_1.
(*b*) If $\epsilon_M = \frac{2}{3}\epsilon_{ult}$, give the ratios σ_1/σ_M and $\epsilon_1/\epsilon_{ult}$, and compare them to those used by Whitney [see Eq. (2.17)].

2.15 A material has a stress-strain relation which can be expressed by the equation

$$\sigma = k\epsilon^n$$

(*a*) Sketch the stress-strain curve for $n < 1$, $n = 1$, $n > 1$, $n = 0$. State which of these are most likely for actual material.
(*b*) Give a general expression for the tangent modulus.
(*c*) For $0 < n < 1$, calculate the initial stiffness corresponding to $\epsilon = 0$.

2.16 The table presents data from a tension test:

σ_1, ksi	ϵ_1, in./in.
0	0
10	0.001
20	0.003

Find a parabolic expression to represent the stress-strain curve of the material.

Note: In Probs. 2.17 to 2.25, try to follow the typical steps of the strength of materials approach outlined. Label each step and look for analogies among the various problems.

2.17 Obtain an expression for the relationship between the torque applied to a shaft of circular cross section of elastic material and the resulting shear stresses and deflection. Identify each one of the steps of a typical strength of materials approach; try to draw a strict analogy with the steps and solutions of Example Problem 2.7. In particular, specify the assumption used on strain distribution.

2.18 Explain why the approach outlined in the previous problem does not serve to describe the torsional behavior of members of noncircular cross section.

2.19 In an elastic shaft of square cross section of side length $2a$, the shear strain components parallel to the X and Y axes through the centroid of the section are specified by

$$\gamma_x = -Ay\left[1 - \left(\frac{x}{a}\right)^2\right]$$

$$\gamma_y = Ax\left[1 - \left(\frac{y}{a}\right)^2\right]$$

By applying the appropriate steps of the strength of materials approach, find the relation between the applied torque and the constant A. (*Hint:* the solution of Prob. 2.7 can be used here.) Find the relation between the applied torque and the maximum shear stress in the section, and locate the point of maximum shear stress.

2.20 A torsion member of circular cross section of radius R is made of a material of the shear stress-shear strain relation

$$\tau = C\gamma^n$$

By making a simple assumption on the distribution of shear strains and following the steps of the strength of materials approach, find relations between the applied torque and the resulting maximum stress and angle of twist. Plot these relations for $n = \frac{1}{2}$.

2.21 A beam can be idealized into a "sandwich beam," in which all bending is resisted only by the narrow flanges, each of area A. Assuming uniform axial deformation of each flange, and following the steps of the strength of materials approach, find relations between the applied moment and the resulting stresses and curvature. Assume elastic material.

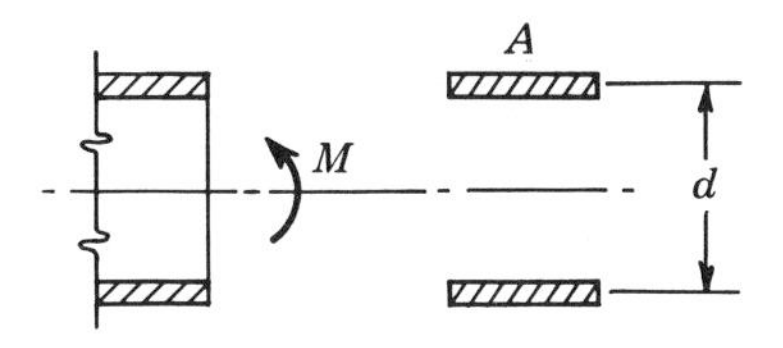

2.22 Repeat Prob. 2.21 but now assume inelastic material of stress-strain relation

$$\sigma = k\epsilon^n$$

2.23 Two elastic plates are fastened by means of a lap joint containing three equally spaced rivets. Assuming the rivets to be rigid, apply strength of materials principles to determine how the load P is transmitted through the three rivets. (*Hint:* Geometry requires that lapped portions of the two plates between rivets must stretch equally.) Compare your result with the conventional assumption.

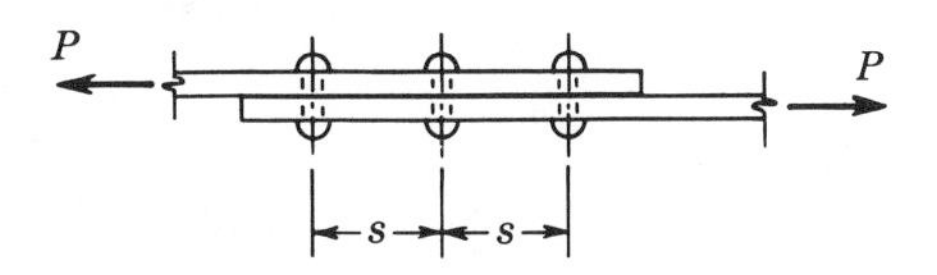

2.24 A bracket is bolted to a member by means of n bolts as shown and subjected to a pure moment M. Apply strength of materials methods (including a suitable assumption on deformations) to find the force transmitted through each bolt.

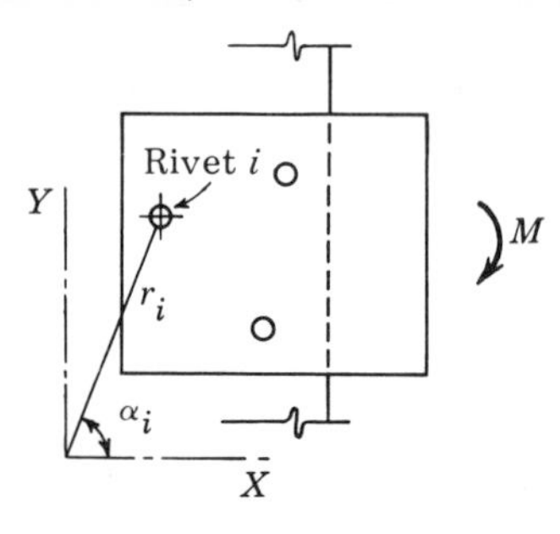

2.25 A flanged coupling is connected by means of two bolt circles, the inner one of radius R_1 and containing N_1 bolts, the outer of radius R_2 and containing N_2 bolts. Apply strength of materials methods to find the force in the inner and outer bolts due to a torque T.

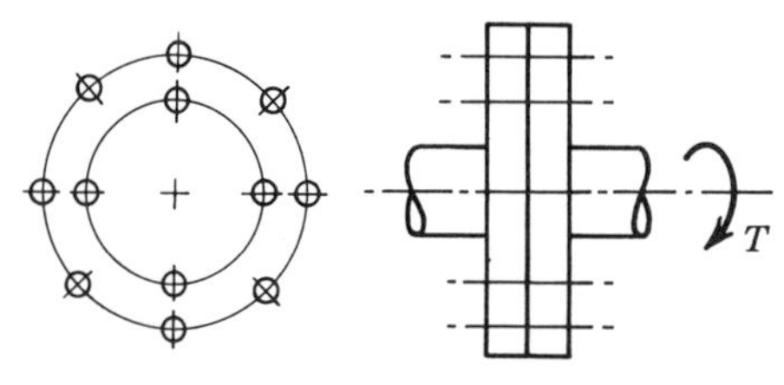

2.26 Consider a continuous elastic body whose elements are subjected to body force components of intensity $\bar{X}$, $\bar{Y}$, and $\bar{Z}$, in pounds per unit volume. These body forces can arise due to the presence of inertia or gravity forces in the body. Modify the governing equations of the problem to take these body forces into account.

2.27 A continuous elastic body, of material whose coefficient of thermal expansion is α, is subjected to a temperature field $T(x,y,z)$. Modify the governing equations of the problem to take the effects of temperature expansions into account.

2.28 Rewrite the set of governing equations (2.29) to (2.32) in cylindrical coordinates R, T, Z suitable for analysis of a circular torsion member. Verify that the stresses and deformations obtained from the strength of materials method for an elastic circular torsion member satisfy all governing and boundary equations and therefore constitute an exact solution of the problem.

2.29 Calculate the critical load on the perfectly rigid bar restrained by an elastic spring of stiffness k.

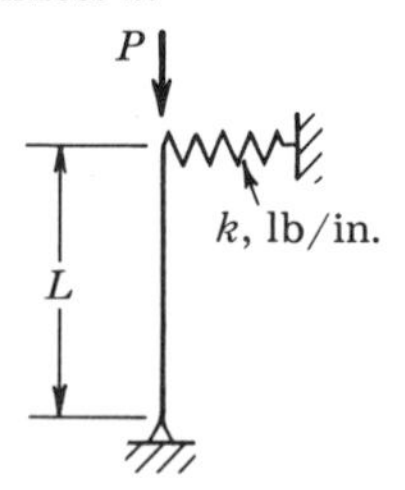

2.30 Calculate the critical load on the cantilevered column consisting of three rigid bars connected by elastic springs of stiffness k in.-lb/rad.

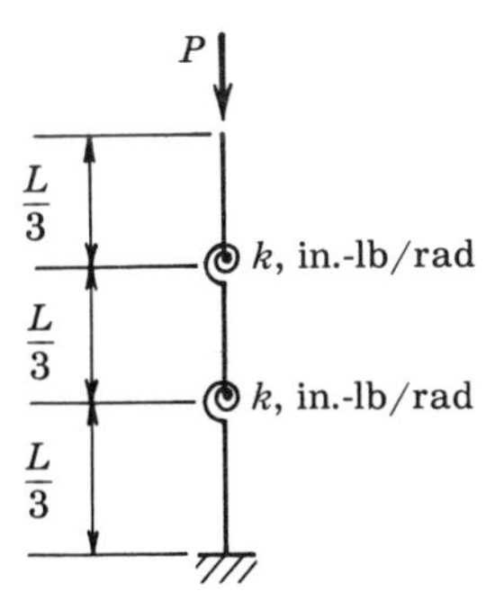

2.31 The rigid member is supported at the origin in such a fashion that it cannot rotate about the X axis. It is elastically restrained from rotation about the Y and Z axes by means of springs of stiffnesses k_y and k_z (in in.-lb/rad) respectively. A load P is applied in such a fashion that as the free end translates and rotates, the load moves with it, that is, the line of action of the applied load always passes along the center line of the section.

Under a sufficiently large load P, the structure will suddenly buckle by excessive rotation about the Y and Z axes; this phenomenon is called *lateral-torsional buckling*. It is required to calculate the critical buckling load P.

To do this, follow the typical steps outlined in this section:

First, consider the possible deformation modes (there are two degrees of rotational freedom), write the equilibrium equations of the deformed structure (vector representation is useful, but not essential), relate the reactive moments to the rotations, and find the nontrivial solutions for the resulting set of homogeneous equations.

Carefully consider procedure and results, and discuss them from a theoretical as well as an engineering viewpoint; in particular, discuss danger of lateral buckling in a beam under load.

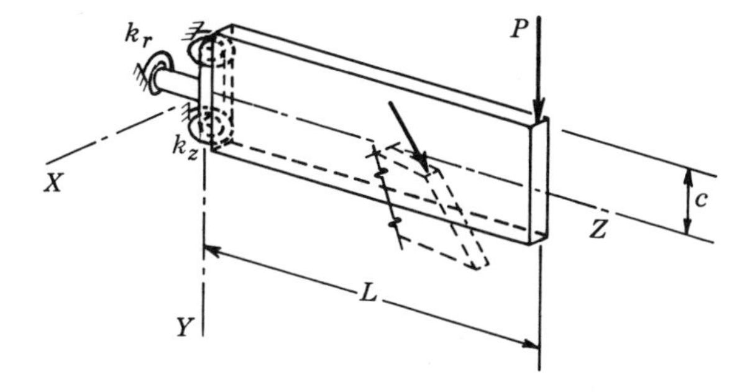

part two

Axially Loaded Members

3

Stress Analysis and Application to Design

3.1 Homogeneous Members

We consider a prismatic member under an axial load P acting at the centroid of the cross section; we then deduce, under the assumption that strain, and therefore stress σ, is uniform over the section, that equilibrium is satisfied if

$$\Sigma F = 0: \qquad P - \int \sigma \, dA = 0$$

or
$$\sigma = \frac{P}{\int dA} = \frac{P}{A} \tag{3.1}$$

The design of such a member involves only the selection of an appropriate cross-sectional area A_{reqd}, which is computed on basis of the allowable stress σ_{allow}, and the axial load P obtained from an analysis:

$$A_{reqd} = \frac{P}{\sigma_{allow}} \tag{3.2}$$

According to the elastic method, this procedure is correct only for prismatic portions of the member without any geometric irregularities. In the vicinity of irregularities such as bolt or rivet holes it is necessary either to compute stresses based on more exact deformations, to apply a stress concentration factor, or to lower the allowable stresses to compensate for zones of overstress.

To illustrate the behavior near irregularities in axially loaded members, we next consider a steel strap containing a small hole, subjected to tension;

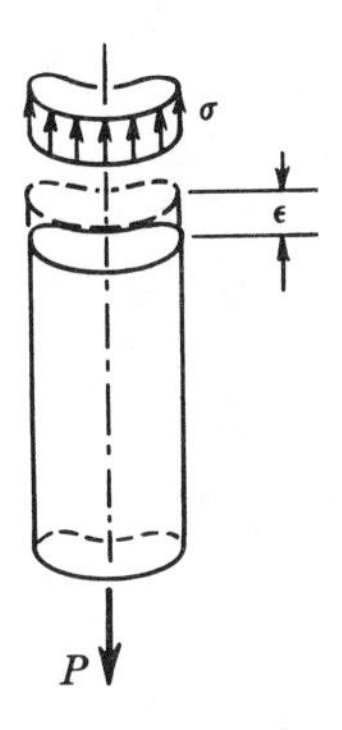

Fig. 3.1

more advanced analysis indicates that in the elastic range the stresses are distributed as shown in Fig. 3.2*a*, which shows that adjacent to the hole the stress peak reaches three times the average value; the stress concentration factor is 3.

In the following analysis, the further behavior of the strap is traced through in a simplified fashion in order to clarify the basic action of a ductile member at a point of stress concentration. The radical simplification which is here made consists of the assumption that the stress-strain relations of Fig. 3.2*b* which apply to a uniaxial state of stress are valid. Actually, a state of biaxial stress exists in the vicinity of the hole, and the stress-strain relations should be adjusted accordingly; an exact elastic-plastic analysis using these exact relations falls in the field of the theory of plasticity. The approximate approach taken here (which yields results for the ultimate strength identical with those due to the precise method) is found justified for the sake of explaining the process by which the redistribution of the stresses can take place.

The elastic behavior shown by curves 1 in the stress and strain plots will prevail till the maximum strain and stress reach their elastic limit values ϵ_{YP} and σ_{YP} respectively, as shown on the elastic–perfectly plastic stress-strain curve of Fig. 3.2*b* applicable to structural steel. We shall next pursue the action beyond the elastic limit.

Under further stretching, the strains will increase beyond ϵ_{YP} in the neighborhood of the hole, as shown by strain distribution 2. But, according

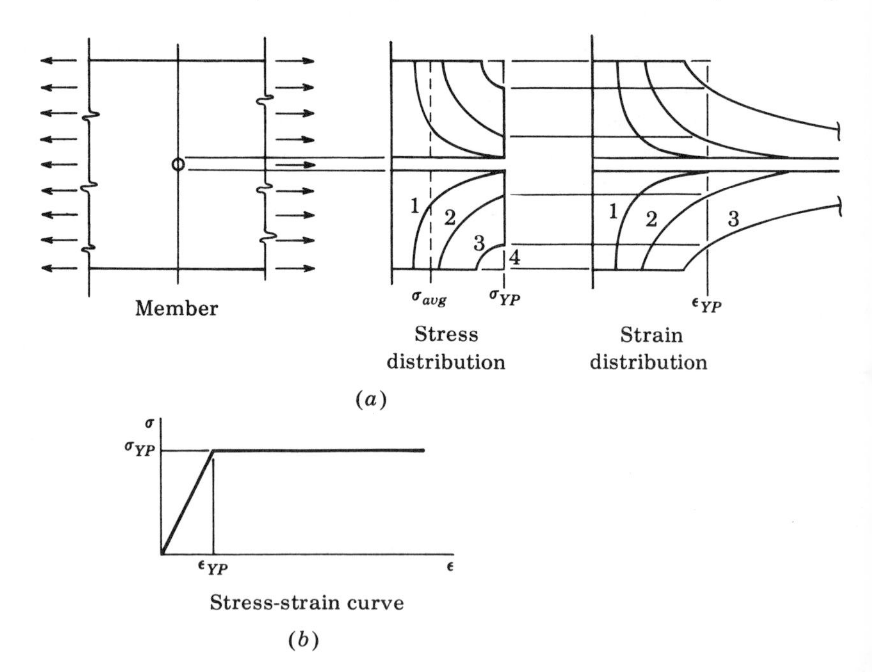

Fig. 3.2 Plastification at point of stress concentration

to the stress-strain curve, to any strain larger than ϵ_{YP} there corresponds a stress σ_{YP}; in the stress distribution, therefore, all regions subject to strains larger than ϵ_{YP} will be stressed to a value σ_{YP}, and this is indicated in curve 2 of the stress plot. Owing to further stretching, strain and stress distributions 3 will be reached. Note that during this stage, a portion of the cross section is still elastic and serves to keep the deformations to a reasonable value. Eventually, at stage 4 the strain everywhere will exceed ϵ_{YP}, and at this stage the entire cross section will have become plastified under a stress σ_{YP}. Only then can unrestricted plastic flow take place, and the strap will break on a line through the hole; this will occur under an ultimate or plastic load

$$P_P = A_{net}\sigma_{YP} \tag{3.3}$$

We note that again, owing to the ductility of the material, the stress was able to redistribute itself to the initially understressed portions of the section. Thus, by following the behavior of the strap into the inelastic range, we deduce that elastic stress peaks in ductile members tend to get ironed out prior to failure. Local elastic overstress then loses its significance, and we may possibly neglect it if we are willing to let the plastic range of the material help out. In practice, this is commonly done, and in the following examples tacit use is made of the ductility of steel.

The analysis just presented is appropriate for tension members; in members subjected to compressive forces, the danger of buckling exists, and the allowable stress is primarily a function of the slenderness ratio L/r, where L is the length of the member and r the radius of gyration of the cross section. Pending a closer look at the phenomenon of buckling, we shall restrict our consideration to tension members and compression members of such stubby proportions that buckling plays no role.

Example Problem 3.1

The Fink-type roof truss spans 40 ft. Total snow and roof dead load results in the forces applied as shown on page 58. It is required to design the lower chord of the truss according to AISC Specifications. Steel to be used is A7 (f_{YP} = 33 ksi). Connections are to be riveted, using $\frac{3}{4}$-in. rivets.

Solution: Once the loads on the structure have been determined, it should be analyzed. For this statically determinate truss, it is a simple matter and can be done by the analytical methods of joints or sections, or graphically. The results are as shown.

The two sections of the bottom chord are subjected to different tensions. If this chord is to be built of one single member, it must be designed to resist the larger force, of value $26.0^{K}T$. If a splice must be provided, then it may be advantageous to design the center portion to resist only $17.3^{K}T$.

The chords of a small roof truss are usually fashioned of two angles set back to back, spaced apart so as to receive the gusset plates. The vertical angle legs are punched for the rivets, and the missing portion of the cross section, of value $\frac{7}{8}$ in. times the angle thickness (see AISC Specifications sec. 1.14), must be deducted from the gross section to obtain the effective area which transmits the force.

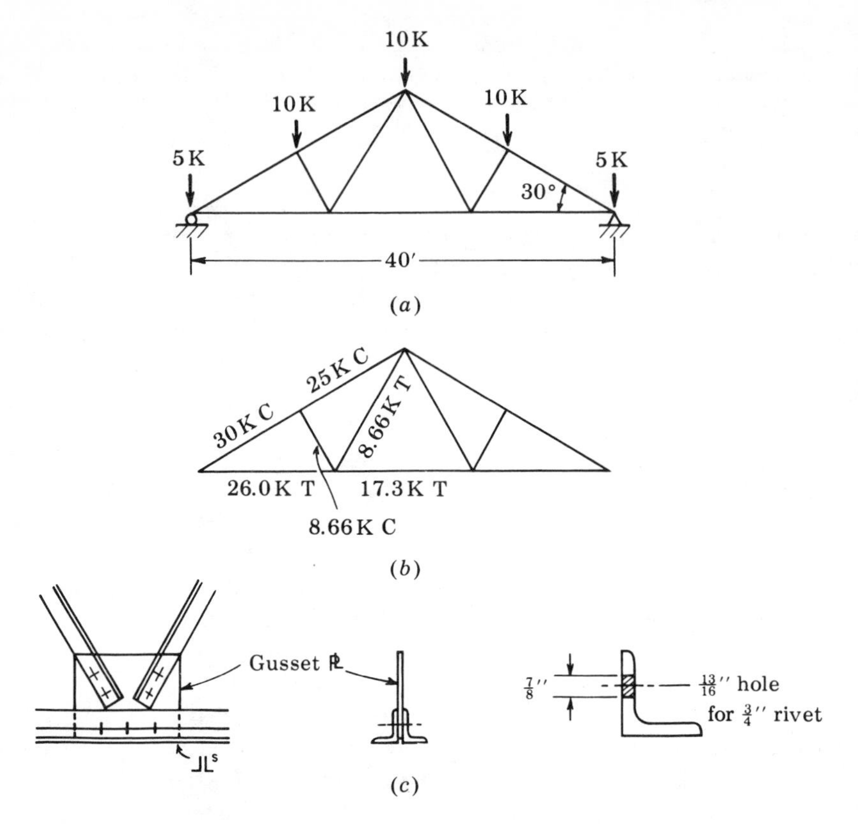

Thus, for alternate 1 (no splice), the required net area, referring for the allowable stress to AISC Specifications sec. 1.5.1.1, according to which

$$F_t = 0.60 \times 33 \text{ ksi} = 20 \text{ ksi}$$

is

$$A_{\text{net}} = \frac{P}{\sigma_{allow}} = \frac{26.0}{20} = 1.30 \text{ in.}^2$$

Assuming $\frac{1}{4}$-in.-thick angles, the gross area of *one* angle must be at least

$$\frac{1.30}{2} + \tfrac{7}{8} \times \tfrac{1}{4} \text{ in.} = 0.65 + 0.22 = 0.87 \text{ in.}^2$$

Page 1-29 of the AISC Handbook indicates that ⅃L s $2\frac{1}{2}$ by $2\frac{1}{2}$ by $\frac{3}{16}$ in. ($A_{furn} = 2 \times 0.90$ in.^{2}) are adequate.

Attention at this point should be directed to AISC Specifications sec. 1.8.4, which prescribe a maximum slenderness ratio (length/least radius of gyration) of 240 for main members. A calculation shows that here this value is

$$\frac{13.3 \text{ ft} \times 12}{0.78} = 205 < 240 \qquad \text{O.K.}$$

Therefore, use ⅃L s $2\frac{1}{2}$ by $2\frac{1}{2}$ by $\frac{3}{16}$ in.

We next investigate alternate 2 in which the possibility of substituting a smaller center member, of net area

$$A_{net} = \frac{17.3 \text{ kips}}{20 \text{ ksi}} = 0.87 \text{ in.}^2$$

Now, assuming $\frac{3}{16}$-in. thickness,

$$A_{gross} \text{ of one angle} = \frac{0.87}{2} + 0.17 = 0.61 \text{ in.}^2$$

Angles ⅃L $2 \times 2 \times \frac{3}{16}$ in. would satisfy this but not AISC Specifications sec. 1.8.4 regarding slenderness. The conclusion is that in either case ⅃L s $2\frac{1}{2}$ by $2\frac{1}{2}$ by $\frac{3}{16}$ in. should be used throughout.

3.2 Effect of Initial Stresses

All commercially available steel members have initial stresses locked in which arise from the plastic deformations occurring during cooling and straightening operations at the mill. These locked-in initial stresses have considerable influence on the behavior of members which will be discussed here.

We consider the behavior under axial load of a member of elastic–perfectly plastic material as shown in Fig. 3.3*b*, with the initial stresses shown in Fig. 3.3*a*. Note that since these stresses exist even when no external load is applied to the member, they must be *self-equilibrating*, that is, they must satisfy equilibrium of any free body. For this reason, tensile and compressive initial stresses must always occur jointly.

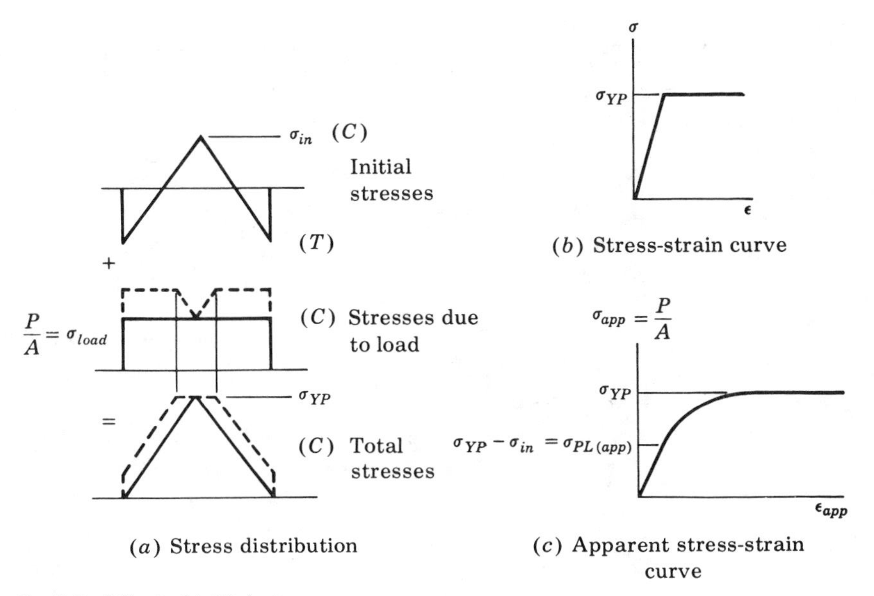

Fig. 3.3 Effect of initial stresses

Now, if, due to an applied compressive load, P/A stresses are added to the initial stresses, that fiber which had been subjected to the highest compressive initial stress will yield first, when the total stress,

$$\sigma_{total} = \left(\frac{P}{A}\right) + \sigma_{in}$$

reaches the yield value σ_{YP}. This, therefore, will occur under a load

$$\frac{P}{A} = \sigma_{YP} - \sigma_{in} \tag{3.4}$$

As soon as the first fiber yields, the load-deflection curve will deviate from a straight line, and as more and more fibers reach their yield stress (as shown by the dashed lines of Fig. 3.3*a*), this curve will become more nonlinear, as shown in Fig. 3.3*c*. The effect of initial stresses is thus to indicate an apparent proportional limit at a stress

$$\sigma_{PL,app} = \sigma_{YP} - \sigma_{in} \tag{3.5}$$

and this fact is of considerable importance in assessing the buckling strength of steel columns. This will be discussed in a later section. Meanwhile, an example problem will illustrate the approach outlined here in determining the apparent stress-strain curve due to residual stresses. We can see that the effect of initial stresses is very similar to that of a stress concentration.

Example Problem 3.2

A flat strap has the initial stresses shown locked in. The mechanical properties of the material are shown in the elastic–perfectly plastic stress-strain curve. Plot a curve of the load P versus the resulting elongation of the strap per unit length, taking into account the effect of the residual stresses. In particular, show that the ultimate strength of the strap is not affected by the residual stresses.

To obtain the load-deflection curve, we use superposition to add the uniformly distributed stresses resulting from the already existent initial stresses. The stresses due to the load P, $\sigma = P/bt$, are uniformly distributed, based on the assumption of uniform deformation and the elastic stress-strain relation

$$\epsilon = \frac{\sigma}{E}$$

The load-strain relation is therefore the well-known expression

$$\epsilon = \frac{P}{EA} = \frac{P}{Ebt}$$

This is valid till the largest total stress reaches the value σ_{YP}, which will occur under the stress distribution shown in (*c*), when the load equals $P_{YP}/2$ (where $P_{YP} = \sigma_{YP}bt$, corresponding to complete plastification of the section). Till this load is reached, the load-deflection curve is linear, of slope

$$\frac{P}{\epsilon} = Ebt = AE$$

As each fiber attains a stress σ_{YP}, it becomes plastic, and as further straining occurs, only those fibers of the section which are below yielding will be able to

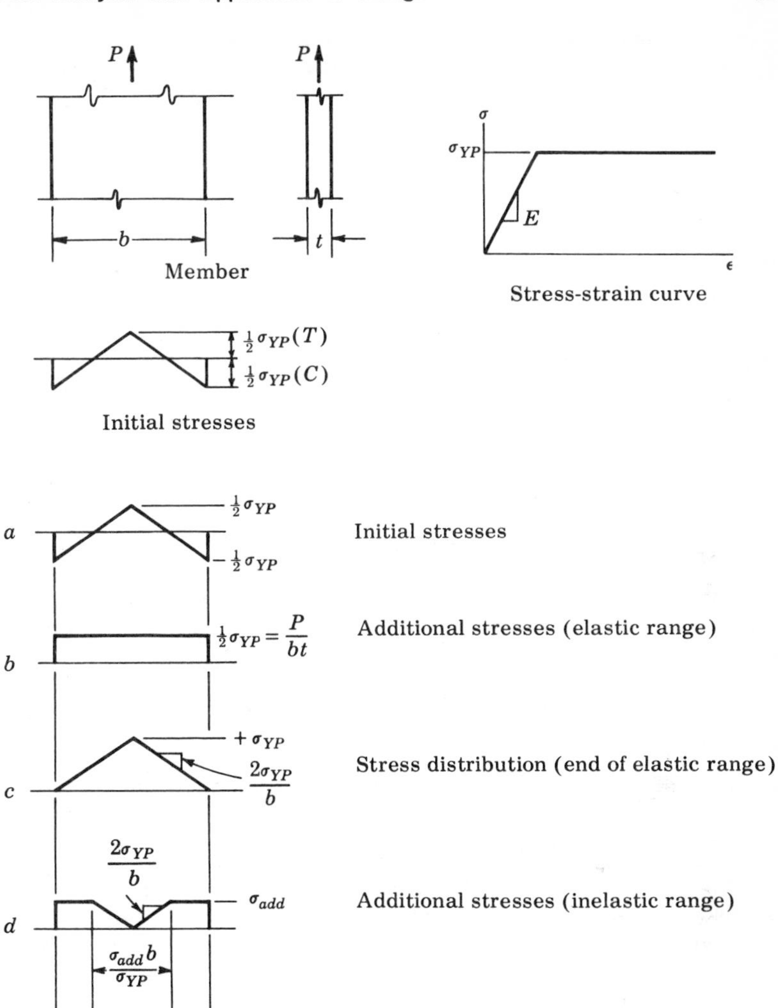
P
P
b
t
Member
σ
σ_YP
E
ε
Stress-strain curve
½σ_YP(T)
½σ_YP(C)
(a)
Initial stresses
a
½σ_YP
−½σ_YP
Initial stresses
+
½σ_YP = P/bt
b
Additional stresses (elastic range)
=
+σ_YP
2σ_YP/b
c
Stress distribution (end of elastic range)
+
2σ_YP/b
σ_add
d
Additional stresses (inelastic range)
σ_add b/σ_YP
=
σ_YP
σ_add
e
Stress distribution (inelastic range)
(b)

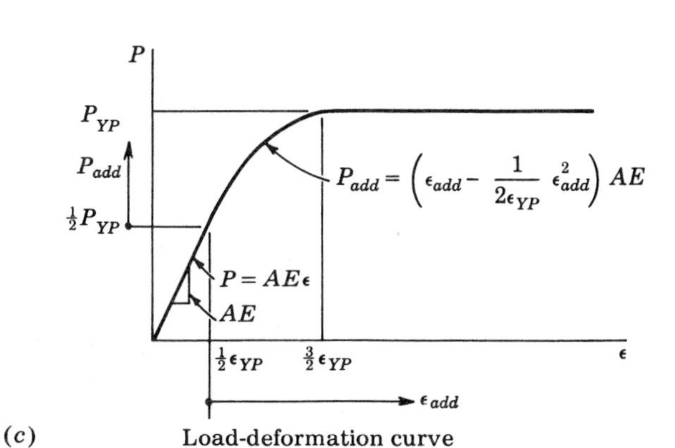
P
P_YP
P_add
½P_YP
P_add = (ε_add − 1/(2ε_YP) ε²_add) AE
P = AEε
AE
½ε_YP
3/2ε_YP
ε
ε_add
(c)
Load-deformation curve

resist additional stresses. These additional stresses are shown in d; the additional stress in the least highly stressed fiber is σ_{add} and the corresponding additional strain is

$$\epsilon_{add} = \frac{\sigma_{add}}{E}$$

The additional load is computed by summing the additional stresses over the area, so that, according to (d) in the figure,

$$P_{add} = \sigma_{add}bt - \frac{1}{2}\left(\frac{\sigma_{add}}{\sigma_{YP}}b\right)\sigma_{add}t$$

$$= \left(\sigma_{add} - \frac{1}{2}\frac{\sigma_{add}^2}{\sigma_{YP}}\right)bt$$

$$= \left(\epsilon_{add} - \frac{1}{2}\frac{\epsilon_{add}^2}{\epsilon_{YP}}\right)AE$$

This expression serves to plot the rest of the load-deformation curve which is shown plotted below.

Note that whereas the ultimate load is not affected by initial stresses, the elastic range of the strap is radically reduced owing to the premature yielding of some fibers. We shall see in a later section that this behavior has an important influence on the buckling strength of steel columns.

3.3 Elastic Behavior of Nonhomogeneous Members

We consider next a prismatic member made of two elastic materials, one of stiffness E_1 and cross-sectional area A_1, the other of corresponding values E_2 and A_2, under an axial load P.

In order to determine the stresses due to this load, it is best to follow the strength of materials approach outlined earlier. Let us then first of all consider the finite free body shown in Fig. 3.4. Equilibrium of vertical forces demands that

$$\int_A \sigma\, dA = \int_{A_1} \sigma_1\, dA + \int_{A_2} \sigma_2\, dA = P \tag{3.6}$$

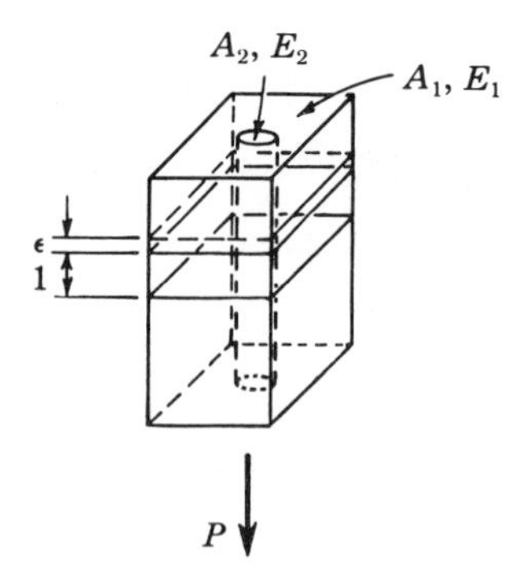

Fig. 3.4

where σ_1 and σ_2 are stresses on the two materials, of areas A_1 and A_2 respectively. To establish the distribution of strains over the cross section, we assume that every fiber of the entire section elongates equally:

$$\epsilon = \epsilon_1 = \epsilon_2 = \text{constant} \tag{3.7}$$

To relate the stresses to these strains, we utilize the elastic relationships

$$\sigma_1 = E_1\epsilon_1 = E_1\epsilon \qquad \text{over area } A_1 \tag{3.8}$$

and $$\sigma_2 = E_2\epsilon_2 = E_2\epsilon \qquad \text{over area } A_2 \tag{3.9}$$

Substituting into Eq. (3.6), and taking the constant quantities outside the integral sign, we find that

$$E_1\epsilon \int_{A_1} dA + E_2\epsilon \int_{A_2} dA = P$$

or $$\epsilon(E_1A_1 + E_2A_2) = P$$

Letting the ratio of the moduli $E_2/E_1 = n$, we can solve for the strain as

$$\epsilon = \frac{P}{E_1(A_1 + nA_2)} \tag{3.10}$$

and substituting this back into the stress-strain relations, we can solve for the stresses:

$$\sigma_1 = E_1\epsilon = \frac{P}{A_1 + nA_2} \tag{3.11}$$

$$\sigma_2 = E_2\epsilon = n\frac{P}{A_1 + nA_2} = n\sigma_1 \tag{3.12}$$

The stresses in the two materials are proportional to their stiffnesses. A larger stress is needed to deform the stiffer material through a certain strain than the more flexible.

Equations (3.11) and (3.12) can be interpreted differently: if the stress in material 1 is desired, we imagine material 2 replaced by an equivalent amount of material 1; since material 2 is n times as stiff as material 1, each unit area of material 2 must be replaced by n times that amount of material 1 in order to have the same resistance to deformation:

$$A_{2_{equiv}} = nA_2 \tag{3.13}$$

The quantity $A_{2_{equiv}}$ is called an *equivalent area;* its substitution will make the section a homogeneous one of material 1, and thus σ_1 can be found as the stress of a homogeneous section:

$$\sigma_1 = \frac{P}{A} = \frac{P}{A_1 + nA_2} = \frac{P}{A_{T1}} \tag{3.14}$$

The denominator A_T is called the *transformed area* in terms of material 1. Likewise, to find σ_2, we can transform the cross section into a fictitious one of the properties of material 2:

$$\sigma_2 = \frac{P}{(1/n)A_1 + A_2} = \frac{P}{A_{T2}} \tag{3.15}$$

The concept of transformed section is a useful one, and widely used in reinforced concrete design where the basic structure is a nonhomogeneous one. Let us apply these methods then to the stress analysis of a reinforced concrete column.

Example Problem 3.3

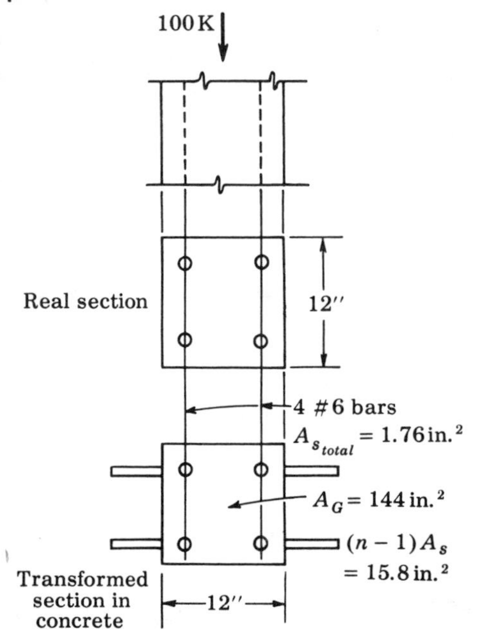

Given: Reinforced concrete column, 12 by 12 in. section, E_c = 3,000 ksi; four No. 6 reinforced bars, E_s = 30,000 ksi (No. 6 bar has $\frac{3}{4}$-in. diameter and cross-sectional area of 0.44 in.²).

Required: Stress in steel and concrete, and allowable load, according to elastic theory.

Solution: To find the concrete stress, we transform the column section into a homogeneous one of an equivalent concrete area. Since steel has n = 30,000/3,000 = 10 times the stiffness of concrete, each unit area of steel must be replaced by 10 times that amount of concrete; the area of the transformed concrete section is then, letting the gross area of the section be A_G, the area of concrete A_c, and the steel area A_s,

$$\begin{aligned} A_T &= A_c + nA_s = (A_G - A_s) + nA_s \\ &= A_G + (n-1)A_s \\ &= 144 + 9 \times 1.76 = 159.8 \text{ in.}^2 \end{aligned}$$

The concrete stress will then be calculated as that of a homogeneous section:

$$\sigma_c = \frac{P}{A_T} = \frac{100 \text{ kips}}{159.8 \text{ in.}^2} = 0.625 \text{ ksi}$$

The stress in the steel can then be determined as n times that in the surrounding concrete:

$$\sigma_s = n\sigma_c = 10 \times 0.625 = 6.25 \text{ ksi}$$

Alternately, the stresses could have been determined by transforming in terms of steel; in reinforced concrete practice it has however become standard practice to transform in tèrms of concrete.

We shall next determine the maximum allowable load which may be applied to the column of the preceding example, assuming that the following are the allowable stresses in steel and concrete:

$$\sigma_{c\ allow} = 1.0 \text{ ksi}$$
$$\sigma_{s\ allow} = 20.0 \text{ ksi}$$

By prorating the previous results, we see that the allowable concrete stress will be reached under a load of

$$P_{allow} = \frac{1.00}{0.625} \times 100 \text{ kips} = 160 \text{ kips}$$

at which time the steel stress is

$$\sigma_s = \frac{160 \text{ kips}}{100 \text{ kips}} \times 6.25 \text{ ksi} = 10 \text{ ksi}$$

We see that under a load at which the concrete capacity is exhausted, the steel stress is still way below its allowable value, and thus not very efficiently used. A *balanced design* is one in which steel and concrete reach their allowable values simultaneously.

To gain further insight in the analysis of nonhomogeneous sections and the use of the transformed-section method, we consider a reinforced concrete section with only one axis of symmetry, as shown in Fig. 3.5, and ask ourselves where an axial load would have to be applied to result in uniform strain over the entire section, that is, in pure compression of the member.

We realize that owing to the symmetry of the previous problem, moment equilibrium was satisfied by symmetric placement of the axial load. In the unsymmetric case considered now, this equilibrium is satisfied by summing moments about a convenient X axis, say one through the point of load application, and setting equal to zero.

$$\Sigma M_x = 0: \qquad \int_{A_s} \sigma_s y\, dA + \int_{A_c} \sigma_c y\, dA = 0 \tag{3.16}$$

Since, for uniform strain, $\sigma_s = n\sigma_c$,

$$\sigma_c\left(n \int_{A_s} y\, dA + \int_{A_c} y\, dA\right) = 0$$

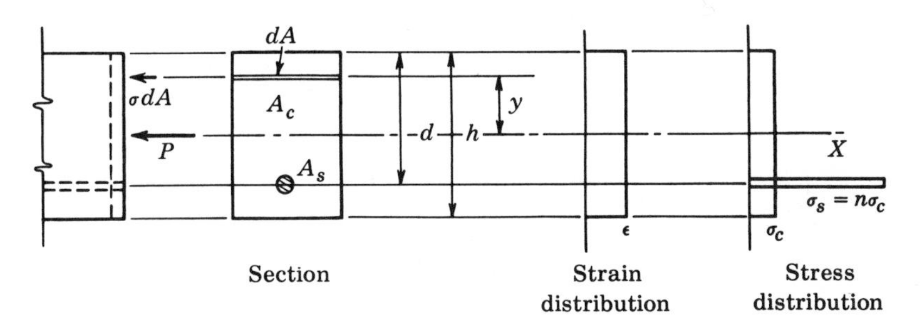

Fig. 3.5

Since σ_c is not zero, the expression in parentheses, which can be identified as the first moment of the transformed section, must be zero. To satisfy this requirement, the axial load must be applied through the centroid of the transformed section; otherwise, bending with resultant nonuniform stress and strain distribution will result.

The foregoing considerations were based on elastic behavior, as is indicated by the ratio n of the *elastic* moduli; the analysis just presented is therefore valid to the extent that the materials act elastically. In concrete, for which with some imagination we may at best visualize an elastic range only for low stresses, and which exhibits creep under long time loads, the use of elastic methods should be looked on with suspicion; the next section will discuss the effect of inelastic behavior on the action of reinforced concrete columns.

3.4 Inelastic Behavior of Nonhomogeneous Members

In the preceding section, stress analysis of nonhomogeneous members was based on the assumption of elastic behavior. If we now consider the stress-strain curves of steel and concrete, we see that the actual behavior of both materials over their full range is far different; by pursuing the behavior of the member using actual (rather than elastic) stress-strain relations we can gain more reliable information concerning the strength of reinforced concrete columns.

Again, conditions of statics demand in any case that

$$\Sigma F = 0: \qquad \int_{A_s} \sigma_s \, dA + \int_{A_c} \sigma_c \, dA = P \qquad (3.17)$$

For axial loading in which stresses are uniformly distributed, this can be reduced to

$$\sigma_s A_s + \sigma_c A_c = P \qquad (3.18)$$

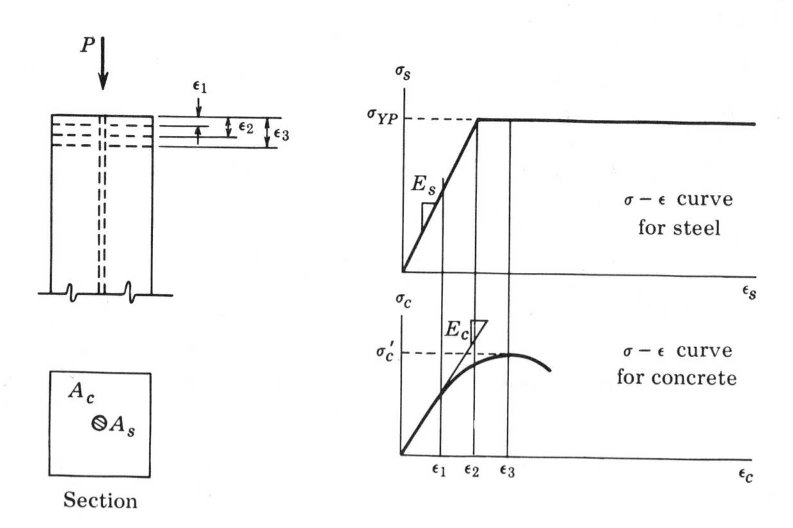

Fig. 3.6

The distribution of strains is again uniform throughout the cross section; under increasing value of the load P, the strain increases through value ϵ_1 (under which strain both materials act elastically, and for which the previous elastic analysis is therefore appropriate) to a value of ϵ_2. To compute the value of load P corresponding to this strain, we find the steel and concrete stresses σ_s and σ_c from the stress-strain curves and enter them into Eq. (3.18), from which we can find the desired value P. We now note that if the strain is further increased, the steel stress will remain constant at a value σ_{YP}, while the concrete stress increases further, till at a strain ϵ_3 the maximum concrete stress σ_c' is reached. At that time, both steel and concrete are stressed to their capacity, and the ultimate column load is, according to Eq. (3.18),

$$P_{ult} = A_s\sigma_{YP} + A_c\sigma_c' \tag{3.19}$$

Further straining will lead to reduction of concrete stress and subsequent crushing of the concrete in compression. The failure load on the column is thus given by Eq. (3.19).

Example Problem 3.4

Given: A reinforced concrete column is made of materials of the mechanical properties shown in the stress-strain curves.

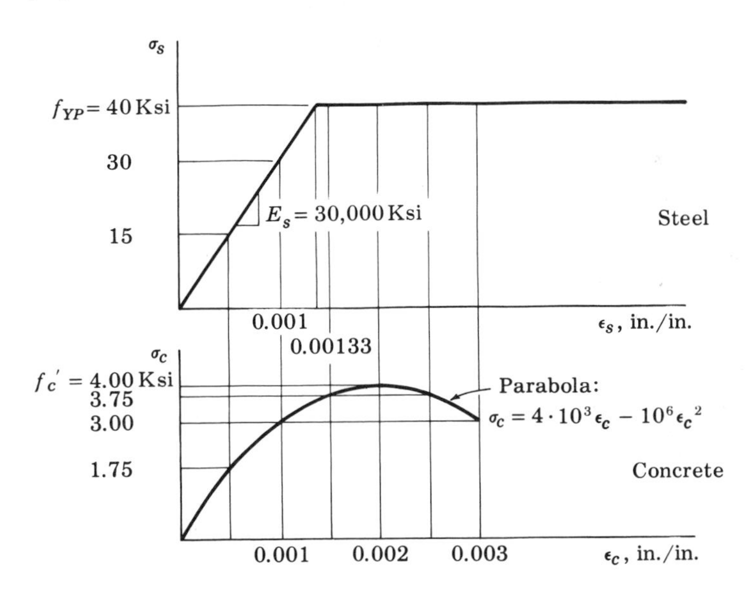

The column is 12 by 12 in. square, with four No. 9 bars (A_s of one No. 9 bar = 1 in.^{2}) as reinforcement, and 18 ft long.

Required: Plot a graph of applied axial load on the column versus the total shortening of the column, based on the actual material properties. Indicate on this plot the portion of the load taken by the steel and that taken by the concrete.

Solution: Assuming a certain value of strain, we can enter the stress-strain curves to find the corresponding stresses. These, multiplied by the areas over which they act, will give the loads carried by steel and concrete:

$$P_s = \sigma_s A_s = \sigma_s \times 4.00 \text{ in.}^2$$
$$P_c = \sigma_c A_c = \sigma_c (A_G - A_s) = \sigma_c \times 140 \text{ in.}^2$$

The total load is obtained by adding the two, and the total shortening by multiplying the assumed strain by the column length:

$$\Delta_{col} = \epsilon \times 216 \text{ in.}$$

The entire computation is done in tabular fashion:

ϵ, *in./in.*	σ_s, *ksi*	P_s, *kips*	σ_c, *ksi*	P_c, *kips*	P_{total}, *kips*	Δ, *in.*
0	0	0	0	0	0	0
0.0005	15	60	1.75	245	305	0.108
0.0010	30	120	3.00	420	540	0.216
0.0015	40	160	3.75	525	685	0.324
0.0020	40	160	4.00	560	720	0.432
0.0025	40	160	3.75	525	685	0.540
0.0030	40	160	3.00	420	580	0.648

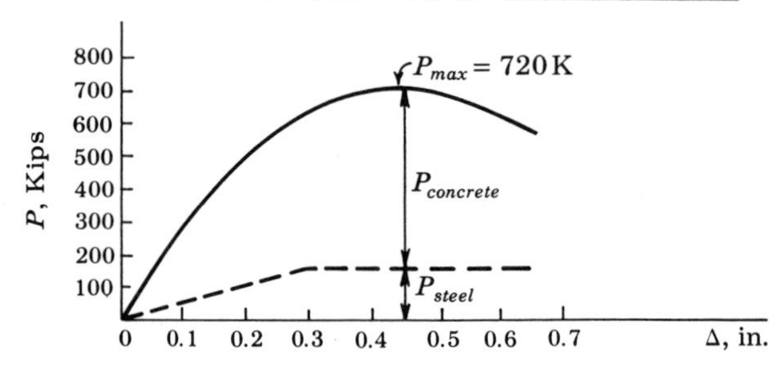

An alternate, and more elegant, procedure would be to express the stresses in Eq. (3.18) as functions of strain, that is, for the steel,

$$\sigma_s = E_s\epsilon = 30{,}000\epsilon \qquad \text{in the elastic range}$$

and

$$\sigma_s = f_{YP} = 40 \text{ ksi} \qquad \text{in the inelastic range}$$

while

$$\sigma_c = 4 \times 10^3\epsilon - 10^6\epsilon^2$$

for the full range of concrete strains. Equation (3.18) will then become

$$P = 4.0 \times 30{,}000\epsilon + 140(4 \times 10^3\epsilon - 10^6\epsilon^2)$$

for the elastic range of the steel, that is, for $\epsilon \leqq 0.00133$ in./in., and

$$P = 4.0 \times 40 + 140(4 \times 10^3\epsilon - 10^6\epsilon^2)$$

for strains above that value. Plotting this expression will again result in the previously drawn graph.

3.5 Code Provisions and Design of Reinforced Concrete Columns

If we compare the action of steel and concrete at failure to that under elastic loads (at which it was seen that the steel was greatly understressed), we see that under inelastic action a redistribution of stresses has taken place which has enabled the full strength of the steel to be brought into play before failure. A consideration of elastic stresses alone does not therefore lead to a realistic appraisal of the strength of the member. An analysis for the collapse load is therefore indicated, and this approach is followed in the ACI Code (chap. 14). According to this procedure, Eq. (3.19) is rewritten[1] in terms of the "steel ratio" $p_G = A_s/A_G$ as

$$P_{ult} = A_G\,(\sigma_c + p_G\sigma_{YP}) \tag{3.20}$$

The allowable load is then obtained by applying appropriate safety factors: for steel, 40 percent of the yield stress, and for concrete, because of variability resulting from field placement, only 25 percent of its ultimate value. The allowable column load, as given by eq. 14-1 of the ACI Code, is then

$$P_{allow} = A_G\,(0.25\sigma_c' + p_G \times 0.40\sigma_{YP}) \tag{3.21}$$

It should be noted that the stress values in Eq. (3.21) do not represent stresses actually prevailing under the applied load P_{allow} but have been obtained by consideration of conditions at collapse of the member. This criterion, therefore, is an application of the *ultimate load* or *collapse* method of design.

The safety factors leading to Eq. (3.21) are considered appropriate for "spiral columns," in which the vertical reinforcing bars are arranged circularly and spirally wrapped with a steel rod, as shown in Fig. 3.7.

[1] In Eq. (3.20) A_G should, more precisely, be A_c; the difference between the two is usually slight.

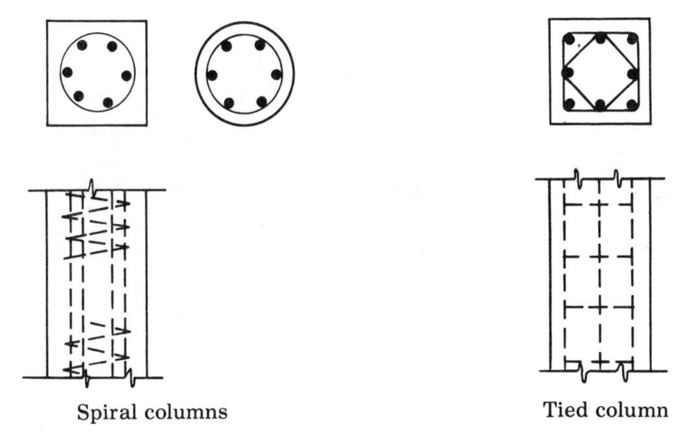

Fig. 3.7

Columns in which the vertical bars are arranged in a rectangular array, and stayed laterally by individual rectangular hoops or "ties," are called *tied columns.* Probably because of lack of polar symmetry, this type of column is found to be weaker than spiral columns, and, according to sec. 1403 of the ACI Code, the allowable load on tied columns is only 85 percent of that given by Eq. (3.21).

Example Problem 3.5 Using sec. 1403 of the ACI Code, find the allowable load on the tied column of Example Problem 3.4 and its safety factor against collapse.

$$P_{allow} = 0.85 \times 144(0.25 \times 4.0 \text{ ksi} + 0.40 \times 40 \text{ ksi} \times 0.0278)$$
$$= \underline{177 \text{ kips}}$$

$$\text{Safety factor} = \frac{P_{\text{failure}}}{P_{allow}} = \frac{720 \text{ kips}}{177 \text{ kips}} = \underline{4.07}$$

While the preceding discussion of reinforced concrete columns has been concerned mainly with *analysis*, Eq. (3.21) is equally well suited to *design*. If a column is to be designed, the allowable stresses and the load to be resisted are known, while the gross area A_G and the steel ratio p_G are unknowns. We assume one of these and solve for the other. Since the steel ratio, according to ACI sec. 913, must fall between the values of 1 to 8 percent (0.01 to 0.08), an average value, say 0.04, can be assumed, the corresponding A_G determined, and the column dimensions and steel bars set. Note that the higher the steel ratio, the smaller the column, but the spacing requirements of ACI sec. 804 may be hard to satisfy with a large amount of steel.

A number of secondary design considerations are of importance and will be briefly mentioned here (sections refer to 1963 ACI Code):

1. Section 912 provides for minimum column sizes. To avoid buckling of the compression member, the rules of secs. 915 and 916 (which prescribe reductions of allowable load for slender columns) should be followed.
2. Amount of longitudinal steel, bar sizes, and spacing requirements are laid down in secs. 913 and 804. Steel ratio should fall between 1 and 8 percent. A minimum of six bars of No. 5 minimum size is specified; they must be spaced far enough apart so that good bond between the inner concrete core and the outer shell prevails; this spacing is therefore related to the maximum size of aggregate used.
3. The longitudinal reinforcing bars will tend to buckle under axial compression, thus leading to premature failure by spalling off the outer concrete shell, as shown in Fig. 3.8. This is prevented by providing lateral reinforcement, spirally wound for spiral columns, lateral ties for tied columns. Amounts and spacing requirements for this reinforcing are laid down in ACI secs. 806 and 913. Since this lateral reinforcing is of vital importance to the strength of the column, these provisions should be strictly adhered to.

4. The slenderness of columns affects their strength, and sec. 916 gives correction factors R for slender concrete columns. Slenderness ratio and end-fixity conditions are variables; while the section provides for increase of design loads resulting from slenderness, it is equally appropriate (and possibly clearer) to decrease the strength by multiplying the allowable load of eq. 14-1 by this R factor (which is always less than unity).

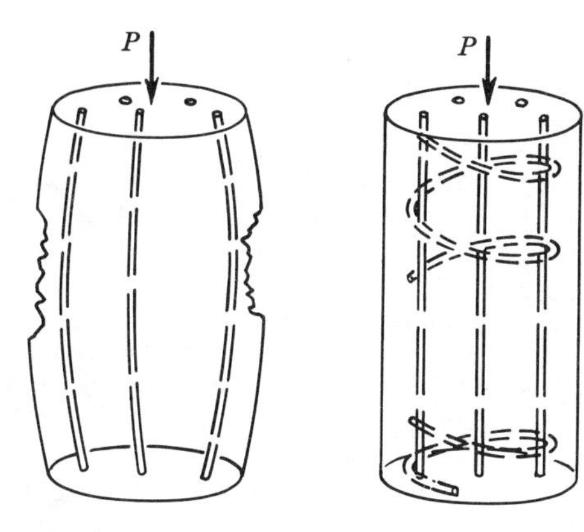

Fig. 3.8

Example Problem 3.6 Design a circular spirally reinforced concrete column 15 ft long to support an axial load of 500 kips, following the ACI Code. Economy demands a small amount of reinforcing steel. Consider ends fixed against rotation and displacement. Concrete strength: $f'_c = 3.75$ ksi; steel strength: $f_{YP} = 50$ ksi.

Solution: Since a small amount of reinforcing is called for, let us assume $p_G = 0.015$. Then Eq. (3.21) can be solved for A_G. We assume that the column may turn out to weigh 10 kips, and this dead load must be added to the superimposed load.

$$A_{G,\,reqd} = \frac{500 \text{ kips} + 10 \text{ kips}}{0.938 \text{ ksi} + 0.015 \times 20 \text{ ksi}} = 412 \text{ in.}^2$$

Use 23-in. diameter column: $A_{G,\,furn} = 415 \text{ in.}^2$

Section 916*a* of the ACI Code requires a slenderness correction in case the ratio of column height h to radius of gyration r exceeds 60; here,

$$\frac{h}{r} = \frac{15 \text{ ft} \times 12}{5 \times 75 \text{ in.}} = 31$$

so no correction is required for slenderness. The actual column dead weight is 7 kips.

To find the actual amount of longitudinal steel required, we subtract that portion of the load which can be safely taken by the concrete from the total load:

$$P_s = A_{s,\,reqd} \times 20 \text{ ksi} = 507 \text{ kips} - (415 \text{ in.}^2 \times 0.938 \text{ ksi})$$
$$A_{s,\,reqd} = 5.90 \text{ in.}^2$$

Use six No. 9 *bars:* $A_{s,\,furn} = 6.00$ in.²

$$p_{G,\,act} = \frac{6.00}{415} = 0.014$$

Spacing requirements (sec. 804*d*) are satisfied.

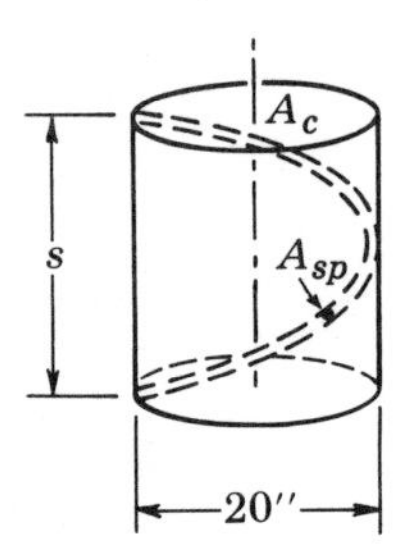

The design of the spiral steel follows eq. 9-1 of sec. 913, and the requirements of sec. 806*a*.

We first calculate the cross-sectional area of the column core A_c. The diameter of this core is determined by deducting from the overall diameter twice the clear concrete cover (according to ACI sec. 808*c*, $1\frac{1}{2}$ in.): [23 in. − 2($1\frac{1}{2}$ in.)] = 20 in.

$$A_c = \frac{\pi}{4} \times 20^2 = 314 \text{ in.}^2$$

Turning to eq. 9-1, we see that the spiral steel ratio p_s (which, according to the definition of sec. 900, is the ratio of the *volume* of steel to the *volume* of concrete contained in the column core, that portion, of area A_c, of the column section contained within the spirals) must have a minimum value

$$p_{s,\,reqd} = 0.45\left(\frac{415 \text{ in.}^2}{314 \text{ in.}^2} - 1\right)\frac{3.75 \text{ ksi}}{50.0} = 0.0108$$

To determine the amount of spiral steel per length s of column corresponding to this ratio, we set

$$p_{s,\,reqd} = 0.0108 = \frac{\text{vol. of spiral}}{\text{vol. of conc. core}} = \frac{\text{vol. of spiral}}{314 \text{ in.}^2 \times s}$$

or

$$\text{Volume of spiral} = 3.40s \text{ in.}^3$$

The spiral size must be selected so that the spacing will conform to the provisions of sec. 806*a*. Taking a spiral of cross-sectional area A_{sp}, and considering one spiral turn, of pitch s, gives a spiral volume of approximately

$$\text{Volume} = A_{sp} \times \pi \times 20.0 = 62.8A_{sp} = 3.40s$$

Solving for the pitch s gives

$$s = 18.5A_{sp}$$

The following table gives the required pitch for various-sized spirals:

Size: dia, in.	A_{sp}, *in.*^{2}	s_{reqd}, *in.*	s_{act}, *in.*
$\frac{1}{4}$	0.05	0.92	
$\frac{3}{8}$	0.11	2.03	2
$\frac{1}{2}$	0.20	3.70	

Note that according to sec. 806*a*, the maximum pitch can be one-sixth of the core diameter, or $20.0/6 = 3.33$ in., or 3 in. plus the diameter of the cross section of the spiral steel. The minimum pitch (predicated on the requirement of good contact between core and outer concrete shell) is to be $1\frac{3}{8}$ in. plus the cross-sectional spiral diameter. Note that only the $\frac{3}{8}$-in.-diameter spiral satisfies these requirements.

Final design:

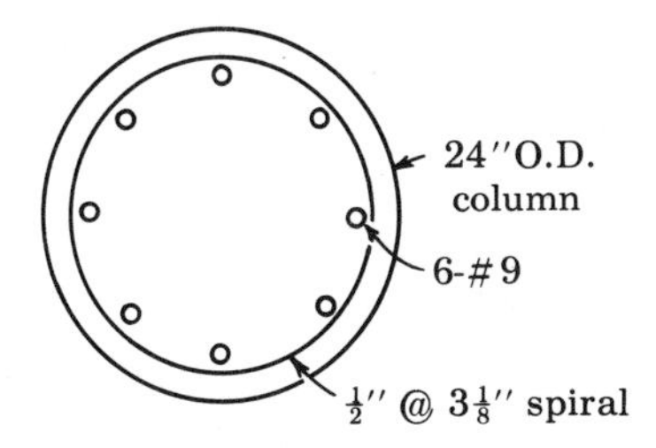

Composite and combination columns Composite columns are those in which structural steel members are encased in reinforced concrete; again inelastic considerations are used to arrive at an allowable load given by eq. 14-2 of ACI sec. 1404. Additional requirements are also laid down in that section.

Combination columns are composed of steel members and concrete; a commonly used combination column is a steel pipe filled with concrete; such members may be designed according to secs. 1405 and 1406 of the ACI Code.

3.6 Elastic Buckling of Ideal Columns

The nature of the buckling phenomenon was discussed in Sec. 2.7, and the analysis of this problem outlined there consisted of three steps: equilibrium of the buckled shape, force-deformation relation, and eigenvalue solution of a set of homogeneous simultaneous equations.

Here, we shall consider the classical case of an elastic, pin-ended column under axial load (the so-called "Euler column"), of bending stiffness EI and length L, as shown in Fig. 3.9. For step 1, we consider the typical element of the buckled member of Fig. 3.10 and write the appropriate equilibrium conditions:

$$\Sigma F_y = 0: \qquad (V + dV) - V = 0 \qquad dV = 0$$
$$\Sigma M_0 = 0: \qquad M - (M + dM) - V\,dx - P_{cr}\,dy = 0$$

or
$$\frac{dM}{dx} + V + P_{cr}\frac{dy}{dx} = 0 \tag{3.22}$$

Differentiating once more with respect to x, we get

$$\frac{d^2M}{dx^2} + P_{cr}\frac{d^2y}{dx^2} = 0 \tag{3.23}$$

Step 2 relates the internal moment M to the deflection:

$$M = EI\frac{d^2y}{dx^2}$$

and substituting this into Eq. (3.23), we get

$$\frac{d^4y}{dx^4} + \frac{P_{cr}}{EI}\frac{d^2y}{dx^2} = 0 \tag{3.24}$$

This is a fourth-order, linear, homogeneous differential equation with constant coefficients, the solution of which (using, for instance, operational methods) is

$$y = C_1 + C_2x + C_3 \sin\sqrt{\frac{P_{cr}}{EI}}\,x + C_4 \cos\sqrt{\frac{P_{cr}}{EI}}\,x \tag{3.25}$$

We need four boundary conditions, two at each end of the member, in order to solve for the four constants of integration. Considering, for

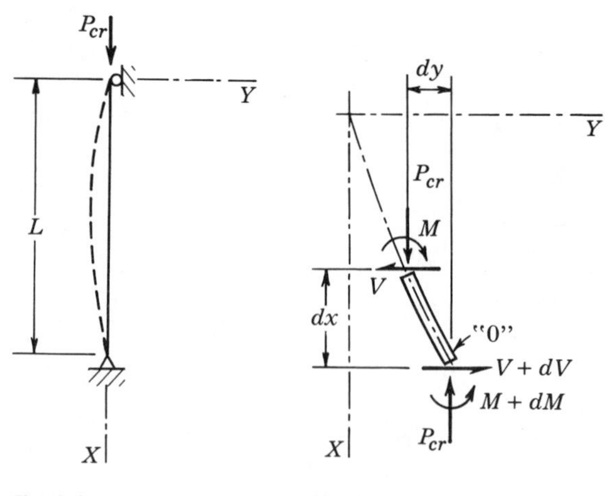

Fig. 3.9

Fig. 3.10

example, the pin-ended so-called Euler column of length L shown in Fig. 3.13a, we see that at both ends, that is, at $x = 0$ and $x = L$, the deflection y and the moment $M = EI\,(d^2y/dx^2)$ vanish:

$$\begin{aligned}
y\Big|_{x=0} &= 0: \quad C_1 + 0 + 0 + C_4 = 0 \\
\frac{d^2y}{dx^2}\Big|_{x=0} &= 0: \quad 0 + 0 + 0 - \frac{P_{cr}}{EI} C_4 = 0 \\
y\Big|_{x=L} &= 0: \quad C_1 + LC_2 + \sin\sqrt{\frac{P_{cr}}{EI}}\,LC_3 + \cos\sqrt{\frac{P_{cr}}{EI}}\,LC_4 = 0 \\
\frac{d^2y}{dx^2}\Big|_{x=L} &= 0: \quad 0 + 0 - \frac{P_{cr}}{EI}\sin\sqrt{\frac{P_{cr}}{EI}}\,LC_3 - \frac{P_{cr}}{EI}\cos\sqrt{\frac{P_{cr}}{EI}}\,LC_4 = 0
\end{aligned} \tag{3.26}$$

Equations (3.26) constitute a set of homogeneous simultaneous equations (that is, the right-hand side of all equations is zero), and such a set can have a solution only if the determinant of the coefficients of the unknowns is zero; accordingly in step 3 we equate this determinant to zero:

$$\begin{vmatrix}
1 & 0 & 0 & 1 \\
0 & 0 & 0 & -\dfrac{P_{cr}}{EI} \\
1 & L & \sin\sqrt{\dfrac{P_{cr}}{EI}}\,L & \cos\sqrt{\dfrac{P_{cr}}{EI}}\,L \\
0 & 0 & -\dfrac{P_{cr}}{EI}\sin\sqrt{\dfrac{P_{cr}}{EI}}\,L & -\dfrac{P_{cr}}{EI}\cos\sqrt{\dfrac{P_{cr}}{EI}}\,L
\end{vmatrix} = 0 \tag{3.27}$$

Note that the only unknown in this equation is P_{cr}/EI, and if we expand the determinant to solve for this unknown, we find the "characteristic equation"

$$L\left(\frac{P_{cr}}{EI}\right)^2 \sin\sqrt{\frac{P_{cr}}{EI}}\,L = 0$$

Since presumably $L\left(\frac{P_{cr}}{EI}\right)^2$ is not zero, then

$$\sin\sqrt{\frac{P_{cr}}{EI}}\,L = 0 \tag{3.28}$$

which can only be if the argument (or angle)

$$\sqrt{\frac{P_{cr}}{EI}}\,L = 0, \pi, 2\pi, \ldots, n\pi, \ldots \tag{3.29}$$

from which

$$P_{cr} = \frac{(n\pi)^2 EI}{L^2} \qquad n = 0, 1, 2, \ldots \tag{3.30}$$

Substituting this value back into the equation for the deflected shape, Eq. (3.25), and the boundary conditions, Eqs. (3.26), we find that the lateral deflection of the column at buckling will be in the form of a sine curve:

$$y = C_3 \sin \frac{n\pi x}{L} \tag{3.31}$$

where the maximum deflection C_3 is indeterminate. Plotting the shapes given by Eq. (3.31) for various values of n, as in Fig. 3.11, we see the meaning of the solution for the different n's; the lowest meaningful critical load (also called the *Euler load*) corresponds to $n = 1$ and applies in cases where the column is unrestrained against deflection over its full length. If by any means, such as lateral bracing, the column can be forced to buckle into two half waves, as in Fig. 3.11*c*, then the buckling load is four times the Euler value. Obviously a column may be stiffened tremendously against buckling by providing suitable bracing.

The result of the analysis as given by Eq. (3.30) may be put in different form by dividing both sides of the equation by the cross-sectional area A:

$$\frac{P_{cr}}{A} = \frac{\pi^2 E}{L^2} \frac{I}{A}$$

or, calling the average compressive stress at buckling $P_{cr}/A = \sigma_{cr}$, and remembering that the radius of gyration $r = \sqrt{I/A}$,

$$\sigma_{cr} = \frac{\pi^2 E}{(L/r)^2} \tag{3.32}$$

This result can be plotted in the form of a buckling curve which shows the critical stress σ_{cr} as a function of the slenderness ratio L/r. The hyperbola of Fig. 3.12 results. The critical stress varies as the square of the slenderness ratio, the most important column parameter. To determine the buckling load or stress of a column, it is necessary to determine that cross-sectional axis of the member with respect to which the slenderness ratio is a maximum.

We should recall that our analysis was based on elastic considerations, as evidenced by the presence of the modulus E in Eq. (3.32). For this

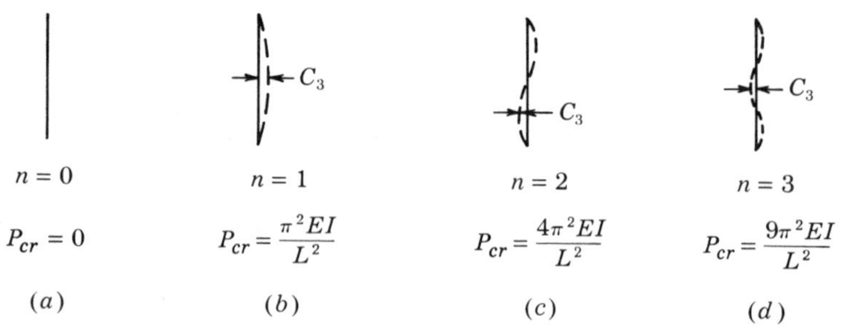

(a)	(b)	(c)	(d)
$n = 0$	$n = 1$	$n = 2$	$n = 3$
$P_{cr} = 0$	$P_{cr} = \frac{\pi^2 EI}{L^2}$	$P_{cr} = \frac{4\pi^2 EI}{L^2}$	$P_{cr} = \frac{9\pi^2 EI}{L^2}$

Fig. 3.11

reason, our analysis ceases to be valid when σ_{cr} exceeds the proportional limit stress, and that portion of the curve of Fig. 3.12 which is dashed does not possess any physical meaning. The particular value of the L/r ratio corresponding to this stress divides the slenderness ratio into the *slender column range*, for which Euler's formula holds, and the *intermediate column range*, for which buckling occurs only after the proportional limit stress has been exceeded. A theory of plastic buckling is required for such intermediate-range columns and will be covered in a subsequent section.

For example, if we consider steel of $E = 30{,}000$ ksi, $\sigma_{PL} = 30$ ksi, then substituting these values into Eq. (3.32) we find that the corresponding L/r ratio is 99. Elastic buckling results cannot be used for steel columns of L/r ratio less than this value.

If a full-size steel member is subjected to an axial load test, it is found that, owing to the presence of initial stresses in some fibers of the member, the stress-strain curve of such a member will deviate from the straight line already at a stress of about half the yield-point value as discussed in Sec. 3.2. As far as overall behavior of the member is concerned, this may then be considered as the proportional limit stress. To find the corresponding L/r ratio (denoted by C_c) limiting the slender-column range, we substitute this value into Eq. (3.32) and find

$$\frac{L}{r} = \sqrt{\frac{2\pi^2 E}{\sigma_{YP}}} = C_c \qquad (3.33)$$

Equation (3.33) appears in the 1961 AISC Specifications. For columns of greater slenderness, this code utilizes the Euler formula, Eq. (3.32), with $E = 29{,}000$ ksi, to which a safety factor of 1.92 has been applied:

$$\sigma_{allow} = \frac{\sigma_{cr}}{1.92} = \frac{\pi^2 \times 29{,}000{,}000}{1.92(L/r)^2} = \frac{149{,}000{,}000}{(L/r)^2} \quad \text{(in psi)} \qquad (3.34)$$

This equation appears as formula 2 in the AISC Specifications.

Our discussion so far has concerned itself only with pin-ended or simply supported columns. The procedure which has been outlined lends itself equally well to the determination of buckling loads for any boundary condi-

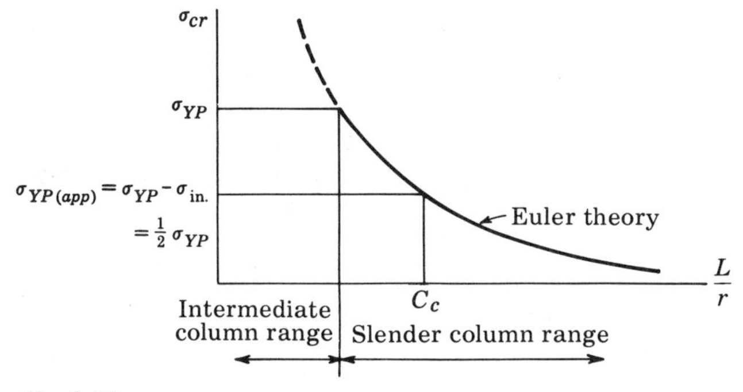

Fig. 3.12

tion. Taking, for instance, the case of a column fixed against rotation at both ends, as shown in Fig. 3.13b, we set

$$y\Big|_{x=0} = 0: \quad C_1 + 0 + 0 + C_4 = 0$$

$$\frac{dy}{dx}\Big|_{x=0} = 0: \quad 0 + C_2 + \sqrt{\frac{P_{cr}}{EI}}\,C_3 + 0 = 0$$

$$y\Big|_{x=L} = 0: \quad C_1 + LC_2 + \sin\sqrt{\frac{P_{cr}}{EI}}\,LC_3 + \cos\sqrt{\frac{P_{cr}}{EI}}\,LC_4 = 0$$

$$\frac{dy}{dx}\Big|_{x=L} = 0: \quad 0 + C_2 + \sqrt{\frac{P_{cr}}{EI}}\cos\sqrt{\frac{P_{cr}}{EI}}\,LC_3 - \sqrt{\frac{P_{cr}}{EI}}\sin\sqrt{\frac{P_{cr}}{EI}}\,LC_4 = 0$$

The determinant of the coefficients to be equated to zero is

$$\begin{vmatrix} 1 & 0 & 0 & 1 \\ 0 & 1 & \sqrt{\frac{P_{cr}}{EI}} & 0 \\ 1 & L & \sin\sqrt{\frac{P_{cr}}{EI}}\,L & \cos\sqrt{\frac{P_{cr}}{EI}}\,L \\ 0 & 1 & \sqrt{\frac{P_{cr}}{EI}}\cos\sqrt{\frac{P_{cr}}{EI}}\,L & -\sqrt{\frac{P_{cr}}{EI}}\sin\sqrt{\frac{P_{cr}}{EI}}\,L \end{vmatrix} = 0$$

and expanding and solving for P_{cr}, we find

$$P_{cr} = 4\frac{(n\pi)^2EI}{L^2} \qquad n = 0, 1, 2, \ldots$$

or, dividing both sides by A, and considering the lowest nontrivial mode $n = 1$,

$$\sigma_{cr} = 4\frac{\pi^2E}{(L/r)^2} = \frac{\pi^2E}{(\frac{1}{2}L/r)^2} \tag{3.35}$$

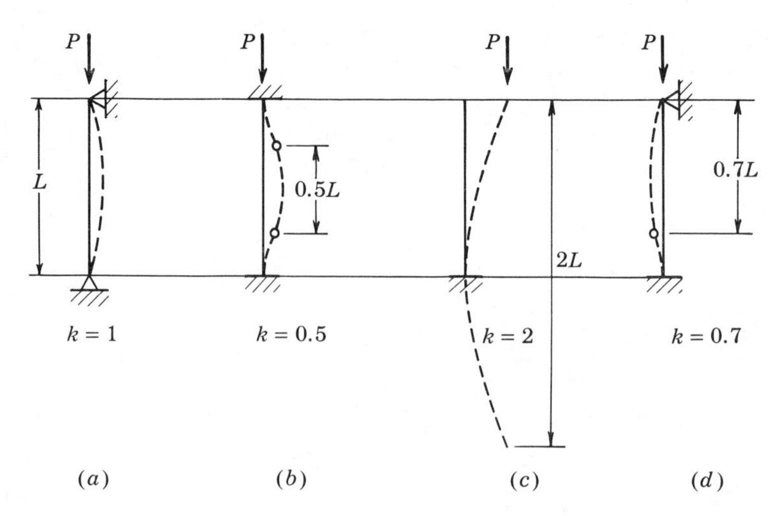

Fig. 3.13

We see that fixing the ends of a column against rotation will increase its buckling strength fourfold.

This result can be interpreted physically by drawing the buckled shape of the column, as in Fig. 3.13*b*, the symmetry of which indicates that the points of inflection will occur at the quarter points, so that the center half (of length $L/2$) will act as a Euler column.

A free-standing column may be considered as acting similarly to one-half of a Euler column of length $2L$; thus, a cantilever column will have one-fourth of the buckling resistance of an equivalent pinned column. Similarly, the fixed-pinned column has an effective length $= 0.7L$.

The effect of end conditions can be expressed generally in terms of the *effective length* kL, where k is a factor to be applied to the actual column length. Values for k are given in Fig. 3.13, in each case representing the ratio of the distance between inflection points to the actual length. For other end conditions, values of k may be assumed by the experienced designer.

The effective length kL may now be substituted for the actual length to account for support conditions other than pinned, so that the effective slenderness ratio to be used in buckling computations becomes kL/r.

3.7 Inelastic Buckling of Ideal Columns

The column curve shown in Fig. 3.12 based on Euler's theory indicates that for columns with low slenderness ratio the average stress will exceed the elastic range of the material prior to buckling, thereby voiding the use of Euler's formula for determining the critical load, since it was based on the elastic properties of the material. A theory of inelastic buckling is required for intermediate-range columns. This is provided by the *tangent-modulus theory* proposed by Engesser (1889) and Shanley (1948).

Figure 3.14*a* shows the stress distribution at the critical column section of a member of material whose stress-strain curve is shown in Fig. 3.14*b*. Up to the instant of buckling, the column remains perfectly straight and is

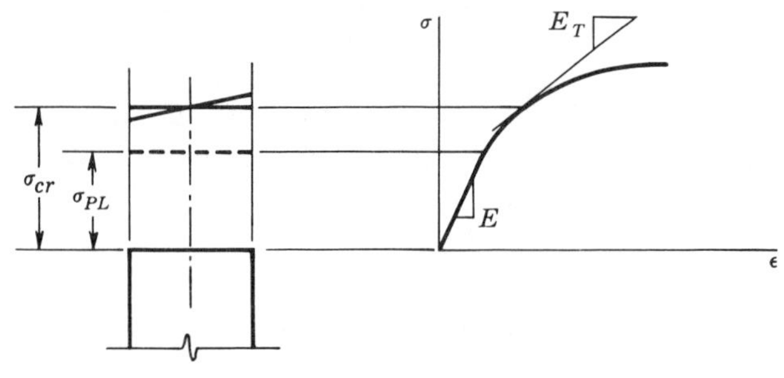

(*a*) Stress distribution in column (*b*) Stress-strain curve

Fig. 3.14

therefore subjected to a uniform stress P/A which, because of the low slenderness of the column, can exceed the proportional limit stress σ_{PL}. Eventually, under a stress σ_{cr}, buckling will impend, and the small bending stresses resulting from the moment Py will occur. The resistance to bending will depend on the stiffness of the material subjected to the inelastic stress σ_{cr}, which is given by the tangent modulus E_T to the stress-strain curve, as shown in Fig. 3.14*b*. Assuming this stiffness to prevail over the entire cross section (a good approximation for small bending stresses), we can write, in analogy with the procedure for elastic stresses:

$$\frac{d^4y}{dx^4} + \frac{P_{cr}}{E_T I}\frac{d^2y}{dx^2} = 0 \tag{3.36}$$

and the critical stress can therefore be written as

$$\sigma_{cr} = \frac{\pi^2 E_T}{(kL/r)^2} \tag{3.37}$$

We see that the results of Euler's formula can be generalized to apply in the inelastic as well as in the elastic range, if the stiffness of the material at the inelastic critical stress is substituted for the elastic stiffness. Figure 3.15*a* shows the column curve extended into the inelastic range according to the tangent-modulus theory, following three different assumed stress-strain curves, as indicated in Fig. 3.15*b*. It is seen that in the case of the idealized flat-topped stress-strain curve, the inelastic strength for intermediate and short columns is constant at σ_{YP}.

We have seen in Sec. 3.2 that the effect of initial stresses in the column material is to lower the apparent proportional limit, as shown in Fig. 3.16*a*, and if buckling occurs at a stress above this value, the calculations should be based on the tangent modulus of the apparent stress-strain diagram[1] of

[1] Since column buckling is really a bending problem, the apparent stress-strain curve should be based on the behavior under bending moment, rather than under axial load as was done in Sec. 3.2. The difference in results due to the two different assumptions is slight.

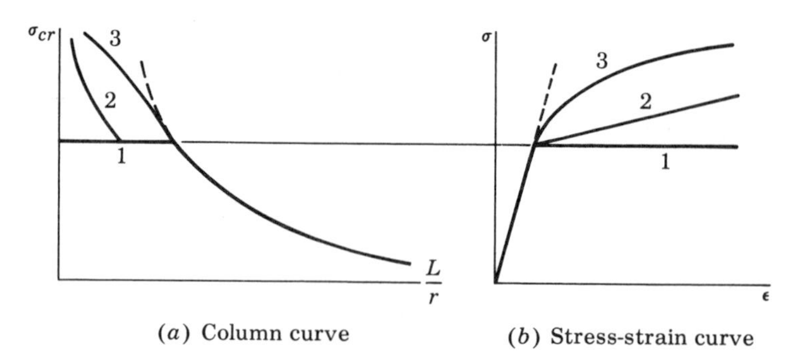

(*a*) Column curve (*b*) Stress-strain curve

Fig. 3.15

Fig. 3.16*a*. The buckling strength of columns is then considerably reduced by the presence of initial stresses, as shown in Fig. 3.16*b*.

Extensive experimental investigations have shown that peak initial stresses in commercial sections are of the order of $\frac{1}{2}\sigma_{YP}$. In this case, the apparent proportional limit becomes, according to Eq. (3.5), equal to $\frac{1}{2}\sigma_{YP}$. And Euler's equation, Eq. (3.32), ceases to be valid when the L/r ratio is less than the value C_c given by Eq. (3.33).

The theoretical buckling strength of intermediate-range columns, of lesser L/r ratio, can then be found by using the tangent-modulus theory, and if this is done, a buckling curve as shown in Fig. 3.17 is obtained. Experimental evidence shown by the data points of Fig. 3.17 closely verifies this analysis.

The analytical curve of Fig. 3.17 varies somewhat depending on both the distribution of the residual stresses and the cross section of the member, but is considered reasonably closely approximated by a parabola becoming tangent to the Euler curve at the point $\sigma_{cr} = \sigma_{YP}/2$, $L/r = C_c$. Such a parabola is expressed by the formula

$$\sigma_{cr} = \sigma_{YP} - \frac{\sigma_{YP}^2}{4\pi^2 E}\left(\frac{L}{r}\right)^2 \tag{3.38}$$

and introducing the value C_c from Eq. (3.33), we can write

$$\sigma_{cr} = \left[1 - \frac{(L/r)^2}{2C_c^2}\right]\sigma_{YP} \tag{3.39}$$

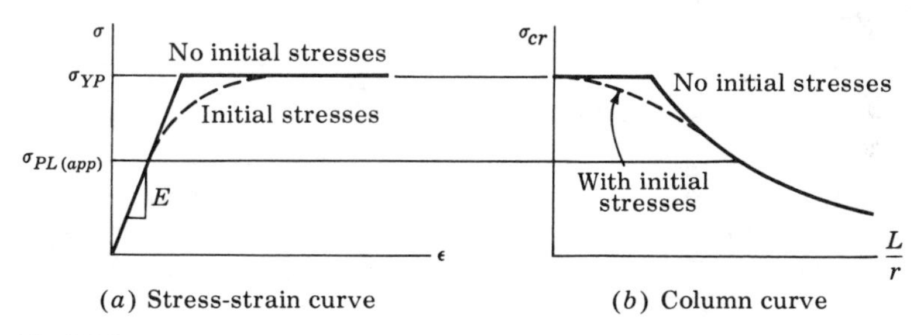

Fig. 3.16

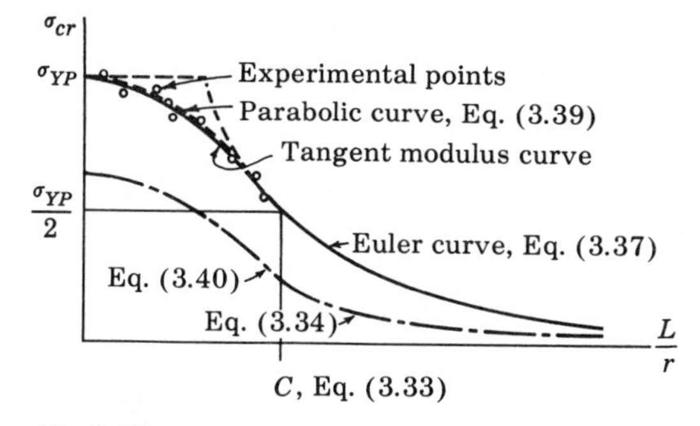

Fig. 3.17

If we divide this critical stress by an appropriate safety factor, we arrive at formula 1 in the 1961 AISC Code:

$$F_A = \frac{\sigma_{cr}}{\text{F.S.}} \tag{3.40}$$

The safety factor as specified there varies from a minimum of 1.67 for $L/r = 0$ (because accidental eccentricities are less detrimental to short than to long columns) to a maximum of 1.92, to match that used with the Euler formula, Eq. (3.34).

Equations (3.34) and (3.40) furnish allowable stresses over the entire range of columns up to $L/r = 120$. Columns of higher slenderness are subject to another reduction of allowable stress given by formula 3 of the AISC Code.

Example Problem 3.7 A rectangular cross-section member is made of elastic–perfectly plastic material which has initial stresses locked in as shown in the figure.

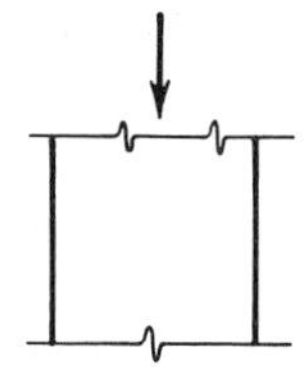

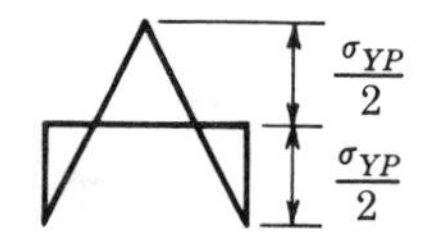

Calculate the buckling stress on this member for the full range of slenderness ratios, and plot the buckling curve.

Solution: The response of this member to axial load was discussed in Example Problem 3.2, and we reproduce here the apparent stress-strain curve based on the results of that problem (compare this to the graph of that example problem).

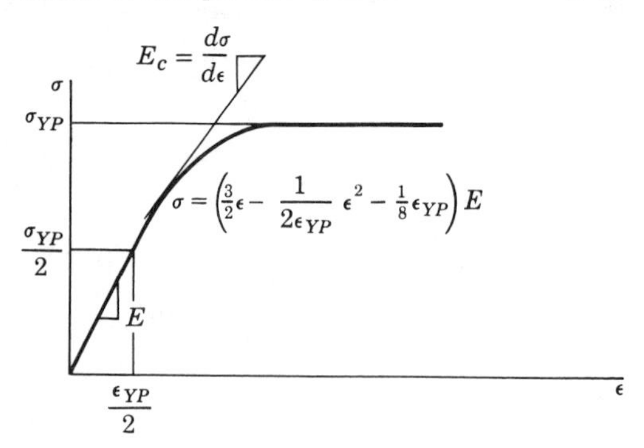

For the linear range of the stress-strain relation, that is, up to $\sigma_{YP}/2$, the critical stress is given by Eq. (3.32):

$$\sigma_{cr} = \frac{\pi^2 E}{(L/r)^2}$$

valid for columns of slenderness ratio

$$\frac{L}{r} \geqq \sqrt{\frac{2\pi^2 E}{\sigma_{YP}}}$$

For higher stresses, the nonlinear portion of the stress-strain curve applies, and accordingly Euler's formula must be generalized to Eq. (3.36):

$$\sigma_{cr} = \frac{\pi^2 E_T}{(L/r)^2}$$

where

$$E_T = \frac{d\sigma}{d\epsilon} = \frac{d}{d\epsilon}\left(\frac{3}{2}\epsilon - \frac{1}{2\epsilon_{YP}}\epsilon^2 - \frac{5}{8}\epsilon_{YP}\right)E$$

$$= \left(\frac{3}{2} - \frac{\epsilon}{\epsilon_{YP}}\right)E$$

Substituting this into the expression for the buckling stress, we get

$$\sigma_{cr} = \frac{\pi^2 E}{(L/r)^2}\left(\frac{3}{2} - \frac{\epsilon}{\epsilon_{YP}}\right)$$

Note that the ϵ in the parentheses of the right-hand side is a function of the σ_{cr} on the left-hand side. It is possible to solve this expression explicitly for σ_{cr}, but for our purposes we will leave it as is and solve in the following tabular fashion:

ϵ, *in./in.*	$\sigma_{add.}$, *ksi*	σ_{cr}, *ksi*	$(L/r)^2$	L/r
$0.5\ \sigma_{YP}$	$0 \quad \sigma_{YP}$	$0.500\ \sigma_{YP}$	$2.0\ \frac{\pi^2 E}{\sigma_{YP}}$	$1.41\sqrt{\frac{\pi^2 E}{\sigma_{YP}}}$
0.6	0.095	0.595	1.51	1.23
0.8	0.255	0.755	0.93	0.96
1.0	0.375	0.875	0.57	0.75
1.2	0.455	0.955	0.31	0.56
1.4	0.495	0.995	0.10	0.32
1.5	0.500	1.000	0	0

Plotting σ_{cr}/σ_{YP} versus L/r, we get the buckling curve shown in the figure.

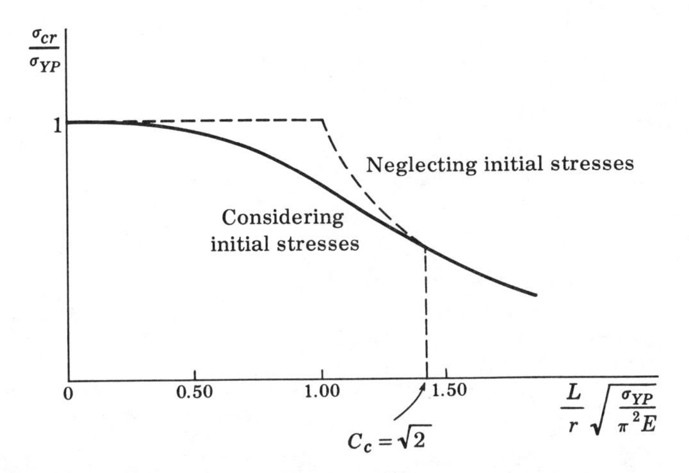

3.8 Design of Steel and Timber Columns

Steel Columns Whereas the allowable stress in members in tension is a fixed quantity, in columns subject to buckling, the critical stress (and therefore the allowable stress, determined by dividing the critical stress by a suitable safety factor) is dependent on the slenderness ratio of the member. For this reason, design of columns is generally a trial-and-error procedure, in which one of the quantities, say the allowable stress, is assumed, the required section is calculated, the L/r ratio is computed, and the actual stress is compared to the allowable. If the actual stress is at, or somewhat less than, the allowable value, the design is satisfactory. If the actual stress is too low, the member is uneconomical, and if too high, it is unsafe. In either case, a second cycle is required which generally will be sufficient to bring actual and allowable stresses in agreement.

Various design aids are available to shortcut the procedure. In particular, Eqs. (3.34) and (3.40) may be plotted (as in Fig. 3.17) or tabulated (as in the AISC Handbook) so that the allowable stress corresponding to a given L/r value may be immediately determined.

It was pointed out in Sec. 3.6 that the degree of end fixity plays a major role in determining buckling strength, and that this effect can be taken into account by providing for an "effective" column length kL.

Figure 3.18, taken from "Commentary on AISC Specifications" (American Institute of Steel Construction, 1961), gives both theoretical and recommended k factors for various end restraints. We note that recommended values are generally on the high side, because usually there is considerable doubt whether the ideal restraint conditions are actually realized in the structure.

In the following design example, a method is proposed which will usually lead to reasonably fast results.

Example Problem 3.8

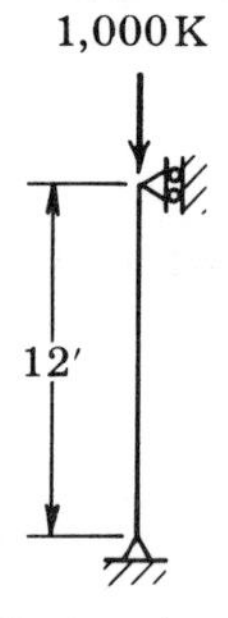

Design the pin-ended column to resist an axial load of 1,000 kips, following the AISC Specifications. Use steel with 36 ksi yield stress.

The column will be a wide-flange section whose weak axis will be critical for buckling. To determine the required size, we resort to a trial-and-error procedure in tabular form. We first assume a likely value of allowable stress, corresponding to an average L/r ratio, and find the corresponding A_{reqd}. Then a member is selected with that value of cross-sectional area and its L/r ratio computed. The allowable stress for this L/r ratio is found, compared with the assumed value, and corrections made if necessary.

F_A *(assumed)*, *ksi*	A_{reqd}, *in.*^{2}	*Member*	r_{min}, *in.*	$(L/r)_{max}$	F_A *(actual)*, *ksi*
16.0	62.50	14 W 219	4.08	35.3	19.5

Note that the allowable stress was assumed too low, and a new trial is made with F_A *(assumed)* at an intermediate value:

18.0	55.5	14 W 193	4.05	35.6	19.5

Another round is indicated:

19.4	51.5	14 W 176	4.02	35.8	19.5 (O.K.)

Therefore, use 14 W 176.

Timber columns The "National Design Specifications for Stress Grade Lumber" base the allowable column stress on the Euler formula, as may be recognized by the formula

$$\left(\frac{P}{A}\right)_{allow} = \frac{\pi^2 E}{2.727(L/r)^2} = 3.619\,\frac{E}{(L/r)^2} \tag{3.41}$$

Buckled shaped of column is shown by dashed line	(a)	(b)	(c)	(d)	(e)	(f)
Theoretical K value	0.5	0.7	1.0	1.0	2.0	2.0
Recommended design value when ideal conditions are approximated	0.65	0.80	1.2	1.0	2.10	2.0

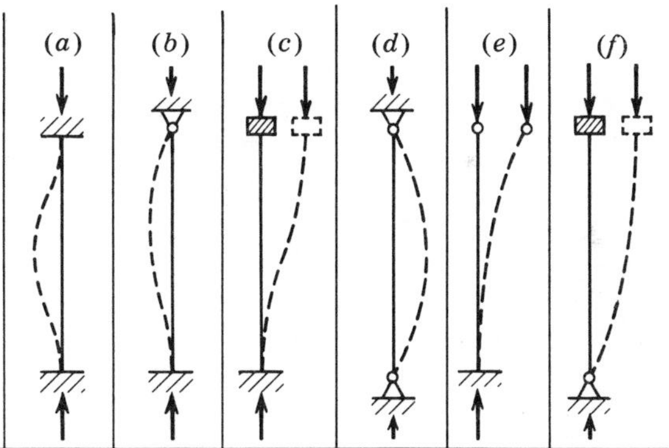

End condition code		
	Rotation fixed	Translation fixed
	Rotation free	Translation fixed
	Rotation fixed	Translation free
	Rotation free	Translation free

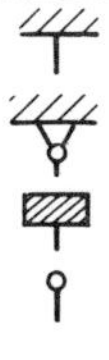

Fig. 3.18 (From "Commentary on the 1963 A.I.S.C. Specifications")

of sec. 401-E-2 of these specifications. We note that, because of lack of uniformity of timber, and because the use of an elastic buckling approach may be questioned, the safety factor is higher than that used for steel members.

For square or rectangular columns,

$$r = \sqrt{\frac{I}{A}} = \sqrt{\frac{bd^3}{12\,bd}} = \frac{1}{3.46d}$$

and Eq. (3.41) becomes

$$\left(\frac{P}{A}\right)_{allow} = \frac{0.30E}{(L/d)^2} \tag{3.42}$$

where d is the column dimension in the direction of buckling. For such simple cross sections, then, we can express the buckling strength in terms of the ratio of length to width of the member.

3.9 Buckling of Compressed Plates: Local Flange Buckling*

The compressive stress in a member with thin flanges may cause these flanges to buckle in the manner shown in Fig. 3.19, with consequent immediate loss of compression strength. Such local flange buckling is to be avoided at all costs, and the following theoretical treatment of plate stability will indicate the critical parameters which influence this phenomenon.

The plate-buckling problem is tackled by means outlined in Sec. 2.7 and applied to other instability problems in Secs. 3.6 and 4.6. Reference to those sections is useful in order to keep in mind the main plan of attack for stability problems of this type. Here, we restate the basic steps used: equilibrium equation of slightly buckled structure, force-deformation relations, eigenvalue solution of a system of homogeneous equations.

The classical theory of elastic plates gives as the governing equation relating the transverse deflection w to the applied transverse load intensity q on the plate element of Fig. 3.20

$$\frac{\partial^4 w}{\partial x^4} + 2\,\frac{\partial^4 w}{\partial x^2\,\partial y^2} + \frac{\partial^4 w}{\partial y^4} = \frac{q}{D} \tag{3.43}$$

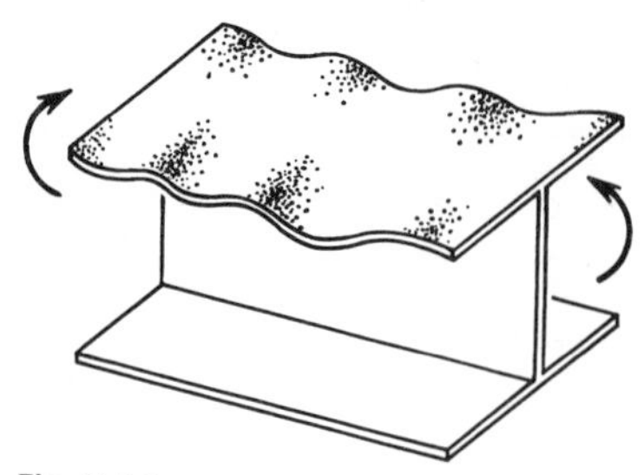

Fig. 3.19

where $D = Et^3/(1 - \mu^2)$ is the elastic plate stiffness.[1] The reader is encouraged to study the derivation of this important equation, which is based on the elements of strength of materials and continuum mechanics outlined in Chap. 2

We now consider the effect of a compressive stress σ_x acting in the middle plane of the deformed element which obeys Eq. (3.43). Figure 3.21*a* indicates a component of the applied σ_x stress which acts normal to the plane of the deformed element, that is, analogous to the load q; its magnitude

[1] S. Timoshenko and S. Woinowsky-Krieger, "Theory of Plates and Shells," 2d ed., p. 79, McGraw-Hill, 1959.

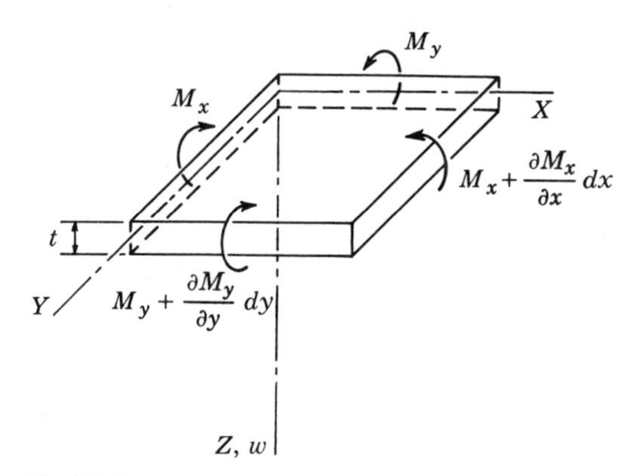

Fig. 3.20

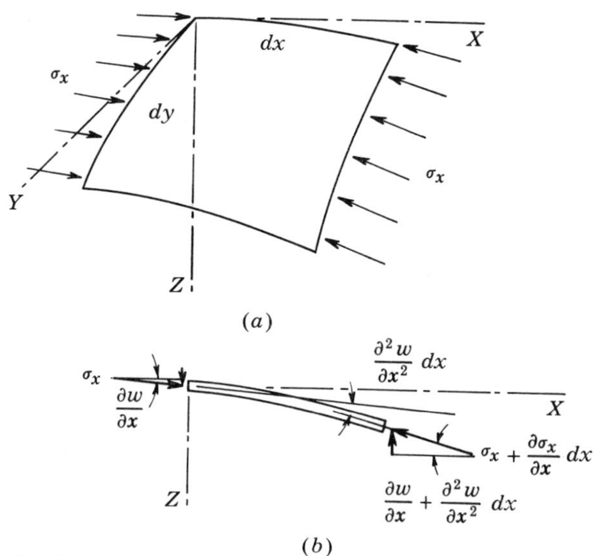

Fig. 3.21

is found by summing the forces of Fig. 3.21*b* along the *Z* axis:

$$\Sigma F_z = (\sigma_x t\, dy)\frac{\partial w}{\partial x} - \left(\sigma_x + \frac{\partial \sigma_x}{\partial x}\,dx\right) t\, dy \left(\frac{\partial w}{\partial x} + \frac{\partial^2 w}{\partial x^2}\,dx\right)$$
$$= -\sigma_x t \frac{\partial^2 w}{\partial x^2}\,dx\,dy$$

after canceling and deleting a higher-order term. Since this is the transverse load component per element area $dxdy$, the unit transverse force is $-\sigma_x t \dfrac{\partial^2 w}{\partial x^2}$, and inserting this into Eq. (3.43), we obtain

$$\frac{\partial^4 w}{\partial x^4} + 2\frac{\partial^4 w}{\partial x^2\,\partial y^2} + \frac{\partial^4 w}{\partial y^4} = -\sigma_x t \frac{\partial^2 w}{\partial x^2} \tag{3.44}$$

The solution, subject to specified boundary conditions, of the homogeneous differential Eq. (3.44) gives us the value $\sigma_{x,cr}$ under which the plate can maintain its deformed equilibrium state, that is, the critical plate-buckling stress.

We consider now the buckling of a rectangular plate of length *a* and width *b*, as shown in Fig. 3.22, with the edges parallel to the *Y* axis simply supported. We assume a solution of the form

$$w = F(y)\sin\frac{n\pi x}{a} \tag{3.45}$$

which satisfies the specified boundary conditions, and proceed by the method of separation of variables; accordingly, we perform the indicated differentiations and substitute Eq. (3.45) into Eq. (3.44). Canceling a common factor

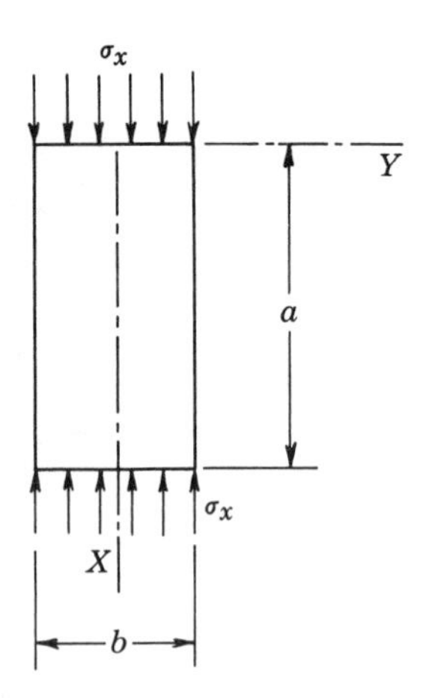

Fig. 3.22

$\sin (n\pi x/a)$, we obtain the ordinary differential equation for $F(y)$:

$$\frac{d^4F}{dy^4} - 2\left(\frac{n\pi}{a}\right)^2 \frac{d^2F}{dy^2} + \left[\left(\frac{n\pi}{a}\right)^4 - \left(\frac{\sigma_x t}{D}\right)\left(\frac{n\pi}{a}\right)^2\right] F = 0$$

The solution of this ordinary, fourth-order, homogeneous differential equation is

$$F(y) = C_1 \operatorname{Sinh} \alpha y + C_2 \operatorname{Cosh} \alpha y + C_3 \sin \beta y + C_4 \cos \beta y \qquad (3.46)$$

where $\alpha = \sqrt{\left(\frac{n\pi}{a}\right)^2 + \sqrt{\frac{\sigma_x t}{D}\left(\frac{n\pi}{a}\right)^2}}$

and $\beta = \sqrt{-\left(\frac{n\pi}{a}\right)^2 + \sqrt{\frac{\sigma_x t}{D}\left(\frac{n\pi}{a}\right)^2}}$

and the transverse deflection is therefore

$$w = (C_1 \operatorname{Sinh} \alpha y + C_2 \operatorname{Cosh} \alpha y + C_3 \sin \beta y + C_4 \cos \beta y) \sin \frac{n\pi x}{a} \qquad (3.47)$$

The constants C_1 to C_4 are found from two boundary conditions each along the two longitudinal edges, that is, at $y = +b/2$ and $y = -b/2$. Let us consider one possible case, that in which these edges are simply supported, that is

$$w(y = -b/2) = w(y = +b/2) = M_y(y = -b/2) = M_y(y = +b/2) = 0$$

For this symmetrical condition with respect to the X axis, we anticipate a displacement w which is symmetric, and for this reason the coefficients C_1 and C_3 of the unsymmetric terms must be zero. The function w is then of the form

$$w = (C_2 \operatorname{Cosh} \alpha y + C_4 \cos \beta y) \sin \frac{n\pi x}{a}$$

and the remaining constants are then found from the boundary conditions

$$w\left(y = \pm\frac{b}{2}\right) = 0: \qquad C_2 \operatorname{Cosh} \frac{\alpha b}{2} + C_4 \cos \frac{\beta b}{2} = 0$$

$$M = D\left(\frac{\partial^2 w}{\partial y^2}\right)\left(y = \pm\frac{b}{2}\right) = 0: \qquad C_2\alpha^2 \operatorname{Cosh} \frac{\alpha b}{2} - C_4\beta^2 \cos \frac{\beta b}{2} = 0$$

To get the nontrivial solution for this set of homogeneous equations, we equate the determinant of the coefficients to zero:

$$\begin{vmatrix} \operatorname{Cosh} \frac{\alpha b}{2} & \cos \frac{\beta b}{2} \\ \alpha^2 \operatorname{Cosh} \frac{\alpha b}{2} & -\beta^2 \cos \frac{\beta b}{2} \end{vmatrix} = 0$$

or $\quad (\alpha^2 - \beta^2) \operatorname{Cosh} \frac{\alpha b}{2} \cos \frac{\beta b}{2} = 0 \qquad (3.48)$

Of these three factors, Cosh $(\alpha b/2)$ is always larger than one; $(\alpha^2 - \beta^2) = 0$ leads to the trivial condition $\sigma_x = 0$; and the necessary condition for a nontrivial solution of the characteristic equation, Eq. (3.48), is therefore

$$\cos \frac{\beta b}{2} = 0$$

or
$$\frac{\beta b}{2} = \frac{\pi}{2}, \frac{3\pi}{2}, \frac{5\pi}{2}, \ldots$$

We take the lowest of these critical values and insert the definition of β:

$$\sqrt{-\left(\frac{n\pi}{2}\right)^2 + \sqrt{\frac{\sigma_x t}{D}}\left(\frac{n\pi}{a}\right)^2}\,\frac{b}{2} = \frac{\pi}{2}$$

from which, solving for the critical stress, we obtain

$$\sigma_x = \frac{D}{b^2 t}\pi^2\left(\frac{1}{n}\frac{a}{b} + n\frac{b}{a}\right)^2$$
$$= \frac{\pi^2 E}{12(1-\mu^2)}\left(\frac{t}{b}\right)^2\left(\frac{1}{n}\frac{a}{b} + n\frac{b}{a}\right)^2$$

This result of the stability analysis can be rewritten as

$$\sigma_{cr} = k\,\frac{\pi^2 E}{12(1-\mu^2)}\left(\frac{t}{b}\right)^2 \tag{3.49}$$

where $k = \left(\frac{1}{n}\frac{a}{b} + n\frac{b}{a}\right)^2$

is called the *plate-buckling coefficient.* This coefficient is a function of the aspect ratio a/b and the number of longitudinal waves n. To obtain some insight in its variation, we plot k versus a/b for various values of n and arrive at the curves of Fig. 3.23. We see that the lowest value of buckling coefficient, of value 4, occurs when the ratio $(a/b)/n$ is unity, suggesting

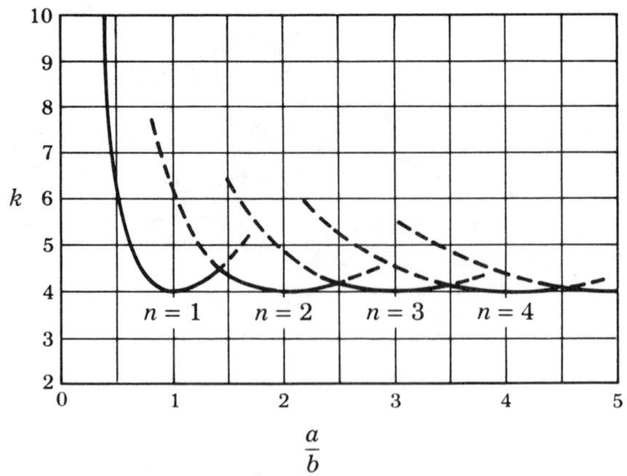

Fig. 3.23 Buckling coefficients for simple-simple plate

that the plate may find it easiest to buckle into n square buckles, each of side length b, as shown in Fig. 3.24.

As the aspect ratio increases, we get into the buckling phenomenon of long plates, and this is the case which applies when local flange or web buckling of rolled steel members must be considered. In this case, k reaches a value of 4, irrespective of aspect ratio, and the critical buckling stress becomes

$$\sigma_{cr} = 4 \frac{\pi^2 E}{12(1 - \mu^2)} \left(\frac{t}{b}\right)^2 \tag{3.50}$$

The critical stress given by Eq. (3.50) is of course valid only for the simple-simple boundary conditions on which these calculations were based. For other boundary conditions, similar calculations lead to the same form of expression, but with different buckling coefficients which are listed in the table that follows (see page 92). But, in any case, the important conclusion can be drawn that the critical plate parameter is the ratio of plate width to thickness b/t. Note the difference between this and the critical parameter in column buckling, the slenderness ratio L/t.

Just as in the case of columns, it is possible that for plates of sufficiently low b/t ratios, yielding may occur prior to buckling, and in this case a plastic theory of plate buckling is called for. The result of such theories is a *plasticity reduction factor* which will further diminish the elastic buckling coefficient.

Based on the foregoing theory, sec. 1.9 of the AISC Specifications lays down definite rules about the maximum width-thickness ratio allowed for thin elements under compression. For the purpose of ensuring stability of

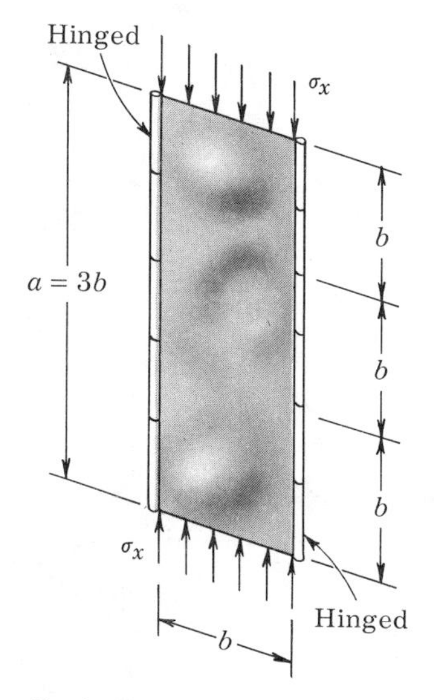

Fig. 3.24

thin flanges, Eq. (3.49) is rewritten in the form

$$\frac{b}{t} = \frac{\sqrt{k\,[\pi^2 E/12(1-\mu^2)]}}{\sqrt{\sigma_{cr}}} = \frac{N_1}{\sqrt{\sigma_{cr}}} \tag{3.51}$$

where N_1 is a numerical factor which depends only on material properties and support conditions. To ensure that the critical buckling stress is larger than the allowable stress (which according to AISC sec. 1.5 is proportional to the yield stress σ_{YP}), a maximum b/t ratio of

$$\left(\frac{b}{t}\right)_{allow} = \frac{N}{\sqrt{\sigma_{YP}}} \tag{3.52}$$

is specified, where N is another numerical factor which incorporates the required safety factor. Note that N factors are specified for the case of elements with one free edge in AISC sec. 1.9.1, and for elements with both edges restrained in sec. 1.9.2.

In plastic design, the possibility of plastic hinge formation must be considered, and in plastified regions the compression flanges of beams must have sufficient stability so that they can yield without premature buckling. An analysis of the plastic plate buckling problem is beyond the scope of this book, but it may be anticipated that more stringent controls on allowable b/t ratios are indicated in this case. AISC sec. 2.6 provides for maximum width-thickness ratios which enable the development of plastic hinges without the danger of premature local buckling.

Plate-buckling coefficients

Boundary condition, longitudinal edges	k
Fixed-fixed	7
Simple-fixed	5.5
Simple-simple	4
Fixed-free	1.2
Simple-free	0.4

General Readings

Elementary theory

Any of the strength of materials texts listed in Chap. 2 covers the fundamentals of Secs. 3.1 to 3.4.

Reinforced concrete columns

Ferguson, Phil M.: "Reinforced Concrete Fundamentals," 2d ed., Wiley, 1965. Chapter 2 presents experimental evidence and current thinking on axially loaded concrete columns.

Column and plate buckling

The texts by Gerard and by Timoshenko and Gere listed in Chap. 2 may be consulted about basic theory; in addition, the following books are valuable:

Bleich, Friedrich: "Buckling Strength of Metal Structures," McGraw-Hill, 1952. Chapter II is a thorough treatment of the mathematical formulation of stability problems, and there is extensive discussion of inelastic buckling.

Column Research Council: "Guide to Design Criteria for Metal Compression Members," 1960. A concise treatment with emphasis on structural engineering applications, including a number of design aids.

Design of steel columns

Gaylord, Edwin H., Jr., and Charles N. Gaylord: "Design of Steel Structures," McGraw-Hill, 1957. This, like most books on structural design, emphasizes procedures and structural details.

Problems

3.1 A mezzanine floor is to be installed of dimensions as shown. It is required to design the hanger rods. According to AISC specifications these rods are to be made of A-7 steel. The live load on the mezzanine is to be 50 lb/ft². (*Note:* Refer to AISC Handbook to determine the net area after threading.)

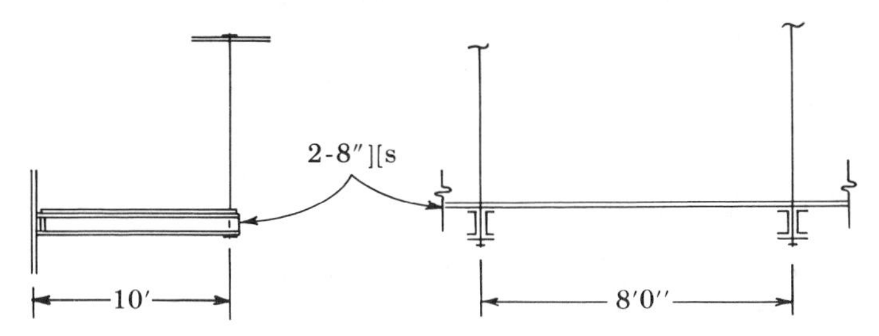

3.2 A two-lane highway bridge has a pair of Warren trusses as main carrying members. The dead weight of the roadway deck and supporting members may be taken at 100 lb/ft². The live load resulting from traffic may be simulated by a uniform load of 640 lb/ft per lane of roadway. The weight of bridge trusses is sometimes predicted on basis of the formula

$$W = \frac{L\sqrt{p}}{20} + \frac{L(b - 16)}{10} + 50$$

where W = weight of truss, lb/ft
L = span length, ft
b = roadway width, ft
p = live load, lb/lin ft of each truss

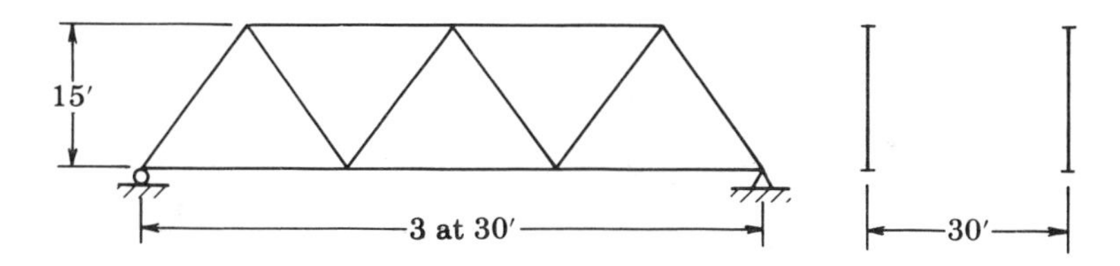

The trusses are to be fabricated of A-7 steel and designed according to the AASHO Specifications, sec. 1.4.2, applicable for highway bridges, which sets the allowable stresses:

Axial tension, structural steel, net section 18,000 psi

(a) Lay out a system of members supporting the roadway deck which is able to transmit the dead and live loads to the lower panel points of the truss, and analyze the truss.
(b) Consider an all-welded alternate for the truss (note that this does not require any holes to be punched into the members). For this alternate, select suitable wide-flange beam members for the bottom chord.
(c) For the welded alternate, select suitable wide-flange beam shapes for those web members which are in tension.

3.3 A riveted alternate is considered for the highway truss of Prob. 3.2. Rivets to be used are $\frac{3}{4}$ in.

(a) Select bottom chord members for this alternate. These members are to be composed of two angles back to back, with space between to accommodate gusset plates.
(b) Select suitable double-angle members for those web members which are in tension.

3.4 Section 1.14.3 of the AISC Code provides for the case of a tension member with a chain of rivet or bolt holes extending through it in a zigzag line. The stress analysis for such an arrangement is complicated, and code provisions are based in the main on test results.

Calculate the critical net section of the $4 \times 4 \times \frac{3}{8}$ in. angle and the allowable tension force on the angle, following the above-mentioned AISC provisions.

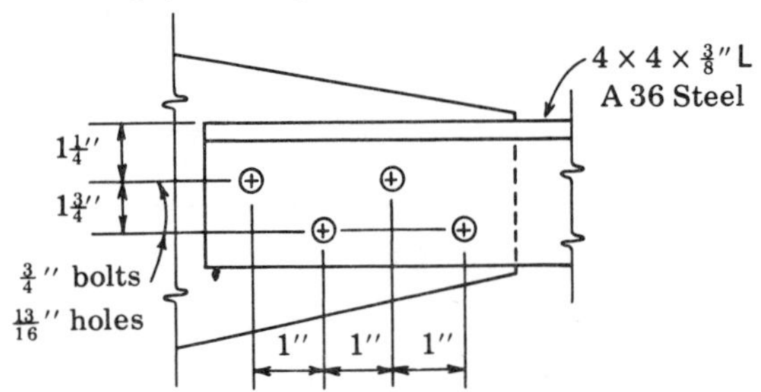

3.5 The riveted splice is to transmit a tensile force of 40 kips. Calculate the net section through the main plate and the required thickness of the member to transmit the load according to the AISC Specifications.

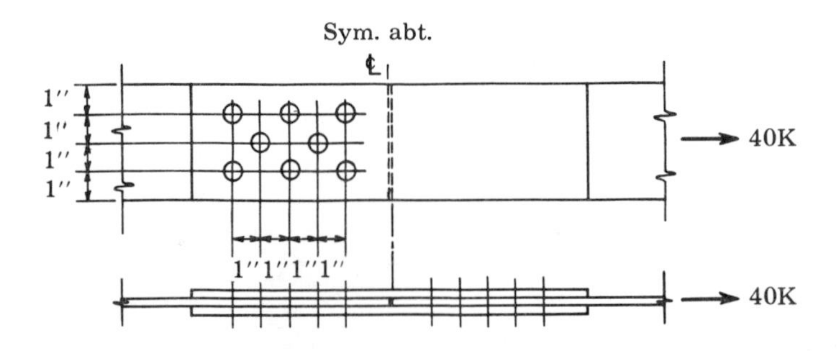

3.6 Design a square timber to transmit a tensile load of 8 kips according to the NDS for stress-grade lumber. Wood to be used is Douglas fir, coast region.

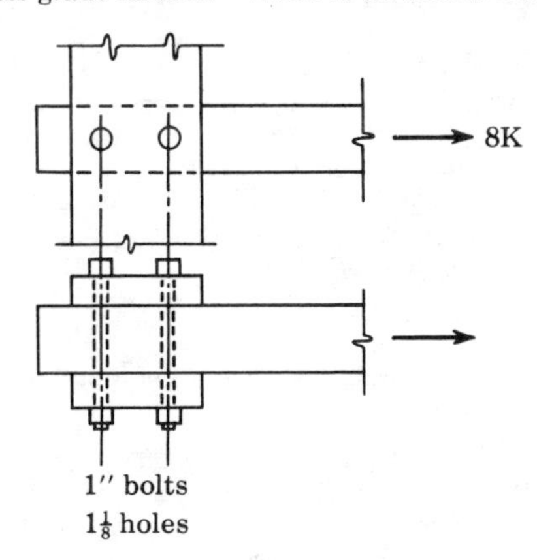

3.7 A rectangular-cross-section member has a set of initial locked-in stresses of the distribution shown. The material is elastic-plastic, with elastic modulus E, yield stress σ_{YP}.

(a) Find the value of the tensile initial stress σ_{in} for a self-equilibrating stress distribution.
(b) Calculate and plot the stress-strain curve, taking into account the locked-in stresses.

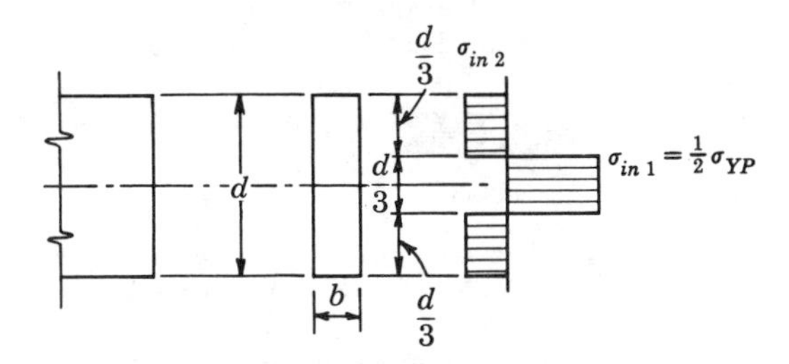

3.8 A circular tension member of radius R, made of elastic–perfectly plastic material of yield stress σ_{YP} and modulus E, has parabolically varying initial stresses, of the magnitude

$$\sigma_{in} = \sigma_0 + ar^2$$

where r is the distance from the centroid of the section.

(a) By integrating and applying statics, find the value of the constant a for a self-equilibrating set of stresses.
(b) Calculate and plot the apparent stress-strain curve for this member, for the case when $\sigma_0 = \sigma_{YP}$.

3.9 A 2-in. standard steel pipe (see AISC Handbook for dimensions) is filled with concrete; $E_s = 30{,}000$ ksi, $E_c = 4{,}000$ ksi.

(a) Find stresses and strains in steel and concrete due to an axially applied load P.
(b) Find the allowable load if the allowable stresses in steel and concrete are 20 and 1.5 ksi respectively.

3.10 A 12 × 12 in. square concrete column (E_c = 3,000 ksi, $f_{c\ allow}$ = 1.0 ksi) is reinforced with four No. 8 bars (E_s = 30,000 ksi, $f_{s\ allow}$ = 26.7 ksi).

(*a*) Find stresses in steel and concrete due to an axial load P.
(*b*) Find the allowable value of the load P.
(*c*) What should be the allowable strength in the concrete so that the allowable stresses in the two materials are attained simultaneously?

3.11 A 4 × 8 in. (nominal size) timber is reinforced with a $\frac{1}{8}$ × 6 in. plate on each side and used as a column. E_{wd} = 1,600 ksi, E_s = 30,000 ksi.

(*a*) Find stresses in steel and concrete due to an axial load P.
(*b*) Evaluate the reduction in stresses and deformation of the wood resulting from the addition of the steel plates.

3.12 A 4 × 8 in. (nominal size) timber is reinforced with one $\frac{1}{8}$ × 6 in. steel plate along one 8 in. side. E_{wd} = 1,600 ksi, E_s = 30,000 ksi.

(*a*) Find the point of application of an axial load P so as to prevent bending of the member.
(*b*) Find stresses in wood and steel due to this load P.

3.13 A member of cross-sectional area $b \times d$ is to resist an axial load P. The member is cast flat, and owing to segregation during the casting process the modulus of the elastic material varies from a value of E_1 at the top to $2E_1$ at the bottom.

(*a*) Where must the axial load be applied to this member to prevent bending?
(*b*) Calculate and plot the stresses due to a load P applied as in part (*a*).

3.14 Consider a member with one axis of symmetry along the Y axis; material stiffness varies as function of the distance y:

$$E = E(y)$$

A load P is to be applied so that no bending results (that is, uniform axial deformation only).

By following the basic steps of the strength of material method, find the required point of load application as

$$c = \frac{\int E(y)y\,dA}{\int E(y)\,dA}$$

where c is the Y coordinate of the point of load application, and identify this point as the centroid of a transformed section.

Also, determine the variation of stresses due to this load as

$$\sigma = \frac{E(y)}{\int E(y)\,dA}P$$

Sketch the transformed section and the stresses for a simple cross-sectional area and variation of stiffness.

3.15 A 2-in. standard steel pipe (see AISC Handbook for dimensions) is filled with concrete. The stress-strain curve of the steel is elastic–perfectly plastic, with E_s = 30,000 ksi, f_{YP} = 40 ksi. That for the concrete is given by the equation

$$\sigma_c\ (\text{ksi}) = 3.5 \times 10^3\epsilon - 0.875 \times 10^6\epsilon^2$$

(*a*) Draw a plot of the applied axial load versus the resulting strain. Compare the stiffness of the member at low loads with that computed in part (*a*) of Prob. 3.9. Calculate the maximum load on the member.
(*b*) Calculate the safety factor if the allowable load is that obtained in part (*b*) of Prob. 3.9.

3.16 The concrete properties in the preceding problem are simplified into an elastic–perfectly plastic stress-strain curve, with modulus of elasticity E_c = 4,000 ksi and yield stress of 3.5 ksi. Under this assumption, redraw the load-strain curve, and calculate initial stiffness (in in./in./kip of load) and ultimate strength of the member.

3.17 The nonhomogeneous compression member consists of materials of the properties shown. The load P is applied concentrically.

(*a*) Write an expression relating the applied load P to the resulting strain.
(*b*) Find the ultimate load on the section.

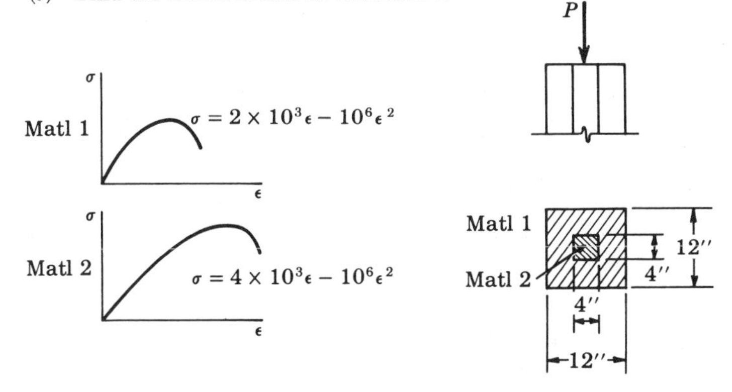

3.18 A composite tension member consists of a tube of 2024-T3 aluminum alloy, of outside diameter 2 in. and inside diameter 1.5 in., into which is inserted a tube of steel of elastic–perfectly plastic properties, of E_s = 30,000 ksi, f_{YP} = 60 ksi. The properties of the aluminum alloy are to be taken from Sec. 2.4, Eq. (2.16). Plot a curve showing the relation between the applied axial load and the resulting axial strain of the member.

Use 1963 ACI Code for Probs. 3.19 to 3.26.

3.19 A reinforced concrete column is 16 by 16 in. in cross section and has eight No. 7 bars of intermediate-grade steel (f_{YP} = 40 ksi) which are suitably tied and spaced. Concrete strength: f'_c = 4.0 ksi.

(*a*) Find the allowable load on the column if it is sufficiently short so that no strength reduction for slenderness is necessary.
(*b*) If the member is 20 ft long and considered hinged at both ends as per sec. 916*a* 2 of the ACI Code, find the allowable load on the column.
(*c*) Find the safety factor of the column of part (*a*).

3.20 A spirally reinforced circular concrete column has 24 in. diameter, eight No. 11 reinforcing bars suitably spaced and wound with spiral; f'_c = 5.0 ksi, f_{YP} = 50 ksi. If the column is restrained at both ends and 24 ft long, what is its allowable axial load?

3.21 A square tied reinforced concrete column is to be designed to safely withstand an axially applied load of 200 kips. It is to be 16 ft long and restrained at the ends in the sense of ACI sec. 916*a* 1; f'_c = 3,000 psi, f_{YP} = 40 ksi. Design the column. Draw the cross section showing dimensions, longitudinal bars, and arrangement and spacing of ties. Find the safety factor against failure.

3.22 A circular spirally reinforced column is to be designed to carry an axial load of 1,000 kips. It is to be 30 ft long and restrained at its ends. $f'_c = 4.0$ ksi, $f_{YP} = 40$ ksi. Design the column. Draw the cross section showing dimensions, amount and size of longitudinal bars, and size, spacing, and arrangement of the spiral reinforcing.

(**3.23–3.26**) The building frame shown is to be designed of reinforced concrete. $f'_c = 4.0$ ksi, $f_{YP} = 40$ ksi. It is required to make a preliminary design of the columns for estimating purposes. For this purpose, the bending moments due to frame action are to be neglected. Only the effects of the floor loads are to be included. This floor load, comprising both live and dead load, is to be taken at 150 lb/ft² of floor area. The axial load in each column may be taken as the load on that portion of the floor area which is tributary to the column stack to be designed. In addition to the floor load, the exterior columns must also carry the weight of the exterior wall, which weighs 30 lb/ft² of wall area and is carried by the periphery of each floor to the adjacent columns.

Problems 3.23 to 3.26 call for the design of column stacks extending from elevation 100 to 145 ft. Since each one of the three columns in a stack has a different axial load, it will be designed separately. It is desirable for ease of forming and standardization to maintain constant gross column dimensions from top to bottom of each stack. To do this, it is advisable to design the bottom column first with maximum steel ratio, and diminish the steel for the upper, less heavily loaded columns.

The load calculation and column design are best carried out in a well-organized tabular manner.

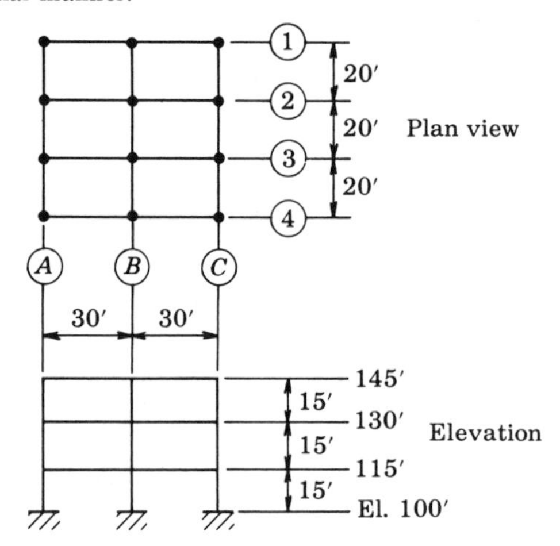

3.23 Design column stack A-1 as square tied columns.

3.24 Design column stack B-2 as square tied columns.

3.25 Design column stack A-1 as circular spiral columns.

3.26 Design column stack B-2 as circular spiral columns.

3.27 Write the boundary conditions for an elastic column of length L which is fixed at one end, pinned at the other. Setting the determinant of the coefficients equal to zero, verify that the characteristic equation for this column is

$$\tan \sqrt{\frac{P_{cr}}{EI}}\, L = \sqrt{\frac{P_{cr}}{EI}}\, L$$

Solve this for P_{cr} and verify that $P_{cr} = \pi^2 EI/(0.7L)^2$.

3.28 Write the boundary conditions for an elastic column of length L which is supported at both ends by elastic restraints of rotational stiffness α. Setting the determinant of the coefficients equal to zero, find the characteristic equation for this case. Find the critical load for various values of α, and plot a curve showing the influence of end restraint on the buckling strength. Verify that the solutions for $\alpha = 0$ and $\alpha = \infty$ check the known buckling loads for pinned and fixed columns.

3.29 A square bar is to serve as a compressively loaded strut to carry 5 kips over a length of 10 ft. Material is high-strength steel of $f_{YP} = 60$ ksi, $E = 30{,}000$ ksi. Calculate the required dimensions of the bar to carry the load with a safety factor of 2.

3.30 It is found that the maximum initial stress in a certain member is 10 ksi. The yield-point stress of a small axial specimen is 30 ksi, $E = 30{,}000$ ksi. Find the lowest L/r ratio for this member for which elastic buckling theory may be applied.

3.31 A strut is to carry 3 kips of compressive load over a length of 5 ft. Material is to be aluminum alloy, $E = 10{,}000$ ksi, and the safety factor is to be 1.50.

(*a*) Design a solid circular rod to safely transmit the load.
(*b*) Design a thin-walled tube of wall thickness equal to $\frac{1}{10}$ of the radius to transmit the load.
(*c*) Calculate the ratio of the amount of material in alternate (*a*) to that of alternate (*b*).

3.32 Write a formula for the minimum slenderness ratio for which elastic buckling theory may be applied in terms of the yield-point stress, the maximum residual stress, and the modulus of elasticity.

3.33 A square column is pin-ended and of such proportions that elastic buckling can occur. Find the percent of material which may be saved by fixing the column ends against rotation.

3.34 A material has a stress-strain curve which can be represented by the equation

$$\sigma = \sigma_{YP}(1 - e^{-k\epsilon}) \qquad (e \text{ is base of natural logs})$$

valid over the entire range of stress and strain.

(*a*) Plot this stress-strain relation.
(*b*) Analytically obtain an expression for the buckling stress σ_{cr}.
(*c*) Plot the buckling curve.

3.35 Plot the buckling curve for aluminum alloy 2024-T3, if its stress-strain relation is given by Eq. (2.16) and the constants of Sec. 2.4.

3.36 A material has the properties shown by the stress-strain curve. Obtain analytical expressions for the buckling stress, and plot the buckling curve, for a rectangular-section column.

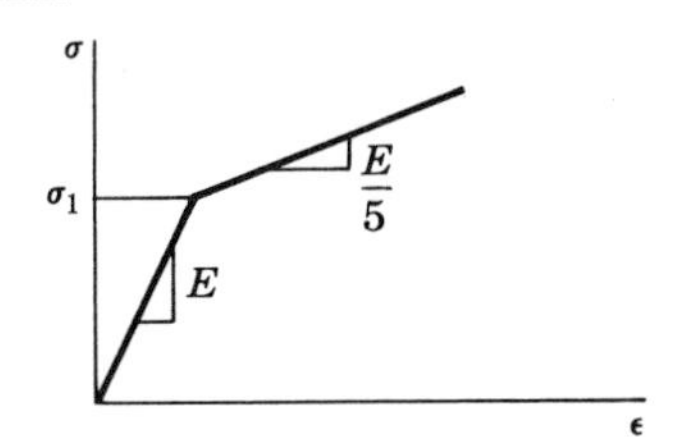

3.37 Plot a column curve which would predict the buckling behavior of a rectangular column with the locked-in initial stresses of Prob. 3.7.

3.38 Show that Eq. (3.39) of Sec. 3.7 represents a parabola which actually passes through the specified boundary points in the manner prescribed.

3.39 Superimpose the parabola given by Eq. (3.38) or (3.39) onto the plot of Example Problem 3.7, and discuss the magnitude of the errors resulting from replacing the actual by the parabolic curve.

3.40 Design a pin-ended steel column of length 20 ft to carry an axial load of 300 kips. Steel to be A-7, member to be selected from wide-flange shapes.

3.41 Calculate the member required for the situation of Prob. 3.40 if it is possible to fix both ends against rotation. Calculate the percent weight saving by providing this fixity.

3.42 The calculated axial load in the column shown is 600 kips. Assume a suitable effective column length and design a wide-flange column of A-36 steel. The floor is restrained against horizontal sway.

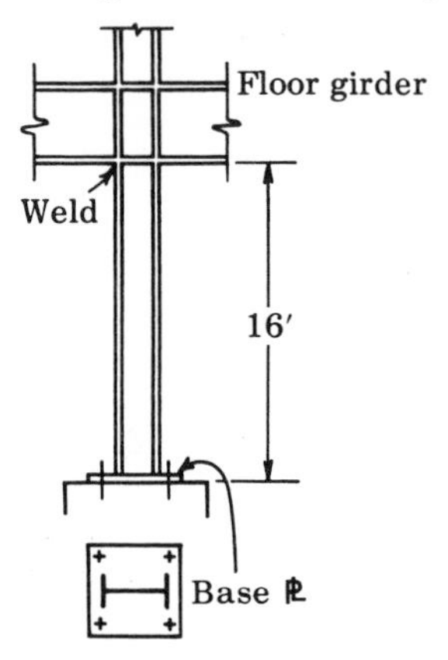

3.43 A column is to be designed of the cross section shown to transmit 200 kips over an effective length of 25 ft. Steel is A-7, and the AISC Specifications are to be followed. Select the standard channel sections and determine the spacing between channel backs.

3.44 The boom for the tower crane of 40 kips capacity is to be designed of four angles of suitable size, laced together by bars to act as a unit. Steel is to be high tensile, of $f_{YP} = 50$ ksi; L/r ratio of the boom is not to exceed 120. Using the AISC Specifications (note that they are not strictly applicable for this type of structure), design the boom. Do not design the lacing bars and end plates.

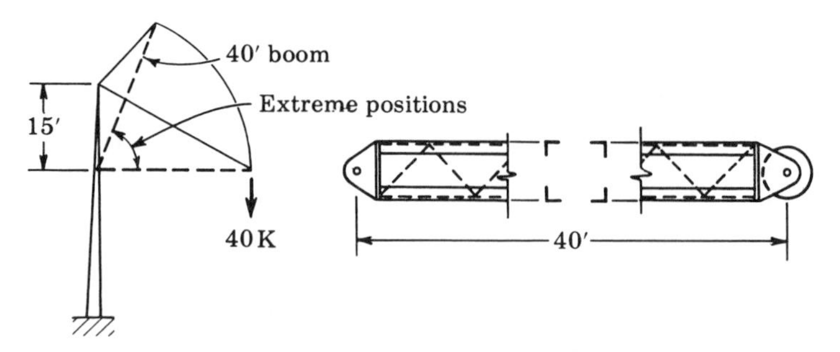

3.45 The roof trusses shown below are spaced at 30 ft on center. The roof load, totaling 60 lb/ft² of roof area, is transmitted through the purlins to the truss joints. The truss is to be built of suitable angles arranged back to back as shown in Sec. A-A. Steel is A-7, and AISC Specifications are to be followed. Design the top chord of the truss. Consider whether it is worth while to arrange a splice at the intermediate top joint, or whether it is more economical to use the same member throughout.

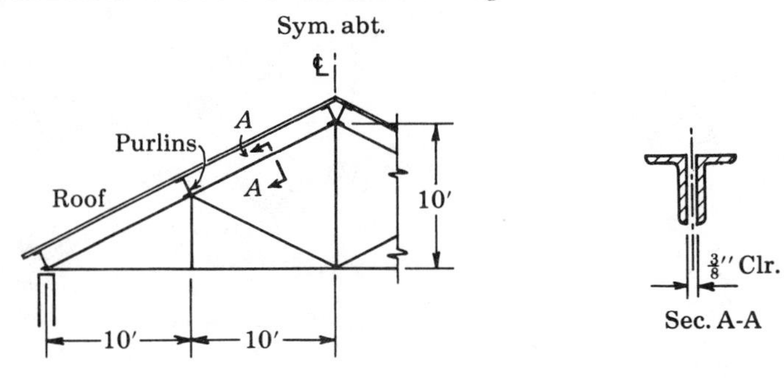

3.46 A wide-flange building column transmits an axial load of 1,000 kips. It is to be diagonally braced at its midpoint in its weak plane. Select an economical and safe column. Steel is A-7.

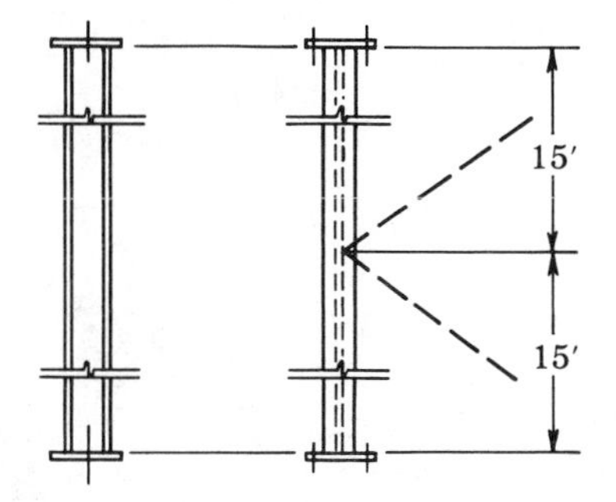

3.47 Design a square column of length 12 ft to transmit a compressive axial load of 120 kips. Material to be Douglas fir, coast range. Follow National Design Specifications for Stress Grade Lumber and Its Fastenings.

part three

Members in Bending

4

Homogeneous Elastic Beams

4.1 Bending Stresses

The analysis of bending stresses of homogeneous elastic beams of symmetric section was treated in Sec. 2.5 as an illustration of the use of the basic steps of the strength of materials approach; the results of that analysis are summarized here:

$$\phi = \frac{M}{EI} \tag{2.25}$$

Since for small deflections the curvature ϕ can be expressed in terms of the beam deflection y as

$$\phi = \frac{d^2y}{dx^2} \tag{4.1}$$

we can therefore write

$$\frac{d^2y}{dx^2} = \frac{M}{EI} \tag{4.2}$$

By successive integration of Eq. (4.2) we can find the deformations of a structure in bending; all the well-known methods of calculating elastic deflections may be derived from Eq. (4.2).

Bending stresses are obtained by use of

$$\sigma = \frac{My}{I} \tag{2.26}$$

This result was based on the assumption of a linear strain distribution and of homogeneous elastic behavior; we must therefore remember that when the maximum stress σ, at the fiber the farthest removed from the neutral axis, exceeds the proportional limit of the material, Eq. (2.26) ceases to be valid.

For the purposes of design, the largest value of bending stress, occurring at the maximum distance $y_{\max} = c$ from the neutral axis, is of importance; we find this by setting

$$\sigma_{\max} = \frac{Mc}{I} \tag{4.3}$$

Now grouping those parts of Eq. (4.3) which depend only on the properties of the member cross section

$$\frac{I}{c} \equiv S \tag{4.4}$$

and calling this quantity the *section modulus*, we can rewrite Eq. (4.3) as

$$\sigma_{\max} = \frac{M}{S} \tag{4.5}$$

For any beam cross section, we can evaluate the section modulus without reference to loading or support conditions, and for standard beam sections, such as commercially available steel beams or standard timber sizes, we can find such values tabulated in various publications.

To calculate the section modulus of a member, we go back to the definition of Eq. (4.4): we need to calculate the location of the neutral axis, the moment of inertia, and the extreme fiber distance c; some examples will show the calculations involved.

Example Problem 4.1

(a) Rectangular section, width b, depth d: By symmetry, the neutral axis is at mid-depth, making the distance $c = d/2$.

$$I = \tfrac{1}{12}bd^3$$

therefore

$$S = \frac{I}{c} = \frac{1}{12}\,bd^3\,\frac{2}{d} = \frac{bd^2}{6}$$

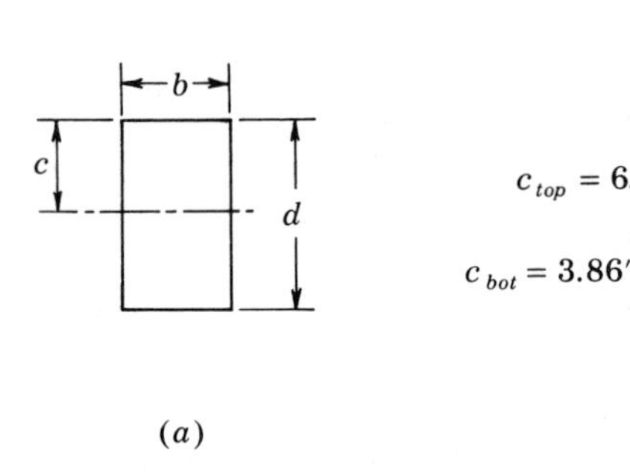

(a) (b)

(b) T section, of dimensions shown; here the neutral axis is unsymmetrically placed, and the distances to top and bottom fibers are therefore different. We use a tabular form to organize our calculations.

Part	A, *in.*^{2}	y', *in.*	Ay', *in.*^{3}	y, *in.*	y^2, *in.*^{2}	Ay^2, *in.*^{4}	$\bar{I}$ *in.*^{4}
1	16	6	96	2.14	4.58	73.3	85.3
2	12	1	12	2.86	8.19	98.2	8.0
	28		108			171.5 + 93.3 = 264.8 in.^{4} = I	

$\bar{y} = \frac{108}{28} = 3.86$ in. $= c_{bot}$
$10 - 3.86 = 6.14$ in. $= c_{top}$

therefore

$$S_{bot} = \frac{I}{c_{bot}} = \frac{264.8}{3.86} = 68.6 \text{ in.}^3$$

$$S_{top} = \frac{I}{c_{top}} = \frac{264.8}{6.14} = 43.1 \text{ in.}^3$$

(**c**) A member of the T section of the previous problem is used as a simple beam, with $M_{\max} = 100$ kip-ft; find the largest tensile and compressive fiber stresses.

$$\sigma_{top} = \sigma_{\max\,comp} = \frac{M}{S_{top}} = \frac{100 \text{ kip-ft} \times 12 \text{ in./ft}}{43.1} = 27.8 \text{ ksi}$$

$$\sigma_{bot} = \sigma_{\max\,tensile} = \frac{M}{S_{bot}} = \frac{100 \times 12}{68.6} = 17.5 \text{ ksi}$$

(**d**) If, for the beam of part (*c*), the allowable tension stress is 20 ksi and the allowable compression stress is 25 ksi, find the allowable moment.

The maximum tensile and compressive stresses are in the ratio c_{bot}/c_{top}:

$$\frac{\sigma_{tensile}}{\sigma_{comp}} = \frac{c_{bot}}{c_{top}} = \frac{3.86}{6.14} = 0.63$$

The allowable stresses are in the ratio

$$\frac{\sigma_{tensile\ allow}}{\sigma_{comp\ allow}} = \frac{20 \text{ ksi}}{25 \text{ ksi}} = 0.80$$

Comparison of these ratios shows that under increasing moment, the allowable compression stress in the top fiber is attained first, at a moment

$$M_{allow} = S_{top}\sigma_{comp\ allow} = \frac{43.1 \text{ in.}^3 \times 25.0 \text{ ksi}}{12 \text{ in./ft}} = 90.0 \text{ kip-ft}$$

This result could also have been attained by prorating the results of part (*c*):

$$M_{allow} = \frac{25.0 \text{ ksi}}{27.8 \text{ ksi}} \times 100 \text{ kip-ft} = 90 \text{ kip-ft}$$

For purposes of design, we can rewrite Eq. (4.5) to solve for a required section modulus to resist a certain moment without exceeding the given allowable stress in the extreme fiber:

$$S_{reqd} = \frac{M}{\sigma_{allow}} \tag{4.6}$$

Having obtained the moment M from an analysis, and having decided on an allowable stress, we can use Eq. (4.6) to find the minimum required section modulus; we can then select a member of cross section satisfying this requirement.

Example Problem 4.2 A floor beam is to be designed to carry a superimposed uniform load of 500 lb/ft; it is supported on masonry walls (Fig. 4.1). The beam could be either of structural steel ($\sigma_{allow} = 20$ ksi) or timber ($\sigma_{allow} = 1.45$ ksi). Maximum beam deflection is to be $\frac{1}{360}$ of the span length. Select the appropriate material and beam size.

We must first decide on the span length and support conditions to be used; it seems reasonable to assume a theoretical support at the center of the walls, leading to a span length L of 20.67 ft; since the masonry walls are unable to resist moment, we should assume simple supports. To the live and floor load of 500 lb/ft we must now add the estimated dead weight of the beam; let us assume this at 50 lb/ft of beam.

With this information we can analyze the beam for shear and moment and obtain the diagrams of Fig. 4.1*b* and *c*; the maximum moment at the midspan is 29.3 kip-ft.

If timber is to be used, the required section modulus is

$$S_{reqd} = \frac{29.3 \times 12}{1.45} = 243 \text{ in.}^3$$

For steel,

$$S_{reqd} = \frac{29.3 \times 12}{20.0} = 17.6 \text{ in.}^3$$

In timber, a 6 × 18 in. beam ($S_{furn} = 281$ in.^{3}) is needed for adequate strength; beams of this size are practically unobtainable, and it therefore seems that the choice will fall on steel. From the AISC Handbook, we find that a 10 W 21 wide-flange beam has a section modulus of 21.5 in.^{3} and thus seems adequate; we now can see that the previously assumed beam weight of 50 lb/ft was excessive, since the beam weighs only 21 lb/ft. Accordingly, the maximum moment, and therefore the required section modulus, should be reduced in the ratio of the assumed to the actual weight:

$$S_{reqd(2)} = \frac{(500 + 21) \text{ lb/ft}}{(500 + 50) \text{ lb/ft}} \times S_{reqd(1)} = 0.947 \times 17.6 \text{ in.}^3$$
$$= 16.7 \text{ in.}^3$$

We see that because of the reduction in weight, the required section modulus has been reduced to the point where an 8 W 20 beam ($S_{furn} = 17.0$ in.^{3}) is adequate for strength; however, before making a final selection, we should check the stiffness of the beam.

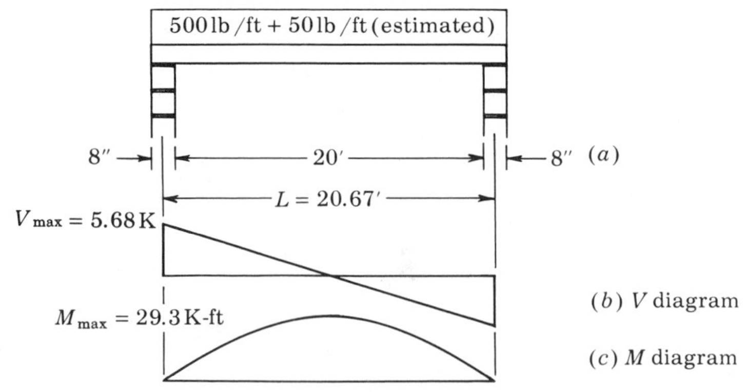

Fig. 4.1

Elastic calculations give for the midpoint deflection of the uniformly loaded simple beam

$$\Delta_{max} = \frac{5}{384} \frac{wL^4}{EI}$$

or

$$\frac{\Delta_{max}}{L} = \frac{1}{360} = \frac{5}{384} \frac{wL^3}{EI}$$

With E for steel at 30,000 ksi and assuming 21 lb/ft for the beam dead load, we can solve for I_{reqd}:

$$I_{reqd} = 360 \times \frac{5}{384} \times \frac{0.521 \text{ kips/ft} \times 20.67^3 \times 144 \text{ in.}^2/\text{ft}^2}{30{,}000 \text{ ksi}}$$
$$= 124 \text{ in.}^4$$

Since the 10 W 21 has a moment of inertia of only 106.3 in.^{4}, we see that this size is deficient in stiffness; going up the tables of wide-flange sections, we see that a 10 W 25 beam has an

$$S_{furn} = 26.4 \text{ in.}^3$$
$$I_{furn} = 133.2 \text{ in.}^3$$

This size is therefore adequate in both strength and stiffness, with sufficient excess capacity to take care of the extra 4 lb/ft beam weight. The final choice is therefore a 10 W 25 steel beam.

Going over the calculations of the preceding example, we draw the following conclusions:

1. The engineer has to decide how best to represent the actual support conditions of the structure.
2. The inclusion of the beam dead weight results in a trial-and-error problem which is usually solved by successive approximation.
3. Elastic beam strength is proportional to the section modulus; elastic stiffness is proportional to the moment of inertia.
4. Due to the trial-and-error nature of the design process, excessive accuracy in the early stages of sizing of members is unwarranted.

4.2 Shear Stresses

Along with the normal bending stresses in beams we must consider the shear stresses; these are a consequence of bending, and may be determined on basis of bending stresses. The steps in the stress analysis involving deformations and stress-strain relations were already used in determining bending stresses, and our further analysis will only involve the equilibrium of an appropriately chosen finite free body, shown in Fig. 4.2.

We consider a free body of length dx along the beam, shown in Fig. 4.2a; the moment on the left side is M, that on the right side slightly different, $M + dM$; in consequence of this difference the bending stresses also vary from My/I on the left face to $(M + dM)y/I$ on the right face. If it is now required to find shear stresses on a horizontal plane a distance y_1 from the neutral axis, we pick a sub-free body to which this plane is exterior, shown in Fig. 4.2b.

Summing forces in the longitudinal direction on this free body, we get:

$$\Sigma F = 0: \quad \int_{\text{Area beyond } y_1} \frac{My}{I}\, dA - \int_{\text{Area beyond } y_1} \frac{(M + dM)y}{I}\, dA + \tau b\, dx = 0$$

or

$$-\frac{dM}{I} \int_{\text{Area beyond } y_1} y\, dA + \tau b\, dx = 0$$

Solving for τ:

$$\tau = \frac{dM}{dx} \frac{\int y\, dA}{Ib} = \frac{V \int y\, dA}{Ib} \tag{4.7}$$

The integral in the numerator denotes the first moment with respect to the N.A. of the part of the cross section beyond the plane on which shear stresses are required; b is the width of the section at this plane. The smaller this width, the smaller the area over which the equilibrating shear force is distributed, and the larger is the shear stress. For a beam of constant width, the first moment takes its largest value at the neutral axis, and it is usually there that we look for the largest value of shear stress.

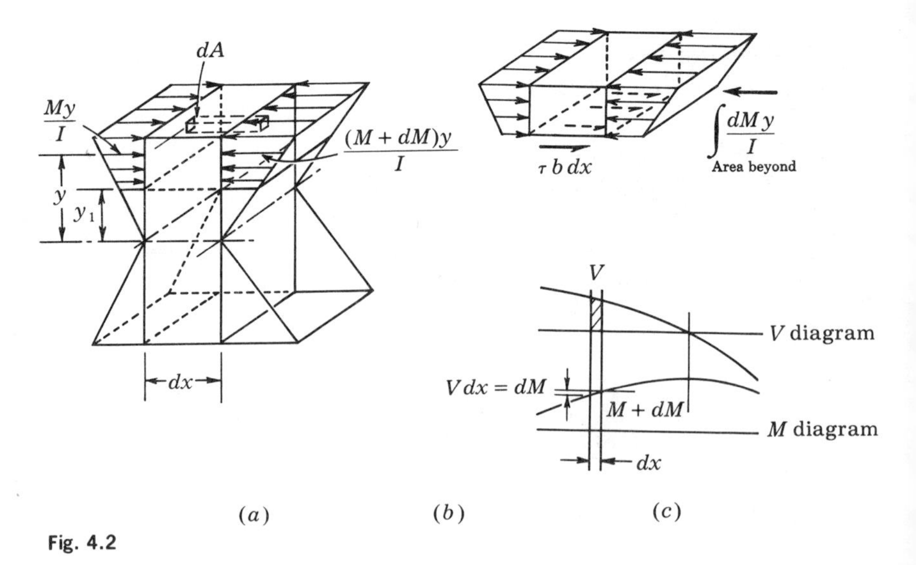

Fig. 4.2

We also see that it is the variation of moment dM from one face of the free body to the other which is responsible for the shear stresses, and since the transverse shear force V is proportional to this change of moment, we should expect the shear stress to be proportional to this shear force, as is indeed shown by Eq. (4.7). This relation is brought out more strongly by considering a transverse plane through the beam, subjected to a total transverse shear force V:

The conjugate shear-stress theory tells us that the shear stress on a longitudinal plane is accompanied by one of equal magnitude acting on a transverse plane, as shown in Fig. 4.3. Summing up all these shear stresses over the transverse face we arrive at the value of the total transverse shear force V:

$$\int_A \tau \, dA = V \tag{2.7}$$

An example of such a calculation was performed in Example Problem 2.2.

It should be noted in passing that the bending and shear stresses at any point may be combined to lead to principal normal and maximum shear stresses on oblique planes. However, since the point of largest bending stress (at the extreme fiber) is one of zero shear stress, and the point of largest shear stress (at the neutral axis) one of zero bending stress, stresses on oblique planes of homogeneous beams will rarely be critical, and only those on transverse planes are considered for design purposes.

Having found the largest values of shear stress, usually at the neutral axis of the section of maximum shear force, we must compare it to some allowable shear stress. For metallic materials, the proportional limit stress in shear is usually about half of the normal proportional limit stress: indeed, this is predicted by "Tresca's" or the "shear-stress failure theory" covered in texts on strength of materials. The allowable shear stress may then be expected to be of the order of half the allowable normal stress.

The situation is different for a nonisotropic material such as timber. Because of the orientation of the fibers of the wood parallel to the beam axis, planes of weakness exist between these fibers, and shear strength of timber beams is slight. Allowable shear stresses in timber are usually less than 10 percent of the corresponding tension strength.

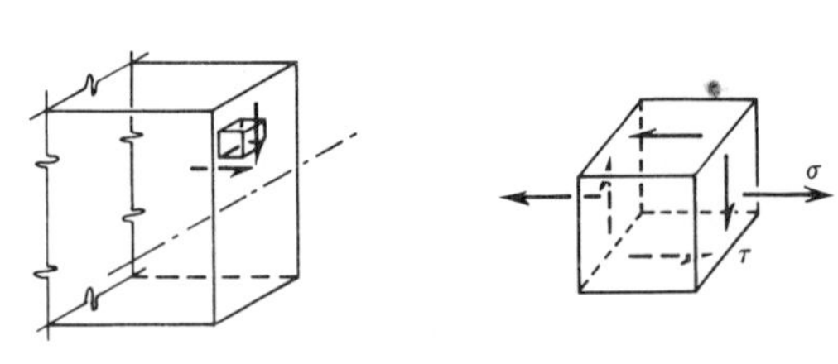

Fig. 4.3

Example Problem 4.3

(**a**) Rectangular cross-section beam, width b, depth d, shear force V. Find shear stresses.

Solution: In this case the width is constant, and the element of area may be taken as

$$dA = b\,dy$$

Equation (4.7) then becomes

$$\tau = \frac{V}{I}\int_{y_1}^{d/2} y\,dy = \frac{V}{2I}\left[\left(\frac{d}{2}\right)^2 - y_1^2\right]$$

It is seen that the shear stresses vary parabolically over the depth, with a maximum value at the neutral axis ($y_1 = 0$) of

$$\tau_{\max} = \frac{V}{2I}\left(\frac{d}{2}\right)^2 = \frac{V}{2(bd^3/12)}\left(\frac{d}{2}\right)^2 = \frac{3}{2}\frac{V}{bd} = \frac{3}{2}\frac{V}{A} \tag{4.8}$$

The maximum shear stress at the neutral axis of a rectangular section is thus $\frac{3}{2}$ of the average value V/A which would be obtained if the total shear force were assumed uniformly distributed; this shows that such simplifying assumptions are on the unsafe side.

At the extreme fiber, $y_1 = d/2$ and the shear stress vanishes. Again this can be predicted, since no traction exists on the free outer surface of the beam.

Now applying Eq. (2.9) we check transverse equilibrium:

$$\int_A \tau\,dA = \frac{Vb}{2I}\int_{y_1=-\frac{d}{2}}^{d/2}\left[\left(\frac{d}{2}\right)^2 - y_1^2\right]dy_1 = \frac{Vb}{2I}\left[\frac{d^2}{4}y_1 - \frac{y_1^3}{3}\right]\Bigg|_{y_1=-\frac{d}{2}}^{d/2}$$

$$= \frac{V}{2I}\frac{bd^3}{6} = V$$

Plot the shear-stress distribution in the rectangular section. (See Figs. 4.4 and 4.5.)

(**b**) 14 WF 87 beam: find shear-stress distribution resulting from shear force V.

Solution: We take the dimensions and properties of the section from the AISC Handbook, page 1-13.

$$I = 966.9 \text{ in.}^4$$

For the shear stresses in the flange, we consider Fig. 4.6.

Going back to basic principles: the unbalanced normal force on a portion of the flange of width y_1 is

$$\int_{\substack{\text{Area}\\\text{beyond}}} \frac{(dM)y}{I}\,dA = dM\,\frac{6.656}{966.9} \times 0.688 \text{ in.} \times y_1 = 0.0069 \times 0.688 y_1\,dM$$

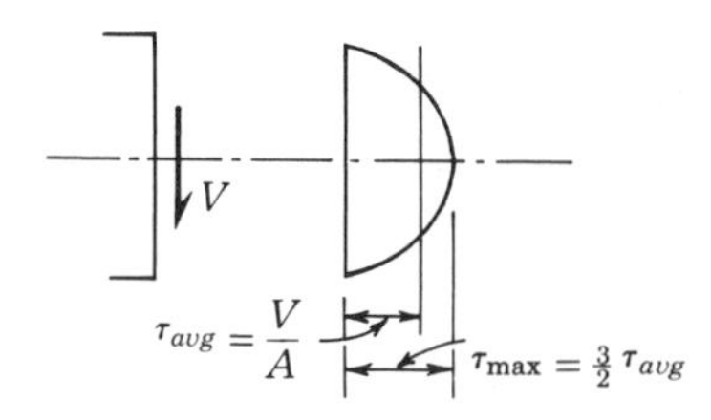

Fig. 4.4

To equilibrate the free body, we set

$$\Sigma F = 0: \qquad \tau \times 0.688 \times dx = 0.0069 \times 0.688 \times dM\, y_1$$

or

$$\tau = 0.0069 \frac{dM}{dx} y_1 = 0.0069V \cdot y_1$$

The shear stress in the flange thus varies linearly from zero at the outer edge to a value of

$$\tau = 0.0069 \times 7.25V = 0.050V$$

at the center of the flange; it is directed horizontally on the transverse plane of the flange. Identical results could of course be obtained by direct application of Eq. (4.7).

In the web, we shall obtain shear stresses by Eq. (4.7).

$$\tau = \frac{V}{966.9 \times 0.420} \left[(14.50 \times 0.688)6.656 \text{ in.} + (0.420y_1)\left(6.312 \frac{y_1}{2}\right) \right]$$
$$= 0.1635 + 0.00653y_1 - 0.000517y_1^2$$

We plot this parabolic stress distribution on Fig. 4.5, and find that the variation is very slight; the maximum value at the neutral axis, corresponding to $y_1 = 6.312$ in., is

$$\tau_{max} = 0.184V$$

It must be noted that only the shear stresses in the web are directed so as to resist the direct shear, and since these stresses are almost uniform, we may try to obtain

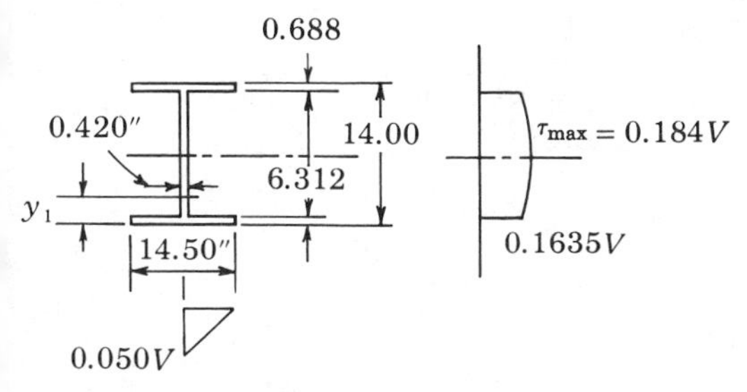

Fig. 4.5

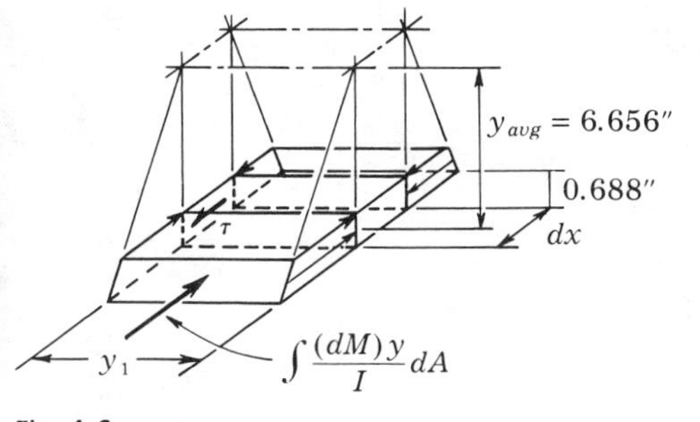

Fig. 4.6

shear stresses in the web by assuming the transverse shear uniformly distributed over the web:

$$\tau_{avg,\ web} = \frac{V}{0.420 \text{ in. } 13.312} = 0.179V$$

In this case, the shear stress based on an average web area is 95 percent of the more rigorously calculated maximum shear stress. In practice, it is customary to calculate shear stress in steel beams by dividing the critical shear by the web area.

(c) A built-up beam section is composed of a 24 WF 100 beam and $\frac{3}{4} \times 16$ in. cover plates top and bottom, riveted by two rows of $\frac{3}{4}$-in. diameter rivets (allowable shear force per rivet is 6.63 kips). Determine the required rivet spacing if the shear in the beam is 160 kips.

Solution: The purpose of the rivets is to keep beam and cover plates from sliding with respect to each other. We calculate the unbalanced shear force on one cover plate per unit length of the beam as

$$\frac{V \int_{cov\ pl} y\, dA}{I}$$

where $\int_{cov\ pl} y\, dA$ is the first moment of one cover plate (this is the "area beyond") $= 12.0 \text{ in.}^2 \times 12.375 \text{ in.} = 148.5 \text{ in.}^3$

$$I = 2987.3 + 12.0 \times 12.375^2 \times 2 + 0.6 = 6{,}658 \text{ in.}^4$$

The shear force per rivet spacing s is s times that per unit length:

$$\text{shear force} = \frac{160 \text{ kips} \times 148.5}{6{,}658}\, s = 3.57s$$

This shear force is resisted by two rivets, of total shear capacity 2×6.63 kips = 13.26 kips; equating actual to safe shear force,

$$3.57s = 13.26$$

and the safe spacing is determined as

$$s = \frac{13.26}{3.57} = 3.72 \text{ in.}$$

The rivet spacing will be taken as $3\frac{5}{8}$ in.

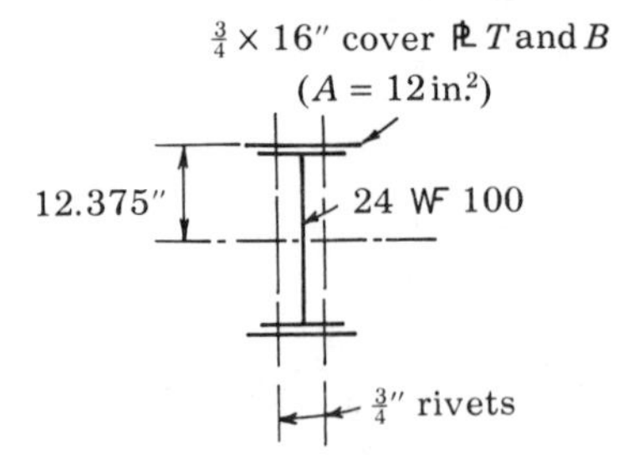

Fig. 4.7

4.3 Deflections

We usually find stiffness requirements imposed on engineering structures, owing to the necessity to prevent excessive sag, sway, springiness, or vibrations. These requirements arise from such diverse considerations as the psychological reaction of users of the structure, noticeable deformations, or the necessity to prevent cracking of window glass resulting from too large deflections of lintels and roofs. It should be noticed that generally strength and stiffness requirements are two different propositions, and a structure may satisfy one without the other; both should be checked to ensure satisfactory performance under working conditions.

Because we generally expect our structures (even those designed by inelastic methods) to resist design loads elastically, we use elastic methods to obtain critical deformations, and the engineer should be in command of methods suitable for calculating this information. An analysis of a loaded structure must always precede the deflection calculation; for commonly occurring simple structures tables are available which give critical values.

Allowable deflections are often at the discretion of the designer; one commonly used criterion for floor beams is a maximum live load deflection of $\frac{1}{360}$ of the span length; this figure arises from the fact that plaster ceilings have been known to crack if hung from more flexible members. For less critical locations, a sag-span ratio of $\frac{1}{240}$ may be indicated.

One way to control deflections of beams is to limit the ratio of the beam depth to span length to a certain value; thus, the AASHO Code (sec. 1.6.11) restricts this ratio to $\frac{1}{25}$ for plate girders and beams.

The relation of this ratio to the beam deformations may be illustrated by consideration of the example of a simply supported beam under uniform load; let the allowable stress be σ_{allow} and the elastic modulus E; if the beam is stressed to its allowable, the curvature ϕ according to the strain distribu-

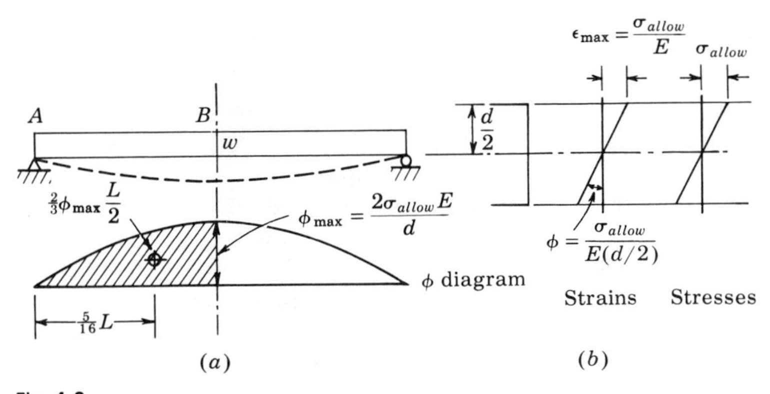

Fig. 4.8

tion of Fig. 4.8*b* will be

$$\phi_{max} = \frac{2\sigma_{allow}}{Ed}$$

at the most highly stressed section, in this case the midspan.

Drawing next the parabolic M/EI, or curvature diagram for this beam, we see that its maximum ordinate under the allowable load is ϕ_{max}. Applying next any of the customary methods of deflection calculation, for instance the "moment area" (or "curvature area") method, we find that

$$\Delta_{max} = \Delta_B = \text{static moment of curvature diagram between points } A \text{ and } B \text{ about point } A$$

$$= \left(\frac{2}{3}\phi_{max}\frac{L}{2}\right)\left(\tfrac{5}{16}L\right) = \frac{10}{48}\frac{\sigma_{max}}{Ed}L^2$$

or

$$\frac{\Delta_{max}}{L} = \frac{10}{48}\frac{\sigma_{max}}{E}\frac{L}{d} \tag{4.9}$$

For the case of a steel beam of $\sigma_{allow} = 20$ ksi, $E = 30{,}000$ ksi, and

$$\Delta_{max}/L = \tfrac{1}{360}$$

the maximum L/d ratio according to Eq. (4.9) would be

$$\frac{L}{d} = \frac{1}{360}\frac{48}{10}\frac{30{,}000}{20} = 20 \tag{4.10}$$

A typical design for critical deflection is shown in Example Problem 4.2 of this chapter. There, the maximum simple beam deflection is set equal to the maximum allowable value, and the required moment of inertia of the beam section is obtained. The beam must then be selected in accordance with this requirement.

In tall structures under lateral loads the amount of side sway should be held to reasonable limits; such considerations are usually checked after the structure has been designed for strength.

4.4 Code Requirements for Steel Beams

As an example of code requirements for steel beams, we consider the pertinent sections of the "Specification for the Design, Fabrication and Erection of Structural Steel for Buildings," 1963 edition, of the American Institute of Steel Construction, henceforth called the "AISC Specifications."

The AISC Specifications, in common with most others, make a somewhat arbitrary decision as to which analysis and design problems are within the knowledge of the code user, and for which applications the code is to provide results and formulas. So, for instance, it is assumed in the AISC Specifications that the determination of maximum bending and shear stresses in beams, as well as deflection calculations, can be readily performed by the designer. Only allowable stresses are provided in sec. 1.5.1.2 for shear, and in sec. 1.5.1.4 for bending:

$$F_v = 0.40F_Y$$

and $F_b = 0.66F_Y$ for cross sections of certain proportions

or $F_b = 0.60F_Y$ for other sections as specified

The reason for the two different bending stresses is apparent when we consider a beam section with extremely thin flanges on the compression side. These flanges will act as thin plates under compression, and will thus be exposed to the danger of buckling instability, characterized by a wavy deformation pattern in the flange, as shown in Fig. 4.9 and discussed in Sec. 3.9 of this book. A certain minimum flange thickness is required to avoid this undesirable action, which would result in rapid loss of bending capacity of the beam. Section 1.9.1 provides for such minimum thickness for projecting elements of members; sec. 2.6 provides for even more stringent limitations on the proportions of flanges, and the allowable stress is accordingly increased to conform with theoretical and experimental predictions. Whereas the code user is expected to be able to calculate the bending and shear resistance of beams, no such knowledge is presupposed for the more involved problem of flange buckling. Nevertheless, an understanding of the buckling phenomenon is highly desirable for intelligent application of the formulas.

Again, the difficult problem of the design of beams to avoid the danger of lateral buckling, a phenomenon whereby the compression side of a sufficiently narrow beam has a tendency to flip sideways under load, thus inducing twisting of the beam, is reduced in the Specifications to the two formulas 4 and 5 of sec. 1.5.1.4.5, which provide a reduction factor for the previously specified stresses under certain conditions. An understanding of the nature of the phenomenon and its analytical treatment is necessary in order to apply these formulas with some intelligence, and Sec. 4.8 in this chapter deals with this in a rational manner.

Other provisions which must be considered when designing steel beams are the following:

Web proportions: Sections 1.10.1 and 1.10.2 and sec. 1.10.6 provide against buckling arising from excessively thin webs of beams and girders. It should be noted that when commercial rolled beam sections are used, it is in general

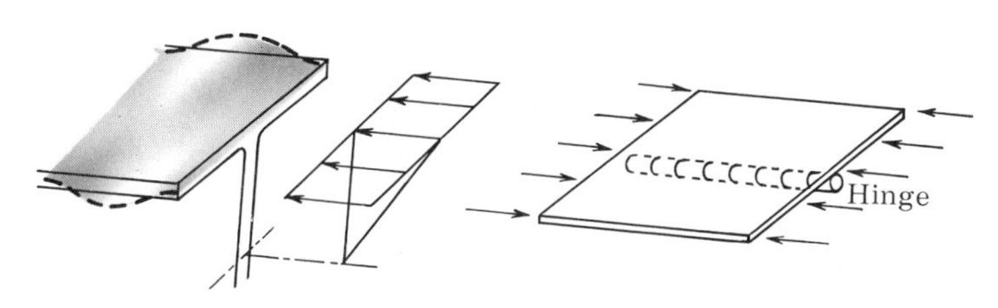

Fig. 4.9

not necessary to check cross-sectional properties, since these sections have been designed sufficiently thick to prevent buckling of flanges or web.

Deflections: Section 1.13 reminds us that deformations may become critical, and provides some criteria for maximum deflections.

Web crippling: At points of support or of concentrated load application, a compressive force is transmitted to the thin beam web which may cause a type of plate buckling called *crippling.* Formulas 13 and 14 of sec. 1.10.6 provide for a maximum compressive web stress based on assumed distribution of the applied force between the 45-deg lines shown in Fig. 4.10.

Stiffeners of plate girders: Section 1.10.5 gives regulations for the proportioning of vertical stiffeners of webs; these are members fastened to thin girder webs at intervals as shown in Fig. 4.11 to prevent shear buckling of the thin web plate as shown in the figure. The elaborate formulas of sec. 1.10.5 are based on a series of theoretical and experimental investigations which corroborated the so-called "tension field" concept. According to this concept, upon shear buckling of the web the shear is transmitted by an action similar to that of a truss, in which the portions of the web parallel to the wrinkles act like tension diagonals, and the stiffeners as compression verticals.

4.5 Code Provisions for Timber Beams

A commonly used set of rules for the design of timber structure are the "National Design Specifications for Stress Grade Lumber and Its Fastenings," published by the National Lumber Manufacturers' Association.

A portion of this code of particular interest in the design of timber beams is table 1, which gives the allowable stresses in bending (F) and in

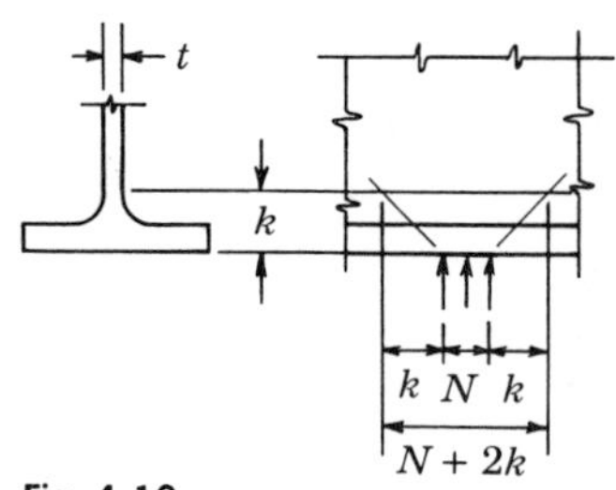

Fig. 4.10

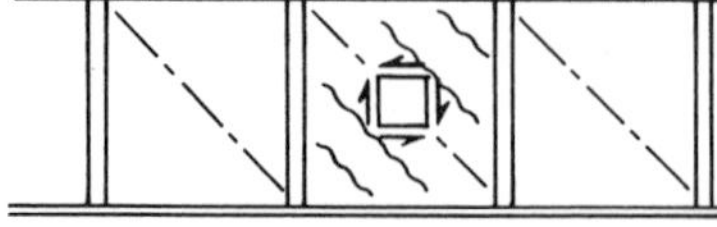

Fig. 4.11

horizontal shear (H) for different commercially available types of lumber. We note that the shear stresses are held very low, owing to the weakness with respect to sliding between cells of wood. It should also be observed that the allowable compression stress perpendicular to the grain is low; therefore, care must be exercised in designing beam supports to provide sufficient bearing area so that the reactions may be distributed to the beam without crushing of the cells.

Design information for beams is given in sec. 400. We note in looking this section over that in this code formulas are provided to determine elementary bending and shear stresses. In all cases, these are identical with those derived here for homogeneous, elastic beams.

4.6 Unsymmetric Bending*

The procedure for beam analysis used in Sec. 4.1 was restricted to beams with at least one axis of symmetry. The reason for this is found in step 1 of Example Problem 2.7 of Sec. 2.5, in which the moment-equilibrium condition about the Y axis was taken care of by symmetry. If this symmetry is absent, an additional equilibrium condition must be written, as shown in the following.

Consider an elastic beam of arbitrary cross section, as shown in Fig. 4.12a, subjected to moments M_X and M_Y about the X and Y axes, respectively. To analyze this section, we again resort to our basic steps:

Equilibrium: Note that now we need three equations:

$$\Sigma F = 0: \quad \int \sigma \, dA = 0 \tag{4.11}$$
$$\Sigma M_Y = 0: \quad \int \sigma x \, dA = M_Y \tag{4.12}$$
$$\Sigma M_X = 0: \quad \int \sigma y \, dA = M_X \tag{4.13}$$

Geometry: We may reasonably assume again that plane sections remain plane, as shown in Fig. 4.12b, but this time we must visualize this plane as being a

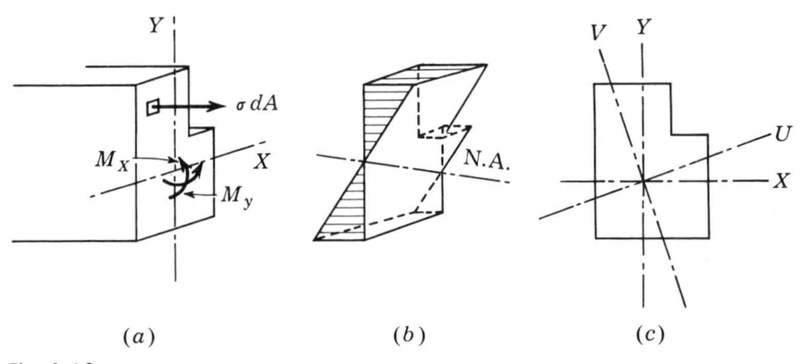

Fig. 4.12

function of both x and y; the equation for the strains then takes the linear form

$$\epsilon = a_1 + \phi_Y x + \phi_X y \tag{4.14}$$

where x and y are coordinates of the X and Y axes and a_1, ϕ_Y, ϕ_X are as yet unknown deformations.

Stress-strain relations: For elastic material, Hooke's law holds:

$$\sigma = E\epsilon \tag{2.12}$$

and substituting Eq. (4.14) into this, we write

$$\begin{aligned} \sigma &= E(a_1 + \phi_Y x + \phi_X y) \\ &= a + bx + cy \end{aligned} \tag{4.15}$$

where now

$$a = Ea_1 \qquad b = E\phi_Y \qquad c = E\phi_X \tag{4.16}$$

Equation (4.15) is now substituted in the equilibrium equations (4.11) to (4.13), to yield

$$\Sigma F = 0: \quad a\int dA + b\int x\,dA + c\int y\,dA = 0 \tag{4.17}$$
$$\Sigma M_X = 0: \quad a\int y\,dA + b\int x^2\,dA + c\int xy\,dA = M_Y \tag{4.18}$$
$$\Sigma M_Y = 0: \quad a\int x\,dA + b\int xy\,dA + c\int y^2\,dA = M_X \tag{4.19}$$

We recognize the integrals as cross-sectional characteristics, namely $\int dA$ is the area A; $\int x\,dA$ and $\int y\,dA$ are first moments, S_X and S_Y, with respect to the reference axes; $\int x^2\,dA$, $\int y^2\,dA$, and $\int xy\,dA$ are second moments of inertia, I_{XX} and I_{YY}, and product of inertia, I_{XY}, respectively.

We can solve the three equations (4.17) to (4.19) for the unknowns a, b, and c; this operation can be vastly simplified by suitable choice of the coordinate axes. So, if the X and Y axes are selected to pass through the centroid, then S_X and S_Y are zero; if further, the axes are oriented so as to form principal axes, then the product of inertia vanishes as well, and only the diagonal terms of the array of equations are retained; denoting quantities measured with respect to the principal axes with the symbols U and V,

$$a = 0$$
$$b = \frac{M_v}{I_v}$$
$$c = \frac{M_u}{I_u}$$

and, substituting these values back into Eq. (4.15), we find the stresses as

$$\sigma = \frac{M_v}{I_v}u + \frac{M_u}{I_u}v \tag{4.20}$$

The curvatures with respect to the principal axes are found, using Eqs. (4.16b) and (4.16c), as

$$\phi_v = \frac{M_v}{EI_v} \tag{4.21}$$

and $$\phi_u = \frac{M_u}{EI_u} \tag{4.22}$$

We note that the developments are analogous to the beam theory for symmetric sections, but must remember that the results hold only if the reference axes are principal axes. The first operation should be to locate these axes by appropriate analytical transformation procedures or by Mohr's circle.

Because any two orthogonal axes, at least one of which is an axis of symmetry, constitute principal axes, the case of symmetric sections becomes identical with the more general case presented here. In that case, the location of the principal axes by transformation becomes unnecessary.

The advantage of the method presented here is the uniformity of the approach for symmetric and unsymmetric sections. The disadvantage is the fact that coordinates must be found with respect to skew axes (see example problem). To avoid this sometimes laborious calculation, we can go back to Eqs. (4.17) to (4.19), and consider convenient centroidal (but nonprincipal) X and Y axes. In this case, the product-of-inertia term must be retained, leading to somewhat more involved equations which take the place of Eqs. (4.20) to (4.22). The development of these equations is left to the student in Prob. 4.25.

It is important to remember that the usual flexure formula must be used with principal axes, and that neglect of this fact may lead to stress calculations which are unsafe. Just as important is the realization that a beam of unsymmetric section will have a deflection component normal to the plane of loading. These facts will be demonstrated in an example problem.

Example Problem 4.4 A beam has a cross section, as shown in the figure, obtained by deducting one-quarter of a square section. Find the bending stresses and curvatures due to a pure moment of value M about the X axis.

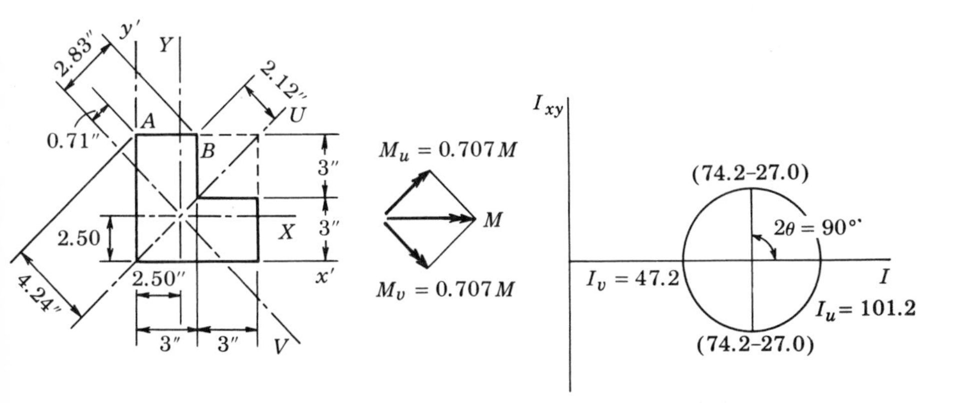

Solution: We first obtain section properties in tabular form, and then apply Mohr's circle to find the principal axes and moments of inertia:

Part	Area, in.^{2}	y', in.	Ay', in.^{3}	x', in.	Ax', in.^{3}	$\bar{x} = \bar{y}$, in.	$A\bar{y}^2 = A\bar{x}^2$, in.^{4}	$\bar{I}_x = \bar{I}_y$, in.^{4}	$I_{xy} = A\bar{x}\bar{y}$, in.^{4}
1	36	3.0	108.0	3.0	108.0	0.50	9.0	108.0	9.0
2	− 9	4.5	− 40.5	4.5	− 40.5	2.00	−36.0	− 6.8	−36.0
	27		67.5		67.5		−27.0	101.2	−27.0
	$y = \frac{67.5}{27} = 2.5$					$I_x = I_y = 74.2 = I_{xy}$			

Having found the moments and product of inertia with respect to the centroidal X and Y axes, we find the principal axes and moments of inertia by Mohr's circle, as shown in the figure. The angle 2θ between the X and Y and the principal U and V axes is 90 deg, indicating that the principal axes are oriented at 45 deg with the X and Y axes. Since the axis of symmetry is at 45 deg, this could have been predicted without calculations. The principal moments of inertia can also be read off Mohr's circle as $I_x \pm I_{xy} = 101.2$ in.^{4} and 47.2 in.^{4}

The principal axes and dimensions are indicated in the sketch, and, having resolved the applied moment M into its U and V components, $M_u = M_v = 0.707M$, we can find the bending stresses and curvatures by Eqs. (4.20) to (4.22):

$$\sigma = \frac{M_u v}{I_u} + \frac{M_v u}{I_v} = 0.707M\left(\frac{v}{101.2} + \frac{u}{47.2}\right)$$

or, for points A and B respectively,

$$\sigma_A = 0.707M\left(\frac{4.24}{101.2} + \frac{0.71}{47.2}\right) = 0.040M$$

$$\sigma_B = 0.707M\left(\frac{2.12}{101.2} + \frac{2.83}{47.2}\right) = 0.057M$$

Stresses at other points can be checked in similar fashion; the stress at point B will be found critical.

The curvatures, by Eqs. (4.21) and (4.22), are

$$\phi_u = \frac{0.707M}{101.2E} = 0.0070\frac{M}{E}$$

$$\phi_v = \frac{0.707M}{47.2E} = 0.0150\frac{M}{E}$$

and finding the components of these vector quantities about the X and Y axes, we calculate

$$\phi_x = 0.707(0.0070 + 0.0150)\frac{M}{E} = 0.0155\frac{M}{E}$$

$$\phi_y = 0.707(0.0070 - 0.0150)\frac{M}{E} = 0.0057\frac{M}{E}$$

thus bearing out an earlier statement that bending about one axis of an unsymmetric section will result in components of deformation about both axes.

4.7 Shear Center*

When a beam whose cross section is unsymmetric with respect to the plane of loading is subjected to transverse shear, the possibility exists that the member may twist under this shear. Such twisting may lead to undesirably high torsional shear stresses and deformations (as discussed in some detail in Part Five), and to avoid this possibility the load should be applied through a certain point of the cross section called the *shear center.*

To show that twisting may occur in a member under transverse shear, we consider the shear stresses resulting from bending in the channel section shown in Fig. 4.13*a*. They are calculated by elementary theory and are plotted in Fig. 4.13*b*. Taking a convenient longitudinal Z axis, we calculate the resisting moment about this axis of these stresses; if this resisting moment is equal to that of the applied shear, no twisting will result. But if resisting and applied moment are not equal, then the member will have to twist till the additional torsional shear stresses equilibrate the unbalanced moment. Here, then, we will try to apply the load in such a fashion that this twisting is avoided.

The shear stress at any point is, by Eq. (4.7),

$$\tau = \frac{V\int y\,dA}{Ib}$$

the shear force on a typical element of the section is this stress multiplied by its area da, the lever arm of this shear force about the Z axis is l, and the total resisting moment about the Z axis is found by integrating the product,

$$M_z = \int_A \tau\,da\,l = \frac{V}{I}\int_A \left(\frac{1}{b}\int y\,dA\right) l\,da$$

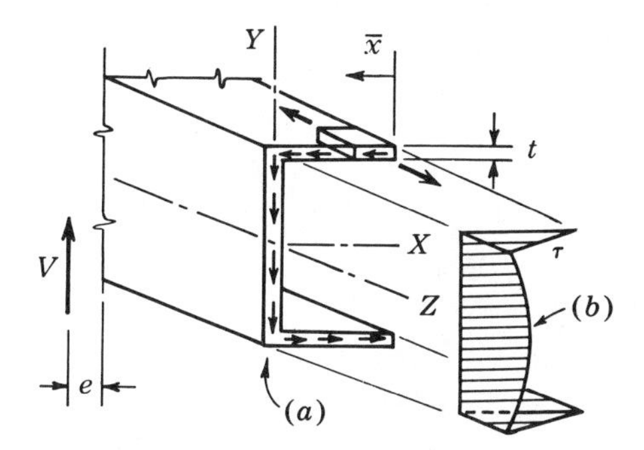

Fig. 4.13

For moment equilibrium about the Z axis, this must be equilibrated by the applied shear times its lever arm e:

$$\Sigma M_z = 0: \qquad Ve = \int \tau \, da \, l = \frac{V}{I} \int \left(\frac{1}{b} \int y \, dA \right) l \, da$$

and, canceling V, we deduce that the location of the shear center e is only a function of the cross-sectional properties:

$$e = \frac{1}{I} \int_A \left(\frac{1}{b} \int y \, dA \right) l \, da$$

To demonstrate the application of these ideas, we consider a semicircular section of constant thickness t subjected to a shear V parallel to the Y axis, as shown in Fig. 4.14.

It is required to find the location of the shear center. To emphasize basic reasoning, we first find the variation of the shear stresses by Eq. (4.7):

$$\tau = \frac{V}{It} \int_{\phi=0}^{\phi=\theta} (tR \, d\phi)(R \cos \phi) = \frac{V}{I} R^2 \sin \theta$$

Thus, the shear stress varies sinusoidally with θ, and the shear force is this stress multiplied by the indicated element, $(\tau)(tR \, d\theta)$. We conveniently pick the center of the semicircle as moment center, so that the lever arm is constant at R. Thus, the resisting moment of the shear forces is

$$\begin{aligned} M_z &= \int_{\theta=0}^{\pi} \tau(tR \, d\theta)R = \frac{V}{I} R^4 t \int_{\theta=0}^{\pi} \sin \theta \, d\theta \\ &= 2 \frac{V}{I} R^4 t \end{aligned}$$

This is to be equilibrated by the moment Ve of the applied shear, thus

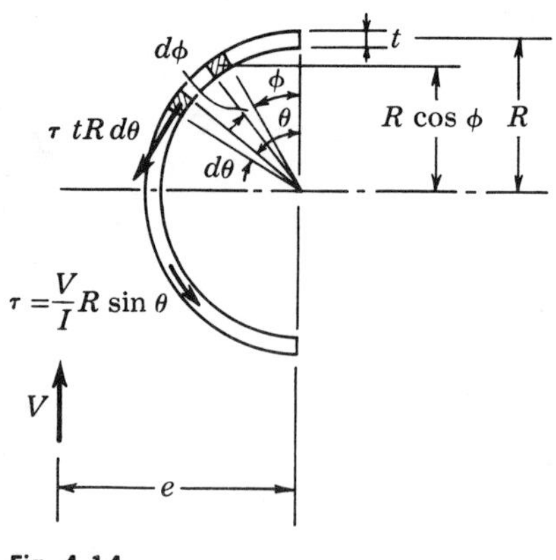

Fig. 4.14

equating,

$$\Sigma M_z = 0: \qquad e = \frac{2R^4t}{I} = \frac{2R^4t}{\frac{1}{2}\pi R^3 t} = \frac{4}{\pi} R = 1.27R$$

We see that the shear center is actually outside the back of the semicircle. Load and support should be attached along this axis to avoid twist of the section.

For channel sections, similar procedures are used; the integration of the moments of the shear stresses can of course be performed by graphical or numerical methods. Thus, in the channel section of Fig. 4.15, the total shear forces in flanges and web are found as the areas under the shear stress diagrams, multiplied by the thickness of the part, and multiplied by the appropriate lever arm. If the moment center O is selected on the web, no moment is caused by the shear force in the web, and the shear center is found by the moment equation

$$\Sigma M_O = 0: \qquad Ve = 2 \times 0.133V \times 3 \text{ in.}$$

or $\quad e = 0.80$ in.

Again, the shear center is outside the channel back.

In sections with very thin webs connecting the flanges, as shown in Fig. 4.16, the flanges may be assumed to supply the total bending resistance. In this case, the shear stresses are constant along the webs, and the calcula-

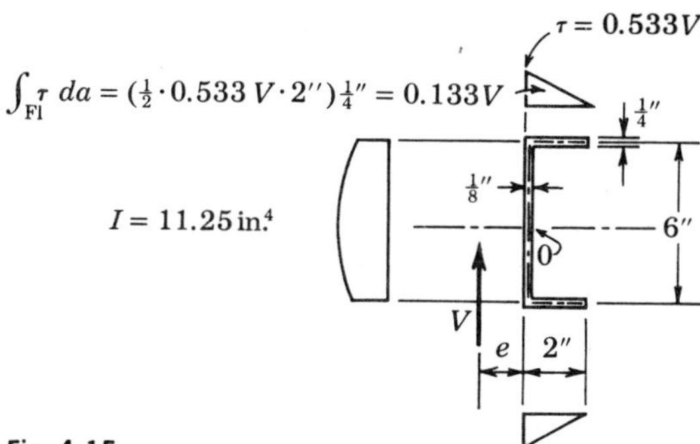

Fig. 4.15

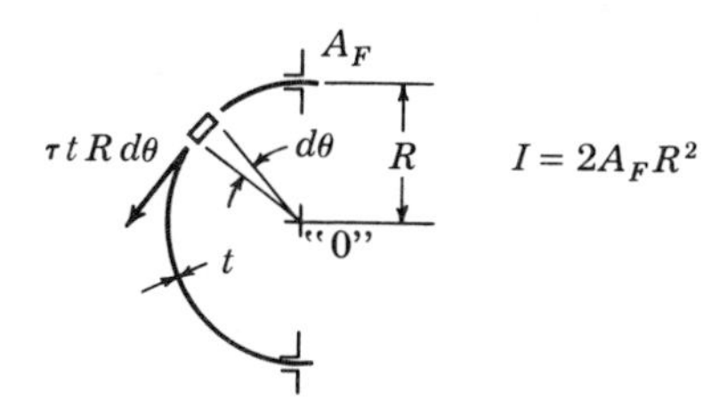

Fig. 4.16

tions for shear center are accordingly simplified. Thus, if the section of Fig. 4.16 is subjected to a shear V, the shear stresses are constant,

$$\tau = \frac{V \int y \, dA}{It} = \frac{V}{2tR}$$

and $\Sigma M_O = 0$: $\quad Ve = \tau \int_{\theta=0}^{\pi} (tR \, d\theta) R = \frac{VR}{2} \pi$

or $\quad e = \frac{\pi}{2} R = 1.57R$

For sections without any symmetry, we remember that bending stresses and shear stresses can be calculated by the familiar methods only if they are referred to principal axes (see Sec. 4.6 and Probs. 4.34 and 4.35). A suggested sequence of operations for finding the shear center in this case is the following:

1. Locate principal centroidal axes of the section.
2. Resolve the applied shear into components along these axes.
3. Treating one component of shear at a time, find the line of application of this component by methods outlined in this section.
4. Locate the shear center as the intersection of the lines found in step 3.

While it is usually desirable to apply loads through the shear center of the section, this sometimes becomes impossible. In such cases, the load should be resolved into a force through the shear center, plus a twisting couple. The stresses due to the former can then be evaluated by elementary bending theory, while the torsional stresses due to the latter must be found by the methods of the torsional theory outlined in Part Five. Shear stresses resulting from torsion and normal and shear stresses resulting from bending can then be superimposed, and principal stresses found by usual methods of stress transformation.

4.8 Lateral Buckling of Beams*

In addition to the possibility of failure of beams due to excessive stresses, we should also be aware of a particular type of instability which can occur in beams of slender cross section. This is the phenomenon of lateral buckling, in which the compression side of the beam flips sideways under load, thus leading to sudden, and possibly catastrophic, collapse.

The rational analysis of this action adheres to the classical methods of elastic stability, and the reader might refer back to Secs. 2.7 and 3.6 to verify the consistency of the approach.

To gain insight into the nature of the problem, we consider an elastic strip of slender rectangular cross section, shown in Fig. 4.17a, subjected to a set of cranked-in end moments, of value M_0. It is required to find the critical value of these moments, M_{cr}, which enables the strip to maintain equilibrium in its deflected shape, shown in Fig. 4.17*b*. The ends of the strip are supported so that rotation about the Z axis is possible, but rotation about the X axis is prevented. We call this type of support *flexurally free, torsionally fixed.* Figure 4.17*c* shows the beam in its buckled shape, and the stress resultant M_0 on a typical section is resolved into its flexural and torsional components along the principal axes of the deformed member. These components are found from Fig. 4.17*d* (neglecting small angles), denoting the displacement parallel to the Z axis as w, as

$$\begin{aligned} M_z &\approx -M_0 \\ M_y &\approx M_0 \frac{dw}{dy} \\ T &\approx M_0 \frac{dw}{dx} \end{aligned} \tag{4.23}$$

Next, we must relate the flexural moment M_y and the torsional moment T

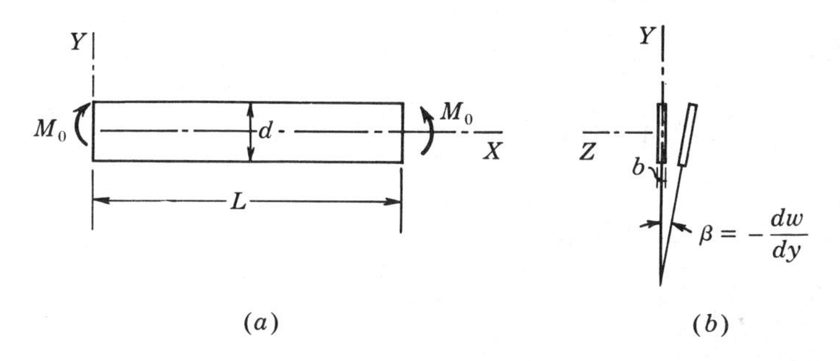

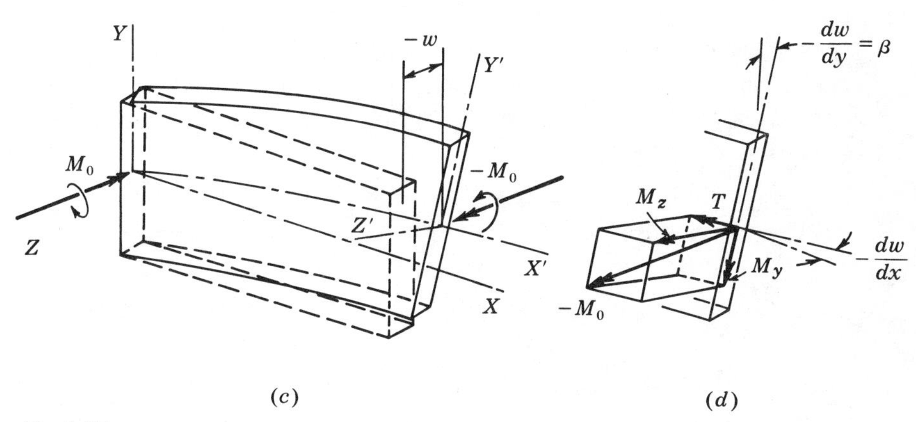

Fig. 4.17

to the deflected shape by elementary theory:

$$M_y = -EI_y \frac{d^2w}{dx^2} = M_0 \frac{dw}{dy} \tag{4.24}$$

$$T = GK \frac{d}{dx}\left(\frac{dw}{dy}\right) = M_0 \frac{dw}{dx} \tag{4.25}$$

In Eq. (4.24), dw/dy is the angle of twist of the buckled shape, and this quantity will henceforth be called β. K is a torsional stiffness factor (analogous to the polar moment of inertia J in a circular section) which depends on the cross-sectional properties only (see Sec. 12.3).

We next differentiate Eq. (4.25) once more with respect to x, and then eliminate the quantity d^2w/dx^2 between Eqs. (4.24) and (4.25), to obtain the differential equation

$$\frac{d^2\beta}{dx^2} + \left(\frac{M_0{}^2}{EI_yGK}\right)\beta = 0$$

or, abbreviating by denoting the coefficient in parentheses by k^2,

$$\frac{d^2\beta}{dx^2} + k^2\beta = 0$$

The solution of this homogeneous differential equation of the harmonic oscillator type is

$$\beta = C_1 \sin kx + C_2 \cos kx$$

The constants C_1 and C_2 are found from the boundary conditions, which call for zero angle of twist at the ends:

$$\beta(0) = 0: \qquad 0 \qquad C_1 + 1 \qquad C_2 = 0 \tag{4.26}$$

$$\beta(L) = 0: \qquad \sin kL\, C_1 + \cos kL\, C_2 = 0 \tag{4.27}$$

Further developments are analogous to those used in connection with axially loaded columns: Eqs. (4.26) and (4.27) constitute a set of homogeneous equations which can have a nontrivial value only if the determinant of the coefficients is zero; therefore

$$\begin{vmatrix} 0 & 1 \\ \sin kL & \cos kL \end{vmatrix} = 0$$

or $\quad \sin kL = 0 \qquad$ or $\qquad kL = 0, \pi, \ldots, n\pi, \ldots$

or $\quad \dfrac{M_0L}{\sqrt{EI_yGK}} = n\pi \qquad n = 0, 1, 2, \ldots$

Solving for the lowest nontrivial value of the critical moment,

$$M_{cr} = \frac{\pi\sqrt{EI_yGK}}{L} \tag{4.28}$$

We see that the critical moment depends on flexural and torsional stiffnesses as well as the length of the beam.

For engineering purposes, it is convenient to introduce a critical stress

$$\sigma_{cr} = \frac{M_{cr}}{I_x}\frac{d}{2} = \left(\frac{\pi\sqrt{EG}}{2}\right)\frac{\sqrt{I_y K}}{I_x}\frac{d}{L} \tag{4.29}$$

Expressing the moments of inertia and the torsional stiffness factor (see Sec. 12.3) in terms of the cross-sectional dimensions, and adopting definite numerical values for a certain material for E and G, we can write the critical stress as

$$\begin{aligned}\sigma_{cr} &= \frac{\pi\sqrt{EG}}{2}\frac{\sqrt{(db^3/12)(db^3/3)}}{(bd^3/12)}\frac{d}{L} \\ &= \frac{C}{Ld/b^2}\end{aligned} \tag{4.30}$$

where C is an appropriate numerical constant for a given material.

The critical lateral buckling stress on a thin strip under constant moment is thus proportional to b^2, representing the weak-axis buckling resistance, and inversely proportional to L and d.

An approximate expression similar to Eq. (4.30) may be derived for lateral buckling of I-section beams, but in this case the weak-axis buckling resistance depends on the cross-sectional area of the compression flange A_f, and in sec. 1.5.1.4.5 of the AISC Specifications, formula 5 gives the allowable compression stress in an unbraced beam as

$$F_b = \frac{12{,}000{,}000}{Ld/A_f} \tag{4.31}$$

This is of the same form as Eq. (4.30).

An approximate, conservative approach to the problem of lateral buckling would be to neglect the torsional resistance of the section and to consider only the lateral buckling strength of the compression flange. The problem is then reduced to one of Euler buckling, and formula 4 of AISC sec. 1.5.1.4.5 results from this method. This formula is similar to AISC formula 1 [Eq. (3.39)], but here the radius of gyration is that associated with lateral flange buckling; a portion of the web is included to account for the restraining effect of the web. A factor C_b accounts for the effect of moment variation along the member, since the danger of lateral buckling is larger for a beam with a region of constant maximum moment than for one with a localized peak moment. This variable is not considered in formula 5.

Lateral buckling is also affected by the manner of load application and end support. Expressions such as Eq. (4.31) or (3.39) give only very approximate solutions to the problem, and in critical cases reference to more exact solutions is in order. It is of course very desirable to restrain the compression side against lateral movement if possible. This can be done by suitable floor arrangement, or by bracing.

General Readings

Elementary beam theory

The basic theory of elastic beams is covered in texts on strength of materials such as those listed in Chap. 2.

Professional design

AISC: "Commentary on the A.I.S.C. Specifications," 1963. Gives explanation and background of 1963 AISC Specifications. Required reading for understanding some of the code provisions. (This is incorporated in AISC Steel Construction Manual, 6th ed.)

Bresler, Boris, and T. Y. Lin: "Design of Steel Structures," Wiley, 1960. This and the previous reference are standard senior-level textbooks in steel design. A number of chapters in both books can serve as supplementary reading.

Gaylord, Edwin H., Jr., and Charles N. Gaylord: "Design of Steel Structures," McGraw-Hill, 1957.

Grinter, Linton E.: "Elementary Structural Analysis and Design," Macmillan, 1965. Beam design, and structural design in general, from a professional viewpoint. Application rather than theory is stressed here for a variety of different materials.

Further topics in beam theory

Murphy, Glenn: "Advanced Mechanics of Materials," McGraw-Hill, 1946. Typical of a number of texts on advanced strength of materials with coverage of topics such as unsymmetric bending, shear center, and curved-beam theory.

Lateral buckling of beams

Harris, Charles O.: "Introduction to Stress Analysis," Macmillan, 1959. Chapter 8 has a simple treatment of lateral buckling similar to the one presented here.

Timoshenko, Stephen P., and James M. Gere: "Theory of Elastic Stability," 2d ed., McGraw-Hill, 1961. Chapter 6 covers lateral buckling of elastic beams.

Problems

4.1 The timber floor system supports a live load of 40 lb/ft² of floor area. Flooring is 2-in. timber decking. Beams are of Southern pine, No. 1 Dense (f_{allow} = 1,600 lb/in.²). Maximum allowable deflection is $\frac{1}{270}$ of the span length. Design a suitable timber beam. E = 1,760,000 lb/in.²

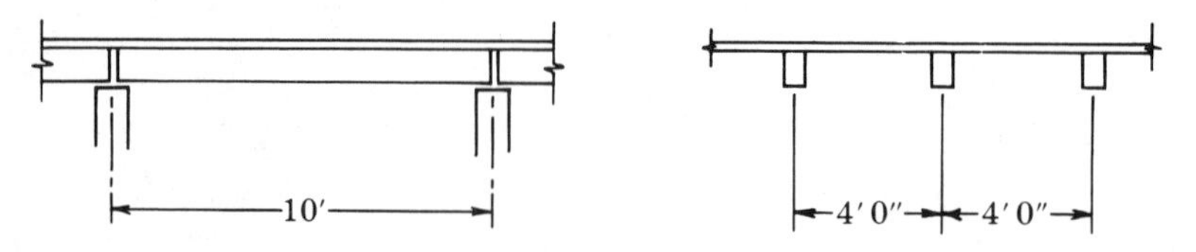

4.2 1-in. sheathing is used on a roof to span across rafters. Material is Douglas fir, inland region (f_{allow} = 1,900 lb/in.², E = 1,760,000 lb/in.²). If the roof live load is 30 lb/ft², what may be the maximum rafter spacing s? Maximum deflection is $\frac{1}{240}$ of the span. (1 in. nominal = $\frac{8}{4}$ in. actual dimension.) Make two assumptions:

(*a*) Sheathing acts as simple beams between rafters.
(*b*) Sheathing acts as continuous beam over rafters.

Comment on appropriateness of assumptions.

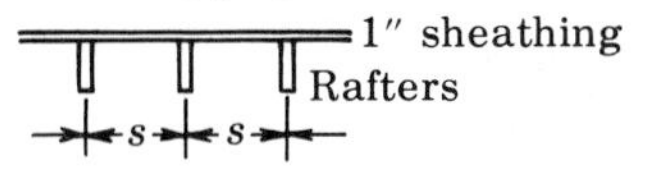

4.3 A 12 × 20 ft platform is to be constructed of Douglas fir timber, coast region, construction grade (f_{allow} = 1,500 lb/in.², E = 1,760,000 lb/in.²). Flooring is to be of 1-in. decking (actual thickness = $\frac{3}{4}$ in.). Live load is 100 lb/ft² of the floor area. Perform the design in the following sequence:

(*a*) Calculate maximum allowable joist spacing (based on capacity of flooring).
(*b*) Design the transverse joists.
(*c*) Design the longitudinal stringers.
(*d*) Make design sketch showing sizes and dimensions.

Remember that in any case, an analysis must be performed before a member can be designed.

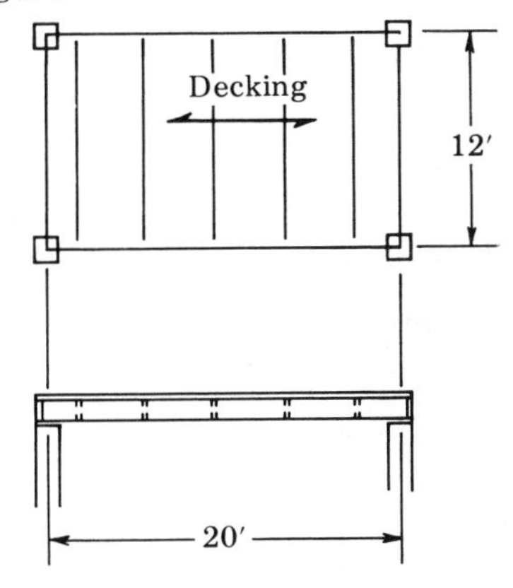

4.4 A floor system is made of 2 × 8 in. joists of Douglas fir (F_{allow} = 1,500 lb/in.²), supporting $\frac{3}{4}$-in.-thick plywood flooring. The joists span over 10 ft, with simple supports.

(*a*) Calculate the allowable floor load if only the joists are assumed to transmit the load across the 10-ft span.
(*b*) Assume that the plywood is rigidly glued to the joists so as to act as a top flange of the beam section. On basis of this assumption, again calculate the allowable floor load. (According to recommended practice, only those plies whose grain runs parallel to the beam axis should be considered in calculating the section modulus. Here, for simplicity, it is permitted to include the entire thickness of the plywood.)
(*c*) Compare results of (*a*) and (*b*) and draw conclusions.

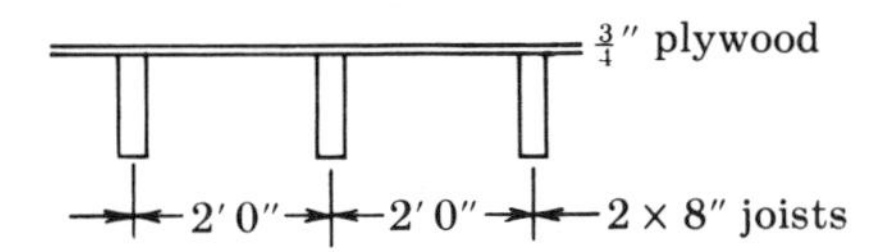

4.5 Four 2 × 8 in. joists are to be combined to form a beam. Consider the following combinations (all suitably fastened together).

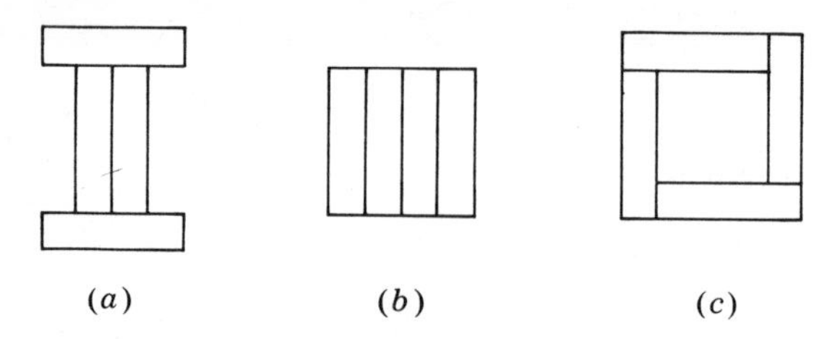

Compare the allowable moments on alternates (*a*), (*b*), and (*c*), and draw conclusions.

4.6 The box beam is glued up of 2 × 6 in. top and bottom flanges connected by ½-in. plywood webs (f_{allow} = 2.05 ksi). Calculate the depth *d* required to resist an applied moment of 40 kip-ft.

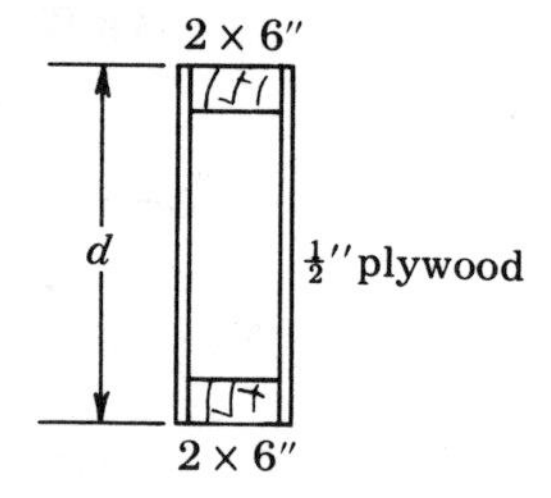

4.7 Use the dimensions from the table "Properties for Designing" of the AISC Handbook for a 14 WF 264 section to compute its cross-sectional area, section modulus, and moment of inertia, and compare these calculated values with those given in the table. Calculate the maximum allowable moment according to the AISC Specifications for this beam made of A-36 steel.

4.8 A beam is made up of a 12 WF 53 beam, to which is welded a 12 [20.7 channel as shown. Calculate the maximum allowable moment according to the AISC Specification for this beam made of A-7 steel.

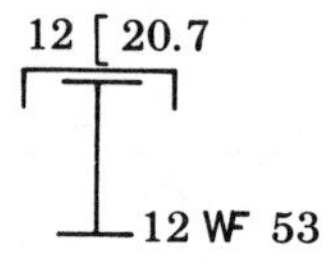

4.9 Dimension the flanges of the welded plate girder to resist an applied moment of 450 kip-ft. Material is A-7 steel, and AISC Code applies.

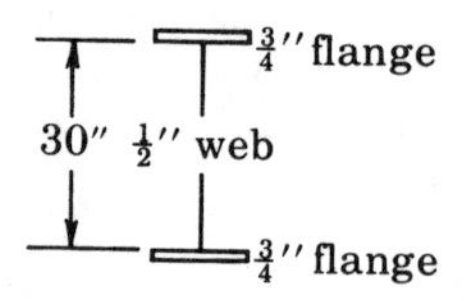

4.10 A framing plan of a typical 24 × 24 ft bay of a building is shown. The beams support a 5-in. reinforced concrete slab (weight of concrete is 150 lb/ft^3) and a live load of 70 lb/ft^2. Select member for floor beams and girders (AISC Specifications and A-7 steel). Consider all members simply supported.

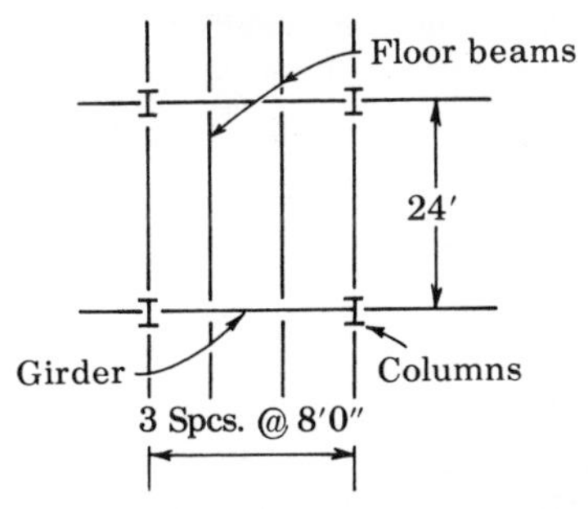

4.11 Two alternate designs are considered for a girder in a building frame: (*a*) consists of a continuous beam over many spans, and (*b*) envisions simple supports between girder and columns. The tributary live- and floor-slab load on the girders is 400 lb/ft. Girders to be de signed of A-7 steel according to AISC Code. Design girders according to (*a*) and (*b*) and draw engineering conclusions.

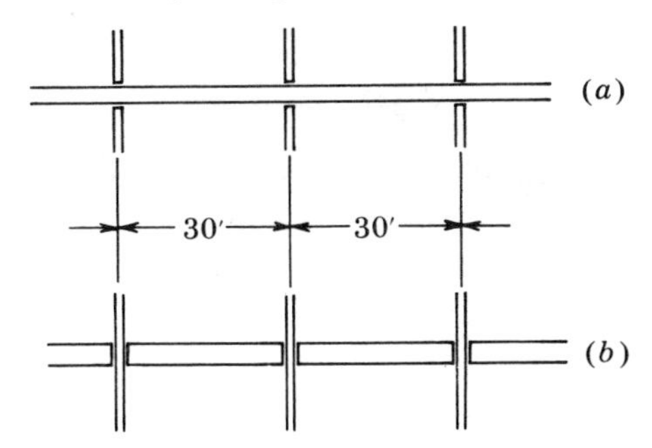

4.12 A built-up plate girder is made of a $\frac{1}{2}$-in. thick web connecting flange plates which are spaced at about 24 in. center to center of flange. The material is A-36 steel.

(*a*) Derive an approximate expression relating the required area of one flange A_F to the applied moment M.
(*b*) The plate girder is used as a simply supported beam across a span of 80 ft and is to carry a uniform load of 1.125 kips/ft. Calculate the number of $\frac{1}{2} \times 12$ in. flange plates required at the point of maximum moment.
(*c*) Determine where each flange plate may be cut off (for required dimensions see figure).

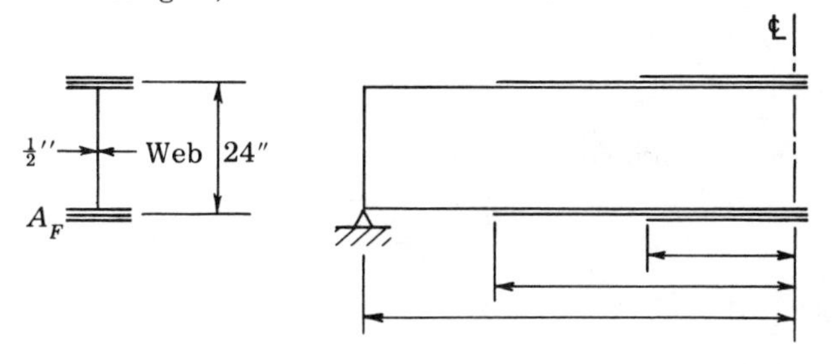

4.13 A 2×10 in. Douglas fir joist is simply supported and uniformly loaded. Allowable stresses are: bending stress $f = 1{,}900$ psi; shearing stress $H = 120$ psi. Calculate the length and load for which bending and shear capacity of the member are reached simultaneously.

4.14 A floor joist as shown is to carry the wall loads shown plus a uniform floor dead and live load of 100 lb/ft. Design the joist of redwood (f = 1,300 psi, H = 95 psi).

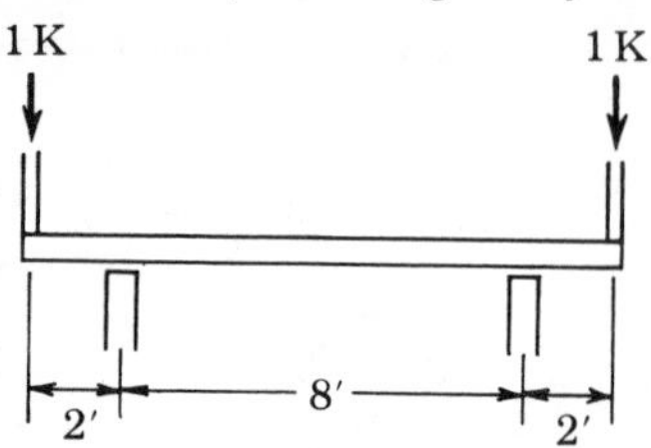

4.15 For a 2 × 8 in. Douglas fir joist (f = 1,200 psi, H = 95 psi), simply supported at its ends, prepare a plot of the safe uniformly applied load versus the span length of the joist.

4.16 The floor system already discussed in Prob. 4.4 envisions coaction of the joists and plywood flooring. The joists span over 10 ft. They are of Douglas fir (f = 1,500 psi, H = 120 psi).

(*a*) Calculate the allowable floor load.
(*b*) Calculate the shear strength of the glue necessary to ensure the composite action.
(*c*) If the joint between joists and plywood is nailed with 10*d* nails (allowable shear force = 94 lb/nail), calculate the required nail spacing.

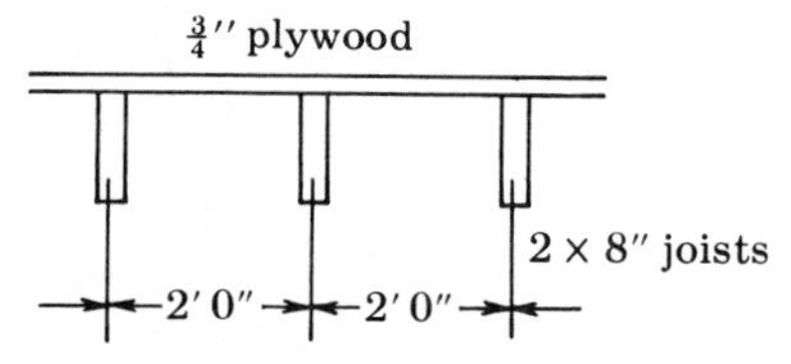

4.17 The composite timber beam is joined by ½ × 6 in. lag bolts (allowable shear force = 740 lb/bolt). Allowable shear stress in timber is 100 psi. Calculate the required lag bolt spacing to make the joint as strong as the timber. Comment on feasibility of detail.

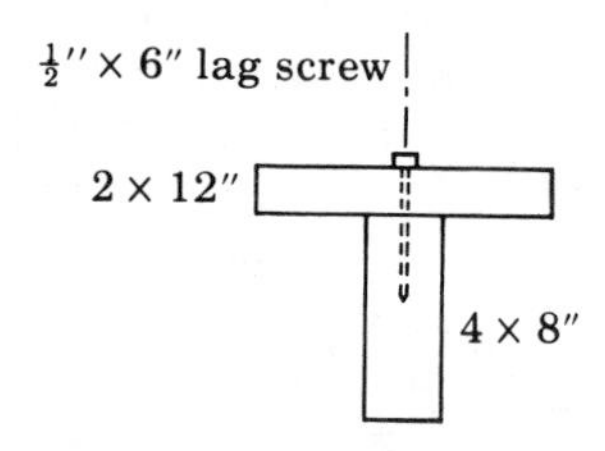

4.18 Calculate and plot the shear stresses in the webs of the box beam due to a shear force of 2 kips. Calculate the strength of glue required to join the component parts of the beam.

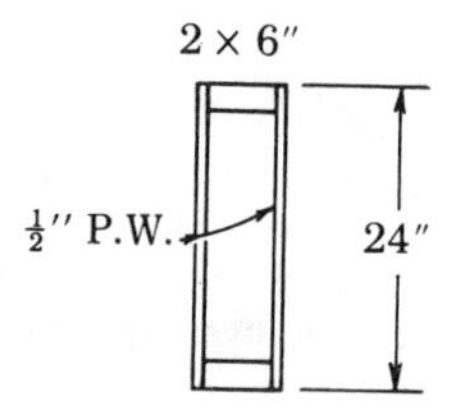

4.19 A built-up section is made of a 24 WF 76 and $\frac{1}{2} \times 12$ in. cover plates, top and bottom. Beam and plates are welded by means of fillet welds, shown separately. The AISC Specifications provide an allowable shear stress in the throat of the weld of 13,600 psi. Calculate the nominal size of weld required to safely resist a shear force of 100 kips.

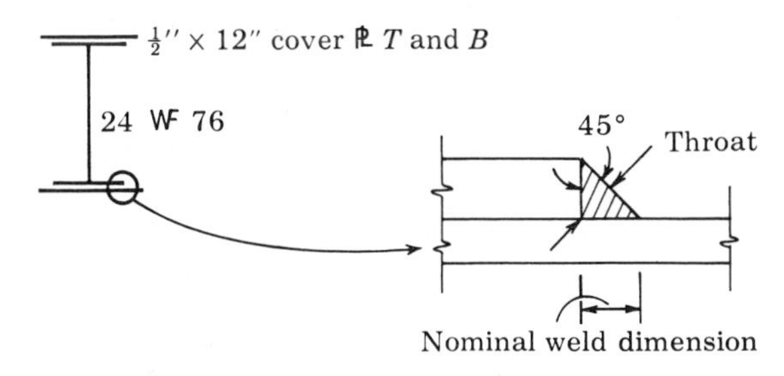

4.20 Calculate the required rivet spacing to resist a shear force of V kips. Allowable rivet shear is 4.40 kips.

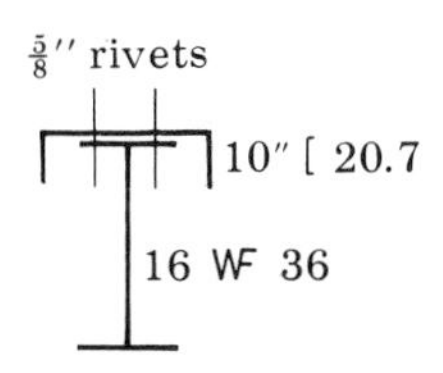

4.21 The built-up girder is to support a uniform load of 12 kips/ft over a span of 20 ft. Ends are simply supported. Calculate required thickness of the web plate to avoid excessive shear and bending stresses. Steel is A-7 ($F_{YP} = 33$ ksi); follow AISC Specifications.

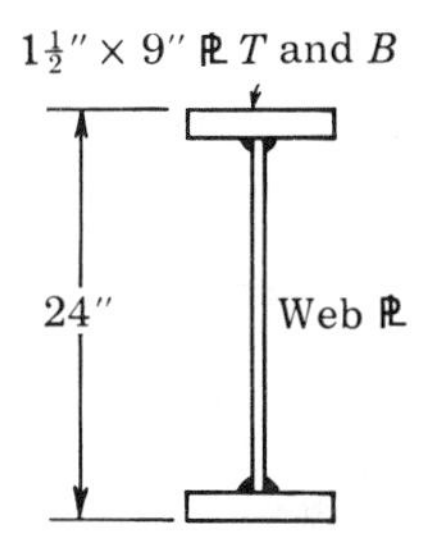

4.22 The beam of Prob. 4.20 is used as a simple beam to support a uniform load of 8 kips/ft over a span of 10 ft. Establish an expression for the rivet spacing s as a function of the distance x from the midspan of the beam. Make a design sketch showing the actual rivet spacing to the nearest $\frac{1}{8}$ in.

4.23 For a simple beam under concentrated midpoint loading, find the maximum allowable L/d ratio if the allowable bending stress is 23.8 ksi and the maximum midspan deflection is $\frac{1}{240}$ of the span length. Repeat the preceding problem if the beam is fixed-ended instead of simply supported.

4.24 Going back to the theory of Sec. 4.6, obtain expressions for the bending stresses referred to centroidal nonprincipal axes X and Y. Use the resulting formulas to check the results of Example Problem 4.4.

4.25 Going back to the theory of Sec. 4.6, obtain expressions for stresses and deformations of members of unsymmetric section under biaxial bending plus axial force.

4.26 For a 12 W 27 beam in bending, plot the variation of the allowable load with the value of angle of inclination α.

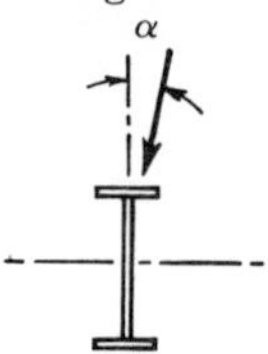

4.27 For an angle $5 \times 3 \times \frac{1}{2}$ in., find directions of the principal axes, and values of the principal moments of inertia. Use cross-sectional dimensions given on page 1-31, AISC Handbook, 6th ed., 1964, and neglect fillets. Note that the information of the last two columns provides a check on your answer.

4.28 An angle $3 \times 3 \times \frac{1}{4}$ in. is loaded parallel to the Y axis, resulting in a moment of value M acting about the X axis; no torsion is present. Calculate the largest value of stress in the section. Section properties are as shown:

$A = 1.44$ in.^{2}
$I_{XX} = I_{YY} = 1.2$ in.^{4}
$x = y = 0.84$ in.
$r_{ZZ} = 0.59$ in.

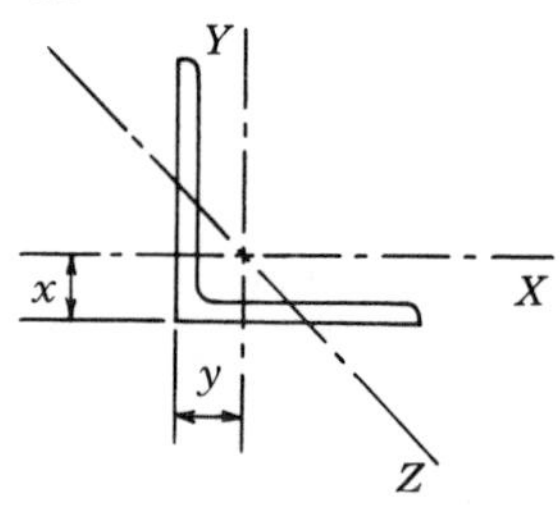

4.29 Refer to page 1-58, AISC Handbook, 6th ed., 1964. The sections shown on that page are used in cases where the main framing consists of W beams to which are fastened wall boards or sheets. The wall itself is fastened to the outside flange of the channel which in this case is called a *girt*. Consider a combination of a 10 W 29 beam and an 8⊏11.5 lb channel. For purposes of this calculation, neglect fillets and rounded-off corners, and use the average thickness of the channel flange. Be sure to use the dimensions of the members given under "Properties for Designing," pages 1-17 and 1-27.

(*a*) For this section, locate the principal centroidal axes.
(*b*) If the member is used as a 20-ft-long simple beam, find the bending stresses due to a uniform load of w kips/ft applied parallel to the Y axis.
(*c*) Find the distribution of shear stresses due to the load of part (*b*).

4.30 (*a*) Calculate and plot the shear-stress distribution due to a transverse shear V parallel to the Y axis.
(*b*) Calculate the location of the shear center.

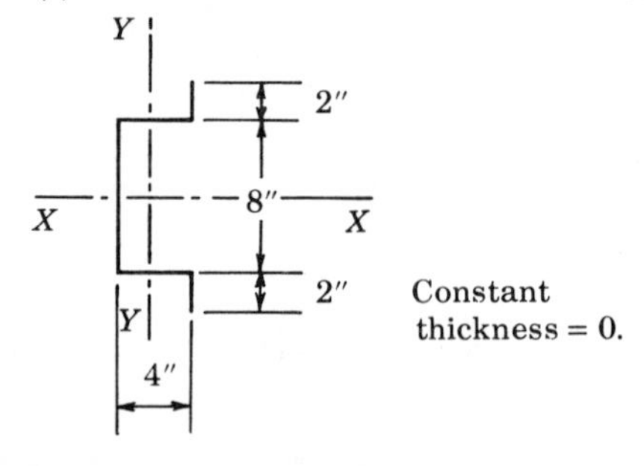

4.31 Find the location of the shear center for any angle section.

4.32 Locate the shear center of the section shown.

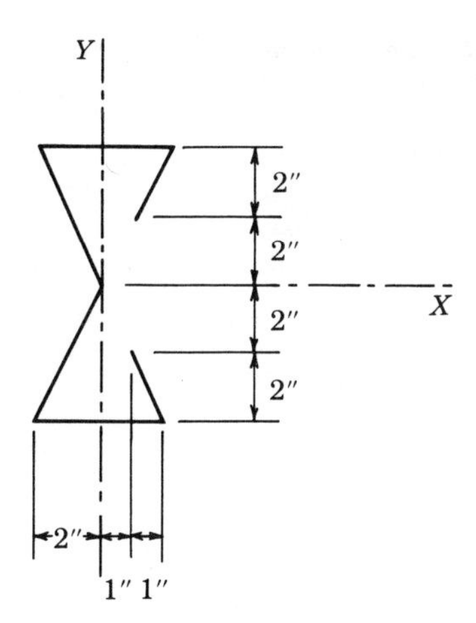

4.33 (*a*) Find the shear center of the 10-in. standard pipe with longitudinal slot whose cross section is shown.

(*b*) For a cantilever beam made of this section, of length 10 ft, calculate the maximum torsional moment due to the dead load of the member.

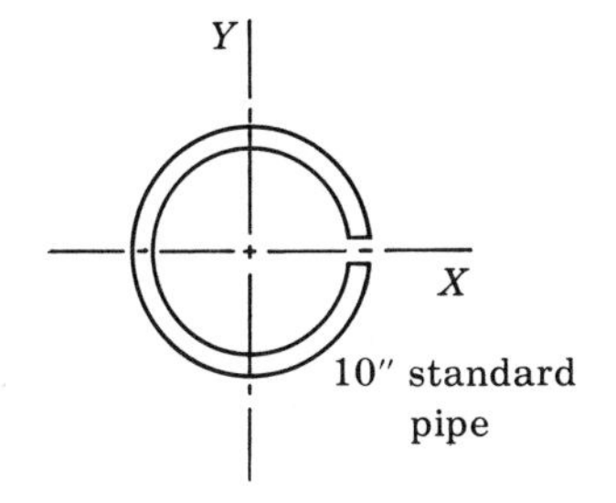

4.34 A 6 × 4 × $\frac{3}{4}$ in. angle is used as a cantilever beam as shown in the sketch. Find:

(*a*) The largest value of bending stress due to the dead load of the angle.

(*b*) The largest value of torsional moment due to the dead load of the angle.

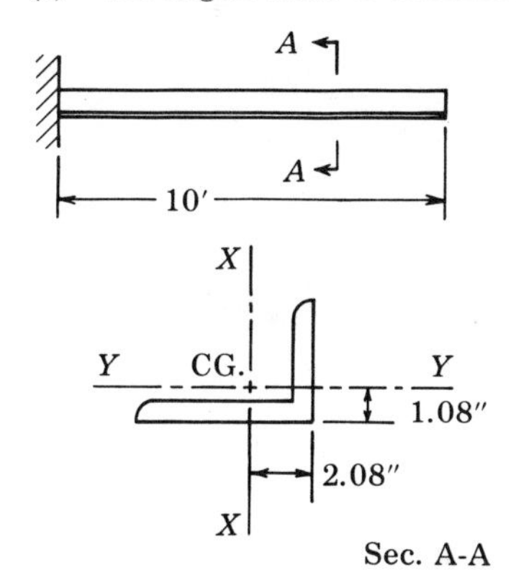

4.35 Consider again the girt section of AISC Handbook, page 1-58, already used in Prob. 4.29.

(*a*) If the member is used as a 20-ft-long simple beam, what is the maximum uniform load which may be applied without exceeding a stress of 20 ksi in the member?

(*b*) Neglecting the effects of lack of symmetry of the section, and utilizing the information on page 1-58 of the AISC Handbook (referring to the XX and YY axes), what would be the maximum uniform load which could be applied to the beam of part (*b*)? What would be the percent error involved in following this procedure?

5

Nonhomogeneous Beams and Composite Design

5.1 General Theory

Nonhomogeneous beams are those in which material properties vary from point to point; in particular, we consider a beam cross section of material (or combination of materials) whose elastic modulus E varies as a function of the depth of the section:

$$E = E(y) \tag{5.1}$$

where y is, as before, the distance from a conveniently chosen reference axis.

We now bring our typical steps of the strength of materials method to bear on this problem:

Equilibrium (as before)

$$\Sigma F = 0: \qquad \int_A \sigma \, dA = 0 \tag{2.18}$$

$$\Sigma M_X = 0: \qquad \int_A \sigma y \, dA = M \tag{2.19}$$

Geometry (as before, we assume that plane sections remain plane)

$$\epsilon = y\phi \tag{2.20}$$

Stress-strain relations [according to the material properties given by Eq. (5.1)]

$$\sigma = E(y)\epsilon = E(y)y\phi \tag{5.2}$$

Equation (5.2) indicates that the stress at any point is proportional to both the material stiffness and the distance from the neutral axis.

Substituting Eq. (5.2) into the equilibrium equations (2.18) and (2.19), we get the governing equations

$$\Sigma F = 0: \qquad \phi \int E(y) y \, dA = 0 \tag{5.3}$$

$$\Sigma M = 0: \qquad \phi \int E(y) y^2 \, dA = M \tag{5.4}$$

As before, the first of these equations serves to locate the neutral axis, the second to find the curvature, which can then be resubstituted into Eq. (5.2) to find the stresses. The difficulty of the integration will depend on the specified variation of E.

If we rewrite the variation of stiffness in terms of some reference value of E_0 multiplied by a dimensionless coefficient or *modular ratio* n, which varies with y, then Eqs. (5.3) and (5.4) can be written as

$$\Sigma F = 0: \qquad E_0\phi\int[n(y)b]y\,dy = 0 \tag{5.5}$$

$$\Sigma M = 0: \qquad E_0\phi\int[n(y)b]y^2\,dy = M \tag{5.6}$$

In this form, the integrals refer to properties of an equivalent cross-sectional area called the *transformed section* which is obtained by multiplying the width b at every point by the modular ratio n. For this transformed section, the governing equations become identical to those for homogeneous beams, and therefore ordinary beam theory can be used. The next section will explain this in greater detail.

5.2 Composite Beams

As an example of a nonhomogeneous, or composite, beam, we may consider a so-called "flitch-plate" beam, consisting of a timber beam and one or more steel plates suitably fastened together; such combinations are used when an existing wooden beam is to be strengthened or stiffened. Figure 5.1a shows the cross section of the beam considered.

To analyze this beam for stresses and deformations due to an applied moment M, we use the ideas of the previous section. The stiffness variation of the cross section in this case is expressed by

$$\begin{aligned} E &= E_{wd} && \text{for the wood} \\ E &= E_s && \text{for the steel} \end{aligned} \tag{5.7}$$

Equilibrium is satisfied by Eqs. (2.18) and (2.19), which for this case are

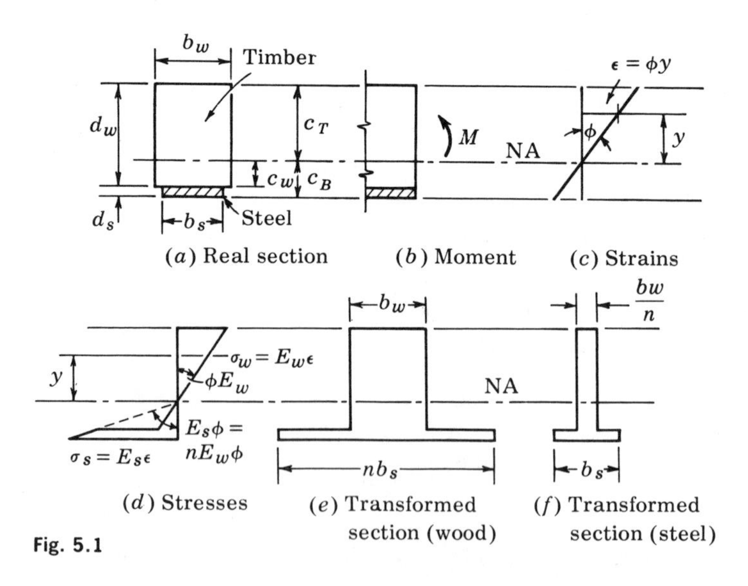

Fig. 5.1

written as:

$$\Sigma F = 0: \qquad \int_{wd} \sigma \, dA + \int_s \sigma \, dA = 0$$

$$\Sigma M = 0: \qquad \int_{wd} \sigma y \, dA + \int_s \sigma y \, dA = M$$

Geometry of the deformed structure is given by Eq. (2.20), which, together with the stress-strain relations (5.7), leads to the stress distribution

$$\begin{aligned} \sigma &= E_{wd}\epsilon = E_{wd}\phi y \qquad \text{for the wood} \\ \sigma &= E_s\epsilon = E_s\phi y \qquad \text{for the steel} \end{aligned} \tag{5.8}$$

and, when substituted into the first equilibrium equation, yields

$$\Sigma F = 0: \qquad \phi\left(E_{wd} \int_{wd} y \, dA + E_{stl} \int_s y \, dA\right) = 0$$

or, invoking the modular ratio $n = E_s/E_{wd}$

$$E_{wd}\phi \left(\int_{wd} y \, dA + n \int_s y \, dA\right) = 0$$

The factor $E_{wd}\phi$ is nonzero. Therefore the neutral axis must be so located that the factor in parentheses vanishes; accordingly,

$$\left(b_{wd} \int_{-c_{wd}}^{c_T} y \, dy + nb_{stl} \int_{-c_B}^{-c_{wd}} y \, dy\right) = 0 \tag{5.9}$$

Turning now to the moment equation:

$$\Sigma M = 0: \qquad \int_{wd} E_{wd} y^2 \phi \, dA + \int_s E_s y^2 \phi \, dA = M$$

$$\text{or} \quad E_{wd}\phi\left(\int_{wd} y^2 \, dA + \int_s \frac{E_s}{E_{wd}} y^2 \, dA\right) = M \tag{5.10}$$

Having located the neutral axis by means of Eq. (5.9), the quantity in parentheses in Eq. (5.10) can be found and the curvature ϕ calculated for any moment M. Having ϕ, the stresses and strains can be found by Eq. (5.8).

Equations (5.9) and (5.10) can be interpreted differently: if the stress in wood is required, we imagine the steel replaced by an equivalent amount of wood; since steel is $n(=E_s/E_{wd})$ times as stiff as wood, each unit area of steel must be replaced by n times that amount of wood in order to have the same resistance to deformation. Let us therefore consider a fictitious "transformed section" of wood, in which the steel is replaced by n times its area of wood, making sure to maintain correct distances from the neutral axis. This transformed section, shown in Fig. 5.1*e*, may now be considered a homogeneous one and analyzed by methods appropriate to such a beam. In particular, the neutral axis is found at the centroid of the transformed section, given by

$$\left(\int_{-c_{wd}}^{c_T} y \, dA + n \int_{-c_B}^{-c_{wd}} y \, dA\right) = 0$$

The validity of this result is verified by comparison with Eq. (5.9). Further, according to Eq. (5.10),

$$E_{wd} I_T \phi = M$$

where I_T is the moment of inertia of the transformed section,

$$\text{or} \quad E_{wd}\phi\left(\int_{wd} y^2\, dA + n \int_s y^2\, dA\right) = M \tag{5.11}$$

as already determined by Eq. (5.10). Equations (5.5) and (5.6) show that strains and stresses in wood can be obtained by applying homogeneous-beam theory to an equivalent section of one material of the stiffness of wood. Stresses in steel can then be found by realizing that, according to Eq. (5.2) and Fig. 5.1*d*, they are n times those in the transformed section.

It is of course permissible to transform the real section into one of one material the stiffness of steel. In this case, each fiber of wood is to be replaced by an equivalent area of a material of the stiffness of steel, given by

$$(dA)_{equiv\ s} = \frac{E_{wd}}{E_s}(dA)_{wd} = \frac{1}{n}(dA)_{wd}$$

The transformed section will then be that of Fig. 5.1*f*, and the stresses obtained will be those in the steel. To get the stresses in the wood, we realize again that wood stresses are related to those in the steel by the factor $1/n$.

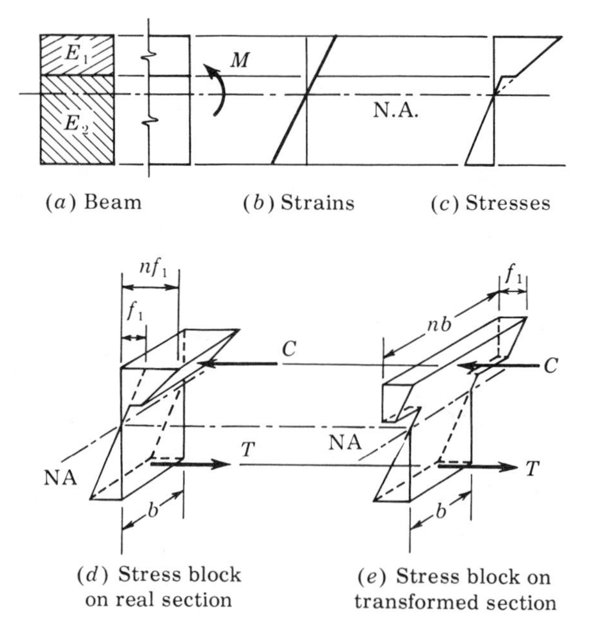

Fig. 5.2

The analogy of the reasoning with that underlying the analysis of nonhomogeneous axially loaded members should be observed. The transformed section of Sec. 3.3 is identical with that discussed in this section.

To summarize the reasoning underlying both the basic and the transformed-section methods for the analysis of nonhomogeneous beams, we consider the beam of two materials shown in Fig. 5.2*a* and subjected to a moment M, which causes the linear strain distribution of Fig. 5.2*b* and the stress distribution of Fig. 5.2*c*. Corresponding to this stress distribution we can draw the stress block of Fig. 5.2*d*, of which the neutral axis must be so located as to equalize tensile and compressive stresses, and whose fiber stresses must be of such magnitude that the internal balances the external moment. Procedures for doing this of course take us back to Eqs. (5.9) and (5.10), but can also be based on calculating resultant forces and moments using the properties of the stress block.

The stress block of Fig. 5.2*e*, corresponding to the transformed section, can be seen to have tensile and compressive resultants identical to those of the first stress block. The neutral axis as well as the stresses will therefore be identical to those obtained from that calculation.

Example Problem 5.1 A composite wood and steel beam of cross section shown is subjected to a moment of 10 kip-ft. $E_s = 30{,}000$ ksi, $E_{wd} = 1{,}760$ ksi. Find the maximum stresses in steel and concrete, and the curvature in the beam. In order to demonstrate both the basic and the transformed-section method, we shall do this problem both ways. First, let us use basic principles. Location of neutral axis: from Eq. (5.9), since $n = 30{,}000/1{,}760 = 17$,

$$\Sigma F = 0: \qquad \left(4.0 \int_{-c_B}^{8.0-c_B} y\,dy + 17 \times 4.0 \int_{8.0-c_B}^{8.5-c_B} y\,dy\right) = 0$$

or

$$\frac{4.0}{2}\left(y^2\Big|_{-c_B}^{8.0-c_B} + 17y^2\Big|_{8.0-c_B}^{8.5-c_B}\right) = 0$$

or

$$(8.0 - c_B)^2 - c_B{}^2 + 17[(8.5 - c_B)^2 - (8.0 - c_B)^2] = 0$$

Solving this for the one unknown c_B (noticing that the terms in c^2 cancel out, so that only a linear equation results), we obtain

$$c_B = 6.18 \text{ in.}$$

Turning now to Eq. (5.10) we solve for the curvature ϕ:

$$\Sigma M = 0: \qquad \phi 1{,}760 \times 4.0\left(\int_{-6.18}^{1.82} y^2\,dy + 17\int_{1.82}^{2.32} y^2\,dy\right) = 120 \text{ kip-in.}$$

or

$$\phi 1{,}760 \times \frac{4.0}{3}[(1.82^3 + 6.18^3) + 17(2.32^3 - 1.82^3)] = 120 \text{ kip-in.}$$

therefore

$$\phi = \frac{120}{1{,}760 \times 470} = 145 \times 10^{-6} \text{ rad/in.}$$

The strains are therefore

$$\epsilon = \phi y = 145 \times 10^{-6} y$$

and the corresponding stresses in the wood

$$\sigma_{wd} = E_{wd}\epsilon = 1{,}760 \times 145 \times 10^{-6} y = 0.256y \text{ ksi}$$

and those in the steel

$$\sigma_s = E_s\epsilon = 30{,}000 \times 145 \times 10^{-6} y = 4.35y \text{ ksi}$$

The stresses are plotted in Fig. 5.3*d*.

Next, we use the alternate, transformed-section method to solve the same problem. First, the transformed section is drawn as shown in Fig. 5.3*e* by replacing the steel by an equivalent amount of wood, of width $n = 17$ times that of the steel strap. Finding now the location of the centroid of this area by taking moments, about the bottom fiber,

$$c_B = \frac{\Sigma A\bar{y}}{\Sigma A} = \frac{32.0 \times 4.0 + 34.0 \times 8.25}{66.0} = 6.18 \text{ in.}$$

as before.

Next, we obtain the transformed moment of inertia, given in its analytical form by Eq. (5.11), by usual methods:

$$I_T = \frac{4 \times 8.0^3}{12} + 32.0 \times 2.18^2 + 34.0 \times 2.07^2 + \frac{68.0 \times (0.5)^3}{12} = 470 \text{ in.}^4$$

a value which we can also recognize in the previous calculations.

In order to find next the stresses in the equivalent homogeneous section, we use the flexure formula, Eq. (2.26),

$$\sigma_{wd} = \frac{My}{I_T} = \frac{120 \times y}{470} = 0.256y$$

and to find those in the steel strap we multiply those in the wood by the modular ratio $n = 17$:

$$\sigma_s = n\frac{My}{I_T} = 17 \times 0.256y = 4.35y$$

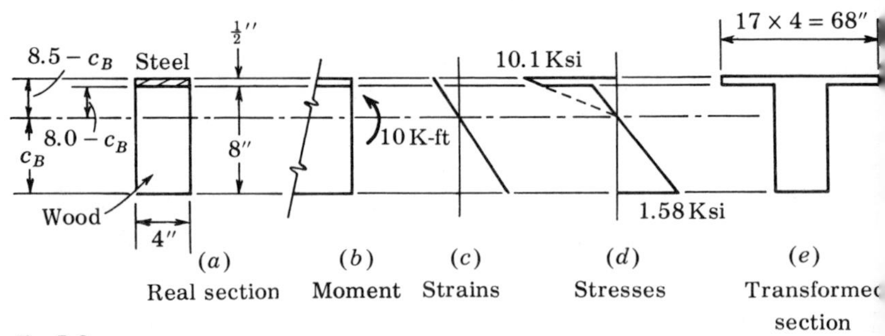

Fig. 5.3

Comparing the calculations involved in the two methods, we conclude that considerable advantage is attached to the use of the transformed section.

5.3 Deflections

To obtain deflections in a nonhomogeneous beam, we can determine the curvature ϕ for the section by use of Eq. (5.10) or by the geometry of the strain distribution of Fig. 5.1*c*, and apply one of the integration procedures based on Eq. (4.2).

5.4 Shear Stresses

To calculate the shear stresses in a nonhomogeneous beam, we recall the basic approach taken earlier, and remember that at any section subject to a moment gradient, equilibrium of an appropriately cut free body requires the presence of shear stresses. Following exactly the procedure in Sec. 4.2, we find that the shear stress in the wood is given by

$$\tau_{wd} = \frac{V \int y \, dA}{I_T b_{wd}} \tag{5.12}$$

and that in steel by

$$\tau_s = n \frac{V \int y \, dA}{I_T b_s} \tag{5.13}$$

The calculation of shear stresses in nonhomogeneous beams thus bears the same relationship to that of bending stresses as does the stress calculation in homogeneous beams.

A specially important shear-stress consideration in beams of several materials is the required strength of bond between the two components to prevent sliding of one with respect to the other. Since the stress at the

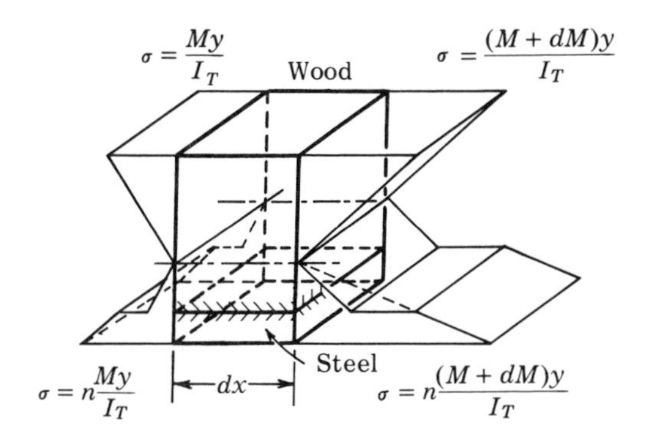

Fig. 5.4

junction of wood and steel in the beam considered is given by

$$\tau = n \frac{V \int_s y \, dA}{I_T b_s}$$

this also represents the minimum allowable shear strength, in kips per square inch, of the bonding agent used.

Generally, a steel plate is fastened to the wooden beam by means of bolts, or lag screws, spaced at intervals. The required bolt size and spacing may then be determined by assigning to each bolt the responsibility for resisting the total shear force over a certain area, as shown in Fig. 5.5:

$$\text{Total shear force/bolt} = \tau \times \text{area}$$

$$= \tau(s \times b) = sn \frac{V \int_s y \, dA}{I_T} = F_{allow} \tag{5.14}$$

where F_{allow} is the allowable shear force which can be transmitted by the bolt and s is the bolt spacing.

Assuming one size of bolt is used along the entire length of beam, then, solving for the required spacing,

$$s = \frac{F_{allow}}{V} \frac{I_T}{n \int y \, dA} \tag{5.15}$$

The spacing thus should be inversely proportional to the shear force along the beam.

An alternate way of accomplishing the selection of bolts and spacing is to start by computing the total number N of bolts required:

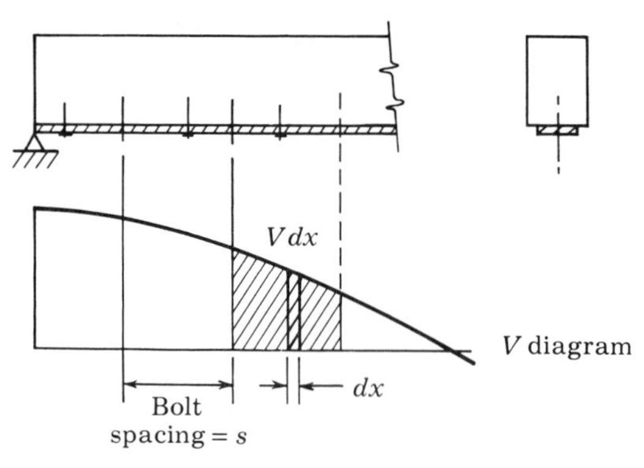

Fig. 5.5

Total shear resistance of all bolts = total horizontal shear force to be resisted:

$$\text{or} \quad NF_{allow} = \int \tau\, dA = \frac{nb \int_s y\, dA}{bI_T} \int_{\text{length}} V\, dx$$

$$= \frac{n \int_s y\, dA}{I_T} \times \text{area under shear diagram} \tag{5.16}$$

Having thus found the total number of bolts, and also realizing, from Eq. (5.16), that the area under the shear diagram is proportional to the shear force transmitted, we can now distribute the bolts so that the areas under the shear diagram tributary to each will be equal. This can be done by eye or by any one of several mathematical or graphical methods.

Example Problem 5.2 The beam of the preceding example problem is used as a simple beam of length 10 ft, to support a uniform load of 0.8 kips/ft. Determine the required spacing of bolts to connect steel plate and timber, if shear capacity of one bolt is 2.0 kips.

We first analyze the beam and draw the shear diagram. We see that the intensity of shear force varies from 4 kips at the ends to zero at the center. Turning next to the beam cross section, we utilize the transformed section to find the shear stress at the junction of wood and steel according to Eq. (5.12).

$$\tau = \frac{V \int y\, dA}{I_T b} = \frac{V \times (34.0 \times 2.07)}{470 \times 4.0} = 0.0375V$$

Following now the procedure of Eq. (5.14), we equate the horizontal shear force per bolt spacing to the capacity of one bolt:

$$\tau b s = 0.0375V \times 4.0 \times s = 2.0 \text{ kips}$$

Solving for the bolt spacing s, we find that this spacing is inversely proportional to the shear force:

$$s = \frac{2.0}{0.15V} = \frac{13.33}{V}$$

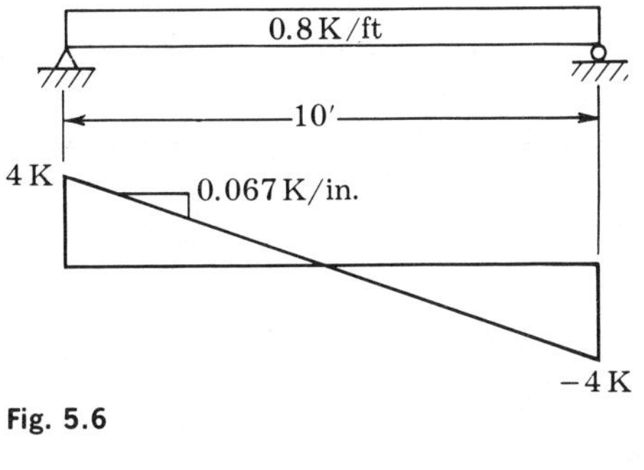

Fig. 5.6

Placing now one bolt at the end of the beam, where $V = 4.0$ kips, we can solve for the first space:

$$s_1 = \frac{13.33}{4.0} = 3.33 \text{ in.}$$

The second bolt will then be spaced $3\frac{1}{4}$ in. from the end. Further calculations of this type can then be carried out in tabular fashion:

V, kips	*Theor. spacing s, in.*	*Act. spacing, in.*	*Dist. from end, in.*
4.00			0
3.78	3.33	$3\frac{1}{4}$	3.25
3.55	3.53	$3\frac{1}{2}$	6.75
	3.76	$3\frac{3}{4}$	10.50
.	.	.	.
.	.	.	.
.	.	.	.

This table can be carried on till the half span of the beam, 60 in., has been reached in the last column.

The total number of bolts required can be verified by solving Eq. (5.16) for the number of bolts N:

$$N = \frac{1}{F_{allow}} \frac{n\int y\,dA}{I_T} \times \text{area under shear diagram}$$

$$= \frac{1}{20} \frac{(34.0 \times 2.07)}{470} \left(\frac{1}{2} \times 4.0 \times 60\right) = 9.0$$

While this indicates that theoretically 9 bolts are required for the half span, the tabular procedure will require somewhat more because each spacing was calculated using the maximum rather than the average value of shear in each increment.

5.5 Balanced Design of Nonhomogeneous Beams

In members of two materials of different allowable strength, it seems appropriate to proportion the section so that the allowable stress in each material will be reached at the same time, thus using each component as efficiently as possible. A design of this type is called a *balanced design.*

To demonstrate how to achieve such a balanced design, we again consider a wood-steel flitch-plate beam, with allowable stresses $\sigma_{wd\ allow}$ and $\sigma_{s\ allow}$ respectively. The requirement that these stresses occur in the extreme fibers simultaneously under the design load leads to the stress distribution of Fig. 5.7*c*, which in turn, by application of the stress-strain relations, Eq. (5.8), can be used to draw the strain distribution, Fig. 5.7*d*.

From this figure, using the plane-section assumption, we can consider the two similar triangles shown to find the location of the neutral axis defined by the relation:

$$\frac{kd}{d} = \frac{\sigma_{wd\ allow}/E_{wd}}{(\sigma_{s\ allow}/E_s) + (\sigma_{wd\ allow}/E_{wd})} = \frac{1}{(\sigma_{s\ allow}/n\sigma_{wd\ allow}) + 1} \tag{5.17}$$

Note that in Eq. (5.17), the d's on the left-hand side can be canceled to yield k, the ratio of the distance from the extreme fiber to the neutral axis and the overall depth of the beam. The expression then depends only on material properties. The next part of the job is to design a transformed section whose neutral axis is at the location specified by Eq. (5.17). Assuming, for the sake of illustration, that the dimensions of the timber are set at width b_{wd} and depth d_{wd}, and the thickness of the steel plate is given at d_s, we wish to find the required width of the steel plate, b_s. To do this we refer to Fig. 5.7e of the transformed section, from which we equate the static moment of the areas about the neutral axis to zero:

$$(b_{wd}d_{wd})\left(kd - \frac{d_{wd}}{2}\right) - (nd_s b_s)(d_{wd} + \tfrac{1}{2}d_s - kd) = 0$$

Solving this for the ratio b_s/b_{wd}, we find

$$\frac{b_s}{b_{wd}} = \frac{d_{wd}\left(kd - \dfrac{d_{wd}}{2}\right)}{(nd_s)(d_{wd} + \frac{1}{2}d_s - kd)} \tag{5.18}$$

Note that, for given depths, the ratio of the widths is to be adjusted so as to put the neutral axis in the proper location, defined by Eq. (5.17). If the actual ratio b_s/b_{wd} is larger than the value specified by Eq. (5.18), the allowable stress will first be reached in the extreme fiber of the wood, and will therefore govern the design, while the steel is not stressed to its capacity. The converse will be true for the ratio b_s/b_{wd} too small.

The next problem in design of composite members is the dimensioning of the member to resist a given design moment, M_{reqd}, under the assumption of balanced design. We assume that headroom limitations determine the values of d_{wd} and d_s. The problem is to select the width of the beam.

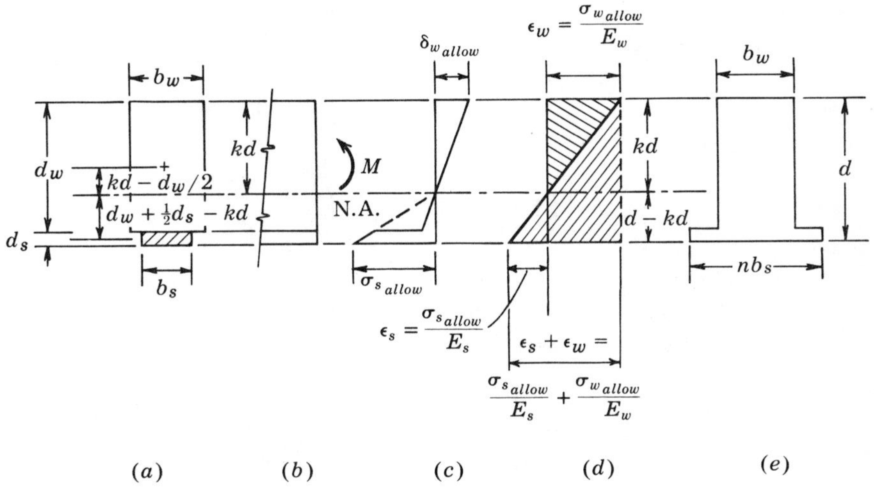

Fig. 5.7

We write, following Eq. (2.26) applied to the transformed section,

$$M_{reqd} = \frac{\sigma_{wd\ allow} I_T}{kd}$$

where, for the balanced section, k is given by Eq. (5.13), and Eq. (5.14) determines the ratio of the widths. Since I_T is proportional to the widths, they can be solved for directly, as shown in Example Problem 5.3.

Very often procedures such as the one outlined above may be shortened considerably by suitable simplifications; for instance, if the steel strap is relatively thin, we can consider all of its material concentrated at the bottom fiber of the timber. Then

$$d_s b_s = A_s = \text{area of steel}$$
$$d_s = 0$$
$$d = d_{wd}$$

We first realize that Eq. (5.17) for the location of the neutral axis for balanced design is still valid, since it depends only on material properties; then following previous reasoning, we determine the properties of the section so as to locate the centroid of the transformed section at the proper place:

$$(bd)\left(kd - \frac{d}{2}\right) - nA_s\,(d - kd) = 0$$

or $$\frac{A_s}{bd} = \frac{\text{area of steel}}{\text{area of wood}} = \text{steel ratio} = p_{bal} = \frac{k - \frac{1}{2}}{n(1 - k)} \tag{5.19}$$

Under this simplification then, the ratio of the amounts of the materials for balanced design again depends only on the material properties. We complete the balanced design to resist a given moment M_{reqd} by using Eq. (2.26):

$$M_{reqd} = \frac{\sigma_{wd\ allow} I_T}{kd}$$
$$= \frac{\sigma_{wd\ allow}\left[I_W + A_W\left(kd - \frac{d}{2}\right)^2 + nA_s(d - kd)^2\right]}{kd}$$
$$= \sigma_{wd\ allow} bd^2\left[\left(1 - \frac{5}{12k}\right) + np_{bal}\left(\frac{1}{k} - 1\right)\right]$$

To find the required beam section, we solve for the only unknown, bd^2 (realizing that the bracketed coefficient depends only on the material constants and allowable stresses):

$$bd^2 = \frac{M_{reqd}}{\sigma_{wd\ allow}\left[\left(1 - \frac{5}{12k}\right) + np_{bal}\left(\frac{1}{k} - 1\right)\right]} \tag{5.20}$$

Any beam whose bd^2 is equal to this value will be adequate to resist the specified moment. We see that we have still considerable choice of proportions; often, headroom limitations will dictate the value of d that is to be used.

After the dimensions b and d are set, the area of the strap is determined by Eq. (5.19).

It must be pointed out that simplifications of the type just discussed may lead to errors the magnitude of which depends on the validity of the assumption, and an exact analysis is necessary in order to perform an error analysis. Nevertheless, it is often possible to ascertain without calculation whether the assumption is conservative or dangerous. So, in this case, by assuming the steel closer to the neutral axis than it actually is, we reduce the internal lever arm below its actual value. Thus, we underestimate the moment resistance of the beam and are therefore bound to arrive at conservative results. A beam designed on the basis of Eq. (5.20) will therefore be slightly larger than theoretically required.

The rather involved equations of this section are derived in order to demonstrate that analytical design formulas can be found. These are of importance when it is desired to make up design aids such as tables or graphs. For our purposes, it is just as quick and much clearer to go back to basic theory, and this is done in the example problem.

Example Problem 5.3 A flitch-plate beam is to be designed to resist a moment of 10 kip-ft. Allowable stresses are: $f_s = 20$ ksi, $f_{wd} = 2.0$ ksi, $E_s = 30{,}000$ ksi, $E_{wd} = 1{,}760$ ksi. The timber to be used of reasonable proportions. Provide a balanced design.

Following the outlined procedure, we first draw the desired stresses and then the corresponding strain distribution. This locates the neutral axis, by similar triangles, at

$$kd = \frac{0.00114}{0.00114 + 0.00067} d = 0.630d$$

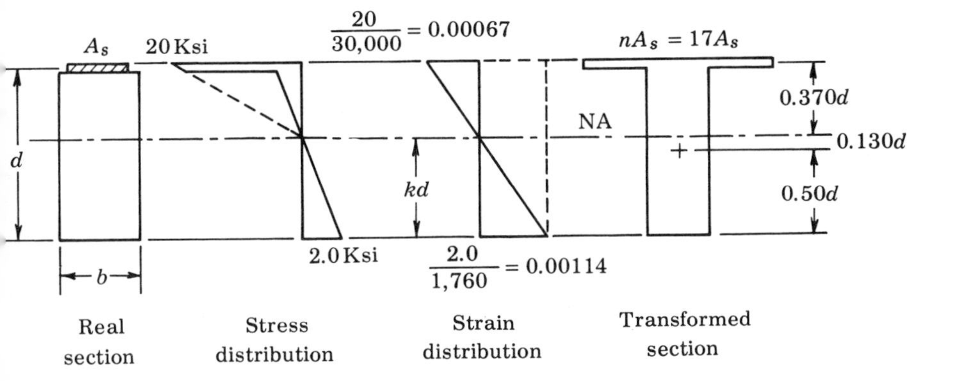

Next, the section is to be proportioned so as to ensure that the neutral axis will be located there: summing first moments of the transformed section about the neutral axis, we get

$$(bd) \times 0.130d - 0.370d \times 17A_s = 0$$

or

$$\frac{A_s}{bd} = 0.0207$$

Lastly, the required moment resistance must be ensured. Using either allowable steel or wood stress (since they are to be reached simultaneously):

$$f_{wd} = \frac{My}{I_T} \qquad \text{or} \qquad 2.00 = \frac{(10 \times 12) \times 0.63d}{I_T}$$

or

$$I_T = 37.8d$$

To furnish this I_T, the cross section must satisfy

$$I_T = (bd)(0.13d)^2 + \frac{bd^3}{12} + 17A_s(0.370d)^2 = 37.8d \text{ in.}^4$$

Since, from before, $A_s = 0.0207bd$, we substitute this and solve for bd^2:

$$(0.0169 + 0.0833 + 0.0483)bd^2 = 37.8$$
$$bd^2_{reqd} = 255 \text{ in.}^3$$

Choose a 4 × 8 in. (actual dimensions) timber:

$$bd^2_{furn} = 256 \text{ in.}^3$$
$$A_{s\ reqd} = 0.0207bd = 0.663 \text{ in.}^2$$

Use a $\frac{3}{16} \times 3\frac{1}{2}$ in. strap:

$$A_{s\ furn} = 0.656 \text{ in.}^2$$

Note that the balanced-design procedure resulted in a considerably smaller steel strap than we were led to believe from Example Problem 5.1. A stress analysis should follow this design to verify the adequacy of the member.

Often the criterion of balanced design for bending stresses furnishes a suitable guideline for dimensioning of the members. Sometimes, however, other considerations may govern—for instance, excessive shear stresses or deflections, availability of certain materials, or relative cost of the materials.

5.6 Analysis of Composite Construction

Many types of structures, among them bridges and buildings, are constructed of a steel frame or beams supporting a reinforced concrete deck or slab, as shown in Fig. 5.8. In the past, the steel and concrete portions of such structures were designed separately, entirely neglecting any mutual interaction which might be achieved by suitable connection between the two materials. Thus, the slab was considered only as a load on the beams, rather than a potential flange.

In recent years it has been realized that considerable economies might be achieved by designing such steel and concrete combinations for interaction, and in this case they are called *composite construction.* We recognize that such composite design is an application of nonhomogeneous-beam theory as discussed in the preceding section. To achieve this composite action careful attention must be given to the joint between steel and concrete so that the shear can be transmitted between the two materials without allowing any sliding to take place between them.

Whereas elementary beam theory assumes that all points in a cross section equidistant from the neutral axis are equally stressed, yet advanced analysis[1] (which does not rely on the plane-section assumption) indicates that owing to the presence of shear strains, beam flanges become progressively less effective as they get wider. Thus, while the stresses in the middle surface of the slab according to strength of materials theory should be constant, as shown by the dotted line in Fig. 5.9, actually they vary, as shown by the solid line. Accordingly, in practice only those portions of the slab which are reasonably close to the beams are considered to be acting as flanges according to beam theory. The width of these portions is called the *effective width.* According to the AASHO Specifications, sec. 1.9.3, this

[1] Stephen P. Timoshenko and J. N. Goodier, "Theory of Elasticity," 2d ed., p. 171, McGraw-Hill, 1951.

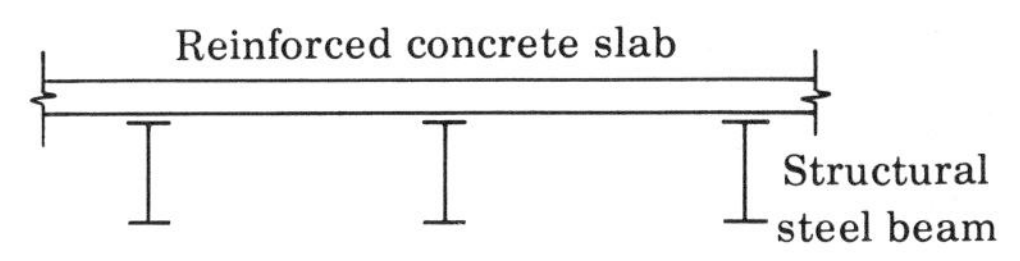

Fig. 5.8

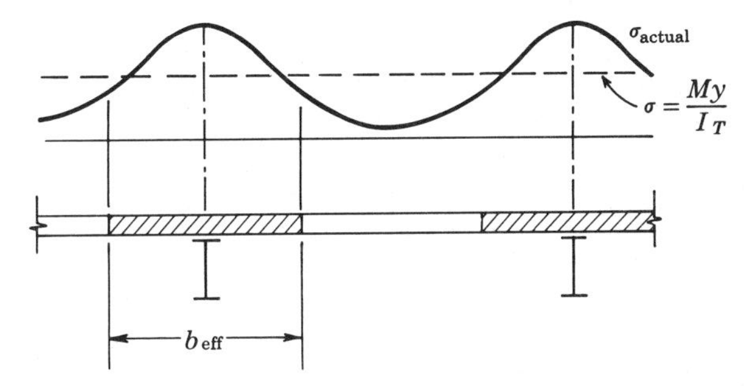

Fig. 5.9

effective width is not to exceed the following:

1. One-fourth the span length of the beam
2. The center-to-center distance between beams
3. Twelve times the least slab thickness

With this information, we can approach the problem of stress analysis of composite beams, using either basic or transformed-section theory. One important fact to remember is that concrete can only take compression stress; if ever the concrete slab should be on the tension side of the beam, it must be considered cracked and completely discounted.

Example Problem 5.4

Concrete: $f'_c = 3{,}000$ psi

$f_{c\,allow} = 1{,}200$ psi

Steel: $f_{s\,allow} = 18{,}000$ psi

$$\frac{E_s}{E_c} = n = 10$$

Find the maximum safe moment which may be applied to a typical 4-ft-wide section of the bridge deck shown.

Solution: We resort to a transformed-section solution, in which we substitute for the concrete an equivalent amount of a material of the stiffness of steel, and then apply homogeneous-beam theory. The effective width is the smaller of:

1. Center-to-center distance between beams = 48 in.
2. Twelve times the slab thickness = 12 × 6 = 72 in.

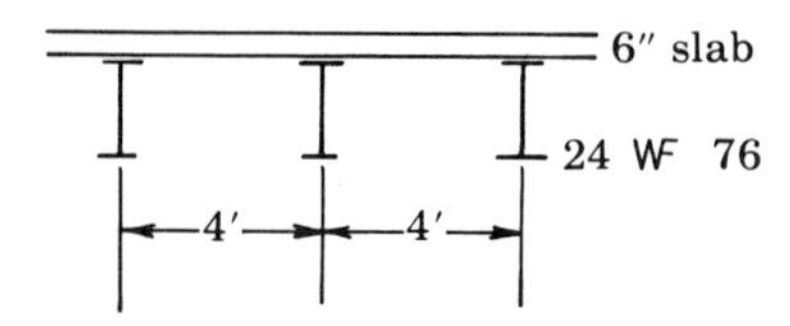

Fig. 5.10

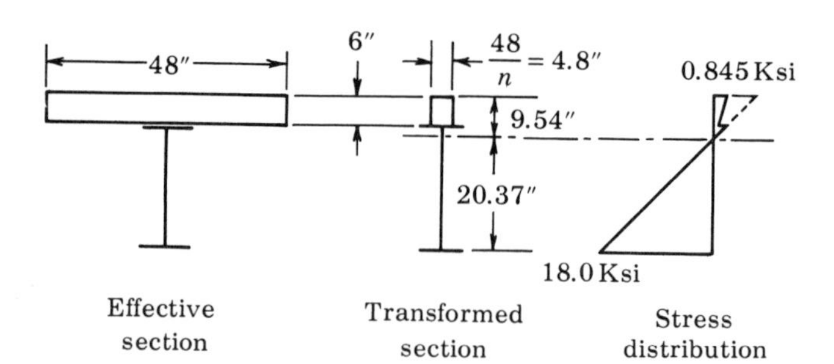

Fig. 5.11

Properties of transformed section

Part	Area, in.^{2}	y'	Ay'	y	y^2	Ay^2	$\bar{I}$
Slab	28.80	14.96	431.0	6.54	42.7	1,230	86
Beam	22.37	0	0	8.41	71.0	1,588	2,096

$\Sigma A = 51.17 \qquad \Sigma Ay' = 431.0 \qquad I_T = 2{,}818 + 2{,}182 = 5{,}000 \text{ in.}^4$

$$\bar{y} = \frac{\Sigma Ay'}{\Sigma A} = 8.41 \text{ in.}$$

Critical material

$$\frac{f_c}{f_s} = \frac{\frac{1}{10} \times 9.54}{20.37} \qquad \text{or} \qquad f_s = 21.3 f_c$$

$$f_{s\ allow} = \frac{18.0}{1.2} f_{c\ allow} = 15 f_{c\ allow}$$

Therefore steel stress is critical.

Maximum moment

$$f_{s\ allow} = \frac{M_{allow} \times 20.37}{5{,}000} = 18.0 \text{ ksi} \qquad \text{or} \qquad M_{allow} = 369 \text{ kip-ft/4 ft width}$$

The maximum concrete stress is 18.0 ksi/21.3 = 0.845 ksi.

Note that the beam does not represent a balanced design, since the concrete is understressed under the allowable moment.

The shear force which must be transmitted at the junction is a critical quantity and will be calculated according to elastic theory, using the transformed section:

$$(\tau b) \text{ kips/in.} = \frac{V \int y\, dA}{I_T} = \frac{V(38.8 \times 6.54)}{5{,}000} = 0.0378V$$

Shear connectors must be provided to transmit this amount of force per unit length of beam.

5.7 Shear Connectors

Mechanical devices which are used to ensure that steel beam and concrete slab deform together include spirals, structural steel channels, and studs, all of which are welded to the top flange of the steel beam (Fig. 5.12).

The maximum shear force which may be applied to one of these connectors before appreciable slip between steel and concrete takes place is called Q_{uc}, and has been determined experimentally. The values of Q_{uc} (in pounds) which are called for in sec. 1.9.5 of the AASHO Specifications are given in Eqs. (5.21) to (5.23).

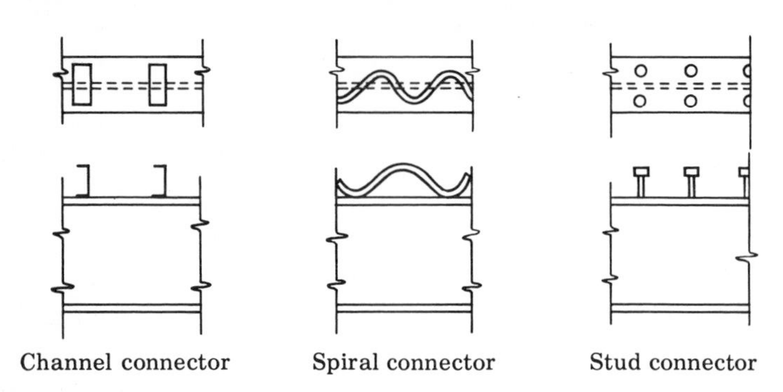

Fig. 5.12

For channels

$$Q_{uc} = 180(h + \tfrac{1}{2}t)W\sqrt{f'_c} \tag{5.21}$$

where h = maximum thickness of channel flange
t = web thickness of channel connector
W = length of channel connector
f'_c = ultimate concrete strength, psi

For welded studs

$$\begin{aligned} &\text{For } H/d > 4.2 \qquad Q_{uc} = 330d\sqrt{f'_c} \\ &\text{For } H/d < 4.2 \qquad Q_{uc} = 80\frac{H}{d}\sqrt{f'_c} \end{aligned} \tag{5.22}$$

where H = height of stud
d = stud diameter

For spirals, per one turn

$$Q_{uc} = 3{,}840d\sqrt[4]{f'_c} \tag{5.23}$$

where d = diameter of spiral rod

For design purposes, these so-called "useful capacities" must be divided by a safety factor which may be taken at 4.

Example Problem 5.5 Design stud connector for the composite beam of Example Problem 5.4 at a section where the shear has a value of 50 kips/4 ft width.

Solution: According to previous calculations, the shear force to be transmitted between steel and concrete is

$$\tau b = 0.0378 \times 50 \text{ kips} = 1.89 \text{ kips/in.}$$

Using four rows of $\frac{3}{4}$-in. studs, 4 in. long ($H/d = 5.33$), we find the design capacity of one stud to be

$$Q = \frac{Q_{uc}}{4.0} = \frac{330}{4.0} \times 0.75\sqrt[2]{3{,}000} = 2{,}550 \text{ lb}$$

Equating the resisting shear of four studs to the shear to be transmitted, we set

$4 \times 2{,}550 \text{ lb} = s \times 1{,}890 \text{ lb/in.}$

where s is the spacing or pitch between studs. Solving, we find this pitch to be

$s = 5.40$ in., say $5\frac{3}{8}$ in.

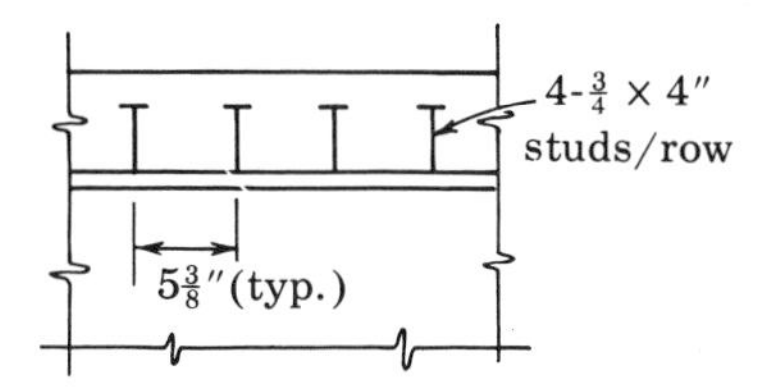

5.8 Design of Composite Beams

Theoretically, it would be desirable to design composite beams of balanced section, in which steel and concrete reach their allowable stresses simultaneously. However, since the concrete slab dimensions are usually already determined by other considerations (such as the slab design for transverse bending), this does not necessarily lead to an economical structure. More often, the size of steel beam is determined which would be required if slab interaction were neglected, and then a trial design of the composite section is made, using a steel beam several sizes smaller. The stresses are determined and the steel beam size adjusted if necessary. In usual cases, the allowable stress in the bottom flange of the steel beam is critical.

Another complicating factor in composite design has to do with the method of construction. As the concrete slab is placed on its formwork supported by the steel beam, all of this dead load must be carried by the steel beam, since the concrete is still in its fluid condition. Thus no composite action is invoked at this stage, and all load leads only to stresses in the steel beam. Only a load which is placed on the structure after the concrete has set is carried by composite action. It is best to compute stresses with and without composite action separately and then use superposition to arrive at critical values.

The necessity of having the steel beam carry the dead load alone can be avoided by supporting it during construction by temporary intermediate shores. If sufficient shores are used, the steel and concrete dead loads may be considered transferred directly to the shores with only minimal bending of the steel beam. When the shores are removed subsequent to hardening of the concrete, all load is transferred to the composite beam and thus carried by composite action.

Example Problem 5.6 A simple composite deck is to be designed consisting of a 4-in. concrete slab and steel beams at 8 ft centers. Span of the deck is 32 ft. The slab weighs 50 lb/ft² and is to support a live load of 200 lb/ft². Construction is to be unshored. $f_c' = 4.0$ ksi, $f_{c\,allow} = 1.8$ ksi, $f_{s\,allow} = 20$ ksi, $n = E_s/E_c = 8$.

Solution: The total uniform load per 8-ft typical section is (200 + 50) lb/ft² × 8 ft = 2.0 kips/ft, to which we add 0.1 kips/ft for beam weight. The maximum positive moment is then

$$M_{\max} = \frac{wL^2}{8} = \frac{2.1 \times 32^2}{8} = 269 \text{ kip-ft}$$

Were we to select a steel beam to carry this load alone, we would require

$$S_{reqd} = \frac{269 \times 12}{20} = 162 \text{ in.}^3 \qquad \text{(use 24 WF 76)}$$

We base our composite section on a smaller section, say a 21 WF 62. The cross section (using the AASHO effective width requirements) will then appear as in Fig. 5.13*a*, and its transformed section as in Fig. 5.13*b*. The section modulus of the steel beam alone is 126.4 in.³ The section properties of the transformed section are computed next:

Part	*Area, in.²*	*y′, in.*	*Ay′, in.³*	*y, in.*	*y², in.²*	*Ay², in.⁴*	*Ī, in.⁴*
Slab	24.00	2.00	48.0	5.40	29.2	701	32
Beam	18.23	14.50	264.5	7.10	50.4	920	1,327

$\Sigma A = 42.23 \qquad \Sigma Ay' = 312.5 \qquad I = 1{,}621 + 1{,}359$

$\bar{y} = \dfrac{312.5}{42.23} = 7.40 \text{ in.} \qquad = 2{,}980 \text{ in.}^4$

The moment due to beam and slab dead weight is

$$M_{DL} = \frac{0.462 \times 32^2}{8} = 59.2 \text{ kip-ft}$$

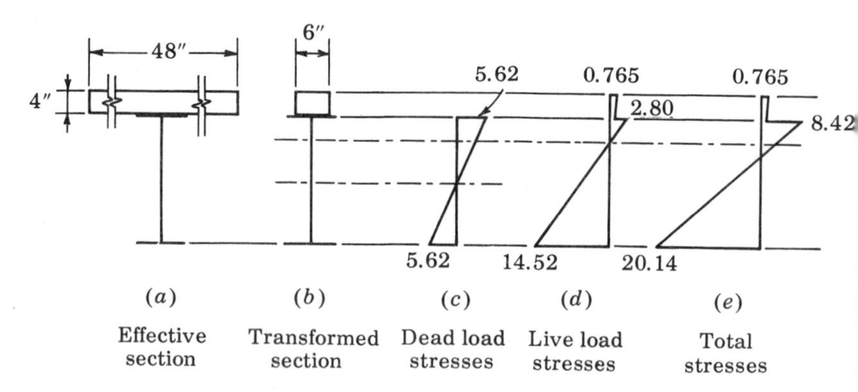

Fig. 5.13

It is resisted entirely by the steel beam, and causes the extreme fiber stresses

$$\sigma = \frac{M}{S} = \frac{59.2 \times 12}{126.4} = \pm 5.62 \text{ ksi}$$

shown in Fig. 5.13*c*.

The moment due to live load is

$$M_{LL} = \frac{1.6 \times 32^2}{8} = 205.0 \text{ kip-ft}$$

It is resisted by composite action, and causes the extreme fiber stresses

$$f_{\text{top of steel beam}} = \frac{205.0 \times 12 \times 3.40}{2{,}980} = 2.80 \text{ ksi}$$

$$f_{\text{bottom of steel beam}} = \frac{205.0 \times 12 \times 17.60}{2{,}980} = 14.52 \text{ ksi}$$

$$f_{\text{top of slab}} = \frac{1}{8} \times \frac{205.0 \times 12 \times 7.40}{2{,}980} = 0.765 \text{ ksi}$$

These live-load stresses are plotted in Fig. 5.13*d* and added to the dead-load stresses to give the resultant stresses shown in Fig. 5.13*e*. We see that the maximum steel stress, of value 20.14 ksi, is 0.7 percent above the allowable value, which is probably all right. The concrete is very much understressed. The assumed section, consisting of a 4-in. slab and 21 WF 62's at 8-ft centers, will therefore be used.

We note that due to composite design, a saving in steel of

$$\frac{76 - 62}{76} = 18.4 \text{ percent}$$

was achieved. Part of these savings have to be expended in installing shear connectors. We shall design these as welded studs, $\frac{3}{4} \times 3$ in., three rows mounted on the top flange of the steel beam. For three studs, the safe shear capacity is

$$\frac{3Q_{uc}}{4.0} = \frac{3}{4.0} \times 80 \times \frac{3.0}{0.75} \sqrt{4{,}000} = 15.2 \text{ kips}$$

The shear to be transmitted from beam to slab is only that due to live load, due to which the maximum shear is 25.6 kips. To establish the required spacing, we equate shear capacity to shear stress times area:

$$15.2 \text{ kips} = \tau b s = \frac{V \times (24 \times 5.40)}{2{,}980} \times s$$

or, solving for the pitch s,

$$s = \frac{350}{V}$$

Proceeding now in tabular fashion as in the previous section:

V, kips	*s, in.*	*Dist. from end of beam, in.*	*Act. spacing*
25.6	13.7	0	3 groups at 14 in.
20.0	17.5	42	3 groups at 17.5 in.
12.6	27.8	94.5	

Since the maximum suggested pitch is 24 in., we shall continue the studs at this spacing across the beam. The final stud spacing is as shown:

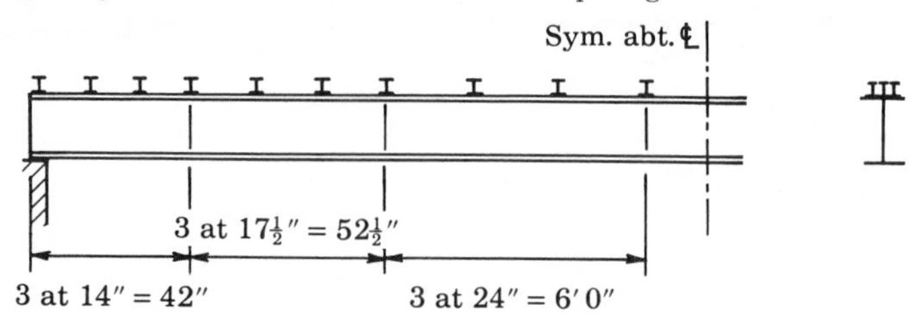

This completes the composite design.

General Readings

The basic ideas of nonhomogeneous-beam analysis are covered in standard texts on strength of materials, and applications are presented in some of the texts on structural design. A more detailed book on composite construction is the following:

Viest, Ivan M., R. S. Fountain, and R. C. Singleton: "Composite Construction in Steel and Concrete," McGraw-Hill, 1958. This book includes a number of design aids.

Problems

5.1 A rectangular beam of width b and depth d is of material whose modulus varies linearly from a value E at its top edge to $2E$ at its bottom edge. It is subjected to a moment M.

(*a*) Calculate curvature and stresses by basic principles.
(*b*) Draw transformed section and check results of part (*a*) by use of transformed section.

5.2 A composite wood timber beam has a cross section as shown. Allowable stresses are $f_{wd} = 1.9$ ksi, $f_s = 20$ ksi; stiffnesses are $E_s = 30{,}000$ ksi, $E_{wd} = 1{,}760$ ksi. Find the maximum allowable moment on this beam.

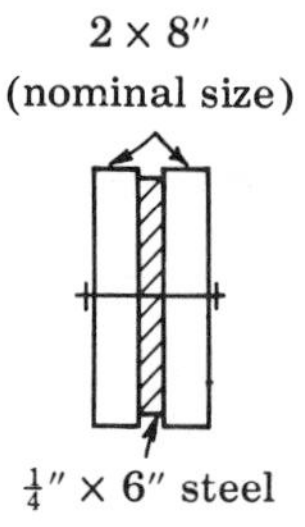

5.3 If the amount of steel used in the beam of Prob. 5.2 is instead used as top and bottom plates, what will be the allowable moment on the beam? Compare with that of Prob. 5.2 and draw appropriate conclusions.

5.4 The beam of Prob. 5.2 is used to carry a uniform load of 0.50 kips/ft transmitted to the 2 × 8's. It is required to determine the bolt spacing. The capacity of each bolt is governed by bearing between bolt and 2 × 8, and is 400 lb in bearing against one 2 × 8. Draw appropriate free bodies to support your reasoning.

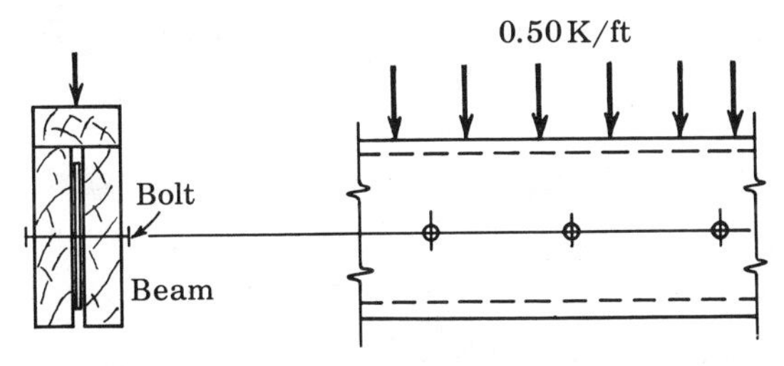

5.5 The composite beam of Douglas fir, construction grade, and A-7 steel is subject to a moment M. Find the stresses in steel and timber. Select appropriate allowable stresses and find the allowable moment.

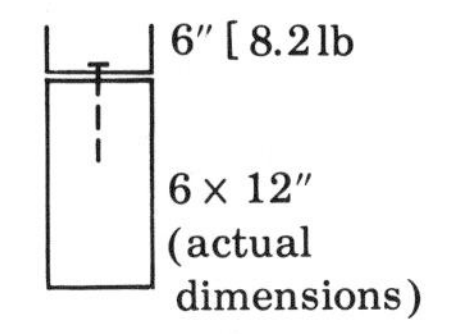

5.6 If the bolt used in Prob. 5.5 to connect steel and timber has a shear capacity of 800 lb, find the required spacing as a function of the shear force V.

5.7 A bimetallic strip is in a thermostat. It is composed of a tungsten strip (E = 51,000 ksi) of cross-sectional dimensions 0.25 by 0.02 in., bonded on its flat side to a similar strip of cadmium (E = 10,000 ksi). Find the ratio of applied moment to resulting curvature for this bimetallic strip.

5.8 A composite beam is made of a 24 WF 100 steel beam of 20 ksi allowable stress and a 6 × 48 in. slab of reinforced concrete of allowable stress f_c = 2,250 psi and stiffness 5,000 ksi. Find the stress in steel and concrete due to an applied moment M. Also find the allowable moment.

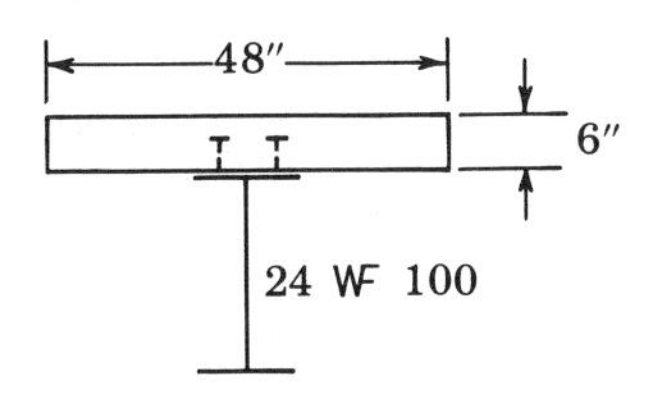

5.9 In the composite beam of Prob. 5.8, coaction between steel and concrete is to be ensured by providing stud-type shear connectors. Find the allowable shear strength of the studs per foot of length of the beam to resist a shear force of 40 kips.

5.10 The beam of Probs. 5.8 and 5.9 is used on a simple span of 20 ft to carry 8.0 kips/ft total uniform load. If the capacity of one stud is 4 kips in shear, and two rows of studs are provided, determine the required stud spacing over the entire beam. Show actual dimensions on a sketch.

5.11 Redesign the beam of Prob. 5.8 to provide a balanced design. Do this by altering the *width* of the concrete slab only. Determine the corresponding moment.

5.12 A composite beam is to be designed to resist an applied moment of 20 kip-ft. It is to consist of a rectangular Douglas fir timber (E = 1,760 ksi, F_{allow} = 1.9 ksi), and a flat steel strap, of allowable stress 20 ksi, bolted to its top. Determine the cross section of the beam to provide a balanced design.

5.13 Determine the bolt spacing required to join steel and timber of Prob. 5.12 to resist a shear force of 2 kips. Shear resistance of one bolt is 0.5 kips.

5.14 A 6 × 12 in. timber beam (E = 1,760 ksi, F_{allow} = 1.5 ksi) is to be reinforced with a flat steel plate bolted to its bottom to resist a uniform load of 0.6 kips over a simple span of 16 ft. Allowable stress of steel = 24 ksi.

(*a*) Determine the portion of the timber beam over which steel reinforcing is required.
(*b*) Determine the amount of steel reinforcing.
(*c*) Determine the bolt spacing (shear strength of one bolt = 0.3 kips).

5.15 The effective section of a composite beam is shown in the figure. Assuming E_s/E_c = 10, calculate and plot the stresses due to a moment of 400 kip-ft, assuming

(*a*) No composite action
(*b*) Composite action between steel and concrete

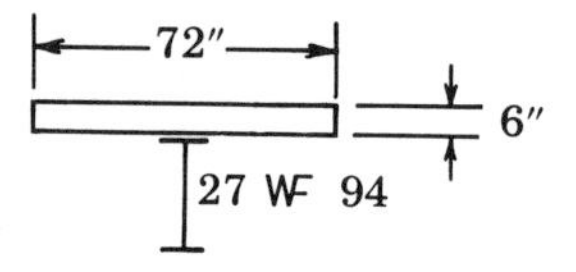

5.16 For the composite beam of Prob. 5.15, plot the shear stresses due to a shear of 32 kips and calculate the required resisting shear force of the shear connectors per foot of length of beam.

5.17 Select appropriate welded stud connectors and spacing to satisfy the requirements of Prob. 5.16, following the provisions of the AASHO Code.

5.18 A simply supported bridge, of 50-ft span, has a cross section as shown. Per typical girder, the traffic live load may be approximated as 1.35 kips/ft of uniform load. Calculate and plot the stresses at midspan, remembering that only the live load can be carried by composite action.

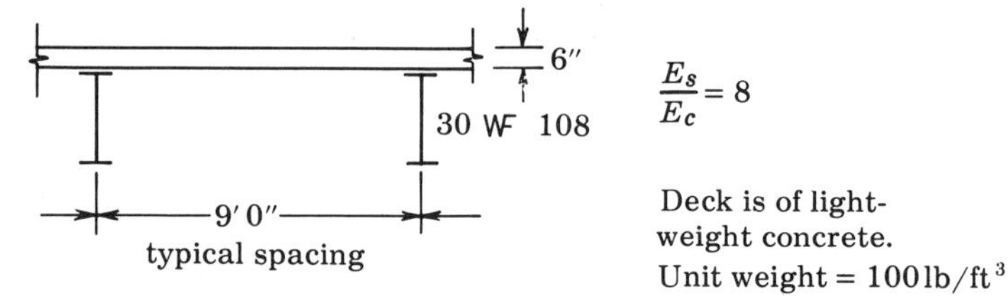

5.19 Assume that in constructing the bridge of Prob. 5.18, the concrete slab (but not the steel beam) is supported by temporary shores which are removed after the concrete slab has hardened. Calculate and plot the stresses at midspan, using superposition, and compare with those obtained in Prob. 5.18.

5.20 Assume that during construction the steel beam of Prob. 5.18 is supported by one temporary shore at midspan. It is required to determine and plot the stresses at midspan. Use superposition of

(*a*) Dead-load stresses on shored structure
(*b*) Stresses due to removal of shore
(*c*) Live-load stresses, considering in each case whether or not composite action is involved

Compare final stresses with those of Probs. 5.18 and 5.19 and draw engineering conclusions.

5.21 If $\frac{3}{4}$ × 4 in. welded studs are to be used to ensure composite action in the structure of Prob. 5.18, determine and show on a design sketch the required spacing of these studs, following the AASHO Specifications.

5.22 $f'_c = 4.0$ ksi, $f_{c\ allow} = 1.6$ ksi, $f_{s\ allow} = 18.0$ ksi;
unit weight of concrete = 150 lb/ft³,
unit weight of asphalt = 100 lb/ft³.

A simple-span highway bridge of length 40 ft is to be designed. The traffic live load can be represented by a uniform load of 3.0 kips/ft per 10-ft traffic lane.

(*a*) Design the bridge, neglecting composite action.
(*b*) Design the bridge, relying on composite action. Assume that no temporary shoring is used. Verify the validity of the design by displaying critical stress distribution. Select suitable shear connectors and determine required spacing. Make sketch of final design.
(*c*) Compare economies of alternates (*a*) and (*b*) and draw conclusions.

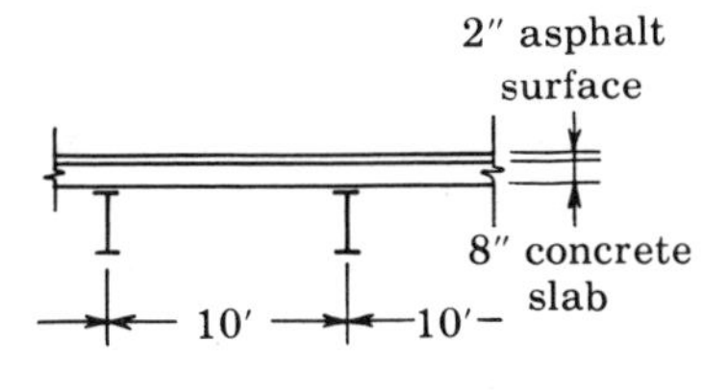

$f'_c = 4.0$ Ksi

$f_{s_{allow}} = 1.6$ Ksi

$f_s = 18.0$ Ksi

5.23 For the bridge of Prob. 5.22, determine whether any savings can be obtained by suitable temporary shoring during construction. Verify your conclusions by selecting steel members and displaying critical stresses. Write engineering conclusions.

5.24 A building floor framing plan is shown. The floor beams, with simple end connections, are to be designed to carry a uniform live load of 100 lb/ft², without exceeding the allowable stresses $f_c = 1.35$ ksi, $f_{s\ allow} = 20$ ksi, $E_s/E_c = 9$. The slab is of lightweight concrete, 100 lb/ft³.

(*a*) Design the floor beams, neglecting composite action.
(*b*) Design the floor beams, considering composite action. Decide whether shoring is to be used during construction, and calculate accordingly. Select suitable shear connectors and determine required spacing. Make design sketch.
(*c*) Compare efficiency of composite and noncomposite alternates. Discuss effects of using temporary shoring, and draw engineering conclusions.

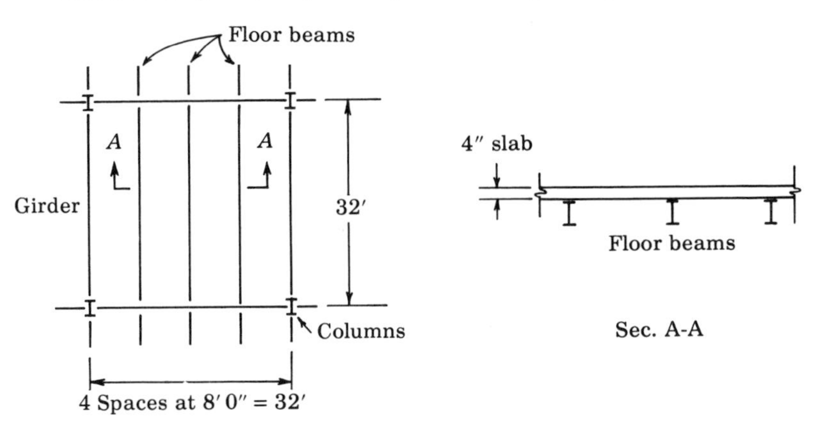

6

Analysis and Design of Reinforced Concrete Beams, Working Stress Theory

6.1 Analysis for Bending Stresses

As an important application of the elastic theory of nonhomogeneous beams, we consider the analysis of reinforced concrete beams in their elastic range. A theory valid within this range is called *working stress theory.*

We must first of all look at the stress-strain relations of the component materials; we refer to Figs. 2.5 and 2.8 and redraw the relevant portions of the stress-strain curves as Fig. 6.1. To a tensile concrete strain there corresponds zero concrete stress, owing to the fact that cracking occurs under tension; the only part of the concrete section effective in resisting moment is that on the compression side.

Next we consider a cross section of the concrete beam of width b, overall depth h, and steel of area A_s at a depth d, under a moment M, as shown in Figs. 6.2*a* and *b*; it is required to find stresses in the beam.

Using our strength of materials procedure, we write the equilibrium and geometrical conditions as before:

$$\Sigma F = 0: \qquad \int \sigma \, dA = 0 \qquad \text{or} \qquad b \int_c \sigma \, dy + A_s f_s = 0 \tag{6.1}$$

$$\Sigma M_{NA} = 0: \qquad \int \sigma y \, dA = 0 \qquad \text{or} \qquad b \int_c \sigma y \, dy + A_s f_s (d - kd) = M \tag{6.2}$$

Plane sections remain plane, as shown by the strains of Fig. 6.2*c*:

$$\epsilon = \phi y$$

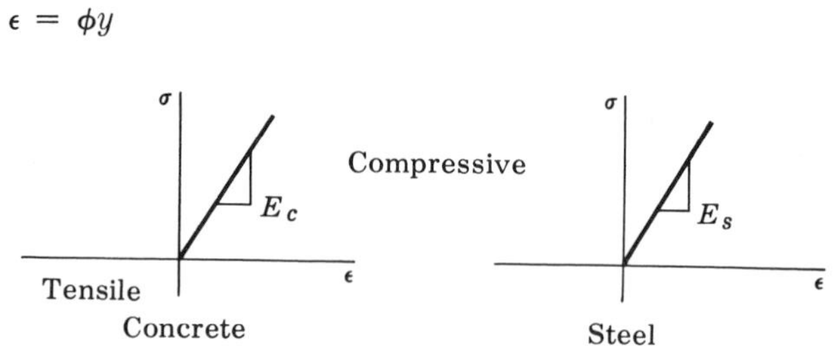

Fig. 6.1

and in particular, the steel strain is

$$\epsilon_s = -\phi(d - kd)$$

Now we introduce the stress-strain relations for concrete and steel:

$$\begin{aligned} \sigma_c &= E_c\epsilon = E_c\phi y && \text{for } \epsilon \text{ compressive} \\ \sigma_c &= 0 && \text{for } \epsilon \text{ tensile} \\ f_s &= E_s\epsilon_s = -E_s\phi(d - kd) && \text{for the steel} \end{aligned} \tag{6.3}$$

Inserting these values into the equilibrium equation, Eq. (6.1), we solve for the location kd of the neutral axis by summing forces:

$$bE_c\phi \int_{y=0}^{kd} y\, dy - A_sE_s\phi(d - kd) = 0$$

or
$$\phi\left[\left(\frac{kd}{2}\right)^2 - \frac{A_s}{b}\frac{E_s}{E_c}(d - kd)\right] = 0 \tag{6.4}$$

Canceling ϕ, dividing through by d^2 and canceling, and introducing the modular ratio $n = E_s/E_c$ and the steel ratio $p = A_s/bd$, we rewrite this quadratic equation as

$$k^2 + 2pnk - 2pn = 0 \tag{6.5}$$

and solve for the ratio k defining the location of the neutral axis:

$$k = -pn + \sqrt{(pn)^2 - 2\,pn} \tag{6.6}$$

With the location of the neutral axis found, we solve for the curvature ϕ by the moment equilibrium equation, Eq. (6.2), into which we insert Eqs. (6.3):

$$bE_c\phi \int_{y=0}^{kd} y^2\, dy + A_sE_s\phi(d - kd)^2 = M$$

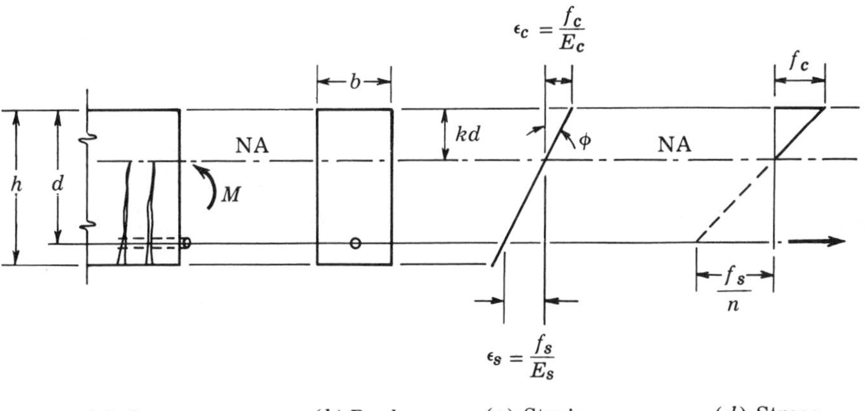

Fig. 6.2

from which

$$\phi = \frac{M}{E_c\{[b(kd)^3/3 + nA_s(d - kd)^2]\}} \tag{6.7}$$

This value can now be resubstituted into Eqs. (6.3) to find the concrete stresses

$$\sigma_c = \frac{M}{\{[b(kd)^3/3 + nA_s(d - kd)^2]\}}\, y \tag{6.8}$$

and the steel stress

$$f_s = n\frac{M}{\{[b(kd)^3/3 + nA_s(d - kd)^2]\}}\,(d - kd) \tag{6.9}$$

The stress distribution is plotted in Fig. 6.2*d*.

The formal integrations can be avoided by drawing stress blocks as shown in Fig. 6.3 and using resultants to satisfy equilibrium; we call this approach the *internal force method.* The compressive resultant of the concrete stresses is (calling the maximum compression stress f_c)

$$C = \frac{1}{2}f_c bkd$$

and the tension force (using the strain relations) is

$$T = A_s f_s = A_s E_s \epsilon_s = A_s E_s \frac{d - kd}{kd}\frac{f_c}{E_c}$$

The internal lever arm is called jd:

$$jd = d - \frac{kd}{3}$$

As usual, we sum horizontal forces to locate the neutral axis:

$$\Sigma F = 0: \qquad C = T \qquad \text{or } \tfrac{1}{2}bkd = A_s n\frac{d - kd}{kd}$$

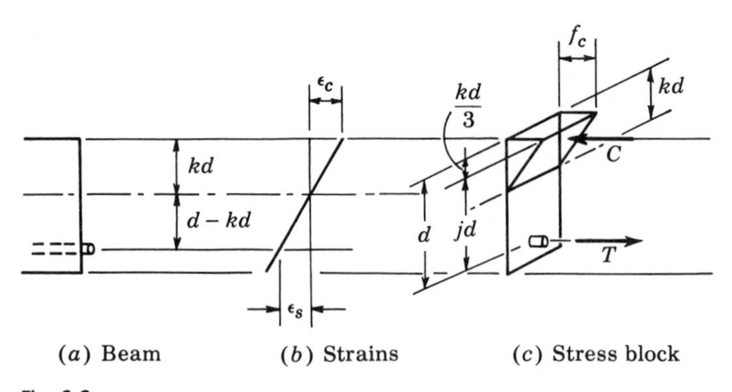

Fig. 6.3

from which we obtain the previous quadratic Eq. (6.4) for k. Summing moments about the tension steel, we solve directly for the maximum compression stress:

$$\Sigma M_T = 0: \qquad Cjd = M$$

from which, substituting for C, we obtain

$$f_c = \frac{M}{\frac{1}{2}bkdjd} = \frac{M}{\frac{1}{2}bd^2k[1 - (k/3)]} \tag{6.10}$$

The tension stress can be found by summing moments about the compressive resultant:

$$\Sigma M_c = 0: \qquad Tjd = M \qquad \text{or} \qquad f_s = \frac{M}{A_s jd} \tag{6.11}$$

This completes the stress analysis by the internal-force method; its use is demonstrated in the following example problem.

Example Problem 6.1 The beam of cross section shown is made of gravel concrete (unit weight = 150 lb/ft³) of compressive strength f'_c = 2.5 ksi. It is subjected to a moment of 125 kip-ft. Determine the stresses in steel and concrete, and the curvature of the beam by the internal force method.

We first determine the ratio of the stiffnesses of steel and concrete, which, according to table 1002*a* of the 1963 ACI Building Code, can be taken at 10. With this information, and determining the cross-sectional area of four No. 9 bars as 4.00 in.², we can draw the stress blocks, shown in Fig. 6.3*b*. Determining the location of the neutral axis of this section, we set

$$\Sigma F = 0: \qquad C - T = 0 \quad \text{or} \quad \tfrac{1}{2}\,15 \text{ in.} \times kd \times f_c - 10\,\frac{25 - kd}{kd} \times 4.0 f_c = 0$$

or

$$(kd)^2 + 5.33(kd) - 133.3 = 0$$
$$kd = -2.67 + \tfrac{1}{2}\sqrt{28.5 + 533.3} = 9.21 \text{ in.}$$

This in turn leads to $C = T = 69.1 f_c$.

$$\Sigma M_T = 0: \qquad Cjd = 69.1 f_c \times 21.93 \text{ in.} = 125.0 \times 12$$

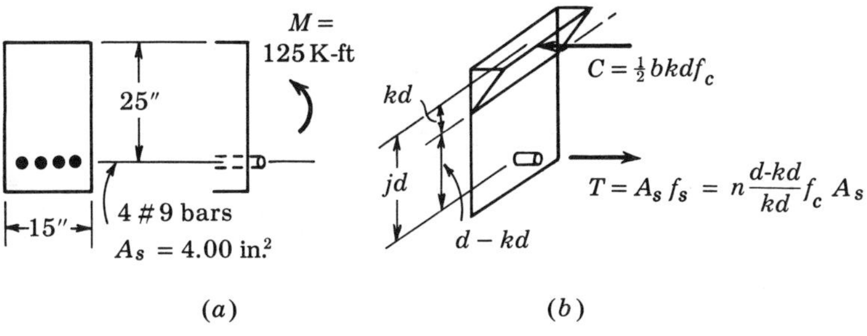

Fig. 6.4

or $f_c = \underline{0.990 \text{ ksi}}$

$\Sigma M_C = 0$: $\quad Tjd = 4.0 \times f_s \times 21.93 = 125.0 \times 12$

or $f_s = \underline{17.1 \text{ ksi}}$

For the curvature, we visualize the strain distribution (Fig. 6.2*c*) and write

$$\phi = \frac{\epsilon_c}{kd} = \frac{f_c}{E_c kd} = \frac{0.990}{3{,}000 \times 9.21} = 35.8 \times 10^{-6} \text{ rad/in.}$$

The transformed-section method is applicable to reinforced concrete as it is to other nonhomogeneous beams. In Fig. 6.5 we transform the effective section to a homogeneous one, of material the stiffness of concrete, by multiplying the steel area by the modular ratio n.

We now use homogeneous elastic beam theory to locate the neutral axis at the centroid of the transformed section:

$$\Sigma_{\substack{\text{Areas} \\ \text{about N.A.}}} = 0: \qquad b\frac{(kd)^2}{2} - nA_s(d - kd) = 0$$

which is identical to the earlier Eq. (6.4).

Curvature and stresses are found using the transformed moment of inertia I_T as

$$\phi = \frac{M}{E_c I_T}$$

$$f_c = \frac{M}{I_T}kd$$

$$f_s = n\frac{M}{I_T}(d - kd)$$

where $I_T = b\dfrac{(kd)^3}{3} + nA_s(d - kd)^2$

a value which we recognize as the denominator of Eqs. (6.7) to (6.9); the results of the transformed-section analysis are thus identical with the earlier results by basic theory.

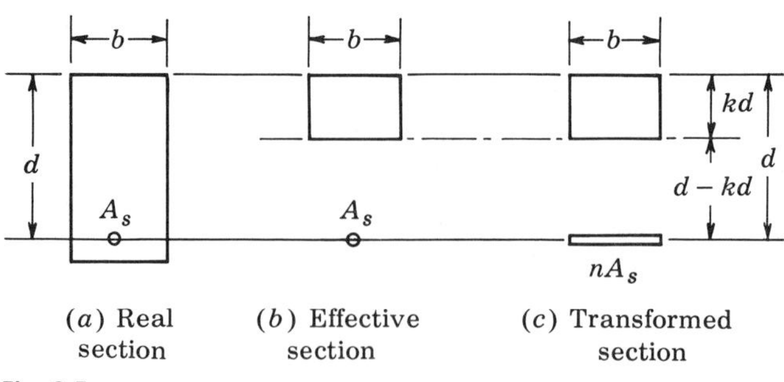

(*a*) Real section (*b*) Effective section (*c*) Transformed section

Fig. 6.5

Example Problem 6.2 Calculate the stresses and curvature in the beam of Example Problem 6.1 by the transformed-section method.

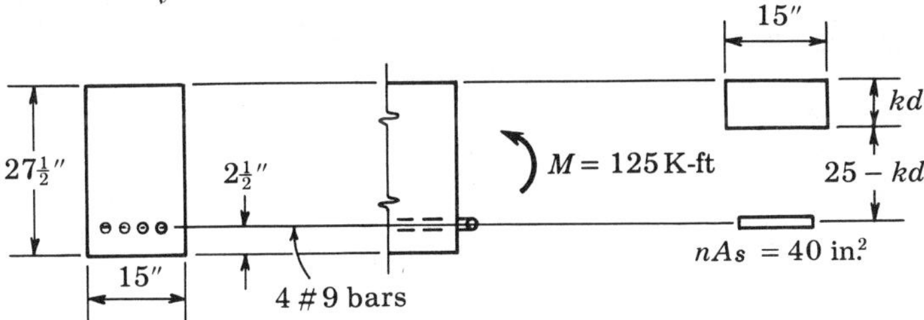

We draw the transformed section and locate its centroidal axis:

$$\Sigma M_{\substack{\text{Area}\\ \text{about N.A.}}} = 0: \qquad (15kd)\frac{kd}{2} - 40.0(25 - kd) = 0$$

$$kd = 9.21 \text{ in.}$$

Treating now the transformed section as a homogeneous section, we calculate the moment of inertia I_T:

$$\begin{aligned} I_T &= \tfrac{1}{3}b(kd)^3 + (nA_s)(d - kd)^2 \\ &= \tfrac{1}{3} \times 15 \times 9.21^3 + 40.0 \times 15.79^2 \\ &= 13{,}870 \text{ in.}^4 \end{aligned}$$

and then find stresses in the usual fashion, remembering to multiply the steel stress by the modular ratio n:

$$f_c = \frac{Mkd}{I_T} = \frac{(125.0 \times 12) \times 9.21}{13{,}870} = 0.998 \text{ ksi}$$

$$f_s = n\frac{M(d - kd)}{I_T} = 10\frac{(125.0 \times 12)15.79}{13{,}870} = 17.1 \text{ ksi}$$

The curvature can be found from the usual formula for homogeneous beams, $\phi = M/EI_T$, as

$$\phi = \frac{125.0 \times 12}{(30{,}000/10) \times 13{,}870} = 36.1 \times 10^{-6} \text{ rad/in.}$$

There is little doubt that for the simple case of a tensilely reinforced beam presented here, the internal-force method has the advantages of closer contact with physical reality as well as quicker computation. Nevertheless, for more involved situations the use of the transformed-section method may become preferable. We shall encounter such situations in the near future.

Another problem in analysis is the one in which we are given the beam and must find the allowable moment which may be applied without exceeding the given allowable stresses.

To do this, we can use first Eq. (6.6) to locate the neutral axis, then Eq. (6.10) to solve for the moment corresponding to the maximum allowable concrete stress:

$$M_{allow} = f_{c\,allow}\, bd^2 \left[\frac{1}{2}k\left(1 - \frac{k}{3}\right)\right] \tag{6.12}$$

This of course presupposes that under increasing moment on the beam the concrete will reach its allowable stress before the steel will reach its allowable stress. To test this hypothesis, we can take the moment-equilibrium condition, about the compressive resultant, so as to express the moment in terms of the steel stress:

$$\Sigma M_C = 0: \qquad Tjd = (A_s f_s)d\left(1 - \frac{k}{3}\right) = M$$

or $$M_{allow} = f_{s\ allow} A_s d\left(1 - \frac{k}{3}\right) \tag{6.13}$$

The lower of the allowable moments according to Eqs. (6.12) and (6.13) will control. If that according to Eq. (6.13) is the lower, it indicates that contrary to the initial assumption, the steel reached its allowable stress first.

It seems irrational to have to evaluate two allowable moments for one beam. We can use the relation between the stresses, Fig. 6.2*d*, to check whether steel or concrete controls: if

$$f_{s\ allow} < n \frac{1 - k}{k} f_{c\ allow}$$

then steel controls, and we speak of an *under-reinforced* beam. If

$$f_{s\ allow} > n \frac{1 - k}{k} f_{c\ allow}$$

then concrete controls, and we speak of an *over-reinforced* beam. If the two sides happen to be equal, then steel and concrete will reach their allowable stresses simultaneously and we speak, as earlier, of a *balanced* beam.

Example Problem 6.3 The beam in Example Problem 6.2 has the following allowable stresses: $f_{c\ allow} = 1.125$ ksi, $f_{s\ allow} = 24$ ksi. Find the allowable moment which may be resisted by the beam.

Having in the preceding example problem found the neutral axis at $kd = 9.21$ in. we can now compute the ratio of the steel and concrete stresses, utilizing the strain distribution:

$$\frac{\epsilon_c}{\epsilon_s} = \frac{9.21}{15.79} = \frac{f_c}{f_s/10} \qquad \text{or} \qquad f_s = 17.1 f_c$$

Since $f_{s\ allow} = \dfrac{24.0}{1.125} = 21.3 f_{c\ allow}$, under increasing moment the allowable stress will first be reached in the extreme fiber of the concrete, under a moment, according to Eq. (6.9), of

$$M_{allow} = C_{allow} jd = \frac{69.1 \times 1.125 \text{ ksi} \times 21.93}{12} = 142 \text{ kip-in.}$$

Since the compressive capacity of the concrete is exhausted before the strength of the steel, the section is over-reinforced.

6.2 Design of Singly Reinforced Beams

Whereas in the *analysis* of beams we are given all characteristics of the beam and its component materials, and are asked to find the stresses or the allowable moment capacity, in *design* we start with a given moment, material properties, and allowable stresses, and wish to arrive at a suitable beam section able to resist the specified moment without exceeding the allowable stresses.

Let us assume that it is desired to design a balanced section which will enable the allowable steel and concrete stresses to be reached simultaneously. We then start with the stress distribution and determine the location of the neutral axis from Fig. 6.6 given by

$$k_{bal} = \frac{kd}{d} = \frac{f_{c\ allow}}{\dfrac{f_{s\ allow}}{n} + f_{c\ allow}} = \frac{1}{\dfrac{f_{s\ allow}}{nf_{c\ allow}} + 1} \tag{6.14}$$

as already found in Eq. (5.13) of an earlier section. The next step is to apply the moment equation, Eq. (6.12), to solve for the required cross-sectional quantity,

$$bd^2_{reqd} = \frac{M}{f_{c\ allow}(\frac{1}{2}kj)} = \frac{M}{f_{c\ allow}[\frac{1}{2}k_{bal}(1 - k_{bal}/3)]} \tag{6.15}$$

We note that this approach is entirely analogous to the familiar concept of the section modulus; there are an infinite number of beams which satisfy Eq. (6.15), and we may use other requirements, such as critical shear stresses or deflections, depth limitations, or cost considerations, to arrive at a decision regarding the properties of the section.

Having decided on the effective depth d and width b of the beam, we are ready to determine the steel area A_s required, using moment condition

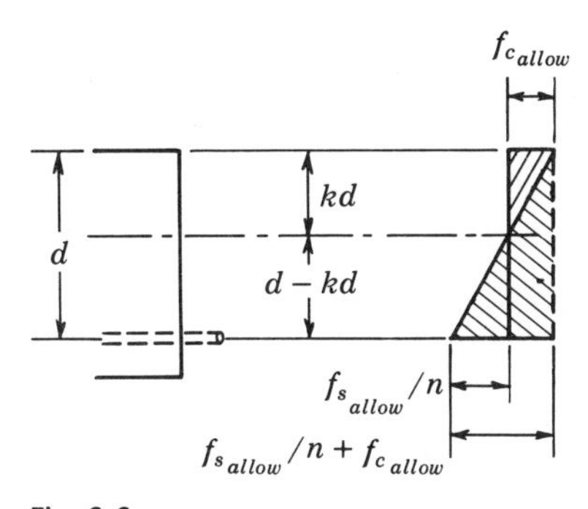

Fig. 6.6

Eq. (6.13):

$$A_{s\ reqd} = \frac{M}{f_{s\ allow}jd} \tag{6.16}$$

Having obtained the required steel area, we can select a combination of commercially available bars totaling up to the requirement. We recall that our bending theory is based on a symmetrical beam section, and therefore are careful to ensure a symmetrical arrangement of bars.

A definite proportion of reinforcing steel to concrete is necessary to ensure a balanced design. This can be verified from the generally valid force equilibrium equation (6.1), rewritten as

$$\Sigma F = 0: \qquad T = A_s f_{s\ allow} = C = \tfrac{1}{2} f_{c\ allow} bkd$$

or $$p_{bal} = \frac{A_s}{bd} = \frac{1}{2}\frac{f_{c\ allow}}{f_{s\ allow}} k_{bal} \tag{6.17}$$

We note that this balanced steel ratio depends only on the materials and allowable stresses.

Example Problem 6.4

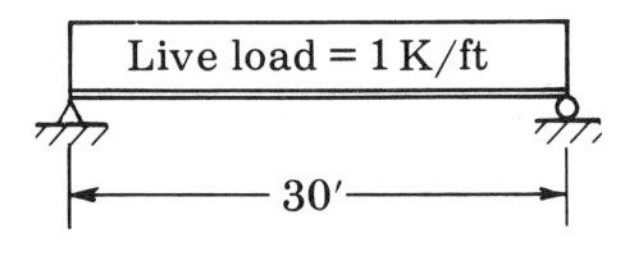

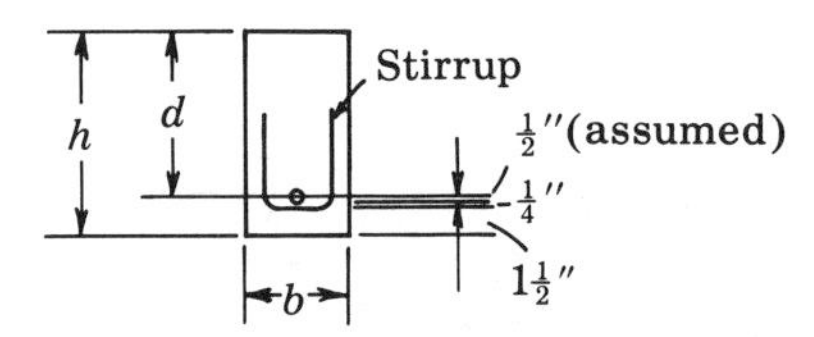

A reinforced concrete beam is to be designed to support the uniform superimposed load of 1 kip/ft across a simple span of 30 ft, following the provisions of the ACI Code. $f'_c = 3.0$ ksi, $f_{s\ allow} = 20$ ksi; unit weight of concrete is 150 lb/ft³.

We must first assume the beam dead weight. Let us anticipate 0.4 kips/ft, so that the total unit load is 1.4 kips/ft. The maximum moment is then

$$M_{\max} = \frac{wL^2}{8} = \frac{1.4 \times 30^2}{8} = 157.5 \text{ kip-ft}$$

The allowable concrete stress is $0.45f'_c = 1.35$ ksi, and, according to table 1002*a* of the ACI Code, $n = 9$.

According to Eq. (6.15), we first find the location of the neutral axis for balanced design by Eq. (6.14) given by

$$k_{bal} = \frac{1}{(f_{s\ allow}/nf_{c\ allow}) + 1} = 0.379$$

The internal lever arm is then given by

$$j_{bal} = 1 - \frac{k}{3} = 0.874$$

Finding next the required concrete section by Eq. (6.15),

$$bd^2 = \frac{M}{\frac{1}{2}f_{c\ allow}kj} = \frac{157.5 \times 12}{0.5 \times 1.35 \times 0.379 \times 0.874} = 8{,}448 \text{ in.}^3$$

We can now select one of many sections, and some possible ones are listed in the table:

b, *in.*	d, *in.*	h, *in.*	*Beam weight, kips/ft*	A_s, *in.*^{2}
12	26.6	29	0.36	4.07
13	25.6	28	0.38	4.23
14	24.6	27	0.39	4.40

Note that the deeper and narrower beam sections are lighter, and also that the amount of tension steel area, calculated by Eq. (6.16) as

$$A_s = \frac{M}{f_s j d}$$

is less for the deeper sections because of the longer internal level arm. In this calculation, the moment M can be corrected for the actual beam weight, and also d can be recalculated, based on the actual overall depth h which is rounded off to the whole inch. The resulting values are entered in the last column of the table.

The beam 14 by 27 in. seems to have reasonable proportions and is selected. The furnished area of three No. 11 bars is 4.68 in.^{2}, and these bars are selected. These bars satisfy the spacing requirements of ACI sec. 804. With these bars and the required cover, the actual effective depth d is somewhat less than predicted, but this is compensated for by the excess of steel area, as can be verified by a stress analysis of the section as designed.

It should be understood that it is not common practice to compute the table of alternate sections. This was done here only for the sake of discussion. We might also keep in mind that it may be possible to delete steel bars at sections of the beam subjected to less than the maximum moment. We shall discuss the cutting off of reinforcing bars in a later section.

We complete the design by showing a design drawing of the critical section:

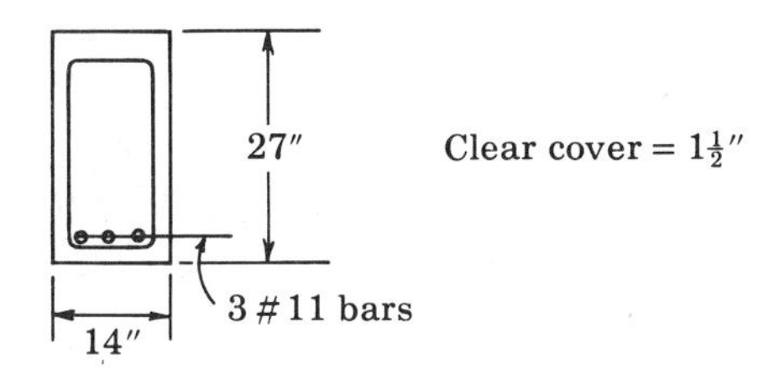

6.3 Aids for Balanced Beam Design

We summarize the design procedure of the preceding section for balanced beams by enumerating the two required steps: To find the concrete section, we use

$$bd^2_{reqd} = \frac{M}{f_{c\,allow}\frac{1}{2}k_{bal}j_{bal}} \equiv \frac{M}{K} \tag{6.15}$$

where $$K \equiv f_{c\,allow}\tfrac{1}{2}k_{bal}j_{bal} = f_{c\,allow}\tfrac{1}{2}k_{bal}\left(1 - \frac{k_{bal}}{3}\right) \tag{6.18}$$

is a *balanced design constant* depending only on the materials selected. It may thus be calculated once and for all for one entire job and used for the design of every member, as long as the same materials are used throughout the job. Note that K has the units of stress; we must be sure to make the units of numerator and denominator of Eq. (6.15) consistent in order to come out with units of in^3 (the same as for section modulus).

Having selected the concrete section, we find the required steel by

$$A_{s\,reqd} = \frac{M}{f_{s\,allow}j_{bal}d} = \frac{M}{ad} \tag{6.16}$$

where $$a \equiv f_{s\,allow}j_{bal} = f_{s\,allow}\left(1 - \frac{k_{bal}}{3}\right) \tag{6.19}$$

is another balanced design constant, independent of loads or proportions of member.

Likewise, we can deduce that, besides K and a, k_{bal}, j_{bal}, and p_{bal} are design constants which will usually be invariant from member to member. Though easy enough to calculate, the values of these constants are tabulated for various combinations of materials in table 1 of the RCD Handbook.

Looking through the tabulated values in this reference, we find that the magnitude of the internal lever arm ratio $j = 1 - (k/3)$ is rather insensitive to changes in modular ratio and allowable stresses, since only its second term (which is small compared to 1) is affected by such changes; in fact, its tabulated values vary only between

$$0.770 \leq j_{bal} \leq 0.914$$

Therefore designers often neglect this variation as secondary, and use an average value for j equal to $7/8 = 0.875$. Whether this approximation is satisfactory must be left to the judgment of the individual analyst. A comparison of beams designed on the basis of an exact j_{bal} with those based on an average j will aid formulation of this judgment.

Indicative of other tabulated design aids is table 2 of the RCD Handbook, which solves Eq. (6.15) for balanced design of a typical strip of a reinforced slab 12 in. wide. In this case, $b = 12$ in., and Eq. (6.15) can be solved directly for the required effective depth:

$$d_{reqd} = \sqrt{\frac{M}{bK}} \tag{6.20}$$

Tables 3 and 5 in the same handbook help in the selection of reinforcing bars for slabs and beams.

It should be realized that the use of design aids without a thorough understanding of their theoretical background and limitations can be disastrous. We should be more than reluctant ever to use any numerical coefficients unless we can check them, starting with basic principles. On the other hand, a thorough knowledge and understanding of available information of this type is necessary for the efficient professional designer, and in well-run offices slack times are often used to compile tables and graphs which will speed up work during rush periods. An understanding of how to compile design or analysis information efficiently is required of the competent engineer.

Allowable stresses While sometimes situations arise where the designer must decide on allowable stresses to be used, generally they are specified by applicable codes. For instance, looking at sec. 1003 of the ACI Code, we find that, depending on grade of steel, bar size, and use, allowable steel stresses vary from 18,000 to 30,000 ksi. In no case is the allowable stress more than half of the yield-point stress. Table 1002*a* of sec. 1002 of this code calls for an allowable compressive concrete stress in bending of $f_{c\ allow} = 0.45f'_c$, where f'_c is the compressive strength of 28-day-old test cylinders of the same batch of mix.

The same section of the code gives values for the modular ratio n, derived from the empirical formula

$$n = \frac{E_s}{E_c} = \frac{30{,}000{,}000}{w^{1.5} \times 34\sqrt{f'_c}}$$

where w is the unit weight of the concrete.

Example Problem 6.5 Redesign the beam of the previous example problem, using available design aids.

Having found the maximum moment as 157.5 kip-ft, we turn to table 1 of the RCD Handbook, where we find, corresponding to $f_s = 20{,}000$ psi, $f_{c\ allow} = 1{,}350$ psi:

$$K = 223 \qquad k = 0.379 \qquad j = 0.874 \qquad a = 1.44$$

Entering the value of K into Eq. (6.15), we get

$$bd^2 = \frac{M}{K} = 8{,}448 \text{ in.}^3$$

leading to the same result as before. The steel is then calculated by

$$A_s = \frac{M}{ad}$$

and the bars selected in the same fashion as before.

6.4 Over- and Under-reinforced Beams

We have seen that a balanced beam requires a definite ratio of steel to concrete, given by p_{bal} according to Eq. (6.17). On the other hand, the costs of materials may be such that an economical beam calls for a different ratio; it may be advantageous to save steel, even at the expense of excess concrete. At other times, architectural requirements may call for a limited concrete section which must be achieved by a disproportionate amount of steel. In either case, an unbalanced beam is indicated, and this section will discuss the design of such beams.

Going back to basic principles, we collect the equations which must be satisfied from Sec. 6.1. Given a moment M to be resisted, modular ratio n, and allowable steel and concrete stresses $f_{s\,allow}$ and $f_{c\,allow}$,

$$\Sigma F = 0: \qquad T = C \qquad \text{or} \qquad A_s f_s = \tfrac{1}{2} kbdf_c \tag{6.1}$$

$$\Sigma M_T = 0: \qquad Cjd = M \qquad \text{or} \qquad \tfrac{1}{2} kbd^2 f_c \left(1 - \frac{k}{3}\right) = M \tag{6.2}$$

$$\text{or} \quad \Sigma M_C = 0: \qquad Tjd = M \qquad \text{or} \qquad A_s f_s d \left(1 - \frac{k}{3}\right) = M$$

where steel and concrete stresses are related by

$$\frac{f_c}{f_s} = \frac{1}{n} \frac{k}{1 - k} \tag{6.3}$$

Equations (6.1) and (6.3) can be combined to give the location of the neutral axis as

$$k = -np + \sqrt{(np)^2 + 2np} \tag{6.6}$$

To the three equations (6.1), (6.2), and (6.6) there correspond the four unknowns A_s, b, d, and k. With more unknowns than equations we have a certain amount of freedom regarding beam proportion. Let us for instance assume that, in order to save steel, an under-reinforced beam is required, with a given steel ratio $p = A_s/bd$ which is only a fraction of the balanced value.

With this given p, Eq. (6.6) furnishes the value of k. We also recall that in an under-reinforced beam, steel will reach its allowable stress $f_{s\,allow}$ first, while the concrete stress is below its allowable. We can thus substitute $f_{s\,allow}$ for f_s in Eq. (6.1), and solve for f_c. Next, we can solve Eq. (6.2) for bd^2, select b and d, and lastly find the required steel area $A_s = (bd)p$.

An analogous procedure could be used for the design of an over-reinforced beam. In either case, the method is lengthy and not too well suited to office practice. We shall therefore consider suitable simplifications.

Consider, for instance, the possibility of substituting for the actual value of the lever arm ratio j (corresponding to the unbalanced beam) the value j_{bal} corresponding to balanced design. In view of the earlier remarks about the insensitivity of j this seems a reasonable approximation. j_{bal}, of course, is an easily calculated design constant.

In the case of an under-reinforced beam, in which $f_s = f_{s\,allow}$ (steel is critical because there is too little of it to balance the design), we can then rewrite Eq. (6.2) approximately as

$$p_{\text{under}} = \frac{A_{s\,reqd}}{bd} = \frac{M}{f_{s\,allow} j_{bal} bd^2} \tag{6.21}$$

and, for an assumed under-reinforced steel ratio p_{under}, find the required value of bd^2. Having selected the dimensions b and d, we can find $A_{s\,reqd}$.

Comparing Eq. (6.21) with Eq. (6.16), we see that bd^2_{reqd} is related to the quantity bd^2_{bal} for a corresponding balanced beam by the ratio

$$bd^2_{reqd} \approx \frac{p_{bal}}{p_{\text{under}}} bd^2_{bal}$$

We can reason similarly in the case of an over-reinforced beam of given steel ratio p_{over} in which $f_c = f_{c\,allow}$, and again assuming that $k_{\text{over}} \approx k_{bal}$ and $j_{\text{over}} \approx j_{bal}$, find from Eq. (6.2) that

$$bd^2_{reqd} \frac{M}{f_{c\,allow}\frac{1}{2}k_{bal}j_{bal}} = \frac{M}{K} \tag{6.10}$$

identical to the balanced value. We see that no savings in concrete section results from over-reinforcement.

To get a feel for the nature of the assumptions made, we should, as usually, compare approximate and exact results numerically. However, in this case we can easily show generally that the substitution of balanced for unbalanced design constants leads to results on the safe side. To do this, we shall first compare the strain distribution of the balanced (dotted line) and the under-reinforced (solid line) beam in Fig. 6.7*a*. Since, because of insufficient steel, the steel stress will reach its allowable while the concrete stress is still safe, we conclude that the neutral axis of the under-reinforced beam will be higher than that of the balanced. The internal lever arm of the under-reinforced section is then larger than that of the balanced. Use of the too small value of j_{bal} in the denominator of Eq. (6.21) results in an amount of steel slightly larger than actually required. Since the steel is the critical part of the under-reinforced beam, the design is on the safe side.

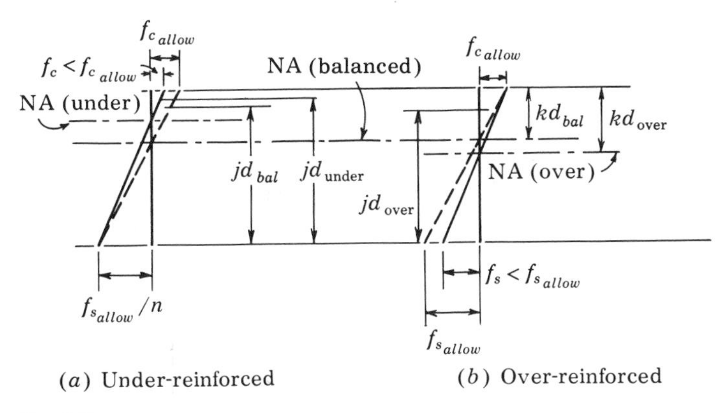

Fig. 6.7

Similar reasoning for the over-reinforced beam shows, referring to Fig. 6.7*b*, that the neutral axis of the over-reinforced beam is lowered, thus increasing the value of k, while lowering the value of j. But since the variation of k predominates over that of j, we conclude that the approximate quantity $k_{bal}j_{bal}$ in the denominator of Eq. (6.10) is smaller than the exact quantity $k_{over}j_{over}$. The resulting approximate value of bd^2_{reqd} will thus be larger than the exact value. Since in an over-reinforced beam concrete controls, the section is on the safe side.

We therefore conclude that the substitution of balanced for unbalanced design constants leads in any case to safe results.

6.5 Design of Longitudinal Reinforcing

We should remind ourselves above all that reinforcing steel belongs on the tension side of the beam, since its primary purpose is to resist the tensile force of the moment couple. Equation (6.16) also tells us that the cross-sectional area of steel A_s must be proportional to the moment M and inversely proportional to the effective beam depth d. The latter of course implies that the larger the internal lever arm, the smaller need be the internal force to resist a given moment. A deep beam therefore usually leads to a saving of steel.

The design of a beam usually starts at the point of maximum absolute value of moment. The required quantity bd^2 is selected for this point, and the dimensions b, d, and h and the steel area A_s are picked. At other points of the beam, the moment will usually have a lesser value, and therefore the moment resistance at those sections may also be less. This can be achieved in two ways: by diminishing the beam section, which leads to a nonprismatic or haunched beam, or by maintaining a constant beam section and diminishing only the amount of reinforcing steel. Usually, the latter course is chosen in order to simplify the formwork, and we shall consider ways of choosing points at which steel bars may be deleted in conformity with the shape of the moment diagram.

If the section to resist the maximum moment has been designed balanced, then at other points of the beam where steel is deleted while the concrete section remains constant, the sections will be under-reinforced; while the use of balanced design constants at such sections is not theoretically correct, we shall nevertheless, in the light of the discussion of Sec. 6.4, use them in order to simplify the design procedure considerably.

Before demonstrating an efficient method for determining steel cutoff points, we shall point out some provisions of the ACI Code pertaining to this matter. Section 918*b* provides for a certain extension of steel bars beyond their theoretical cutoff point. Parts *e* and *f* of this section require that at least one-third of top and bottom steel be extended under certain conditions even though not required by flexural theory.

Example Problem 6.6

Live load = 1 K/ft

30′ 10′

$f'_c = 3.0$ ksi, $f_{c\ allow} = 1.35$ ksi, $f_s = 20$ ksi.
Design the beam and longitudinal reinforcement according to the ACI Code.

Solution: Assume the beam dead load at 325 lb/ft, so that the total uniform load is 1.325 kips/ft. The moment diagram is then drawn and the maximum moment determined as 118 kip-ft. Using then the balanced design constants (for $n = 9$)

$$k_{bal} = 0.379 \qquad j_{bal} = 0.874 \qquad K_{bal} = 0.223$$

the required section must satisfy the requirement

$$bd^2 = \frac{M}{K} = \frac{118 \times 12}{0.223} = 6{,}350 \text{ in.}^3$$

We select the section $b = 14$ in., $d_{reqd} = 21.2$ in. The actual cross section is shown. Note that if the overall depth is chosen to the nearest inch, and if provisions for stirrups are provided, $d_{act.}$ becomes 21.5 in. The beam weight is now 0.35 kips/ft. The maximum moment can then be prorated to $(1.350/1.325) \times 118 = 120$ kip-ft, which will not affect the choice of the section.

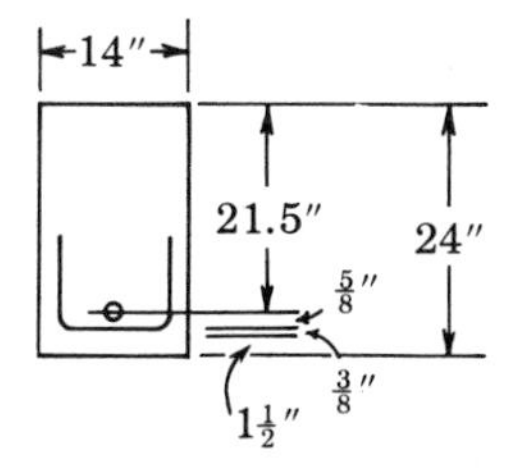

We next plot and draw the moment diagram. Since this will be used to determine cutoff points, it is essential that this be drawn accurately to a large scale, as shown in Fig. 6.8. Since parabolas are easy to construct, it is convenient to superimpose these on straight lines to obtain the final diagram. We next determine the maximum amounts of top and bottom steel as

$$A^+_{s\ reqd} = \frac{120 \times 12}{20 \times 0.874 \times 21.5} = 3.84 \text{ in.}^2$$

Bottom steel: use four No. 9 bars: $A_{s\ furn} = 4.00$ in.^{2}

$$A^-_{s\ reqd} = \frac{67.5 \times 12}{20 \times 0.874 \times 21.5} = 2.16 \text{ in.}^2$$

Top steel: use three No. 8 bars: $A_{s\ furn} = 2.37$ in.^{2}

We now determine the point along the beam where we can afford to delete two of the four bottom bars. At this point, we will have a resisting moment due to the two remaining bars ($A_s = 2.00$ in.^{2}), of value $(2.00/3.84) \times 120$ kip-ft $= 62.5$

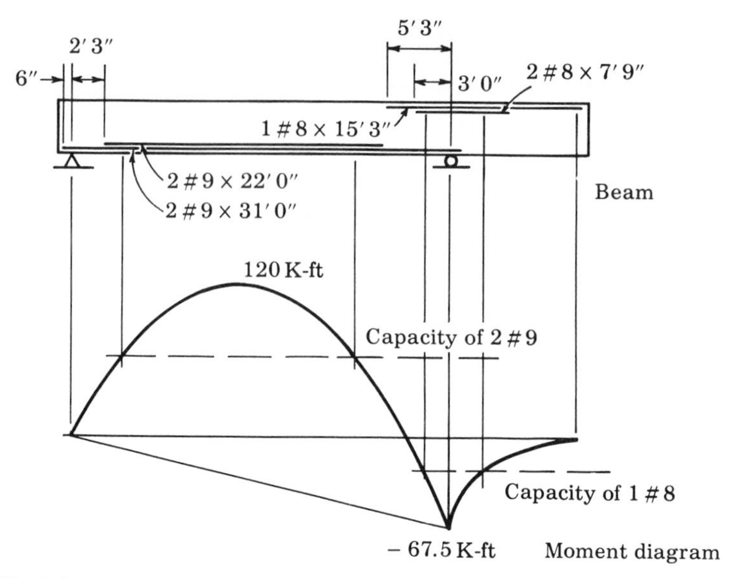

Fig. 6.8

kip-ft. The theoretical cutoff point is then at the point where the moment drops to this value, and we obtain this location by scaling the moment diagram. From these points we conveniently project the theoretical cutoff point in accordance with ACI sec. 918*b*. If the scale has been taken sufficiently large, the dimensions of the bars may be scaled directly. We proceed in a similar fashion for the top bars, deleting two of the three bars at a point where the negative moment drops to a value $(0.79/2.16) \times 67.5 = 24.7$ kip-ft.

Before the beam design may be considered satisfactory, bond and diagonal tension stresses must be investigated. This will be discussed in the next section.

6.6 Bond and End Anchorage

We shall first discuss the problem of end anchorage in terms of a simple example: consider a cantilever beam as shown in Fig. 6.9*a*. Under load, the top steel has a tendency to pull out of the supporting structure. To prevent this possibility (which would of course lead to immediate failure) we must embed or anchor a sufficient length of the bars (called the *embedment* or *anchorage length L*) in the concrete of the support. To calculate the required

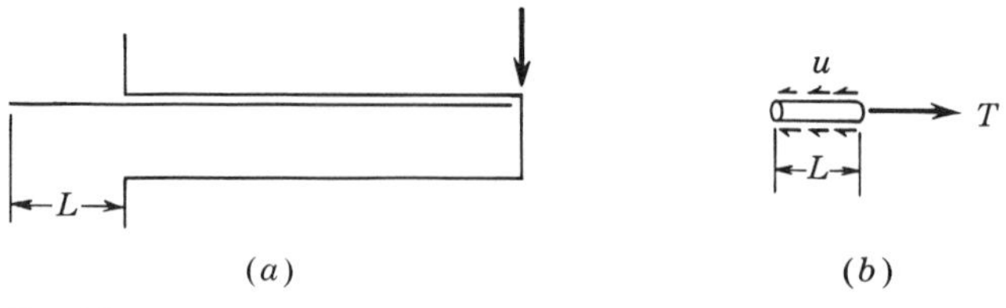

Fig. 6.9

anchorage length, we resort to the free body of the steel bar shown in Fig. 6.9*b*; at the face of the support, a tensile force T is acting (which can be calculated from the moment at this point), and to maintain equilibrium it must be resisted by the *bond stress* u (in pounds of shear per unit of surface area) between steel and surrounding concrete. Assuming that this bond stress is uniformly distributed along the embedment length, we can calculate this length by considering equilibrium of longitudinal forces:

$$\Sigma F = 0: \qquad u\Sigma_0 L = T = f_s A_s \tag{6.22}$$

where Σ_0 is the perimeter of the cross section of the reinforcing bar in inches (this quantity is given for commercial bars in Sec. 1.5).

If the allowable value of the bond stress u_{allow} is specified, then the required length

$$L = \frac{T}{\Sigma_0 u_{allow}} = \frac{f_s A_s}{\Sigma_0 u_{allow}} \tag{6.23}$$

Allowable values of the bond stress are given in ACI sec. 1301, where we see that the bond strength is influenced by several factors:

1. The compressive strength: the allowable bond stress is proportional to the square root of f'_c.
2. The bar diameter: the allowable bond stress is inversely proportional to the bar diameter. If bond is critical, it pays to use small bars.
3. The location of the steel: in the course of setting, the concrete around top bars will settle away from the steel so that part of the contact is lost. For this reason the allowable bond stress is reduced for top bars.
4. The ribs or surface deformations of the bars.

The assumption of uniformly distributed bond stress is of course only an artifice to reduce a very complicated situation to manageable proportions. Gross simplifications of this type (note that we satisfied overall equilibrium, but not the internal geometry of the problem) often lead to irrational allowable stresses which must be determined by elaborate series of experiments.

Note that for a well-designed beam the tensile stress f_s is close to, but probably below, $f_{s\,allow}$. If we use the specification value of $f_{s\,allow}$, the tensile bar force T will be, if anything, overestimated, and so will be the required embedment length. This assumption will furnish safe results and make the required embedment length a direct function of the bar diameter. For instance, considering deformed bottom bars, and using the allowable bond stresses of sec. 1301*c*, we determine

$$L_{reqd} = \frac{f_{s\,allow} A_s}{\Sigma_0 (4.8\sqrt{f'_c}/D)} = \frac{f_{s\,allow}}{19.2\sqrt{f'_c}} D^2 \tag{6.24}$$

where the coefficient of D^2 is a constant for the entire job.

Instead of extending the straight tension bar into the support, it is sometimes bent into a "standard hook" as described in ACI sec. 801. The anchorage strength of such a hook against pulling out is described in sec. 918h. Based on extensive tests, the pullout strength of hooks may safely be assumed to be such that the additional straight embedment length is one-half of that required without hooks. This is achieved by reducing $f_{s\ allow}$ to one-half of its value in Eq. (6.24).

Example Problem 6.7

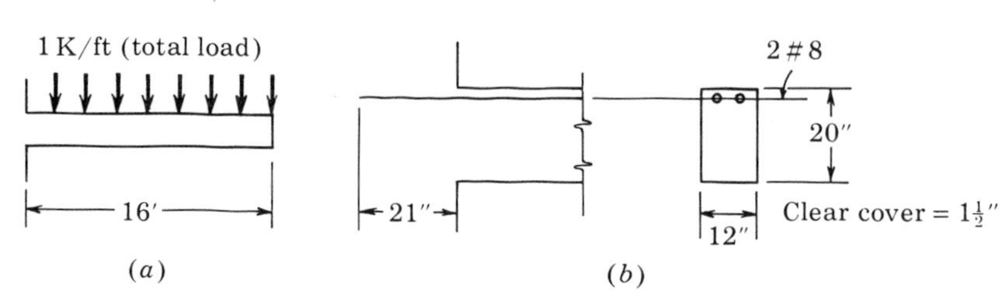

$f_{s\ allow} = 20$ ksi, $f'_c = 5.0$ ksi, $f_{c\ allow} = 2.25$ ksi, $n = 7$.

Design the critical section of the cantilever beam and determine embedment length of bars.

Solution: At face of support:

$M = \frac{1}{2}wL^2 = 128$ kip-ft; $V = 16$ kips

Balanced design constants

$$k = \frac{1}{(f_s/nf_c) + 1} = 0.440$$

$$j = 1 - \frac{k}{3} = 0.853$$

$$K = \tfrac{1}{2}f_{c\ allow}kj = 0.422$$

Gross section

$$bd^2_{reqd} = \frac{128 \times 12}{0.422} = 3{,}640 \text{ in.}^3$$

Use $b = 12$ in., $d = 17.4$ in., $h = 20$ in., $d_{act.} = 17.5$ in.

Steel

$$A_{s\ reqd} = \frac{128 \times 12}{20 \times 0.853 \times 17.5} = 1.58 \text{ in.}^2$$

Use two No. 8 bars; $A_{s\ furn} = 1.58$ in.

Anchorage

$$u_{allow} \text{ (for top bars)} = \frac{3.4\sqrt{f'_c}}{D} = 240 \text{ psi}$$

$$L_{reqd} = \frac{20 \times 1.58}{6.3 \times 0.248} = 20.9 \text{ in., say 21 in.}$$

A similar, but usually not so critical, situation prevails at the end of every bar subjected to a calculated stress. To develop this stress, the force must be transferred to the bar from the surrounding concrete, and this is to be accomplished without exceeding the allowable bond stress. The anchorage provisions of ACI sec. 918 which provide for extension of bars beyond their theoretical cutoff points take care of this development.

We must guard against the possibility of slippage of the reinforcing steel with respect to the surrounding concrete not only at points of support, but throughout the beam. That this danger of slipping exists may be seen from the free bodies of Fig. 6.10. We consider a slice, dx long, of the beam. The left side, at a point where the moment has a value M, has corresponding internal forces C and T acting on it; the right face, subjected to a slightly different moment $M + dM$, has similarly slightly different resultant forces acting, $C + dC$ and $T + dT$, with internal lever jd. Taking moments about the location of the compressive resultant, we find

$$\Sigma M_C^+ = 0: \qquad M - (M + dM) + [T - (T + dT)]jd = 0$$

$$\text{or} \quad dT = \frac{dM}{jd}$$

$$\text{or} \quad \frac{dT}{dx} = \frac{dM}{dx}\frac{1}{jd} = \frac{V}{jd} \tag{6.25}$$

Note that the rate of change of tension force (and compression force) is proportional to the rate of change of moment along the beam, which, as we know, is equal to the shear at that point of the beam.

We consider next a different free body, namely that of Fig. 6.10*b*, representing the steel bar only, of length dx. Equation (6.25) tells us the difference between the tension forces acting on the opposite end of the piece. We realize that in order to maintain equilibrium additional bond stresses u between steel and concrete must be acting. Summing horizontal forces, we find

$$\Sigma F^+ = 0: \qquad (T + dT) - T - u\,dx\,\Sigma_0 = 0$$

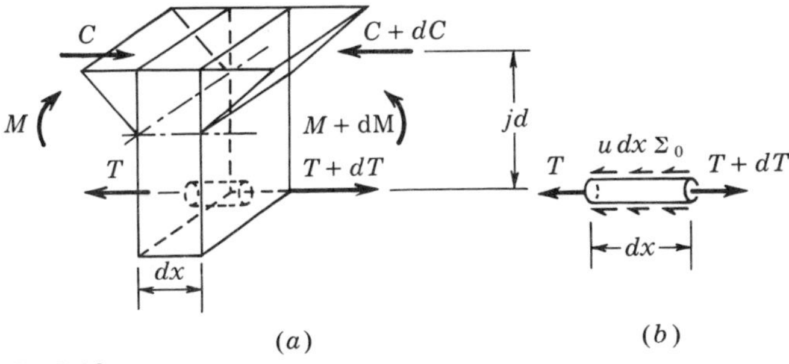

Fig. 6.10

or, using Eq. (6.25),

$$u = \frac{dT}{dx}\frac{1}{\Sigma_0} = \frac{V}{\Sigma_0 jd} \tag{6.26}$$

Equation (6.26) is eq. 13-1 of the ACI Code.

Bond stresses of magnitude given by Eq. (6.26) must be resisted at every beam section in order to keep the steel in equilibrium. The nature of these bond stresses is the same as that of the shear stresses discussed earlier for different types of beams, and the reader might at this time refer back to Secs. 4.2 and 5.3 to convince himself that the method of analysis used here is analogous to that used earlier.

For design purposes, it is essential to select bars of sufficient surface area so that the actual bond stress u does not exceed the allowable values given in ACI sec. 1301b and discussed in connection with end-anchorage problems. To do this, we solve Eq. (6.26) for the required perimeter:

$$\Sigma_{0\ reqd} = \frac{V}{u_{allow} jd} \tag{6.27}$$

Note that a maximum of perimeter is required where the shear is greatest. In simply supported beams, this will be at the supported ends, where the moment is zero. Thus, while reinforcing for such beams will usually be selected to provide the amount of area to resist the maximum moment, we must keep in mind while determining cutoff points that sufficient bar surface area must be retained at the ends to satisfy Eq. (6.27).

Example Problem 6.8

(a) Calculate the critical bond stress due to shear in the cantilever beam designed in the previous section, and compare it to the allowable value according to the ACI Code.

Solution: At point of maximum shear:

$$u_{\max} = \frac{V}{\Sigma_0 jd} = \frac{16 \text{ kips}}{6.3 \times 0.853 \times 17.5} = 0.170 \text{ ksi}$$

Section 1301b, ACI Code:

$$u_{allow} \text{ (top bars)} = \frac{3.4\sqrt{f'_c}}{D} = 0.240 \text{ ksi}$$

Design is O.K. for bond.

(b) Check the beam design of Example Problem 6.6 for bond stresses and revise design of reinforcing if necessary.

Solution: We begin by drawing the beam as designed and also the shear diagram in order to determine critical sections.

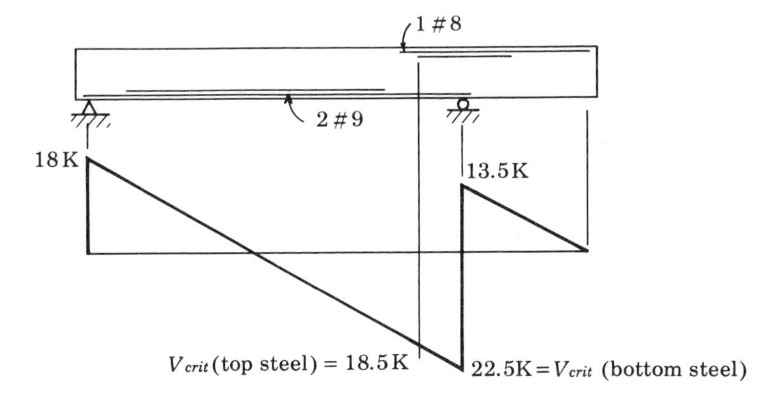

Fig. 6.11

Considering first the bottom steel, we find that the critical point is just to the left of the right support, where maximum shear combines with minimum amount of steel. At this point,

$$\Sigma_{0\,reqd} = \frac{V}{u_{allow}jd} = \frac{22.5 \text{ kips}}{0.243 \times 0.874 \times 21.5} = 4.9 \text{ in.}$$

$\Sigma_{0\,furn}$ (two No. 9) = 7.1 in.; design is O.K.

The critical section for top steel is at the cutoff point of the top bars, where one remaining No. 8 bar must resist 18.5 kips of shear. Here,

$$\Sigma_{0\,reqd} = \frac{V}{u_{allow}jd} = \frac{18.5 \text{ kips}}{0.192 \times 0.874 \times 21.5} = 5.1 \text{ in.}$$

$\Sigma_{0\,furn}$ (1 No. 8) = 3.1 in.; design is inadequate

Insufficient perimeter is available. The top steel is to be revised to provide extension of two No. 8 bars (Σ_0 = 6.2 in.) beyond the point of inflection.

6.7 Shear and Diagonal Tension

Whereas the resistance to transverse shear forces of homogeneous beams is well understood and can be easily determined, the behavior of reinforced concrete beams under shear is not at all clear at this time. While in the past a rational approach has been taken to the solution of this problem, in recent years, making use of the usual principles of structural mechanics which we emphasize in these notes, a critical comparison of test results with theory has revealed the influence of several factors not accounted for in present theory. Much work and thought will have to be expended before the interactions are clearly understood and a fully rational analysis can be attempted.

While a seemingly rational approach will be discussed in the following, we must keep in mind that the validity of this approach, which is based on the questionable assumption of elastic behavior, is not borne out by the actual behavior of beams. To make up for this, the allowable shear stress specified in ACI sec. 1201*d* depends not only on the actual strength of the concrete, but is a function of several additional variables. As has been mentioned before, when the analysis becomes unduly simplified, then the allowable values must become more complicated to keep the results in tune with reality.

It is generally understood that failure of concrete beams under excessive shear force occurs not by exceeding the shear strength but by the tensile strength of the material. This arises from the fact that the strength of concrete in pure shear (which can be determined by testing specimens under torsion and compression) is much greater than that in tension. It follows that fracture will occur on planes of maximum tension stress which accompany the shear and which can be determined by the theory of combined stresses.

We begin our analysis with the determination of shear stresses on horizontal and vertical planes, following earlier procedures. As already in Sec. 4.2, we first consider a section of the beam, dx long, subject to a moment M on one face, $M + dM$ on the other. We recall from Eq. (6.10) that the extreme concrete compression stress resulting from a moment M is

$$f_c = \frac{2M}{bd^2kj}$$

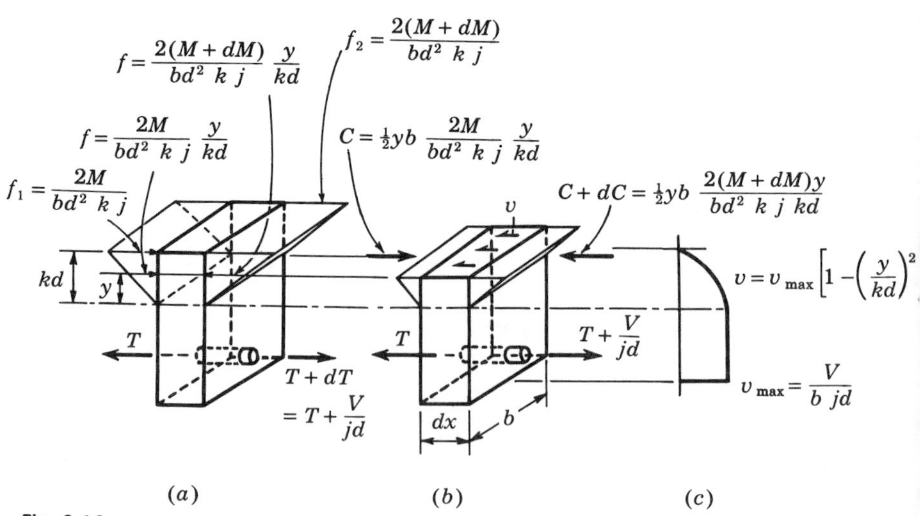

Fig. 6.12

and the increment of tensile force, according to Eq. (6.25), is

$$\frac{dT}{dx} = \frac{V}{jd}$$

Cutting then an appropriate free body which contains the horizontal plane to be investigated for shear stresses as an exterior face, as shown in Fig. 6.12*b*, we establish horizontal equilibrium of all forces acting on the body:

$$\Sigma F^{+} = 0: \qquad \left(T + \frac{V\,dx}{jd}\right) - T + \frac{y^2}{jd(kd)^2}\left[M - (M + dM)\right] - vb\,dx = 0$$

Solving for the required stress on the horizontal plane,

$$v = \frac{V}{bjd}\left[1 - \left(\frac{y}{kd}\right)^2\right] \tag{6.28}$$

Note that the distribution is parabolic, with the maximum value, corresponding to $y = 0$, at the neutral axis:

$$v = \frac{V}{bjd} \tag{6.29}$$

Between the neutral axis and the tension steel, the shear stress remains constant at this value, as may be ascertained from an appropriate free body. The shear-stress distribution is then as in Fig. 6.12*c*.

The shear stress given by Eqs. (6.28) and (6.29) acts on vertical and horizontal planes in the beam. The vertical (or transverse) planes are also subject to axial stress f resulting from bending of the beam. This state of combined stress on a typical element is shown in Fig. 6.13*a*.

Since, as previously discussed, concrete is weak in tension, it stands to reason that one must first look for the largest tensile stress at the point, and this we can find, using any of the techniques of stress transformation,

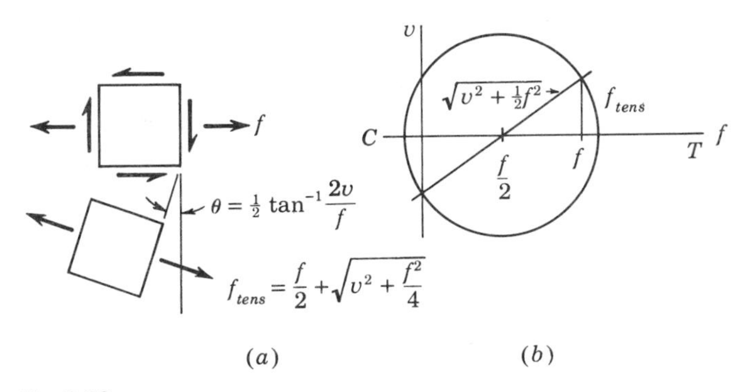

Fig. 6.13

such as the Mohr's circle shown in Fig. 6.13*b*, as

$$f_{\max} = \frac{f}{2} + \sqrt{v^2 + \frac{f^2}{4}} \tag{6.30}$$

acting at an angle of

$$\theta = \tfrac{1}{2} \tan^{-1} \frac{2v}{f} \tag{6.31}$$

with the vertical plane.

This relation between shear, bending, and maximum tensile stress is valid at any point of the beam. It now remains to look for points at which a critical value of $f_{\max}$ might prevail. Our attention is immediately drawn to the tension side of the beam, where the largest value of shear stress [given by Eq. (6.29)] acts. To determine the actual value of tensile bending stress f is a difficult matter, intimately connected with the crack formation in the beam. To avoid the question entirely, we consider a simple beam which has its maximum shear, and therefore shear stress, at the end where the moment, and therefore f, is zero.

At this location then, since $f = 0$, Eqs. (6.30) and (6.31) reduce to

$$f_{\max} = v = \frac{V}{bjd} \tag{6.32}$$

and $\theta = \tfrac{1}{2} \tan^{-1} \infty = \tfrac{1}{2} 90 \text{ deg} = 45 \text{ deg}$ (6.33)

The maximum tension stress is thus equal to the shear stress and acts on planes at 45 deg to vertical and horizontal planes.

Indeed, the cracks forming near the ends of beams under excessive shear are inclined at the angle predicted by this theory, as shown in the beam of Fig. 6.14. The maximum tensile stress, which according to Eq. (6.32) can be determined identically with the maximum shear stress (a fact which the ACI Code reminds us of by talking of "shear as a measure of diagonal tension"), may not exceed the value given in ACI sec. 1201*d* in the case of beams without web reinforcing. This value is rightly held very low (less than 71 psi for 5,000-psi concrete) in view of the low tensile strength of concrete.

In practice, the concrete cross section of the beam is usually designed

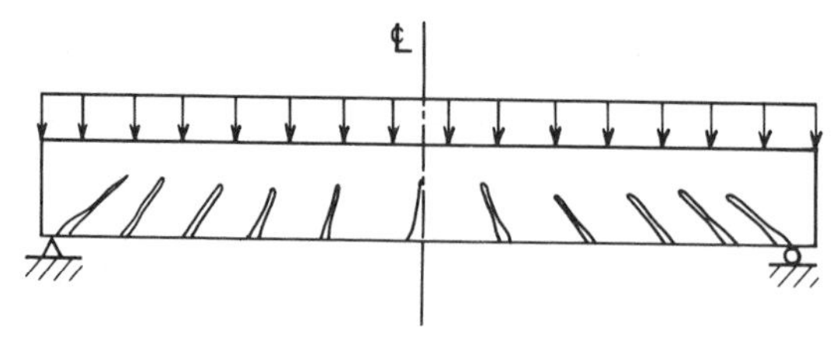

Fig. 6.14

on the basis of a bending analysis and then checked for shear stress at points of critical shear. If the maximum tensile stress determined by Eq. (6.32) (or by eq. 12-1 of sec. 1201*a* of the ACI Code, which simplifies the matter slightly by setting j equal to unity) exceeds the allowable shear stress v_c, then we can either change to a bigger concrete section or resort to web reinforcing consisting of steel bars so placed as to prevent, or minimize, the tensile cracking on oblique planes due to the shear.

Example Problem 6.9 Check the diagonal tension stresses at the critical sections of the beam of Example Problem 6.6. Compare with allowable ACI values and comment on the safety of the beam against diagonal tension failure.

Solution: We redraw the previously designed beam and its shear diagram. At the critical section for shear, the maximum diagonal tension stress according to Eq. (6.32) is

$$v = \frac{V}{bjd} = \frac{22.5}{14 \times 0.874 \times 21.5} = 0.086 \text{ ksi} = 86 \text{ psi}$$

According to ACI eq. 12-1, in which $j = 1.0$,

$$v = \frac{V}{bd} = 75 \text{ psi}$$

Since the allowable diagonal tensile stress for $f'_c = 3.0$ ksi is (according to ACI table 1002*a*) 60 psi, web reinforcing is required at this section.

Turning next to the left end, where $V = 18.0$ kips, we can find the stress (according to ACI) by proportion as

$$v = \frac{18 \text{ kips}}{22.5 \text{ kips}} \times 75 \text{ psi} = 60 \text{ psi}$$

No web reinforcing is therefore required at the left end.

The design of the web reinforcement will be taken up in the next section.

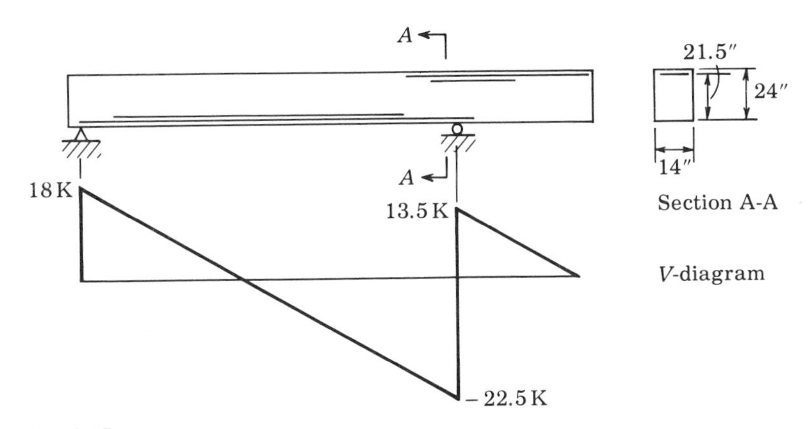

Fig. 6.15

6.8 Web Reinforcing

To prevent collapse of the concrete beam even though subjected to diagonal tensile cracks, we resort to web reinforcing. The basic idea underlying the selection and placing of this reinforcing is identical to that followed in the design of longitudinal bars: we must provide sufficient steel to transmit those tensile stresses across cracks which are necessary to keep every part of the beam in equilibrium.

We consider a portion of the beam which contains web reinforcing consisting of vertical bars or "stirrups," and turn our attention to the equilibrium of a typical free body ABC between two diagonal tension cracks a distance s apart, as shown in Fig. 6.16*a*. The forces acting on this free body are the shear stress v acting on the top face of area bs, the tension force T in a vertical steel bar, or stirrup, and a compression force C exerted by the concrete prism between the cracks, both of the latter at 45 deg with the beam axis, as shown in Fig. 6.16*b*. Equilibrium here is easily established by use of the force polygon of Fig. 6.16*c*, according to which

$$T = vbs = \frac{Vs}{jd} \tag{6.34}$$

The allowable tension force is based on the allowable tension stress in the stirrup f_v acting on the cross-sectional area of the stirrup A_v so that

$$T = A_v f_v = \frac{Vs}{jd} \tag{6.35}$$

For design purposes, Eq. (6.35) can be used in two ways: given the beam section, the allowable stresses, and the shear V at any point in the beam, we can either specify the bar size A_v and solve for the required stirrup spacing

$$s_{reqd} = \frac{A_v f_v jd}{V} = \frac{A_v f_v}{vb} \tag{6.36}$$

or, given the spacing, we can find the required bar size A_v. The former

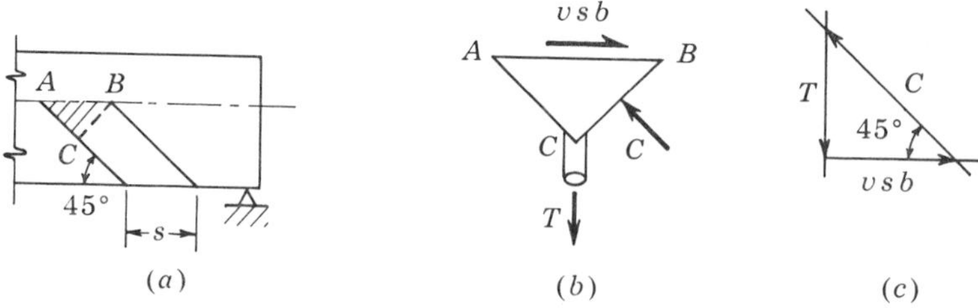

Fig. 6.16

of these alternates is used in practice where it is easier to adjust stirrup spacing to conform to varying shear force than to use different sizes of stirrups.

An equation similar to Eq. (6.36) is given in sec. 1203 of the ACI Code. The difference lies in the deletion of the coefficient j and in the use of the "excess shear force" V' instead of the total shear force V. The excess shear force is that portion of the shear force acting at the section which is above and beyond the value of shear V_c corresponding to the allowable diagonal tensile stress v_c which may be carried by the concrete without help of web reinforcing. In other words, the web steel is responsible only for the part of the total shear not carried by concrete. This reasoning is not in accordance with our previous notions, because the vertical stirrups can be under stress only after diagonal cracks are present. Since of course no diagonal tensile stress can be transferred in the concrete after cracking, the steel should logically be designed to carry the entire shear. Nevertheless, beams designed according to the ACI procedure seem to be safe.

Often inclined stirrups are used, making an angle α with the longitudinal beam axis. The design of such reinforcing follows the same procedure as for vertical bars. From Fig. 6.17:

$$\Sigma F_{x'} = 0: \qquad vsb \times \frac{1}{\sqrt{2}} - T_{x'} = 0$$

$$T \cos (\alpha - 45°) = T[\cos \alpha \cos 45° + \sin \alpha \sin 45°] = vsb \frac{1}{\sqrt{2}}$$

or $$T = \frac{vsb}{\sin \alpha + \cos \alpha} = A_v f_v$$

or $$A_v = \frac{Vs}{f_v jd(\sin \alpha + \cos \alpha)} \tag{6.37}$$

Equation (6.37) is used in the same way as Eq. (6.36). In comparing Eq. (6.37) with eq. 12-7 of the ACI Code, the same comments as above apply.

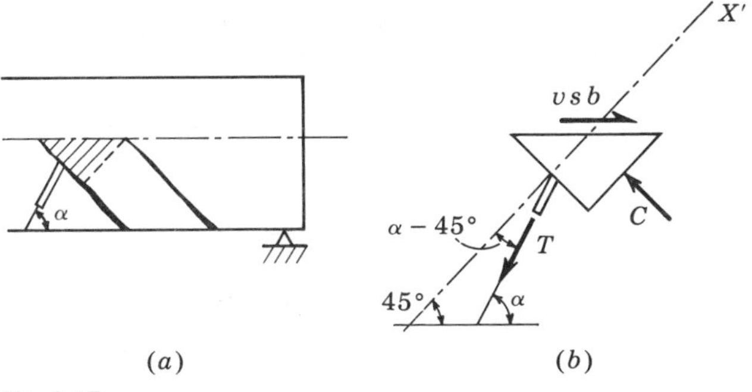

Fig. 6.17

Beams are often reinforced with stirrups at 45 deg. Indeed, in view of the assumption that the principal stress directions (and therefore the cracks) are at 45 deg, this seems a logical arrangement. In this case, Eq. (6.37) reduces to

$$A_v = \frac{1}{\sqrt{2}} \frac{Vs}{f_v jd} \tag{6.38}$$

A comparison of Eq. (6.38) with Eq. (6.35) for vertical stirrups shows that less web steel may be required if the bars cross the cracks at right angles, a conclusion which is to be expected in view of the previous discussion.

For vertical bars, $\alpha = 90°$, and the general equation, Eq. (6.37), reduces to Eq. (6.35).

Among the various provisions and restrictions of chap. 12 of the ACI Code, only one additional important one will be mentioned: it is apparent that if the spacing s between stirrups is so large that diagonal cracks can develop between bars, then the beam may fail in shear regardless of the size of the web bars. Accordingly, sec. 1206*a* of the ACI Code sets a maximum stirrup spacing so that every potential 45-deg crack is crossed by at least one reinforcing bar. For vertical bars this means that the maximum spacing of stirrups is to be not greater than about one-half the beam depth.

We should also remind ourselves of the importance of end anchorage of web bars. In order to develop their safe tensile capacity, their ends must be embedded for a length which will enable them to transfer the tensile force to the surrounding concrete without exceeding the allowable bond stress. The details of how this may be accomplished are given in ACI sec. 919. Because embedment space is scant, recourse to hooks, bends, or welding to the longitudinal reinforcement is often taken in practice.

Various analytical and graphical methods are available for efficient determination of the stirrup spacing. In our example we shall demonstrate only direct use of Eqs. (6.36) or (6.37) to calculate the number and arrangement of stirrups.

Example Problem 6.10 Design web reinforcing for the beam of Example Problems 6.6 and 6.9; $f_v = f_s$. We again reproduce the beam as designed, and also the "shear stress" diagram, proportional to the previously drawn shear force diagram and related to it by the ratio (according to ACI Code)

$$v \text{ (psi)} = \frac{V}{bd} = \frac{V \times 1{,}000}{14 \times 21.5} = 3.33V$$

Wherever this diagonal tension stress exceeds the value v_c (= 60 psi) which may be carried by the concrete alone, we must provide stirrups for the excess shear stress. The spacing of these is given by Eq. (6.36) (assuming No. 2 U stirrups):

$$s = \frac{A_v f_v}{vb} = \frac{0.10 \times 20{,}000}{v \times 14} = \frac{143}{v \text{ (psi)}}$$

Similar to the earlier problem of determining stud spacing in a composite beam, we can again evaluate this spacing in tabular fashion. Near the right support:

v_{excess}, *psi*	*s*, *in.*	*Act. spacing*, *in.*	*Dist. from end*, *in.*
15	9.5	9	9
12	11.9	10	19

Note that the maximum allowable stirrup spacing according to ACI 1206*a* is $d/2$ or 10.5 in. Accordingly, all further stirrups to a distance 40 in. + d = 61 in. will be spaced at 10 in. The table is completed with

		5 @ 10	69

The stirrup spacing is indicated in Fig. 6.18. Note that at all other portions of the beam the diagonal tension stresses are sufficiently low so that no stirrpus are required.

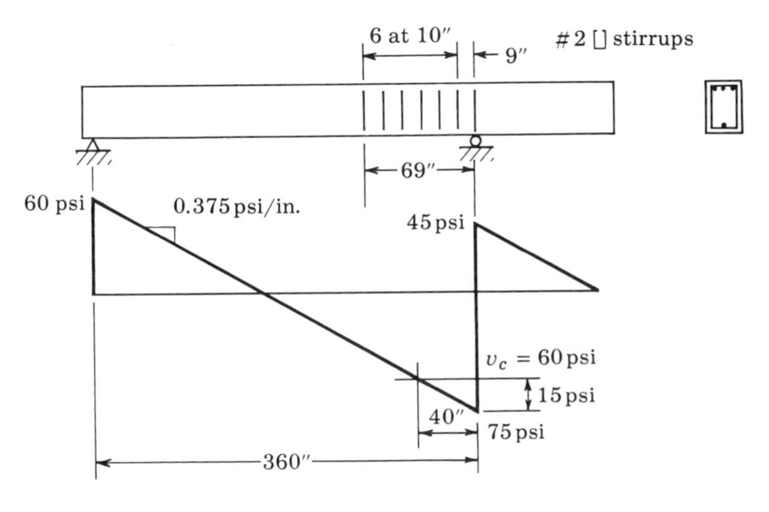

Fig. 6.18

6.9 Doubly Reinforced Beams

Because of constructive considerations, or because of limitations on the gross concrete section, beams will often contain steel in both tension and compression sides. In such doubly reinforced members, the compression steel will help carry part of the internal compression force. Given such a member, we shall first consider the stress analysis for a specified moment.

Analysis We are given the cross section shown in Fig. 6.19, with tension steel area A_s and compression steel area A_s' located as shown. It is required to find elastic stresses in steel and concrete resulting from an applied moment M.

To do this, we have as before two methods available. In order to focus attention on basic principles, we shall first use the internal-force method of analysis. Considering the all-important conditions of statics, we write

$$\Sigma F = 0 \qquad \text{or} \qquad C_c + C_s = T \tag{6.39}$$

$$\Sigma M_T = 0 \qquad \text{or} \qquad C_c d\left(1 - \frac{k}{3}\right) + C_s(d - d') = M \tag{6.40}$$

where
$$\begin{aligned} C_c &= \tfrac{1}{2} f_c b k d \\ C_s &= A_s' f_s' \\ T &= A_s f_s \end{aligned} \tag{6.41}$$

(Note that this neglects the hole in the concrete to accommodate the compression steel.) From the linear strain distribution in Fig. 6.19*c*,

$$\frac{\epsilon_c}{kd} = \frac{\epsilon_s'}{kd - d'} = \frac{\epsilon_s}{d - kd}$$

and introducing the stress-strain relations, we find that

$$\frac{f_c}{E_c kd} = \frac{f_s'}{E_s(kd - d')} = \frac{f_s}{E_s(d - kd)} \tag{6.42}$$

or
$$\frac{f_c}{kd} = \frac{f_s'}{n(kd - d')} = \frac{f_s}{n(d - kd)} \tag{6.43}$$

as also shown in the stress distribution, Fig. 6.19*d*. Writing the steel stresses in Eqs. (6.41) in terms of the concrete stresses and substituting into Eqs. (6.39) and (6.40), we find

$$\tfrac{1}{2} f_c b k d + A_s' \frac{n(kd - d')}{kd} f_c = A_s \frac{n(d - kd)}{kd} f_c \tag{6.44}$$

and
$$\tfrac{1}{2} f_c b k d \, d\left(1 - \frac{k}{3}\right) + A_s' \frac{n(kd - d')}{kd} f_c(d - d') = M \tag{6.45}$$

These two equilibrium equations contain two unknowns, f_c and k, and

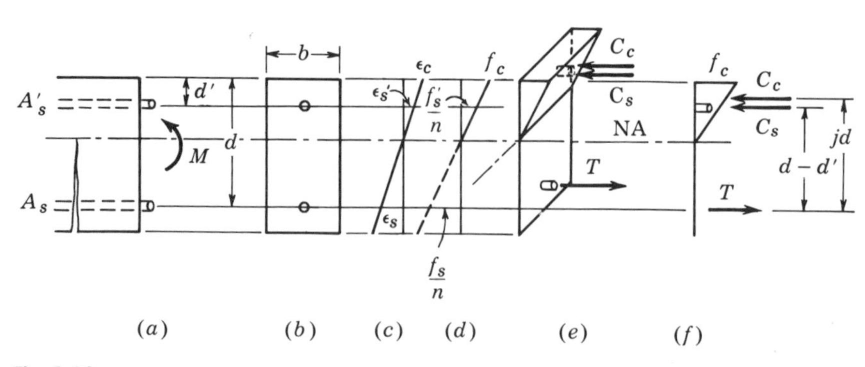

Fig. 6.19

can thus be solved. Note that f_c can be canceled from Eq. (6.44), which then, upon substitution of $p = A_s/bd$ and the compressive steel ratio $p' = A_s'/bd$, can be solved for k:

$$k = -(np + np') + \sqrt{(np + np')^2 + 2[np + np'(d'/d)]} \tag{6.46}$$

Note that this expression for the location of the neutral axis reduces to Eq. (6.6) for singly reinforced beams when $p' = 0$. Having k, we can solve Eq. (6.45) for the concrete stress f_c in the extreme compression fiber. The other stresses can then be found from the plot of the stress distribution, Fig. 6.19d, or the equivalent Eq. (6.43).

As in earlier beam problems involving composite sections, we can make use of the transformed-section method to use homogeneous beam theory. For this purpose, we draw the transformed section in concrete, having transformed both tension and compression steel in the ratio n, as shown in Fig. 6.20b. Solving for the location of the neutral axis as the centroidal axis of the transformed section, we set

$$\Sigma M_{\text{Area } w/r \text{ to N.A.}} = (bkd)\frac{kd}{2} + nA_s'(kd - d') - nA_s(d - kd) = 0$$

Again introducing the tensile and compressive steel ratios p and p' and solving for k, we are led once more to Eq. (6.46).

Once k is found, we can find the transformed moment of inertia I_T and also find the concrete stresses by

$$f_c = \frac{My}{I_T}$$

and the steel stresses by multiplying the surrounding concrete stresses by the modular ratio n.

Note that, more precisely, we should take into account the hole left in the compressive concrete when the compressive steel is transformed. We can plug this hole, shown dotted in Fig. 6.20b, by taking an amount A_s' off the transformed compression steel. This leaves a transformed compression steel area of only $(n - 1)A_s'$, and, accordingly, in Eq. (6.46) the value n which multiplies A_s' should be replaced by $n - 1$. This effect was also

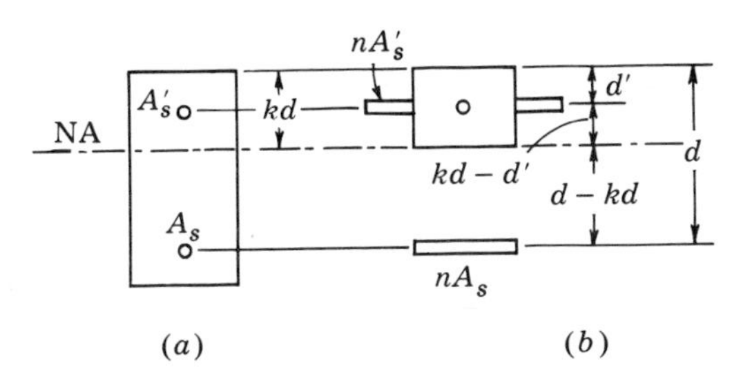

Fig. 6.20

neglected in the first of Eqs. (6.41) for the sake of clarity. However, it can easily be included without increasing the work involved.

A more serious effect is the creep which has been shown to occur in concrete under compression. Creep, or plastic flow, is the continuing deformation of a material under stress. In this case, it results in the compressive concrete flowing out from under the stress, thereby throwing proportionately more force on the compression steel, which is not able to undergo similar flow at ordinary temperatures. A simplified way of evaluating the effects of this creep is to reason that after an appropriate period of time, the ratio of the moduli of compression steel and concrete has increased. The concrete has become more pliable with time.

An appropriate value for this increase of compressive modular ratio with time is a factor of 2. Section 1102*c* of the ACI Code prescribes this value, so that according to this code, the ratio $n = E_s/E_c$ connecting steel and concrete stresses on the compression side should be doubled. If we modify the transformed section in accordance with the two corrections just discussed, the fin representing the transformed compression steel will have an area $(2n - 1)A_s'$, instead of the value nA_s' shown in Fig. 6.20, and the location of the neutral axis will then be given by

$$k = -[np + (2n - 1)p'] + \sqrt{[np + (2n - 1)p']^2 + 2\left[np + (2n - 1)p'\frac{d'}{d}\right]} \qquad (6.47)$$

which can be obtained by suitably modifying Eq. (6.46).

It should be noted that, logically, the n on the tension side should be similarly transformed. This, however, is not usually done. We observe again that Eqs. (6.46) and (6.47) are singularly unsuitable for memorization. A superior plan of attack is to thoroughly understand Figs. 6.19 and 6.20, and remember that any proper solution must satisfy equilibrium, linear strain distribution, and appropriate stress-strain relations.

Example Problem 6.11 The beam of cross section shown is subjected to bending. Allowable stress in steel is 24 ksi, compressive allowable stress is 1.125 ksi; $n = 10$.

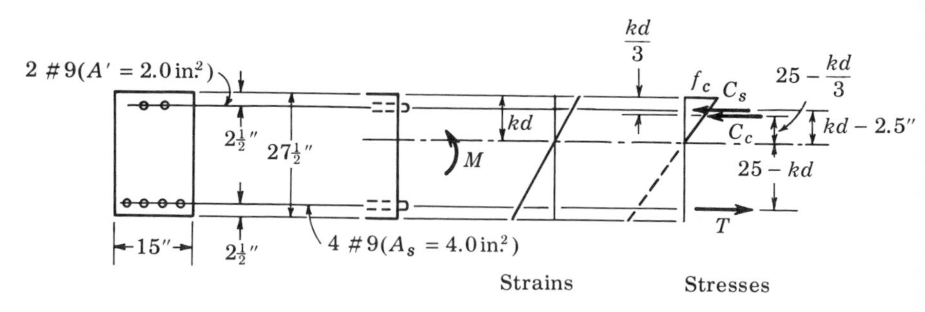

Fig. 6.21

Make suitable allowance for creep of concrete. Find the allowable moment on this beam, using the internal force method.

First, using basic principles, we write the stresses as

$$f'_s = 2n \frac{kd - 2.5}{kd} f_c$$

$$f_s = n \frac{25 - kd}{kd} f_c$$

and the internal forces as

$$C_c = \tfrac{1}{2} f_c \times 15 \times kd$$

$$C_s = A'_s f'_s = 2.0 \times (20 - 1) \frac{kd - 2.5}{kd} f_c$$

$$T = A_s f_s = 4.0 \times 10 \frac{25.0 - kd}{kd} f_c$$

Now writing the equilibrium condition,

$$\Sigma F = 0: \qquad 7.5(kd)^2 + 38.0(kd - 2.5) - 40.0(25.0 - kd) = 0$$

or

$$(kd)^2 + 10.4(kd) - 146.2 = 0$$

from which $kd = 8.0$ in. This value could of course also have been obtained by direct substitution into Eq. (6.47).

The internal forces can now be written as

$$C_c = 60.0 f_c$$
$$C_s = 26.1 f_c$$
$$T = 2.12 f_c$$

and the value of f_c determined by summing moments about the tension steel:

$$\Sigma M_T = 0: \qquad (60.0 f_c)(22.33 \text{ in.}) + (26.1 f_c)(22.50 \text{ in.}) = M$$

or

$$f_c = 0.000519M$$

from which

$$f'_s = 13.75 f_c = 0.00714M$$
$$f_s = 21.2 f_c = 0.0110M$$

Comparing the ratios of the actual to those of the allowable stresses, we see that the compressive concrete stress governs, so that

$$M_{allow} = \frac{f_{c\ allow}}{0.000519} = \frac{1.125}{0.000519} = 2{,}170 \text{ kip-in.} = 181 \text{ kip-ft}$$

Comparing this result to that of Example Problem 6.3, we see that the addition of the top steel (an increase of 50 percent of steel) has raised the allowable moment capacity from 142 to 181 kip-ft (an increase of 27 percent).

Example Problem 6.12 Repeat the preceding example problem, this time using the transformed-section method.

$$\Sigma M_{\text{Areas}} = 0: \qquad (15.0 \times kd)\frac{kd}{2} + 38.0(kd - 2.5) - 40.0(25.0 - kd) = 0$$

from which, as before, $kd = 8.0$ in.

The moment of inertia of the transformed section is

$$I_T = \tfrac{1}{3} \times 15 \times 8.0^3 + 38.0 \times 5.5^2 + 40.0 \times 17.0^2 = 15{,}258 \text{ in.}^4$$

The maximum concrete stress is then

$$f_c = \frac{M \times 8.0}{15{,}258} = 0.000524M$$

and the steel stresses

$$f'_s = 20\,\frac{M \times 5.5}{15{,}258} = 0.00722M$$

and

$$f_s = 10\,\frac{M \times 17.0}{15{,}258} = 0.0111M$$

the same results as before.

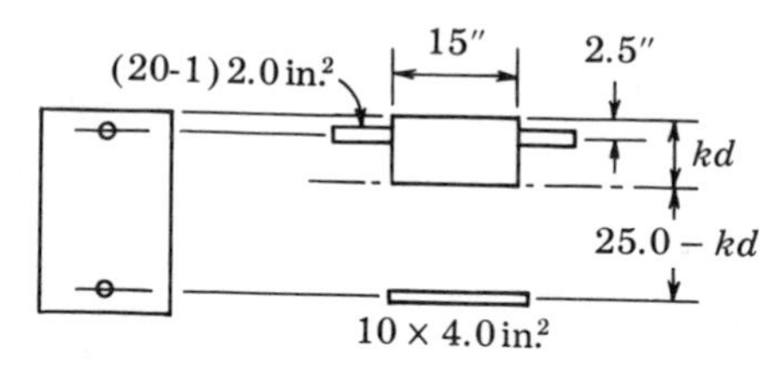

Fig. 6.22

Design of doubly reinforced beams It frequently happens that a beam is called on to resist a certain moment M but that space limitations or other considerations prevent construction of the beam with a value of bd^2 sufficient to satisfy Eq. (6.15), which ensures that the concrete stresses are within their safe values. In such cases, we must provide compressive steel to help carry part of the internal compression force, and it is the purpose of this section to establish guidelines for designing this steel in consonance with our basic principles. In this, we shall follow the provisions of the ACI Code, which calls for a ratio $n' = 2n$ to account for the effect of creep. We shall state our problem as follows:

Given: M_{des}, beam dimensions b and d such that $bd^2 < bd^2_{reqd}$ for simple reinforcement; allowable stresses f_s and f_c, and modular ratio n.

Required: To select steel A_s and A'_s to resist moment M_{des}. To proceed, we use the principle of superposition, and let

$$M_{des} = M_c + M'_s \tag{6.48}$$

where $M_c = Kbd^2$ = moment which the simply reinforced section could resist

M'_s = added couple provided by the resisting moment resulting from compression steel and additional tension steel

Portraying this superposition, we draw Fig. 6.23. We shall first consider the resisting moment M_c which we calculate from Eq. (6.15). Since we aim at simultaneous attainment of $f_{c\,allow}$ and $f_{s\,allow}$, the use of balanced design constants is here perfectly appropriate. We can therefore also calculate the amount of tension steel $A_{s\,1}$ needed to complete the internal couple M_c by equating internal forces, by Eq. (6.16) or Eqs. (6.17).

The required magnitude of the additional couple M'_s is next found by

$$M'_s = M_{des} - M_c \tag{6.49}$$

and moment and force equilibrium is ensured, with reference to Fig. 6.23c, by setting

$$\Sigma M_T = 0: \qquad (A'_s f'_s)(d - d') = M'_s \tag{6.50}$$

and
$$\Sigma M_C = 0: \qquad (A_{s\,2} f_{s\,allow})(d - d') = M'_s \tag{6.51}$$

In these equations, the compressive steel stress can be related to the allowable tensile steel stress by considering the by now familiar stress distribution diagram, Fig. 6.23d.

$$f'_s = 2\,\frac{kd - d'}{d - kd}\,f_{s\,allow} \tag{6.52}$$

and, substituting this into Eqs. (6.50) and (6.51), we can solve for the required compression steel A'_s and the additional tension steel $A_{s\,2}$ necessary to complete the additional couple:

$$A'_s = \frac{M'_s}{\dfrac{2(1 - d'/d)(k - d'/d)}{1 - k}\,f_{s\,allow}\,d} = \frac{M'_s}{cd} \tag{6.53}$$

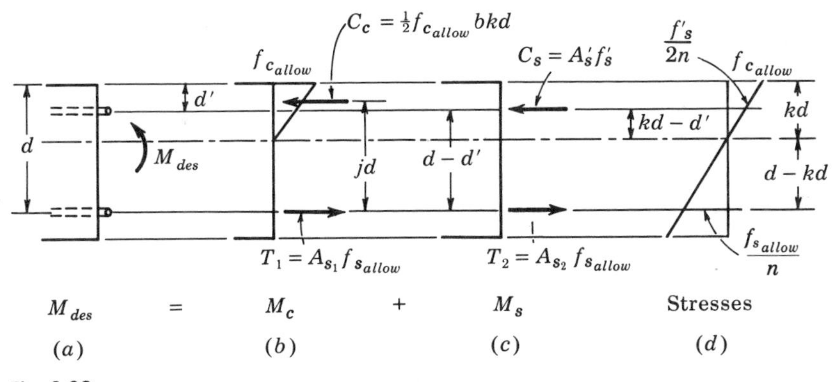

Fig. 6.23

where $$c = \frac{2(1 - d'/d)(k - d'/d)\, f_{s\ allow}}{1 - k} \tag{6.54}$$

and $$A_{s\ 2} = \frac{M_s'}{f_{s\ allow}(d - d')} \tag{6.55}$$

Note that Eq. (6.55) is analogous to the one for $A_{s\ 1}$, except that now the lever arm of the additional couple is $d - d'$ instead of jd. Equation (6.53) is unhandy to memorize or use. Again reference to basic reasoning is indicated here. We note that values for the coefficient c are tabulated in table 7*b* of RCDHB, modified by the addition of a numerical factor to compensate for units, and also by a ratio $(2n - 1)/n$ instead of the factor 2 in Eq. (6.54) to compensate for the hole in the compressive concrete. This table may be considered typical of time-saving design aids which may be made up. Note that as far as possible, the variables are cast in nondimensional fractions to preserve generality of the information.

We note from Eq. (6.52) that when the fraction

$$2\frac{kd - d'}{d - kd} > 1 \qquad \text{or} \qquad \frac{kd - d'}{d - kd} > \frac{1}{2} \tag{6.56}$$

that is, when the distance from the neutral axis to the compression steel is over half that to the tension steel, then the compressive steel stress f_s' exceeds $f_{s\ allow}$. This possibility is expressly prohibited by ACI Code sec. 1102*c*. In such a case f_s' in Eq. (6.50) is taken as $f_{s\ allow}$, and

$$A_s' = \frac{M_s'}{f_{s\ allow}(d - d')} \tag{6.57}$$

Equation (6.57) supersedes Eq. (6.53) if inequality (6.56) prevails.

Example Problem 6.13 Design a beam of overall dimensions 18 by 30 in. to resist a moment of 500 kip-ft; $f_c' = 4{,}000$ psi, $f_s = 20$ ksi; follow ACI Code.

First calculating the value of moment M_c which can be taken without compression steel, we set, allowing 2.5 in. concrete cover,

$$M_c = Kbd^2 = 0.324 \times 18 \times 27.5^2 \times \tfrac{1}{12} = 367 \text{ kip-ft}$$

then

$$A_{s\ 1} = \frac{M_c}{f_s jd} = \frac{367 \times 12}{20 \times 0.861 \times 27.5} = 9.32 \text{ in.}^2$$

Calculating next the added couple, we find

$$M_s' = 500 - 367 = 133 \text{ kip-ft}$$

To resist this added couple, we require tension steel of area

$$A_{s\ 2} = \frac{M_s'}{f_{s\ allow}(d - d')} = \frac{133 \times 12}{20 \times 25.0} = 3.19 \text{ in.}^2$$

and compression steel of area

$$A'_s = \frac{M'_s}{f'_s(d - d')}$$

where

$$f'_s = 2 \times \frac{11.5 - 2.5}{27.5 - 11.5} \times 20 > 20 \text{ ksi}$$

according to ACI sec. 1102*c*; we therefore use $f'_s = 20$ ksi, and

$$A'_s = \frac{133 \times 12}{20 \times 25.0} = 3.19 \text{ in.}^2$$

The total tension steel required is therefore $A_{s\,total} = 12.51$ in.², the compression steel 3.19 in.²

These requirements are satisfied by eight No. 11 bars on the tension side and two No. 11 bars on the compression side. Since only five No. 11 bars will fit in one row in a beam of 18 in. width, according to ACI sec. 804, the tension steel will have to be arranged in two layers. Since this will decrease the available internal lever arm to the centroid of the steel area, one additional No. 11 bar will be added on the compression side to compensate for this effect. The final section appears as shown. As is usual, stresses in the final section may be checked by an analysis as previously described, in which the tension force may be considered concentrated at the centroid of the tension bars.

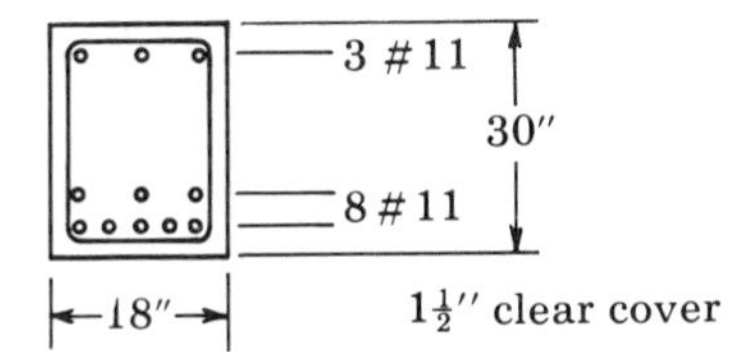

Fig. 6.24

Often it is anticipated that a beam will be doubly reinforced, though no definite beam size is prescribed. This situation can arise from the desire to reduce dimensions and dead load of members. Compressive steel is sometimes called for to minimize deflections resulting from creep, or because of the possibility of reversal of moment, and at the interior points of support of continuous beams (where the moments are often critical) a certain amount of compression steel is bound to be present, because sec. 918*f* of the ACI Code provides for the extension of at least one-third of the maximum area of bottom steel across supports.

We can state this type of design problem in the following way:

Given: Moment to be resisted, M_{des}; allowable stresses, modular ratio.

Required: Doubly reinforced beam section to resist moment M_{des}. Amounts of top and bottom steel to be in reasonable proportion to each other, say, approximately equal in area.

One way of approaching this problem is to assume that the portion M_c represents a certain fraction of the total moment M_{des}, say, $\frac{2}{3}$. Then the added couple M'_s will have to amount to $\frac{1}{3}\, M_{des}$. With this assumption we can proceed along earlier lines, calculating in turn bd^2 from Eq. (6.15), $A_{s\,1}$ from Eq. (6.16), A'_s from Eq. (6.53), and $A_{s\,2}$ from Eq. (6.55). With some experience the proper fractions can be chosen so that the ratio A_s/A'_s falls close to the desired value.

Example Problem 6.14 A continuous beam of uniform span lengths of 20 ft, under uniform total load 1 kip/ft, is to be designed according to the ACI Code. A constant cross section of small overall dimensions is required. Analysis to be according to ACI moment coefficients (ACI sec. 904). $f'_c = 3{,}000$ psi, $f_s = 20$ ksi.

The ACI moment coefficients give values for design moments at critical sections of continuous beams as shown in Fig. 6.25.

Since, according to ACI sec. 918*f*, some bottom steel will be present over all supports, in order to obtain a slender beam we shall design the beam section to resist about 30 kip-ft without double reinforcement, and make sure that sufficient compression steel is provided wherever the moment exceeds this amount.

The section required will be of dimensions such that, with $k = 0.379$, $j = 0.874$,

$$bd^2_{reqd} = \frac{M_c}{K} = \frac{30 \times 12}{0.224} = 1{,}605 \text{ in.}^3$$

Try $\underline{b = 10 \text{ in.}}$, $d = 12.7$ in., $\underline{h = 15 \text{ in.}}$

At the typical interior span section C, single reinforcing is sufficient, and will be

$$A_{s\,reqd} = \frac{M}{f_s jd} = \frac{25.0 \times 12}{20 \times 0.874 \times 12.7} = \underline{1.35 \text{ in.}^2} \text{ (bottom steel)}$$

At the typical interior support section D, double reinforcing is required:

$$A_{s\,1} = \frac{30 \times 12}{20 \times 11.1} = 1.62 \text{ in.}^2$$

$$M'_s = 36.4 - 30.0 = 6.4 \text{ kip-ft}$$

$$f'_s = 2 \times \frac{4.8 - 2.3}{12.7 - 4.8} \times 20.0 = 12.6 \text{ ksi}$$

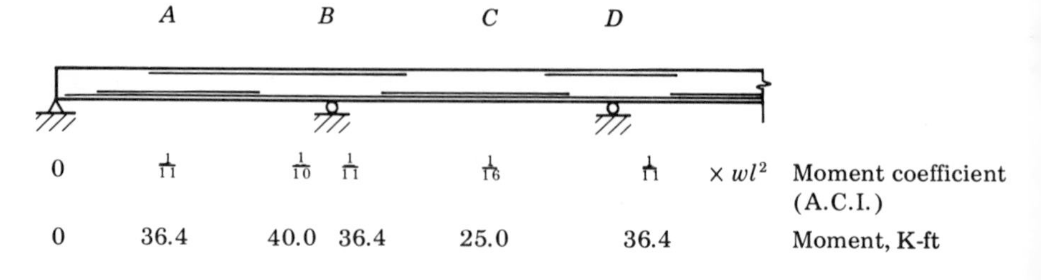

Fig. 6.25

so that

$$A_{s\,2} = \frac{6.4 \times 12}{20 \times 10.4} = 0.37 \text{ in.}^2 \qquad A_{s\,total} = \underline{1.99 \text{ in.}^2} \text{ (top steel)}$$

and

$$A'_s = \frac{6.4 \times 12}{12.6 \times 10.4} = \underline{0.59 \text{ in.}^2} \text{ (bottom steel)}$$

At the exterior span section A, with moment of identical value to that at D, the steel areas required are the same, but top and bottom steel is switched.

Lastly, turning to sec. B, we have $M_{des} = -40$ kip-ft.

$$A_{s\,1} = 1.62 \text{ in.}^2$$
$$M'_s = 40 - 30 = 10 \text{ kip-ft}$$

so that

$$A_{s\,2} = \frac{10 \times 12}{20 \times 10.4} = 0.58 \text{ in.}^2 \qquad A_{s\,total} = \underline{2.20 \text{ in.}^2} \text{ (top steel)}$$

$$A'_s = \frac{10 \times 12}{12.6 \times 10.4} = \underline{0.92 \text{ in.}^2} \text{ (bottom steel)}$$

The next step is to devise an efficient arrangement for the reinforcing steel so that all requirements are fulfilled. Figure 6.26 shows one possible reinforcing scheme.

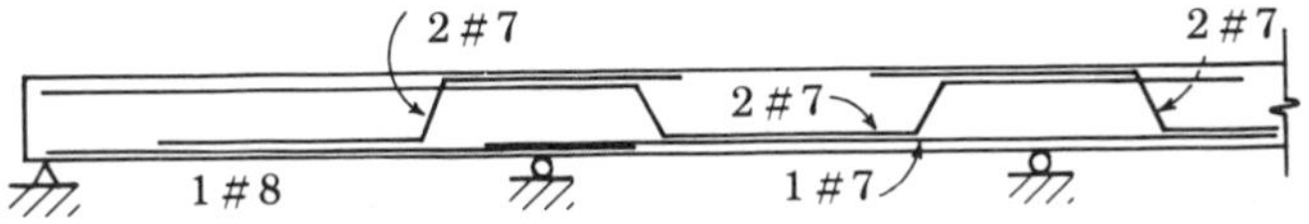

	A	B	C	D	Section,
—	0.59	2.20	—	1.99	A_s top, required, in.^{2}
—	1.99	0.92	1.35	0.59	A_s bot, required, in.^{2}
1.20	1.20	2.40	—	2.40	A_s top, furnished, in.^{2}
0.79	1.99	1.39	1.80	0.60	A_s bot, furnished, in.^{2}

Fig. 6.26

We find that in most doubly reinforced beams the compression steel is not stressed to its allowable value because of its proximity to the neutral axis, and for this reason doubly reinforced beams require a larger amount of steel to resist a certain moment than singly reinforced beams. Doubly reinforced beams are not conducive to saving steel. Of course, when owing to code provisions or constructive details steel is present on the compression side anyway, a good designer will certainly avail himself of its presence.

Because of its compressive stresses, compression steel can be subject to buckling, and to prevent this (with accompanying spalling of the compressive concrete cover) the ACI Code, sec. 806*c*, provides for periodically spaced stirrups or ties which serve to restrain the longitudinal steel.

6.10 T Beams

An extremely common type of reinforced concrete construction is the slab-and-girder system, as shown in Fig. 6.27. Depending on the proportions of the slabs, they may be reinforced to carry the load to the floor beams either by plate or by beam action. The beams, in turn, have to transmit the slab reactions to their supports.

Turning our attention to the floor beam, a section of which is shown in Fig. 6.27*b*, we can assume that a certain width of the slab will act as a flange on the beam, and this so-called "effective width" is given by ACI sec. 906*b* in terms very similar to those already considered for composite construction:

$$\text{Maximum effective width} = \tfrac{1}{4}\text{ span length or beam spacing}$$
$$\text{and}\quad \text{Maximum effective overhang} = 8 \times \text{slab thickness}$$

With this information, we can apply our previous methods (but here, preferably the transformed-section method) to the stress analysis of T sections. Turning our attention to Fig. 6.27*c* and *d*, we see that only under positive moment, in which case the slab is on the compression side, does the T section enter the calculation. Under negative moment the slab is considered cracked and is not part of the effective portion of the section.

An example problem will illustrate the stress analysis of a T section.

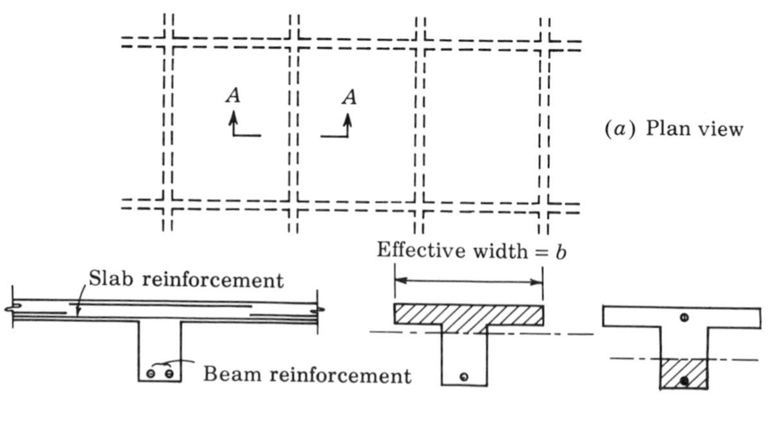

Fig. 6.27

Example Problem 6.15 The section of the floor beam shown in Fig. 6.28 is subjected to a positive moment of 250 kip-ft. For $n = 8$, investigate stresses in steel and concrete. Length of floor beam is 25 ft, beam spacing 12.5 ft.

To draw the effective section, we must first determine whether the neutral axis lies in the slab or in the stem. To do this, we make a trial by assuming that the

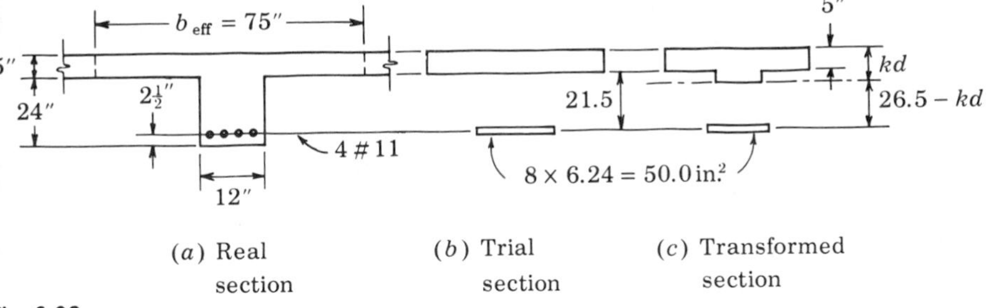

Fig. 6.28

neutral axis lies at the junction of slab and stem. If the static moment of the transformed section is positive when computed on this basis, the conclusion is that the neutral axis has been assumed too low and should therefore lie in the slab, and vice versa.

In this case,

$$\Sigma M_{\text{Areas}} = (75 \times 5) \times 2.5 - 50.0 \times 21.5 < 0$$

N.A. lies in stem, and the transformed section is as shown in Fig. 6.28*c*.

Finding next the centroidal axis of the transformed section:

$$\Sigma M_{\text{Areas}} = 0: \qquad (75 \times 5)(kd - 2.5) + 12 \frac{(kd - 5.0)^2}{2} - 50.0(26.5 - kd) = 0$$

from which $kd = 5.4$ in.

Calculating next the transformed moment of inertia, we set

$$I_T = \tfrac{1}{12} \times 75 \times 5^3 + (75 \times 5)2.9^2 + 50.0 \times 21.1^2 = 26{,}160 \text{ in.}^4$$

so that

$$f_c = \frac{Mkd}{I_T} = \frac{250 \times 12 \times 5.4}{26{,}160} = \underline{0.62 \text{ ksi}}$$

and

$$f_s = n \frac{M(d - kd)}{I_T} = \frac{8 \times 250 \times 12 \times 21.1}{26{,}160} = \underline{19.4 \text{ ksi}}$$

Note that, as might be expected with the large amount of compressive concrete available, the beam is greatly under-reinforced.

The design of T beams as used in slab-type structures is influenced by a number of factors: the slab thickness is usually already established in the course of the slab design. Usually, the floor beams are analyzed as continuous beams, with the largest design moments at the points of negative moment, where only a scant amount of compressive concrete is available. The portion of the bottom steel which must be extended over the supports is then considered as compression steel. Diagonal tension is also a critical factor in T beams with narrow stems.

A more thorough examination of problems of design of such beams is left to books on reinforced concrete design.

6.11 Deformations of Reinforced Concrete Beams

Deformations of structural members have to be calculated for two main reasons: first, to be able to predict excessive flexibility of a structure, and, second, in order to analyze statically indeterminate structures.

In concrete beams, as in others, we start from the strain distribution and obtain the curvature ϕ as the angle of the strain curve with the axis. For instance, this can be obtained from the transformed section by the usual elastic expression

$$\phi = \frac{M}{E_T I_T} = \frac{d^2y}{dx^2}$$

where E_T is the elastic modulus of the equivalent material, in this case the concrete. From here, the usual methods will give us the required deflections.

We note that in the case of even a prismatic concrete beam, we ought to compute a new value of I_T each time the steel changes. Furthermore, in the case of a continuous beam with unequal amounts of top and bottom steel, the value of I_T will be different for regions of positive and negative moments. But the point of inflection can only be determined after the structure has been analyzed, and so the problem becomes highly nonlinear. To avoid such difficulties we take an average value of the moments of inertia for the regions of positive and negative moments (ACI sec. 909*c*).

On the other hand, sec. 905*c* 2 suggests that I may be calculated on the basis of the gross concrete section for purposes of analysis.

Actually, the results are rather insensitive to the nature of the assumption, especially since the continuing creep or plastic flow of the structure during the time of service is a major factor, as shown in sec. 909*d* of the ACI Code. The factors given in this section for the computation of the additional time-dependent deformations indicate that the dead load deflections may eventually reach three times the instantaneous value for singly reinforced beams, and 1.8 times the instantaneous value for beams with heavy compression steel. Obviously, sag can be inhibited by use of double reinforcement.

The creep effects of live loading (which acts only intermittently) are much harder to determine than those due to sustained load. Sec. 909*e* of the Code gives criteria for the short-time allowable deflections resulting from live load only. Section 909*f* is a reminder that the consequence of continuing deformations can be the crushing of partitions and the cracking of glass, and sets limits on the allowable magnitude of total deflections.

We saw in an earlier section that the deflection of elastic beams under given allowable stresses is proportional to the ratio of length to depth, and if deflection calculations must be avoided, a maximum value for this ratio can be used as a criterion of adequate stiffness. Table 909*b* gives maximum slenderness values for beams and slabs of various support conditions. These should be adhered to unless more extensive deflection calculations indicate a sufficiently stiff design.

General Readings

General

ACI Committee 318: Commentary on Building Code Requirements for Reinforced Concrete (ACI 318-63), *ACI Publication SP-10,* 1965. This pamphlet documents and references the provisions of the 1963 ACI Code.

Winter, George, L. C. Urquhart, C. E. O'Rourke, and A. H. Nilson: "Design of Concrete Structures," 7th ed., McGraw-Hill, 1964. Chapters 1 and 2 of this book are a very good summary of the physical and mechanical bases of reinforced concrete design. Chapter 3 deals with elastic and ultimate strength of beams.

The book by Ferguson listed in Chap. 3, as well as a number of others, also covers the subject in detail.

Shear and diagonal tension

ACI-ASCE Committee 326: Shear and Diagonal Tension, *J. ACI,* vol. 59, January–February–March, 1962. This lengthy paper reports the latest thinking and design recommendations on a topic which is still too little understood.

Engineering drawings for concrete structures

ACI Committee 315: "Manual of Standard Practice for Detailing Reinforced Concrete Structures," ACI, 1965. Guidelines for design drawings.

Problems

6.1 A simply supported beam of span length 20 ft carries a total uniform load of 1 kip/ft. Its cross-sectional dimensions are 10 in. wide by 19 in. deep, with reinforcing consisting of three No. 9 bars with concrete cover of 3 in. to the center of the steel. $f'_c = 3.0$ ksi, steel has yield strength of 40 ksi, and $n = 9$. Find maximum stresses in steel and concrete. Determine whether the beam is safe and economical according to the ACI Code. Use both transformed-section and internal-force methods. Draw conclusions regarding appropriateness of each method.

6.2 Calculate the allowable moment on the beam section according to the ACI Code. Draw stress distribution and calculate curvature. State whether beam is over- or under- reinforced. $f'_c = 2.5$ ksi, $f_{YP} = 40$ ksi.

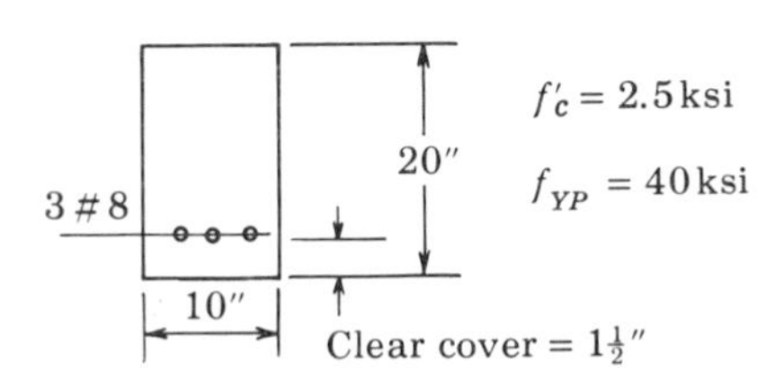

6.3 The beam section shown is subjected to a moment M. $f_c' = 4.0$ ksi, $f_{YP} = 50$ ksi. Calculate stresses in steel and concrete:

(*a*) considering each row of reinforcing in its proper location
(*b*) considering the total reinforcing concentrated at its centroid

Compare results and draw conclusions. Comment on your choice of method of calculation.

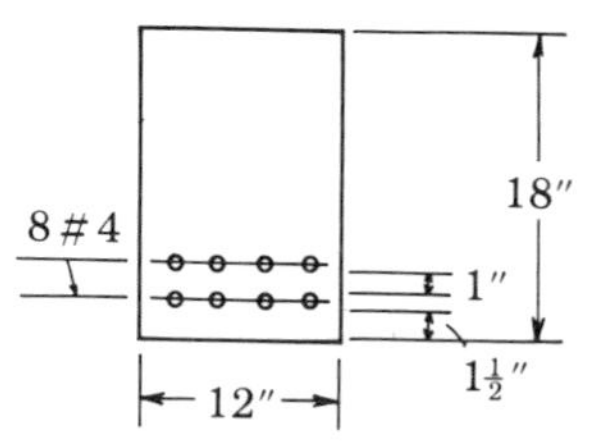

6.4 Calculate the value P of the concentrated loads which may be applied as shown without violating the provisions of the ACI Code. Do not neglect beam dead weight. $f_c' = 4.0$ ksi; steel is structural-grade billet steel; unit weight of concrete = 150 lb/ft³.

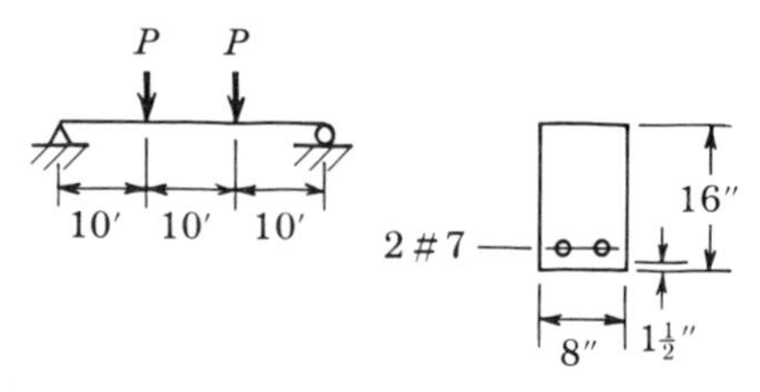

6.5 A typical section through a one-way slab is shown. The slab is a cantilever projecting 6 ft. What is the allowable uniform live load which the slab may carry according to the ACI Code? $f_c' = 3{,}000$ psi, $f_{YP} = 40$ ksi; lightweight concrete, unit weight = 100 lb/ft³.

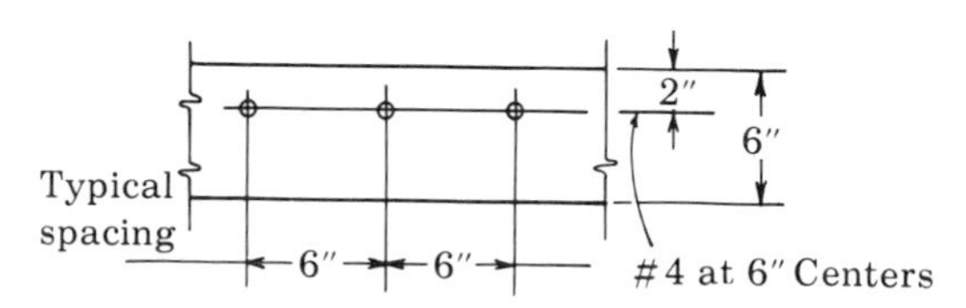

6.6 A common method of constructing floor slabs is to place corrugated sheets over support which act as forms for the concrete and, after the concrete has set, constitute its reinforcing. Using the properties of the cross section of the corrugated sheet, calculate the maximum allowable moment which may be applied to a foot width of the slab according to the ACI Code. $f_c' = 3{,}000$ psi. Properties of corrugated sheet: Corrugation depth $= 1\frac{3}{8}$ in., area = 0.644 in.²/ft width, $f_{s\ allow} = 28{,}000$ psi, $n = 12$.

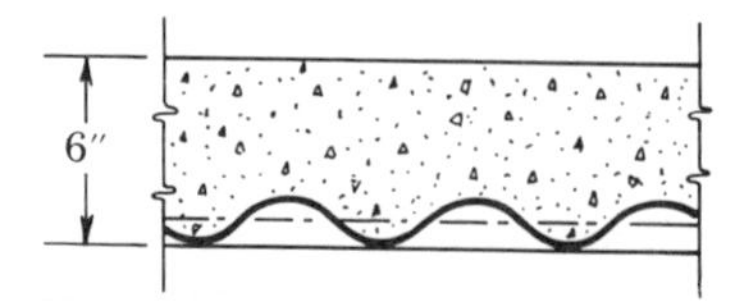

6.7 A cantilever retaining wall as shown is to be checked for stresses. The load on the stem of such a wall can be computed according to Rankine's theory as being similar to hydrostatic pressure, that is, increasing linearly with the depth as shown. The maximum lateral pressure is as shown. $f'_c = 2.5$ ksi, $f_{s\ allow} = 20$ ksi; gravel concrete (150 lb/ft³).

Check the stresses in the bottom section of the stem and comment on its adequacy according to the ACI Code.

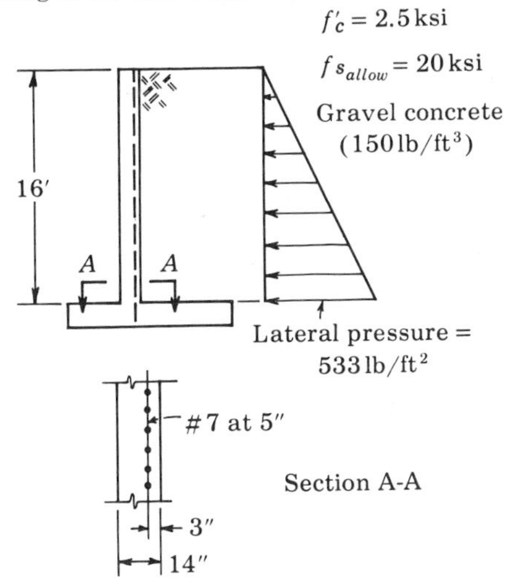

6.8 Design a rectangular singly reinforced beam to resist a moment of 112.5 kip-ft. $f_s = 22$ ksi, $f'_c = 2.5$ ksi; 1963 ACI Code.

(*a*) Use basic elastic theory of cracked sections.
(*b*) Verify by using all available design aids.
(*c*) Select actual steel and dimensions and draw sketch of cross section.
(*d*) Review stresses using actual steel area and dimensions.

6.9 A simple beam of span length 20 ft carries a uniform load of 1.5 kips/ft (not including beam dead weight). Beam is to be made of gravel concrete (unit weight = 150 lb/ft³). $f_s = 20$ ksi, $f'_c = 3.0$ ksi; 1963 ACI Code.

(*a*) Assume beam dead weight and design the critical section. Revise the design as required if the beam dead weight was assumed incorrectly. Select steel and draw the cross section.
(*b*) Check stresses in the section.

6.10 Redesign the beam of Prob. 6.8 as a 20 percent under-reinforced beam. Select cross-sectional dimensions and steel. Check stresses and compare material economies as compared to the balanced design.

6.11 A reinforced concrete slab is to be designed, simply supported over a 16-ft span. It supports a uniformly distributed load of 100 lb/ft² and is made of lightweight concrete weighing 100 lb/ft³, of $f'_c = 4.0$ ksi. Steel is to be of $f_s = 20$ ksi. Follow 1963 ACI Code. Design critical section of the slab, select steel, and draw cross section. Check stresses in steel and concrete. *Hint:* Take typical foot-wide strip. Do not neglect slab weight.

6.12 A cantilever beam 10 ft in length supports a uniform load of 1.5 kips/ft. f'_c = 3,750 psi, f_s = 18 ksi; 1963 ACI Code. Gravel concrete (unit weight = 150 lb/ft³). Design the critical section of the beam; include effect of beam dead weight. Sketch the cross section and check actual stresses.

6.13 Design the sections at midspan and at the end of the fixed-ended beam shown. Make $b = \frac{2}{3}d$. Moment coefficients for fixed-ended beam under uniform load: at ends, $\frac{1}{12}$; at center, $\frac{1}{24}$. Draw cross sections and review stresses in beam as designed. *Hint:* Make beam of constant gross concrete section. Design this section for point of largest moment. At the other section, of lesser moment, retain the concrete dimensions and select steel accordingly, realizing that this section will be under-reinforced. f'_c = 5,000 psi, f_s = 20,000 psi.

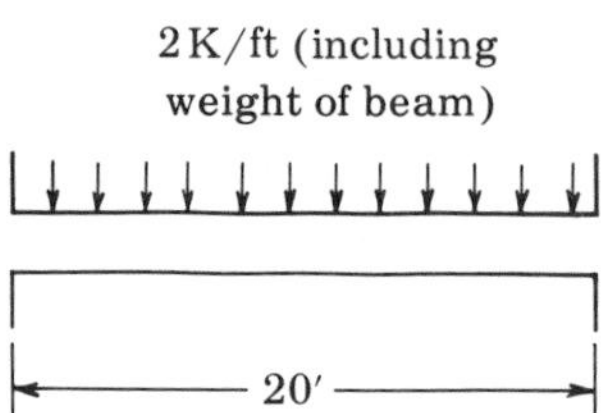

6.14 The stem of the retaining wall is to be designed according to the straight-line theory to resist the lateral soil pressure indicated. The lateral pressure (in lb/ft²) is given by 0.30 wh, where w is the weight of the soil (in lb/ft³) and h is the depth from the surface (in ft). f'_c = 2.0 ksi, $f_{s\ allow}$ = 20 ksi. Weight of soil is 100 lb/ft³. Weight of concrete is 150 lb/ft³.

(*a*) Find required effective depth.
(*b*) Find steel area required per linear foot.
(*c*) Sketch critical cross section as designed and check stresses.

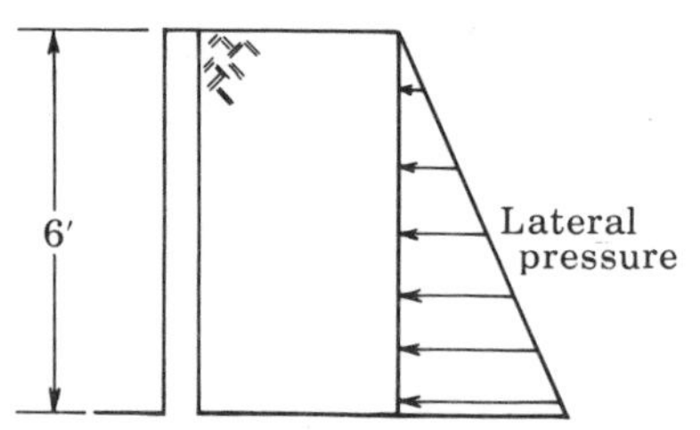

6.15 A cantilever slab as shown is to be designed. Design the critical section, and devise bar cutoff scheme as shown. Indicate bar size and spacing and cutoff dimensions. f_s = 20 ksi, f'_c = 3.0 ksi. Gravel concrete. Uniform live load = 200 lb/ft². Follow ACI Code.

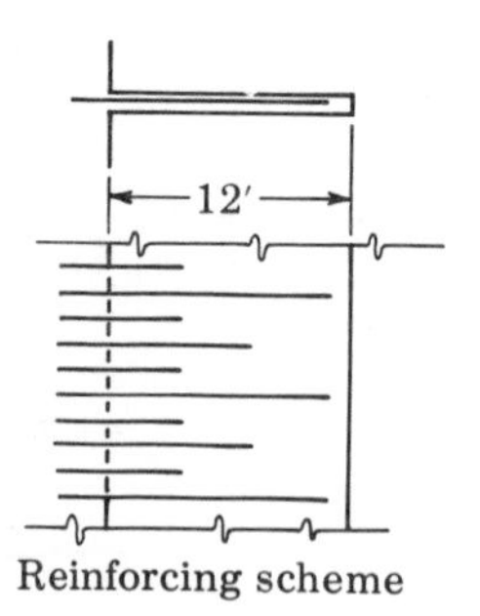

Reinforcing scheme

6.16 A uniformly loaded simple beam is to support a live load of 1 kip/ft over a span of 80 ft. Design the beam and longitudinal reinforcing. Show cutoff points on a design sketch. $f'_c = 4.0$ ksi, $f_s = 20$ ksi. Beam of lightweight concrete. Follow ACI Code.

6.17 Devise a reinforcing scheme similar to the one shown in the sketch of Prob. 6.15 for the retaining wall previously designed in Prob. 6.14. Show bar size, spacing, and cutoff dimensions on a design sketch.

6.18 $f_s = 18$ ksi, $f'_c = 2.0$ ksi. Make $b = d/2$. Gravel concrete. Follow ACI Code.

(*a*) Design cross section of beam (include dead weight). Find A_s and Σ_0 required.
(*b*) Calculate exact tensile force at fixed end, and required embedment lengths for all bars (different bars will have different embedment lengths).
(*c*) Draw scale detail of fixed end of beam, showing reinforcement (1) with straight embedment, (2) with standard hooks.

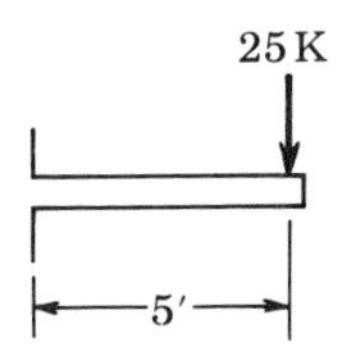

6.19 The fixed-end moment on the end of a fixed-ended concrete beam is 200 kip-ft. Design the end section of the beam and provide end anchorage according to ACI. Make design sketch. $f_s = 22$ ksi, $f'_c = 4.0$ ksi.

6.20 Check the bond stresses at critical points of the beam designed in Prob. 6.16. Revise reinforcing if required, following the provisions of the ACI Code.

6.21 Check the bond stresses at critical points of the cantilever slab designed in Prob. 6.15. Calculate the embedment length required for end anchorage, and the perimeter for bond. Revise steel if necessary, following the ACI Code.

6.22 Draw a shear-stress diagram for the beam of Prob. 6.16. Compare maximum diagonal tension stress to allowable ACI values and provide vertical stirrups if required.

6.23 A simply supported beam of span L, uniformly loaded with intensity w, has width b and effective depth d. It is designed as a balanced beam. Arrive at a formula relating the maximum diagonal tension stress v to the depth span d/L, and draw engineering conclusions regarding the influence of this ratio on the need for web reinforcing. In particular, comment on how this may affect the design for diagonal tension of concrete slabs.

6.24 $f'_c = 2.5$ ksi, $f_s = f_v = 20$ ksi, $b = 13$ in., $d = 21$ in. Design web reinforcing, following the ACI Code. Use No. 3 stirrups. Provide design sketch showing arrangement and spacing of stirrups.

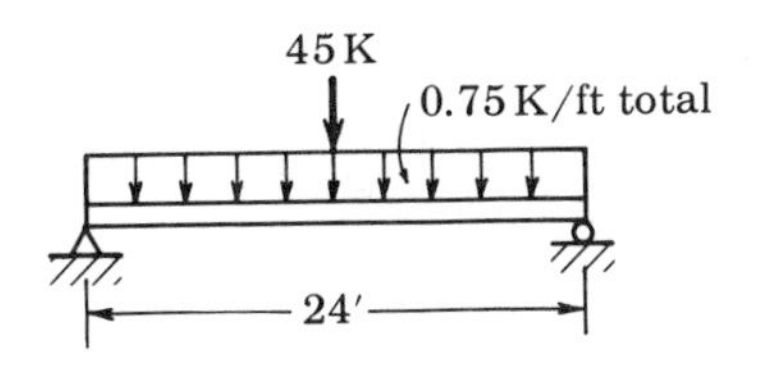

6.25 Repeat Prob. 6.24, this time using No. 4 stirrups. Compare amount of stirrup steel to that used in Prob. 6.24, and draw conclusions regarding efficient size of stirrups.

6.26 $f'_c = 2.0$ ksi, $f_s = 18$ ksi, $b = 12$ in., $d = 24$ in. Design web reinforcing for the beam shown, following the ACI Code. Make design drawing showing arrangement of vertical stirrups.

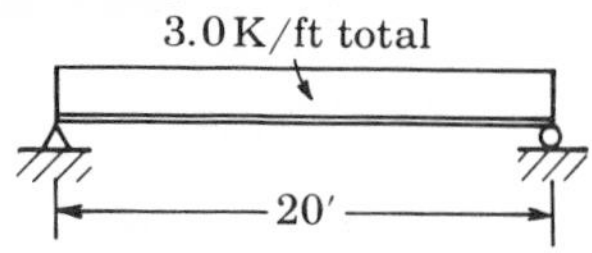

6.27 Redesign the web reinforcement of Prob. 6.26, this time using stirrups inclined at 45 deg. Make design drawing, compare this design to that using vertical stirrups, and draw engineering conclusions.

6.28 The beam of cross section shown ($n = 8$) is subjected to a moment of value 60 kip-ft. Draw the strain and stress distributions, both neglecting and including creep, by

(*a*) the internal-force method
(*b*) the transformed-section method

Compare the values for stress obtained under the two different assumptions.

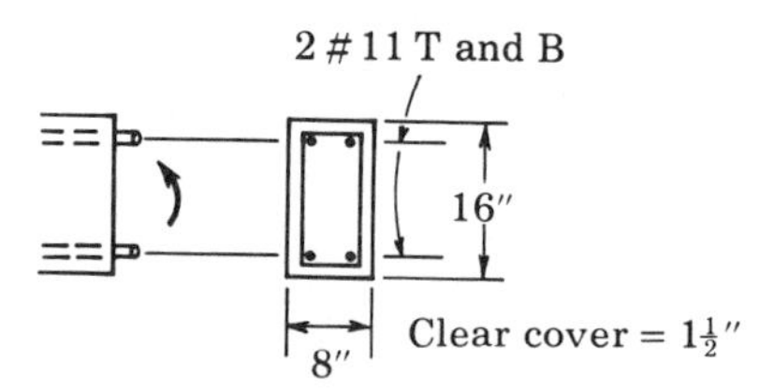

6.29 For the beam section of Prob. 6.28, $f_s = 20$ ksi, $f'_c = 4{,}000$ psi. Find the maximum allowable moment according to the ACI Code.

6.30 $f_s = 22$ ksi, $f'_c = 4{,}000$ psi. ACI Code.

(*a*) Find the allowable moment on the beam section if the top steel is not present.
(*b*) Find the allowable moment including the compression steel.
(*c*) By calculating and comparing the ratio $M_{allow}/A_{s\ total}$ for both parts (*a*) and (*b*), draw conclusions regarding the efficiency of doubly reinforced beams.

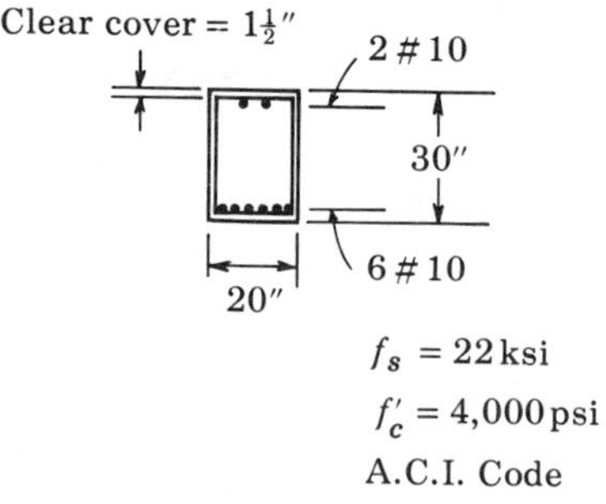

6.31 Design a reinforced concrete beam of width 12 in., overall depth 18 in., to resist a moment of 80 kip-ft. $f'_c = 3.0$ ksi, $f_s = 20$ ksi. Follow ACI Code and sketch cross section.

6.32 A fixed-ended reinforced concrete beam of span length 40 ft is to be designed. It is to resist a uniform total load of 2 kips/ft. For architectural reasons the cross-sectional dimensions are to be held to a minimum, and the beam is therefore to be designed to take structural advantage of the nominal bottom steel which, according to ACI sec. 918*f*, must be extended into the ends. $f'_c = 4.0$ ksi, $f_s = 24$ ksi; ACI Code to be followed. Present complete design sketch, including cutoff points and compression ties.

6.33 $f'_c = 2.5$ ksi, $f_s = 20$ ksi; ACI Code.
The typical floor beam of section shown spans 40 ft and is spaced at 12-ft centers. Calculate the allowable moment on the section, and compute stresses corresponding to this moment.

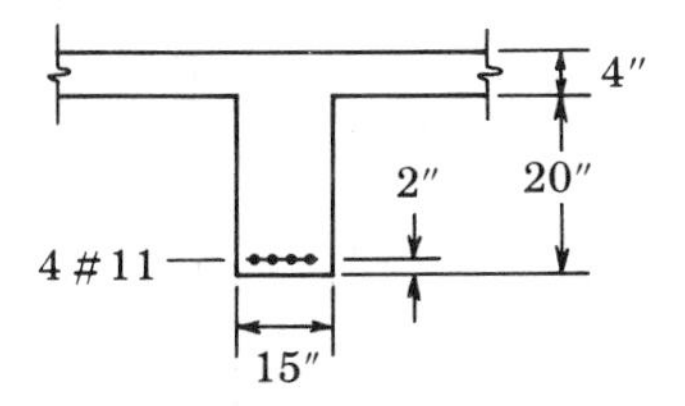

6.34 $f_{c\ allow} = 1.5$ ksi, $f_s = 18$ ksi, $n = 9$.
The beam of section shown is provided with recessed seats for precast slabs. Determine the amount of bottom steel required to result in a balanced design.

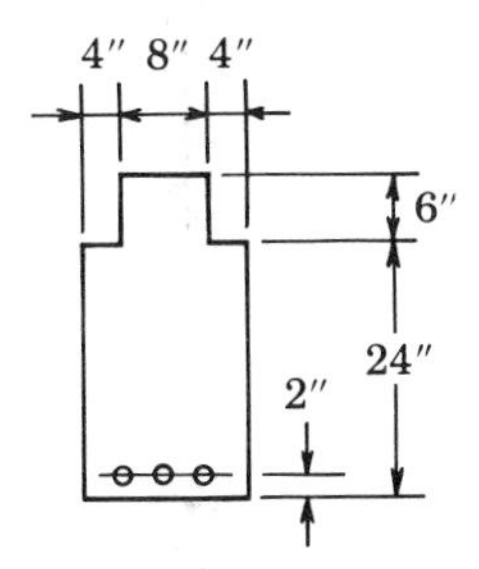

6.35 $f_{c\ allow} = 1.5$ ksi, $f_s = 18$ ksi, $n = 9$.
Determine the allowable moment on the beam shown

(*a*) with top reinforcement
(*b*) without top reinforcement

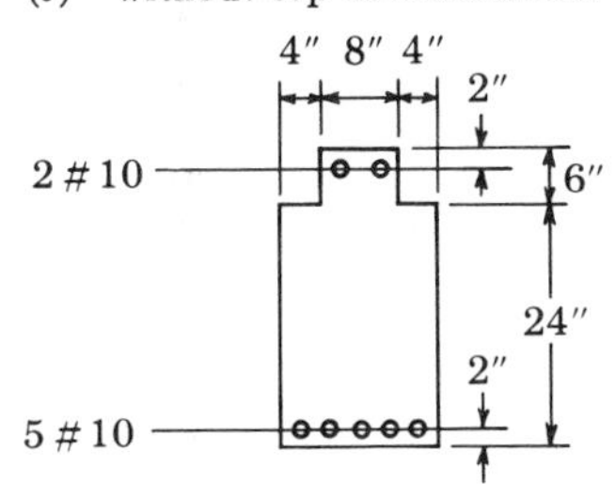

6.36 Calculate allowable negative moment which this section can withstand, following the ACI Code. $f_s = 20$ ksi, $f_{c\ allow} = 2.0$ ksi, $n = 10$.

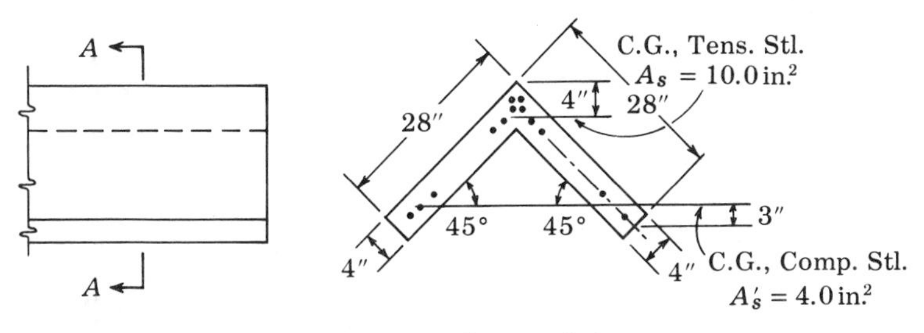

Section A-A

6.37 $f_{c\ allow} = 500$ psi, $f_{s\ allow} = 20$ ksi, $n = 8$. A cylindrical shell has a cross section shown. Assuming this section to act similarly to a beam (an assumption made in the so-called "beam method" of shell analysis), find the allowable moment on the shell section.

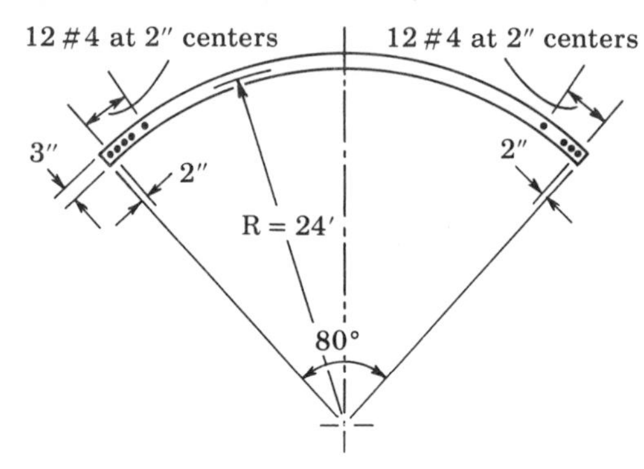

6.38 The view shows a section cut through a doubly reinforced beam subjected to a moment M. Assume that the location of the neutral axis has been determined by conventional methods. An internal-force–type of expression for the extreme fiber compression stress, f_c, is to be derived.

(*a*) Find the magnitude and location of the compressive resultant on the concrete in terms of k, d, b, f_c.
(*b*) Find the magnitude and location of the compression force on the steel in terms of the above plus A_s', d', and n.
(*c*) Find the magnitude and location of the total resultant compression force.
(*d*) Find the value of the internal lever arm.
(*e*) Find the value of f_c in terms of M, k, and the section properties.

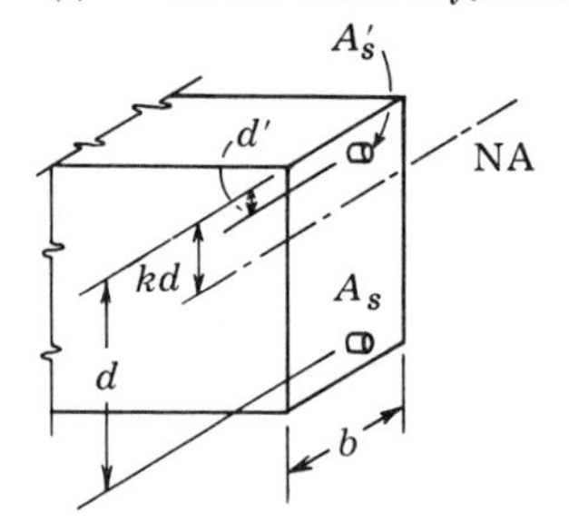

6.39 $f_c' = 2{,}500$ psi, $f_s = 20{,}000$ psi; follow ACI Code.

(*a*) Determine the maximum positive and negative moments which can be resisted safely by the beam section shown.
(*b*) Determine the maximum safe value of shear which may be carried.

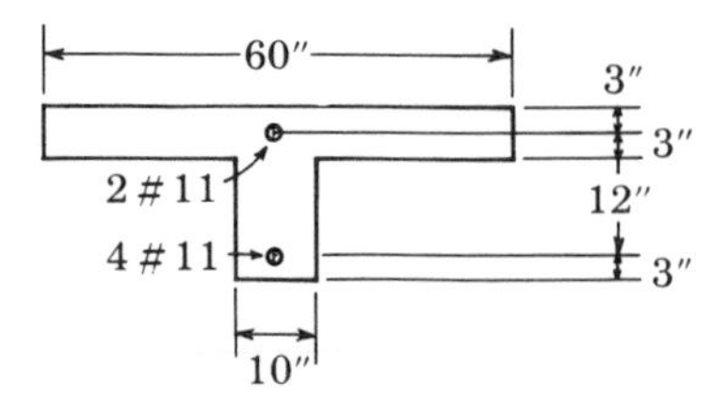

6.40 $f'_c = 3{,}750$ psi, $f_s = 22{,}000$ psi; follow ACI Code.

(*a*) What is the moment capacity of the reinforced concrete I beam shown?
(*b*) What is the largest value of shear which can be safely resisted by this section? Assume that it is provided with properly designed web reinforcement.

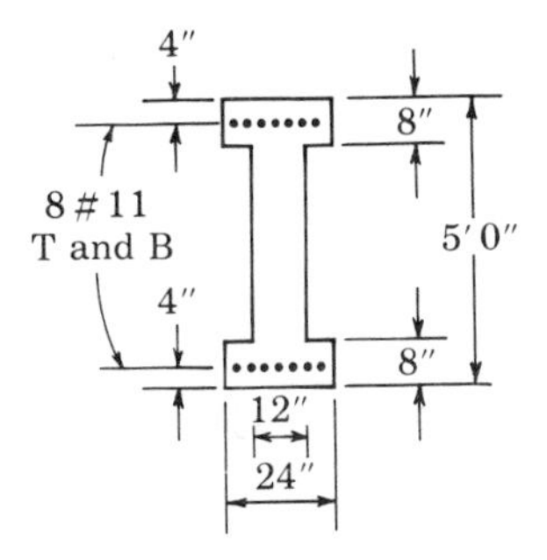

6.41 Design the cantilever beam to support the load shown, following the straight line theory and the ACI Code. Make sure the beam is safe in bending, diagonal tension, bond, and anchorage. Cut bars for maximum steel economy. Neglect the dead weight of the beam itself. Width of beam to be 12 in. $f'_c = 3.75$ ksi, $f_s = 20$ ksi. Available bar sizes are: No. 8 bars: $A_s = 0.79$ in.^{2}, $\Sigma_0 = 3.1$ in.; No. 2 bars: $A_s = 0.05$ in.^{2}, $\Sigma_0 = 0.8$ in.

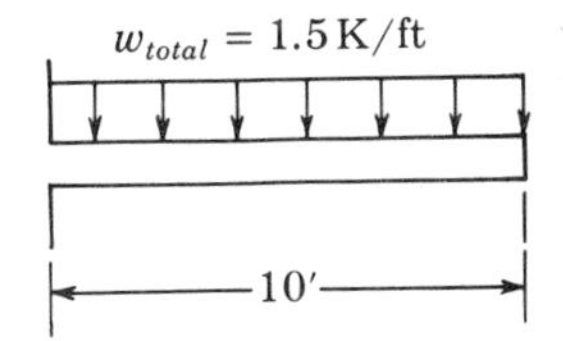

6.42 Design the fixed-ended beam to support the load shown in the sketch. Make sure the beam is safe in bending, diagonal tension, bond, and anchorage. Cut bars for maximum steel economy.

Moment coefficients: At ends $\frac{1}{12}$ at midspan $\frac{1}{24}$. $f'_c = 3{,}750$ psi, $f_s = 22{,}000$ psi. Clear cover $1\frac{1}{2}$ in. Width of beam = 12 in. Assume dead weight of beam 250 lb/ft. Follow ACI Code. Available bars are: No 8 bars: $A_s = 0.79$ in.^{2}, $\Sigma_0 = 3.1$ in.; No. 2 bars: $A_s = 0.05$ in.^{2}, $\Sigma_0 = 0.79$ in. What are the actual stresses in steel and concrete at the center of the beam?

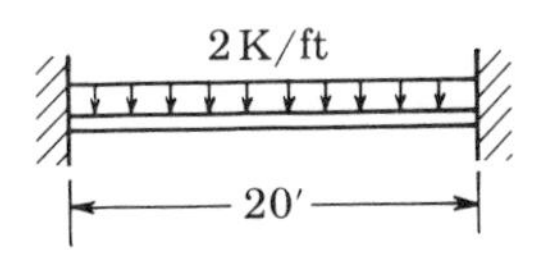

6.43 The beam shown is to be designed to resist the total load shown. A previous analysis of the structure has indicated a negative end moment at the right end of the beam of 80 kip-ft.

(*a*) Completely design the beam, using the requirements of the ACI Code. According to the architect's specifications, the width of the beam should be 14 in. Maximum economy of materials is essential. Allowable stresses are: $f'_c = 3{,}750$ psi, $f_s = 20{,}000$ psi. There are only a few bar sizes available.

For longitudinal reinforcement, use No. 9 bars: $A_s = 1.0$ in.², $\Sigma_0 = 3.54$ in.²
For web reinforcing (if required), use No. 3 bars: $A_s = 0.11$ in.², $\Sigma_0 = 1.18$ in.

(*b*) Suppose plans are afoot to stiffen the vertical supporting member so as to increase the negative end moment at the right end of the beam. What is the largest value of negative moment which could safely be resisted by the section of the beam as designed by you?

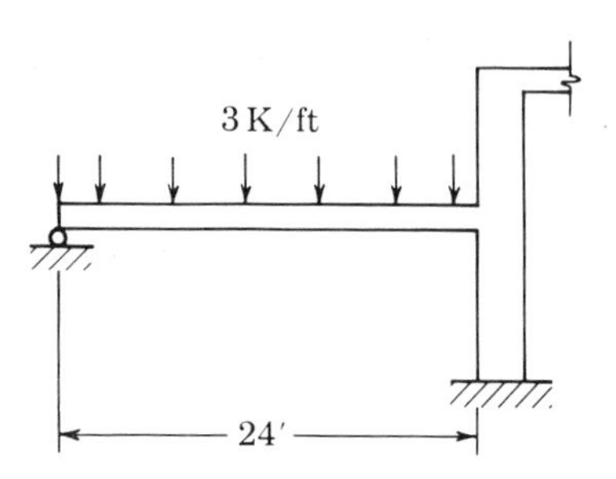

6.44 A proposed building structure is to contain a series of first-story girders as shown. These girders are to support the weight of the first-story floor load and the weight of the brick walls and roof, applied as shown in the sketch. The unit weights of floor and wall loads are as indicated. The columns supporting the girder are relatively slim as compared to the girders, so that their restraining influence on the girders is negligible. The girders may therefore be analyzed as simply supported.

The allowable stresses to be used are: $f_s = 24{,}000$ psi, $f'_c = 3{,}750$ psi; ACI Code to be followed throughout. Required: Complete design of the girders. Design drawings. *Note:* Only No. 8 and No. 2 bars are available from stock. No. 8 bars: $A_s = 0.79$ in.², $\Sigma_0 = 3.1$ in.; No. 2 bars: $A_s = 0.05$ in.², $\Sigma_0 = 0.8$ in.

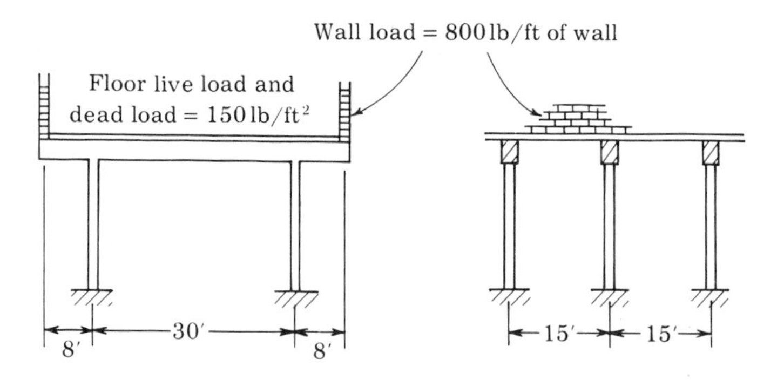

6.45 The practice of supporting elevated roadways on single rows of columns centrally located is becoming more prevalent. The sketches show an elevation and cross section of such a roadway. Two lanes are provided. The live load on each lane is 640 lb/lin ft of roadway. Assume that it is permissible to use ACI moment coefficients (note that in practice a more thorough analysis would have to be made). Design the reinforcing required at support and midspan of the *longitudinal girder*. Calculate only A_s and Σ_0 required. Do not select the bars. Allowable stresses are: $f'_c = 3{,}000$ psi, $f_s = 20{,}000$ psi; follow ACI Code in all respects. Where the code is ambiguous or for points not covered by the code, use your engineering judgment.

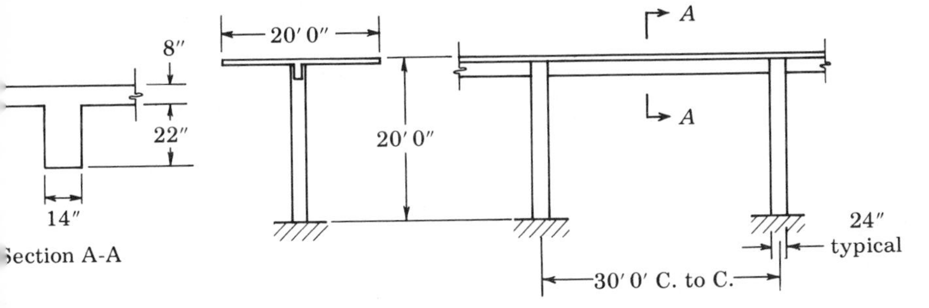

6.46 A footbridge, spanning 50 ft as a simple beam, is to be designed. The width of the walkway is to be 6 ft 0 in.; the live load on the walkway is to be 100 lb/ft² of area. Design is to be as shown.

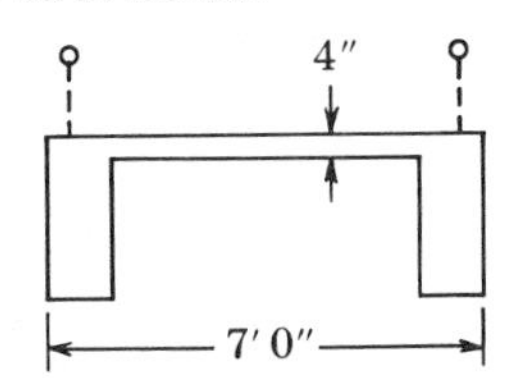

Conventionally reinforced span of the cross section shown, of concrete weighing 150 lb/ft³, of strength f'_c = 4,000 psi, and of intermediate-grade reinforcing steel. Design is to be according to the ACI Code, conventional method. In deciding whether slab may be considered an integral part of the beam, let code regulations on the beams govern. Calculate the required width and depth of the girder webs and the amount of steel at the section of maximum moment. Make sure that the beam is safe in bending and in diagonal tension, and that it can be built (i.e., that steel can be accommodated). Make design sketches as required. Calculate the stresses under working load.

6.47 A portion of a plan of a slab-beam floor is shown, with two sections through a typical interior floor beam, one at the support and one at midspan. The original design computations show that the design was based on the following allowable stresses: f'_c = 2,500 psi, f_s = 18,000 psi. The ACI Code was followed. The live load on the floor is 100 lb/ft². A review is required of the stresses due to the total load. Concrete weighs 150 lb/ft³. Find the stresses in steel and concrete at the points of maximum positive and negative moments. Also determine whether or not stirrups are required, and whether the bond stresses between steel and concrete at the sections considered are within the values allowed by the code.

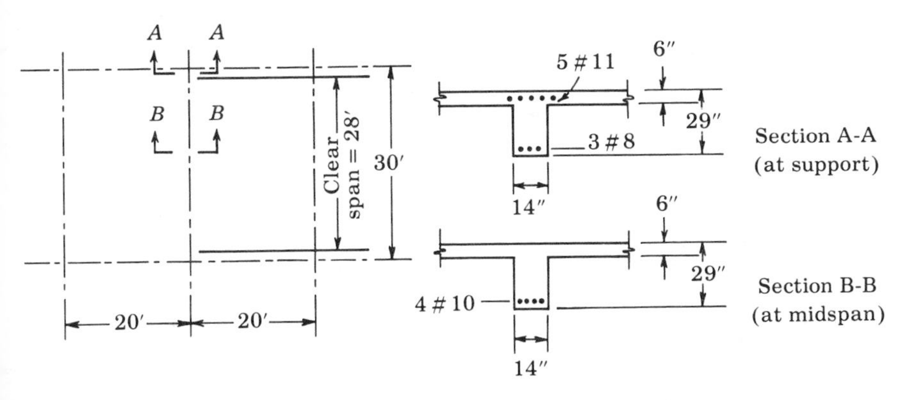

6.48 A cantilever-type retaining wall is to be designed to support an embankment of soil 24 ft high. The soil weighs 120 lb/ft³. Lateral pressure is assumed to vary linearly with the depth, with a ratio of lateral to vertical pressure of 0.3. Allowable stresses are: $f'_c = 2{,}500$ psi, $f_s = 18{,}000$ psi. Allowable soil pressure = 5 kips/ft². No tension allowed between pad and soil. Coefficient of friction between soil and concrete = 0.4. Concrete weighs 150 lb/ft³.

A trial section for this wall is suggested in the sketch. The thicknesses given in this sketch are assumed only for purposes of analysis. They are not necessarily the final thicknesses required by design.

1. Referring to this trial section:

   (*a*) What are the maximum and minimum pressures between pad and soil? Are the allowable values exceeded?
   (*b*) What is the safety factor against overturning (stability ratio)?
   (*c*) What is the factor of safety against sliding?
   (*d*) Draw shear and moment diagrams for pad and stem.

2. Completely design the retaining wall. Be sure it is safe and economical. Make complete design drawings.

Notes

1. The weight of the concrete must be included in the analysis.
2. The volume of soil overlying the pad affects the pressure between pad and soil.
3. Remember that net pressure only causes internal forces and moments.
4. It is possible to show all design details in a typical section through the wall.

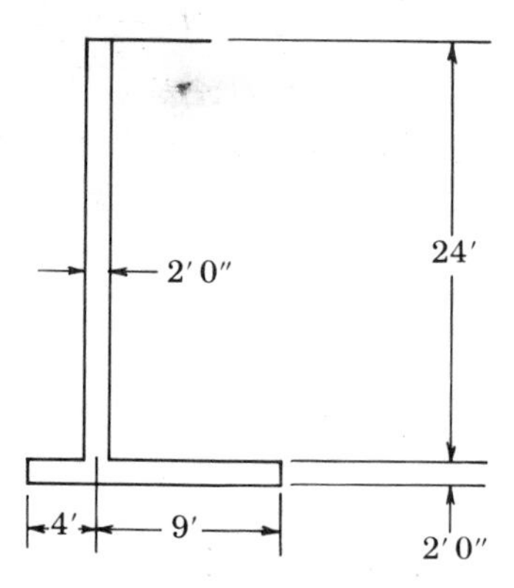

7

Inelastic Beams of Perfectly Plastic Material

7.1 General Theory

To investigate the behavior of beams whose material does not obey Hooke's law, we remember that powerful and general methods were discussed in Sec. 2.5. In that section, four basic steps were outlined for a bending analysis of beams and we recognize that of these four steps the first three, involving equilibrium and geometry of deformation, are independent of the properties of the material; these only enter the picture in step 4, which relates stress and strain.

Accordingly, we summarize the equations of equilibrium

$$\Sigma F_z = 0: \qquad \int_A \sigma \, dA = 0 \tag{7.1}$$

$$\Sigma M_x = 0: \qquad \int_A \sigma y \, dA = M \tag{7.2}$$

and the statement that plane sections remain plane,

$$\epsilon = y\phi \tag{7.3}$$

To relate stress and strain, we are more general than in the previously discussed elastic case, and write, according to Eq. (2.13), that stress is a function of strain:

$$\sigma = f(\epsilon) \tag{7.4}$$

From here we again follow the elastic case in analogous steps: putting Eq. (7.3) into Eq. (7.4), we express stress as a function of its distance from the neutral axis:

$$\sigma = f(y\phi) \tag{7.5}$$

and substituting this in turn into the equilibrium equations (7.1) and (7.2), we get

$$\int_A f(y\phi) \, dA = 0 \tag{7.6}$$

and $$\int_A f(y\phi) y \, dA = M \tag{7.7}$$

which can be integrated as soon as the appropriate stress-strain relation, Eq. (7.4), is known.

Example Problem 7.1 For example, let us consider a rectangular cross-section beam of width b and depth d, of material whose stress-strain curve in both tension and compression can be expressed by Eq. (2.16):

$$\sigma = f(\epsilon) = k\epsilon^n \tag{7.8}$$

where k and n are constants which must be determined by experiment. Note that when $n = 1$, then stress is proportional to strain, and we have the elastic case, and $k \equiv E$.

With this stress-strain relation, Eqs. (7.6) and (7.7) become

$$\int_A k(y\phi)^n \, dA = kb\phi^n \int_{-c_1}^{c_2} y^n \, dy = 0 \tag{7.9}$$

and

$$\int_A k(y\phi)^n y \, dA = kb\phi^n \int_{-c_1}^{c_2} y^{n+1} \, dy = M \tag{7.10}$$

Equation (7.9) has two factors, $kb\phi^n$ and $\int_{-c_1}^{c_2} y^n \, dy$, and since the former is a definite nonzero value, the integral must be zero. By symmetry of beam section and tensile and compressive material properties, this is possible when the neutral axis is at mid-depth of the section, making $c_1 = c_2 = d/2$. The same conclusion is of course obtained when we consider the symmetric stress block of Fig. 7.1d, which shows the condition of force equilibrium.

Having located the neutral axis, we can integrate the moment equation (7.10) between the limits of $c_1 = -d/2$ and $c_2 = +d/2$, and get

$$kb\phi^n \frac{1}{n+2} (y^{n+2}) \Big|_{-\frac{d}{2}}^{+\frac{d}{2}} = M$$

or

$$\phi^n = \frac{(n+2)2^{n+1}M}{kbd^{n+2}}$$

from which the curvature resulting from the moment M is

$$\phi = \sqrt[n]{\frac{(n+2)2^{n+1}}{bd^{n+2}} \frac{1}{k} M} \tag{7.11}$$

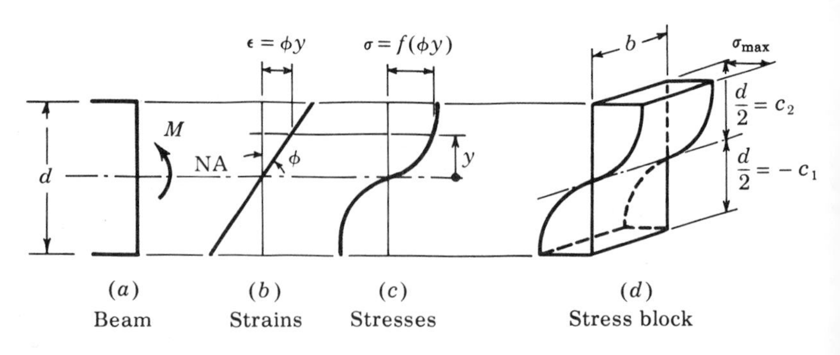

Fig. 7.1

We note that because of the nonlinear stress-strain relation, applied load and resulting deformation are not proportional.

Note that for the special elastic case of $n = 1$, and letting $k = E$,

$$\phi = \frac{12}{bd^3}\frac{1}{E}M = \frac{M}{EI} \qquad (7.12)$$

thus corroborating an earlier result.

7.2 Elastic–Perfectly Plastic Beams

A perfectly plastic material is one which exhibits the properties shown by the dashed line in the stress-strain curve of Fig. 2.5. It has been pointed out in Sec. 2.2 that the properties of mild or structural steel are rather well represented by the assumption of perfect plasticity, if strain hardening is neglected. Since strain hardening contributes to both strength and toughness, its neglect will lead to results which are consistently on the safe side.

Mathematically, an elastic–perfectly plastic stress-strain curve can be expressed by Eqs. (2.15), which are repeated here:

$$\begin{aligned} \sigma &= E\epsilon \qquad 0 < \epsilon \leq \epsilon_{YP} \\ \sigma &= \sigma_{YP} \qquad \epsilon_{YP} \leq \epsilon \end{aligned} \qquad (7.13)$$

where the relationships apply to both tensile and compressive loading.

To illustrate the application of the basic theory to members of perfectly plastic materials, the case of a rectangular beam of such a material is considered.

The usual steps of the strength of materials approach indicate that by symmetry the neutral axis must be at mid-depth of the section. The strain distribution of Fig. 7.2*b* indicates that points of the cross section farther from the neutral axis than a distance y_{YP} are subjected to inelastic strains and, according to the perfectly plastic stress-strain relation (7.13),

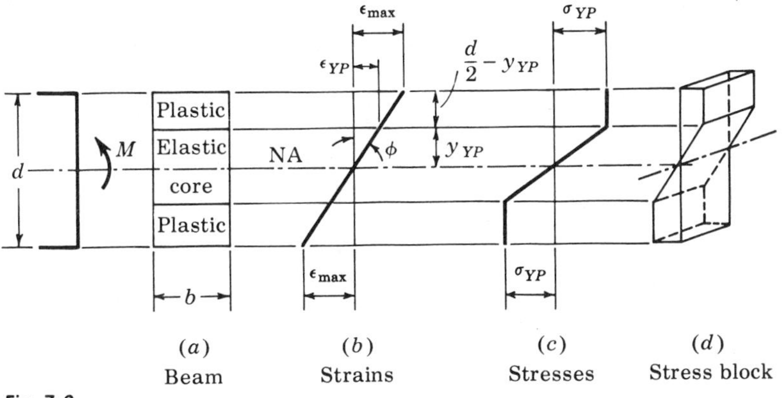

Fig. 7.2

must be stressed to a value σ_{YP}. This is shown in the stress distribution, Fig. 7.2*c*.

Owing to the discontinuous nature of the stress distribution, it is advisable to work with the stress blocks, rather than with Eq. (7.7), which implies analytical integration. To calculate the relationship between applied moment M and resulting curvature ϕ, we set

$$\Sigma M_{NA} = 0: \qquad 2b\sigma_{YP}\left[\left(\frac{d}{2} - y_{YP}\right)\left(\frac{d}{2} + y_{YP}\right)\frac{1}{2} + \frac{y_{YP}}{2}\frac{2}{3}y_{YP}\right] - M = 0$$

therefore

$$y_{YP} = \sqrt{3\left(\frac{d^2}{4} - \frac{M}{b\sigma_{YP}}\right)} \tag{7.14}$$

thus giving the location for the elasto-plastic boundary under any moment

$$M \geq M_{YP}$$

Since from Fig. 7.2*b* the curvature is obtained as

$$\phi = \frac{\epsilon_{YP}}{y_{YP}} = \frac{\sigma_{YP}}{Ey_{YP}} = \frac{M_{YP}}{2EI}\frac{d}{y_{YP}} \tag{7.15}$$

and substituting Eq. (7.14), we obtain as the required relationship

$$\phi = \frac{M_{YP}}{EI}\frac{1}{\sqrt{2}\sqrt{3/2 - (M/M_{YP})}} \tag{7.16}$$

for $M > M_{YP}$. For $M < M_{YP}$, the elastic relationship

$$\phi = \frac{M}{EI} \tag{2.25}$$

will of course apply.

It is of interest to pursue the change in stress distribution throughout the section, and also the increase of curvature with increasing moment. This is done for several values of moment in Fig. 7.3*a* to *c*.

Stress and curvature increase linearly with increasing moment till the applied moment reaches a value of M_{YP}, at which time the most highly stressed fibers reach their proportional limit. As the strain increases further, the perfectly plastic stress-strain curve of Fig. 2.5 is called on to furnish the value of stress corresponding to any value of strain. Thus, it can be seen that a progressively larger number of fibers will be subjected to a strain exceeding ϵ_{YP}, and therefore to a stress equal to σ_{YP}. Figure 7.3*a*, *b*, and *c* shows the stages of strain and stress distribution. From them it can be seen that as the strain increases beyond ϵ_{YP}, there will be a plastic exterior zone and a steadily shrinking elastic core surrounding the neutral axis. As the applied moment increases, an increasingly larger number of fibers will be called upon to contribute their full capacity to the resisting moment, until, when the moment has reached a value of $\frac{3}{2}{}_Y M_P$, finally every fiber is subjected to its yield stress σ_{YP}. The elastic core has disappeared, and the section is therefore free to kink without restraint. This fact is brought out by Eq. (7.16), which for $M = \frac{3}{2}M_{YP}$ indicates an infinitely large curvature, a fact which is confirmed by the strain distribution of Fig. 7.3c. It must, of course, be borne in mind that actually any material will rupture at a definite value of strain, and therefore both the maximum moment value of $\frac{3}{2}M_{YP}$ and the infinite curvature are hypothetical values which form asymptotes to the actual moment-curvature diagram. However, for ductile materials such as structural steel, for which the rupture strain might have a value of possibly 240 times the yield strain ϵ_{YP}, these results yield values close enough for practical purposes.

Several concepts have been brought out in the preceding discussion. The moment under which every fiber is stressed to its maximum capacity is called the *ultimate* or *plastic moment* M_P. The ratio of plastic moment to yield-point moment M_P/M_{YP} is called the *shape factor* and denoted by the symbol f. For the rectangular section, $f = \frac{3}{2}$. The carrying capacity of the rectangular beam is therefore 50 percent higher than the moment which causes first yielding in the extreme fibers. As its name implies, the value of the shape factor varies with the cross section of the member. Thus, beams which have the bulk of their area (and therefore of their reserve strength) near the neutral axis will have a high shape factor. This fact is brought out by Fig. 7.5 (p. 225), which shows plots of moment versus curvature for various section beams.

Whenever a section has become completely plastified and therefore

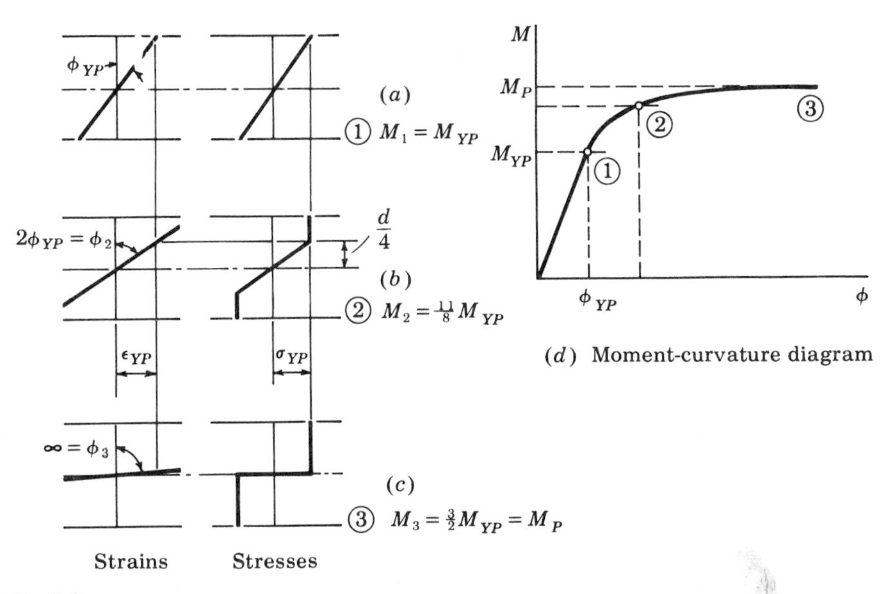

Fig. 7.3

subjected to free yielding, a *plastic hinge* is said to have formed. As can be verified by the moment-curvature diagram, a plastic hinge (resulting from an applied moment M_P) indicates that the beam is able to undergo a large amount of rotation or kink at the particular section under a constant value of moment. The action may thus be likened to a real or mechanical hinge. Plastic hinges are a vital part of plastic analysis of structures. The amount of curvature or kink which a beam can undergo before rupture is called the *rotation capacity* and is to a large extent dependent on the ductility of the material.

A general method of calculating the ultimate moment capacity M_P of any section is obtained from a study of Fig. 7.4, which shows the stress blocks on a section due to M_P.

$$\Sigma F_x = 0: \qquad C = T \qquad \text{or} \qquad \sigma_{YP} A_{comp} = \sigma_{YP} A_{tens}$$

Since σ_{YP} is the same in tension and compression,

$$A_{comp} = A_{tens} \tag{7.17}$$

This shows that the neutral axis at ultimate does not necessarily lie at the centroid of the section; rather, it is located so as to divide the cross section into two equal areas. To find the ultimate moment M_P, we set

$$\Sigma M_{NA} = 0: \qquad \text{thus } M_P = \sigma_{YP} A_{tens} \bar{y} = \sigma_{YP} \frac{A}{2} \bar{y}$$

where $\bar{y}$ is the internal lever arm between the tensile and compressive resultants. By analogy with conventional elastic procedures, we can write

$$M_P = \sigma_{YP} Z \tag{7.18}$$

where $Z = (A/2)\bar{y}$ and is called the *plastic section modulus*.

Moment-curvature diagrams for beams of various cross-sectional shapes have been evaluated by the methods presented and are plotted in nondimensional form in Fig. 7.5.

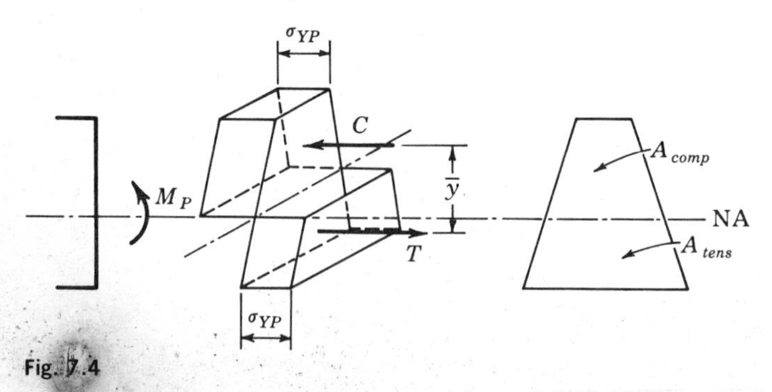

Fig. 7.4

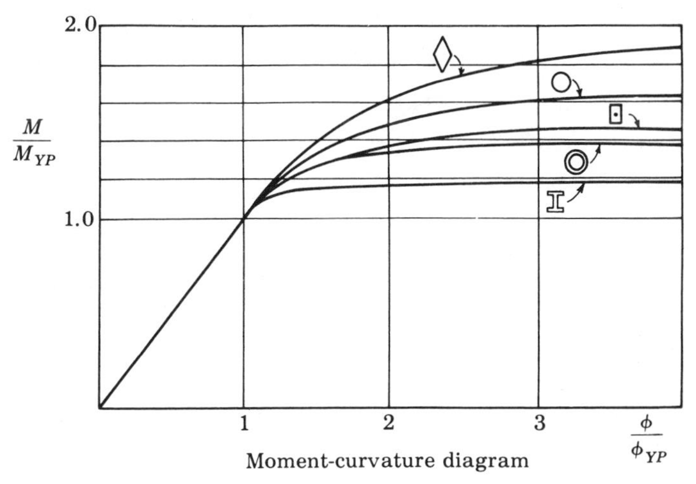

Fig. 7.5

Example Problem 7.2

(a) Find the ultimate moment and the shape factor for the diamond-shaped section shown. Elastic–perfectly plastic material of yield stress σ_{YP}.

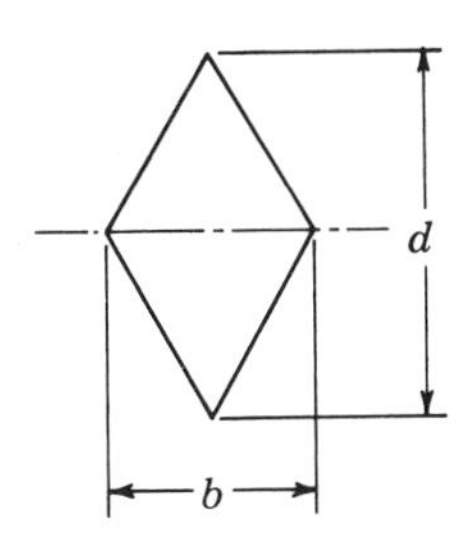

At ultimate:

$$\Sigma M = 0: \qquad \left(\sigma_{YP}\tfrac{1}{2}b\,\frac{d}{2}\right)\left(\tfrac{1}{3}d\right) = \tfrac{1}{12}\sigma_{YP}bd^2$$

At first yielding:

$$\sigma_{YP} = \frac{Mc}{I} = \frac{M_{YP}(d/2)}{bd^3/48} = \frac{24M_{YP}}{bd^2}$$

or

$$M_{YP} = \tfrac{1}{24}\sigma_{YP}bd^2$$

$$\text{Shape factor} = f = M_P/M_{YP} = 2.$$

(b) Find plastic section modulus and shape factor for 12 WF 27 beam. The dimensions shown are taken from the AISC Steel Handbook. Finding the plastic section modulus as the first moment of the half section about the N.A.:

$$Z = 2(6.500 \times 0.400 \times 5.77 \text{ in.} + 5.57 \times 0.240 \times 2.79 \text{ in.})$$
$$= 2(14.97 + 3.72) = \underline{37.38 \text{ in.}^5}$$

Elastic section modulus $= S = 34.1$ in.^{3}; $f = Z/S = \underline{1.09}$.

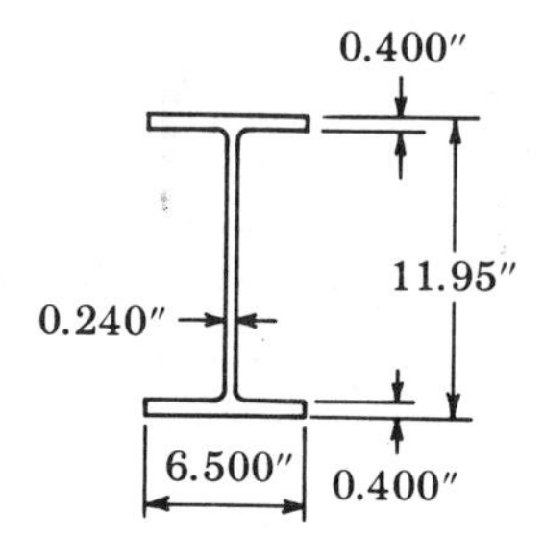

(c) Find the ultimate moment on the T section shown ($\sigma_{YP} = 40$ ksi). To locate neutral axis at ultimate:

$$\Sigma F = 0: \quad [12 + 2y - 2(8 - y)]\sigma_{YP} = 0 \qquad y = 1.0 \text{ in.}$$
$$\Sigma M = 0: \quad [12 \times 2 \text{ in.} + 2.0 \times 0.5 \text{ in.} + 14 \times 3.5]40 \text{ ksi}$$
$$= 2{,}960 \text{ kip-in.} = \underline{247 \text{ kip-ft}}$$

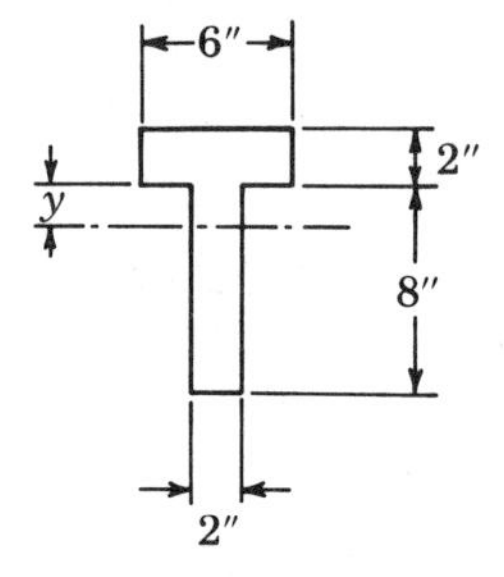

Note that the ultimate strength of perfectly plastic beams is independent of their elastic response. This is a specific indication of one of the important theorems of the theory of perfectly plastic structures.

7.3 Limit Analysis of Perfectly Plastic Structures

Limit analysis concerns itself with the determination of the maximum (or "limit") load which a structure is able to withstand before collapse will take place. The attention, then, is focused upon actual failure, rather than upon the attainment of some assumed allowable stress, as is the case in conventional elastic analysis. Having determined the *collapse* load of the structure, any suitable factor of safety may be applied to determine the safe or *design* load. The problem before us is therefore the calculation of the failure load of any structure. It will be seen that for structures of perfectly plastic material, several convenient methods for doing this are available.

The criteria governing design of a structure will often include the possibility of excessive deflections, resistance to vibratory, repeating, or reversible loads, the need to maintain elastic action, or similar requirements; in many cases of this type, conventional elastic analyses are called for. However, in a great number of cases occurring in structural engineering practice, none of these factors are critical. Collapse will then be the governing consideration.

Conventional elastic analysis and design of structures ostensibly presumes that at no place in the structure is the elastic limit of the material exceeded. Nevertheless, the straightening of accidentally kinked or bent members, the cold bending or cambering of truss chords, and other practices involving inelastic action are accepted. Rolled steel sections as received from the mills may be subjected to initial or locked-up stresses close to the yield point of structural steel. Ductility is the reason why these effects may be neglected without dire results. The effects of redistribution of stresses resulting from ductility have been discussed in a number of examples in these notes.

To begin our discussion, let us consider a simply supported beam of length L subject to a uniform load. It is required to find the value of the uniform load w under which collapse will occur, if the ultimate strength of the member is M_P. Figure 7.6*b* shows the moment diagrams due to two different values of the load w. The maximum moment, by statics, is calculated as $wL^2/8$. Purely elastic action will prevail till the load reaches a value of w_{YP}, at which instant the extreme fibers of the center section, subjected to a moment M_{YP}, will become plastic. The load increases further, until at a value of load w_P the moment at the center section attains a value of M_P. At this time, the center section of the beam becomes completely plastified, and a plastic hinge free to kink has developed. The deflected shape of the beam of Fig. 7.6*a* shows that under this value of load the beam

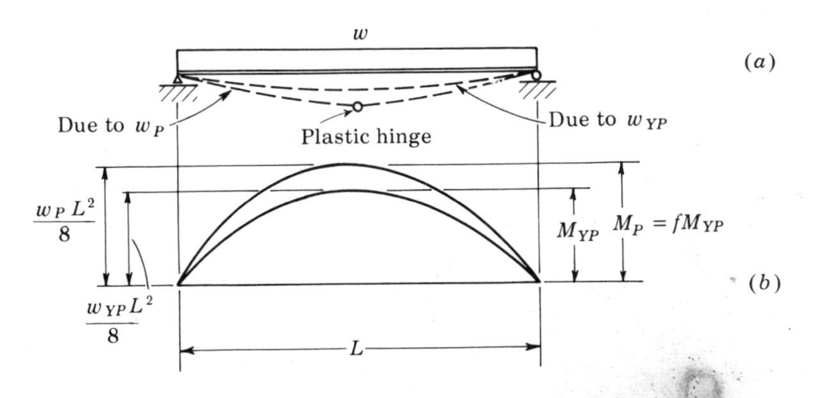

Fig. 7.6

consists of two links able to rotate freely with respect to each other, so that at this time the structure will undergo unrestricted deflection. Therefore, the collapse load is calculated:

$$\frac{w_P L^2}{8} = M_P \qquad \text{or} \qquad w_P = \frac{8M_P}{L^2}$$

The state of collapse just mentioned occurs when the structure has degenerated into a *mechanism*, or *kinematically admissible field*. A kinematically admissible mechanism is defined as a structure able to undergo deformation without the application of any additional load. Generally, then, collapse of a structure will result from the formation of a mechanism, and the determination of the failure load becomes equivalent to finding the load under which the proper mechanism will form.

It should also be noted for further reference that the ratio of the load causing collapse to the load causing first yielding is

$$\frac{w_P}{w_{YP}} = \frac{M_P}{M_{YP}} = \frac{fM_{YP}}{M_{YP}} = f$$

The safety factor against collapse of this structure, had it been designed according to the conventional criterion that any yielding would make the structure unsafe, would have been equal to the shape factor of the section.

For the next example, let us consider the fixed-ended beam of length L and of strength M_P shown in Fig. 7.7*a*. It should be noted that this beam is statically indeterminate to the second degree.

Purely elastic action terminates when the end moment, according to elastic analysis of value $wL^2/12$, reaches a value of M_{YP}. The moment diagram 1, of Fig. 7.7*b*, indicates this stage. As the load is increased, both

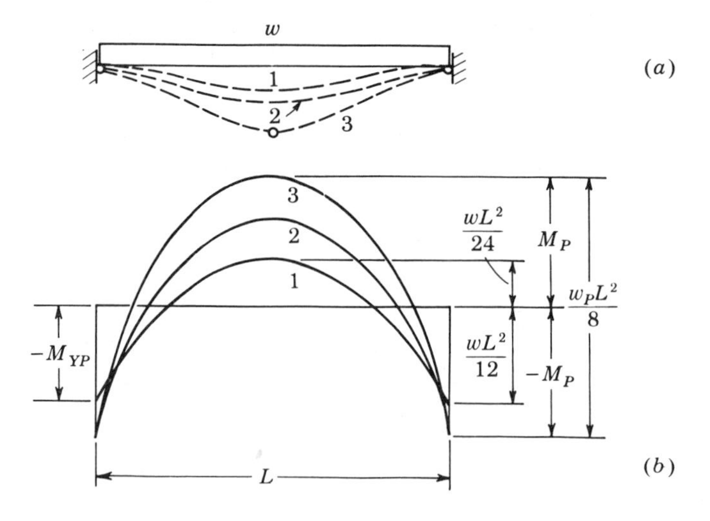

Fig. 7.7

end and center moments will increase, with the end moments eventually attaining a value of M_P. At this instant, plastic hinges form at the ends, as shown in the deflected shape labeled 2 in Fig. 7.7*a*. Under additional loads, since the ends of the beam will be unable to resist further moment, the beam must act as a simple beam, with the ends rotating freely, made possible by the presence of plastic hinges there. This enables the curvature at the midportion to increase, thereby increasing the moment there till, under a load w_P, it has also reached a value M_P. At this instant a plastic hinge forms at midspan, as shown in Fig. 7.7*a*, as stage 3, and the structure has degenerated into a mechanism. Collapse will therefore occur when the moments at the ends and at midspan reach values of M_P.

To calculate the collapse load for this beam, we realize that, by statics alone, the total ordinate of the parabolic moment diagram is equal to $wL^2/8$; since to attain a mechanism, according to moment diagram 3 of Fig. 7.7*b*,

$$\frac{w_P L^2}{8} = 2M_P$$

therefore

$$w_P = \frac{16M_P}{L^2}$$

Since the use of moment diagrams in this fashion is only one way of expressing conditions of statics, the same value could be attained by use of free bodies, as shown in Fig. 7.8:

Free body AB: $\quad \Sigma M_B = 0$: $\quad R_A \frac{L}{2} - 2M_P - \frac{w_P L}{2}\frac{L}{4} = 0$

therefore $\quad R_A = \frac{4M_P}{L} + \frac{w_P L}{4}$

Free body BC: $\quad \Sigma M_B = 0$: $\quad R_C = \frac{4M_P}{L} + \frac{w_P L}{4}$

Free body AC: $\quad \Sigma V = 0$: $\quad R_A + R_C - w_P L = 0$

therefore $\quad \frac{8M_P}{L} + \frac{w_P L}{2} - w_P L = 0 \qquad w_P = \frac{16M_P}{L^2}$

The sequence of formation of the plastic hinges is irrelevant to the determination of the collapse load. The basic approach is to visualize the deflected

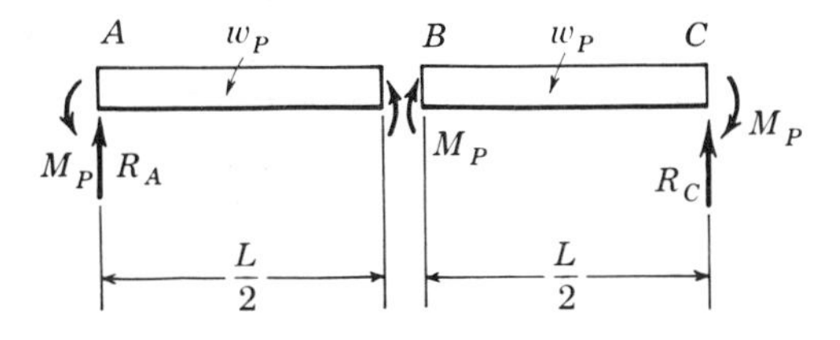

Fig. 7.8

shape leading to collapse, introduce plastic hinges at the appropriate places, and calculate the collapse load by statics alone. Conditions of geometry have become unnecessary at this stage because plastic hinges, thanks to their ability to kink, can accommodate themselves to any curvature.

An important fact demonstrated here is the redistribution of moments prior to collapse. It is noted that, from an initial elastic stage during which the midpoint moment was equal to one-half of the end moments, the midpoint moment at collapse was equal in value to the end moments. This is another demonstration of the principle, stated earlier, that in a structure of ductile material, whenever possible, the less highly stressed portions of a structure will tend to help out the more highly stressed parts.

Finally, let us again compare the load causing failure to the load causing first yielding:

$$\frac{w_P}{w_{YP}} = \frac{16M_P/L^2}{12M_P/L^2} = \frac{4}{3}\frac{fM_{YP}}{M_{YP}} = \frac{4}{3}f$$

For the fixed-ended beam just discussed, then, had it been designed for first yielding as a failure criterion, the safety factor against collapse would have been $\frac{4}{3}f$. This is a higher value than the calculated safety factor of f for the simply supported beam. We conclude that for beams designed according to yield-point stress (or some other elastic criterion), the safety factor against collapse will be variable.

We must again recall that the redistribution of moments prior to collapse depended on the ability of the plastic hinges at the end of the beam to rotate through an angle sufficiently great to enable the midspan moment to reach a value of M_P.

Premature failure will occur if the rotation capacity is impaired because of local buckling, brittle fracture, or other defects. Special attention must therefore be paid to any phenomena which might impair this rotation capacity.

The above examples have also indicated that the redistribution of moments from an overstressed to an understressed portion is possible only in statically indeterminate structures. No such redistribution is possible in statically determinate structures because the occurrence of only one plastic hinge will result in formation of a mechanism with consequent collapse. The principles of limit analysis are therefore in the main applicable to statically indeterminate structures.

Example Problem 7.3 The rigid frame shown is subjected to a lateral concentrated load. Find the value of this load under which the frame will collapse. Strength of all members is M_P; neglect effect of axial forces.

The collapse must occur because of sidesway, as shown in Fig. 7.9. Summing then horizontal forces on the entire frame, we set:

$$\Sigma H = 0: \quad 2\left(2\frac{M_P}{L}\right) - P_P = 0$$

therefore

$$P_P = 4\frac{M_P}{L}$$

We shall next draw the moment diagram for the frame at collapse, as in Fig. 7.9c, and verify that the moments are compatible with the strength of the members.

We note that no more work is involved in finding the collapse load for this three times indeterminate structure than in the case of the fixed beam.

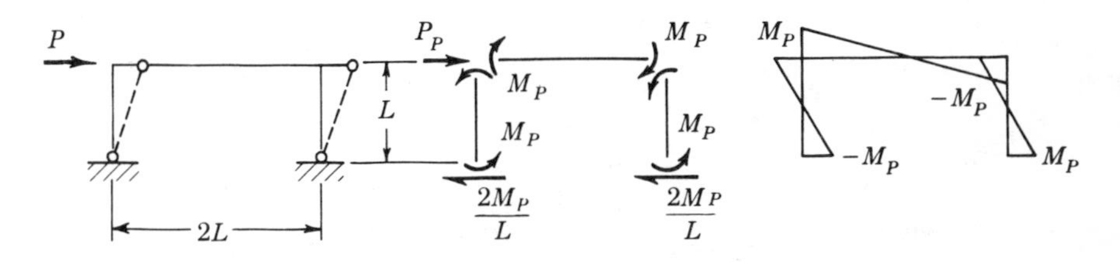

(a) Frame and mechanism (b) Free bodies (c) Moment diagram

Fig. 7.9

The foregoing examples were particularly simple in so far as the location of the plastic hinges at collapse could easily be determined by conditions of symmetry. More often, the analyst encounters structures and loadings for which the location of the plastic hinges is not so easily apparent. Several plans of attack are possible in such cases and will be demonstrated by an example.

Example Problem 7.4

Given: Propped cantilever beam of length L, strength M_P, subjected to a uniform load, as shown in Fig. 7.10a.

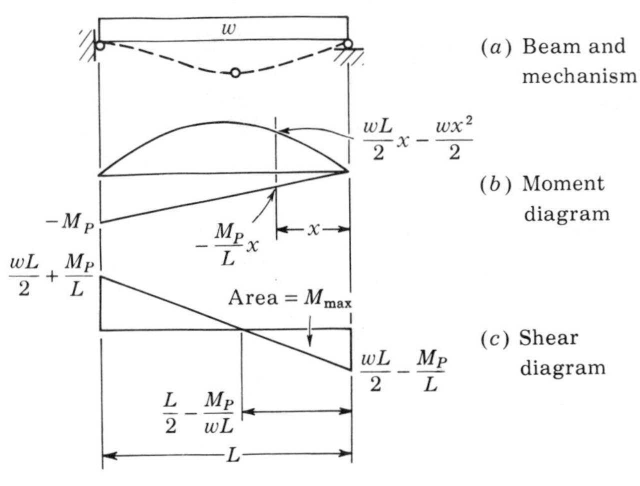

Fig. 7.10

Required: Collapse load w_P.

It is apparent that one plastic hinge will occur at the fixed end A, but the location of the plastic hinge in the span required for the formation of a mechanism is not immediately apparent. We realize, however, that it must occur at the point of maximum moment (in terms of distance x from end B) and we solve for the location of this point by differentiating the moment equation and equating to zero; referring to Fig. 7.10*b*,

$$M = \frac{w}{2}(Lx - x^2) - \frac{M_P}{L}x$$

Setting the derivative equal to zero:

$$\frac{dM}{dx} = \frac{wL}{2} - wx - \frac{M_P}{L} = 0$$

therefore

$$x_{M_{\max}} = \frac{L}{2} - \frac{M_P}{wL}$$

and setting the moment at this point equal to M_P,

$$\frac{wL^2}{8} - \frac{3M_P}{2} + \frac{M_P^2}{2wL^2} = 0$$

from which

$$w_P = \frac{11.66M_P}{L^2}$$

Utilizing the properties of the shear diagram shown in Fig. 7.10*c* (realizing that the well-known relations between shear and moment are based only on statics and can therefore be used), the same value is obtained.

A solution could have been obtained graphically as shown in Fig. 7.11. The statically determinate simple beam moment diagram is drawn, and a skew line is superimposed to represent the effect of the restraining moment at end A. The skew line is rotated about end B till the moment at end A, of value $-M_P$ at collapse, becomes equal to the largest moment within the span (represented by the ordinate subtended between line and curve), of value $+M_P$. The value of M_P is read as a fraction of the maximum ordinate $w_PL^2/8$ of the simple beam diagram. The accuracy of this procedure is limited only by the size and care with which the figure is drawn.

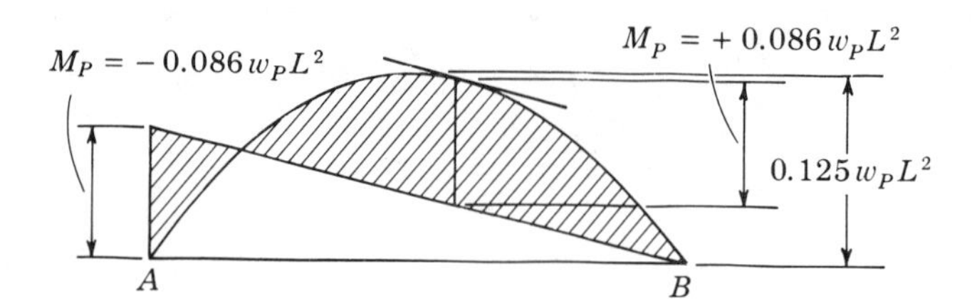

Fig. 7.11

From the discussion and examples of this section we can again deduce that the elastic stiffness of the perfectly plastic structure has no effect on its ultimate strength. Only the strength of the component members affects the fully plastic collapse. This will be discussed further in the next section.

7.4 Upper and Lower Bound Theorems

It has already been seen in the last problem discussed in the previous section that, generally, the correct placement of the plastic hinges presents a major problem of plastic analysis. In structures of the size occurring in structural engineering practice, a large number of possible collapse mechanisms might have to be considered, and often the work involved in obtaining an exact solution becomes prohibitive. In such cases, it is possible to achieve an approximate solution for the collapse load for which the amount of error may be determined. If, for instance, an approximate solution can be obtained quickly with a guarantee that it falls to within a certain percentage, say 5 percent, of the correct value, a more time-consuming exact analysis may become unnecessary.

The tools which enable us to perform this approximate analysis are the *upper and lower bound theorems*, and before these can be explained it is necessary to define two concepts (one of which has been mentioned before).

1. *Kinematic admissibility*—this refers to the ability of a structure to deform without application of additional loads. We understand that this concept requires perfect plasticity of the material, and that it is primarily associated with a sufficient number of plastic hinges in the structure. Any solution in plastic analysis which permits this kind of collapse of the structure is called a *kinematically admissible solution.*
2. *Static admissibility*—this refers to the requirement that under no conditions may the strength of the structure at any point be exceeded. For example, if the fully plastic strength of a beam is M_P, it cannot ever be subjected to a moment larger than this. Any solution in plastic analysis which is based on internal forces which do not exceed the strength of the member is called a *statically admissible solution.* In particular, an elastic design will generally lead to a statically admissible structure because its internal forces, calculated by the laws of statics, are held below the member strength.

Let us demonstrate the use of these concepts by referring to the propped cantilever beam under uniform load of Example Problem 7.4 of the previous section.

To avoid the problem of locating the plastic hinge in the span, an approximate solution could have been attempted by assuming this plastic hinge to be located at, say, midspan of the beam. Based on this assumption the approximate collapse load can be found by statics, referring to the free bodies of Fig. 7.12*b*, as

$$w_P = 12 \frac{M_P}{L^2}$$

a value somewhat higher than the exact value. If, again using statics, we draw the complete moment diagram, we see that over a portion of the beam its strength is exceeded. By our previous discussion, this approximation represents a kinematically admissible but not a statically admissible solution.

It is clear that the moment diagram of Fig. 7.12*c* represents a physical impossibility. Under increasing load, the plastic hinge would have formed earlier in the portion of the beam marked "statically inadmissible region," and thus collapse would have taken place under a load *smaller* than the one predicted by our approximate solution. This reasoning leads us to the important general "upper bound theorem":

The load computed on basis of an assumed kinematically admissible mechanism will always be larger than, or equal to, the correct collapse load of the structure.

Note again that such a kinematically admissible or "upper bound" solution can always be found by assuming a number of plastic hinges sufficient to cause collapse, then applying statics to solve for the corresponding upper bound load. However, such an approximation may be unsafe because it may overestimate the capacity of the structure; it should therefore only

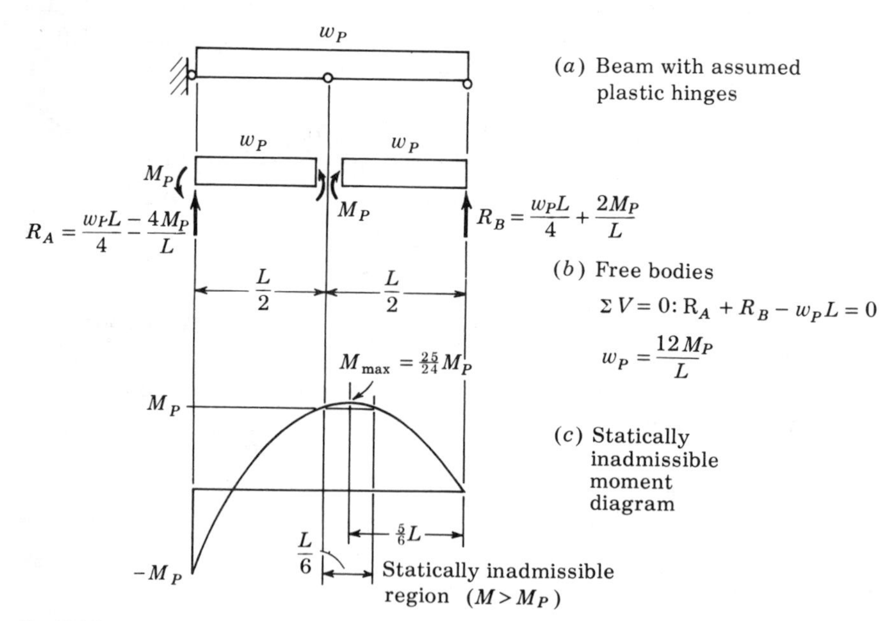

Fig. 7.12

be used with great caution, and mainly in conjunction with a lower bound solution which will serve to estimate the error.

To discuss the establishment of such a lower bound solution we go back to the previously discussed graphic method of analysis of the proposed cantilever beam under uniform load.

Supposing that means for accurate graphical construction were not at hand, it would be possible to attain conservative results by locating the skew base line so as to ensure that the largest moment in the span had a value somewhat below the value of the redundant $-M_P$ at end A. This is done in Fig. 7.13.

We set the maximum moment in the span at, say, 90 percent of the fully plastic value M_P, and scale this value as

$$0.9M_P = 0.08w_P L^2 \qquad \text{or} \qquad w_P = 11.25\,\frac{M_P}{L^2}$$

As should be expected (because no mechanism has formed yet), the load is somewhat below its correct value. Actual collapse would occur under a load *larger* than the one predicted here. We also note that, according to previous definition, the moment diagram on which our approximate analysis has been based is a statically admissible one, and generalize this demonstration to frame the following "lower bound theorem":

The load computed on the basis of an assumed statically admissible set of internal forces will always be smaller than, or equal to, the correct collapse load of the structure.

We see, then, that even though it might be a difficult task to find the exact collapse load for a complicated structure, it will always be possible to bracket this exact result, and often sufficiently closely for practical purposes. Furthermore, as a result of this discussion it may also be said that a solution for collapse load of a structure will be the correct one if three conditions are satisfied:

1. A collapse mechanism must have formed (kinematic admissibility).
2. At no point may the ultimate strength of the members be exceeded (static admissibility).
3. The structure must be in static equilibrium.

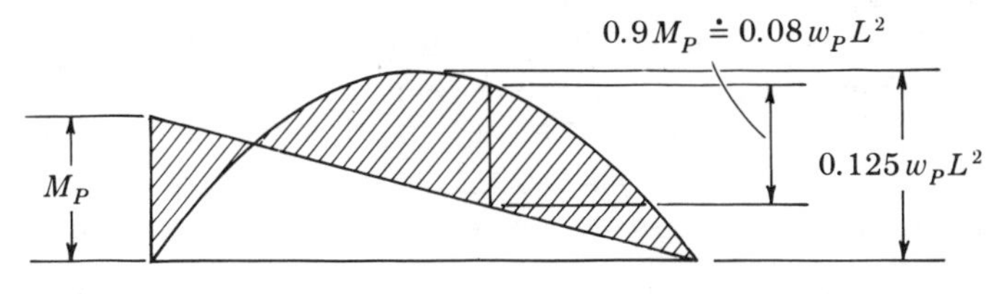

Fig. 7.13

If conditions of both static and kinematic admissibility are satisfied by a solution, then it must be a simultaneous upper and lower bound, indicating that the correct value of the collapse load has been found.

Conversely, the correctness of any limit analysis should be checked by verifying the static admissibility (i.e., that the maximum moments are not too large), for kinematic admissibility (i.e., that the number of plastic hinges is sufficient to allow the structure to collapse), and for satisfaction of the conditions of statics.

The reader should realize that the preceding coverage does not constitute a proof of the upper and lower bound theorems. For such proof, reference should be made to the book by Hodge listed in General Readings for this chapter.

7.5 Mechanism Method of Analysis

To analyze a structure by the mechanism method in order to obtain a kinematically admissible upper bound solution, the collapse mechanism is assumed first. At each section at which a plastic hinge is assumed, the moment must have a value equal to the known moment capacity of the member at that point, M_P; each of these known moments reduces the degree of static indeterminacy of the structure by one; if sufficient hinges for collapse have been assumed, the structure may be analyzed by statics alone, and this has been done in the preceding examples.

An alternate way, here specially convenient, of establishing equilibrium conditions is based on the *principle of virtual displacement*,[1] which may be stated as:

> If a system of forces in equilibrium is subject to a virtual displacement, the work done by the external forces is equal to the work done by the internal forces.

The derivation of this principle shows that it is valid for any structure in equilibrium, no matter whether elastic or inelastic, also that the virtual displacement is restricted only by the support conditions of the structure. For our purpose, it is convenient to assume a virtual displacement which corresponds to the assumed mode of collapse, and in all our problems, the assumed virtual displacement will be identical with the assumed collapsed shape of the structure. The important thing to remember is that the principle of virtual work is only an easy way of writing equilibrium conditions. The external work considered is equal to the constant external forces needed to deform the mechanism, multiplied by their path of travel along their line of action. The internal work is due to the constant values of M_P times the rotation each plastic hinge undergoes. As in all other conventional methods,

[1] Ivan S. Sokolnikoff, "Mathematical Theory of Elasticity," 2d ed., p. 383, McGraw-Hill, 1956.

the displacements are assumed small so that the dimensions of the undeformed structure can be used for all calculations.

Since this method is based on an assumed mechanism, it leads necessarily to results which form an upper bound to the true collapse load. In other words, if the correct collapse mechanism is assumed, the procedure will yield the correct answer. For any other assumed mechanism, the answer will be larger than the correct value.

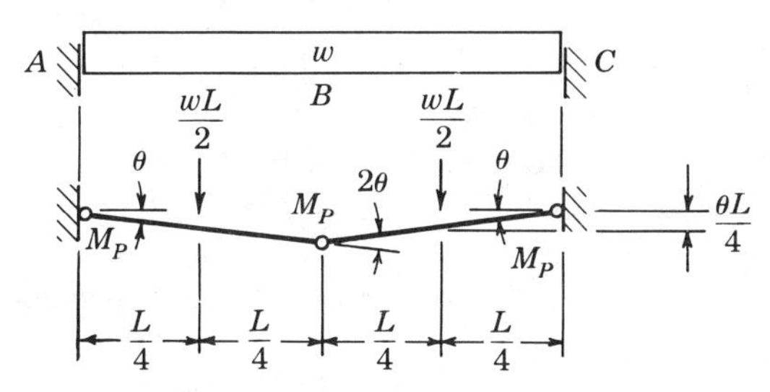

Fig. 7.14

The previously solved, uniformly loaded fixed-ended beam will be utilized to demonstrate this method of analysis. Figure 7.14*a* shows the structure, and Fig. 7.14*b* its mode of failure, which is here used as the virtual displacement, corresponding to a rotation of value θ at point A. The internal work is

$$U_{int} = M_P(\theta + 2\theta + \theta) = 4M_P\theta$$

The external work is due to the vertical motion of the resultants of the uniform load (only loads over any one rigid link may be lumped together into a resultant), and is

$$U_{ext} = 2\frac{wL}{2}\frac{\theta L}{4} = \frac{wL^2}{4}\theta$$

Equating,

$$U_{int} = U_{ext}: \qquad 4M_P\theta = \frac{wL^2}{4}\theta$$

therefore

$$w_P = \frac{16M_P}{L^2}$$

as previously.

The preceding example was exceptionally simple in so far as the locations of the plastic hinges were evident. Now, a pin-ended rectangular frame, shown in Fig. 7.15*a*, subjected to vertical and lateral loads, is to be analyzed for collapse load. There are several collapse mechanisms possible in this frame, all of them shown in Fig. 7.15*b*. Since each of these represents a kinematically admissible solution, each one will lead to a collapse load which is an upper bound. Only the lowest of these upper bounds has a chance of being correct.

Taking up each possible mechanism in turn, and applying the principle of virtual displacements, the calculations given in Fig. 7.16 give the ultimate load for each case.

It is seen that the combined mechanism is the one occurring first, when the load reaches a value of M_P/L. Some experience would have obviated the task of investigating all three mechanisms. The correct one could have been picked up by inspection, making use of the fact that the mechanism combining the smallest amount of internal work with the largest amount of work of the applied forces will be the critical one. Physically speaking, this means that each of the applied loads will want to contribute to the collapse, while the correct collapse mechanism will show a relatively small number of plastic hinges. We note that of the mechanisms under consideration, only the combined one allows both lateral and vertical loads to be effective. When it is believed that the correct mechanism has been chosen, the moment diagram is drawn, as in Fig. 7.17. If it is found that nowhere is the ultimate moment exceeded, then the collapse load and its resulting mechanism must

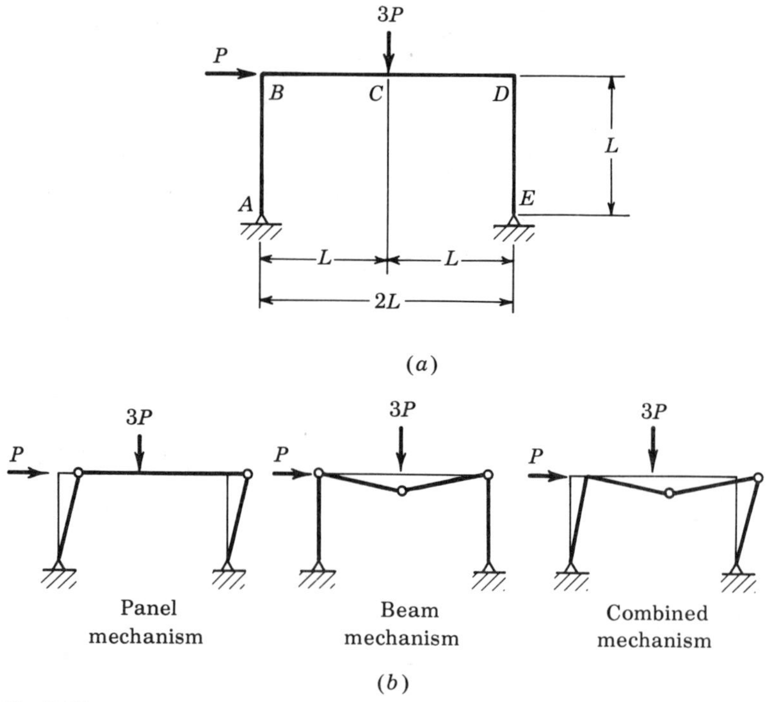

Fig. 7.15

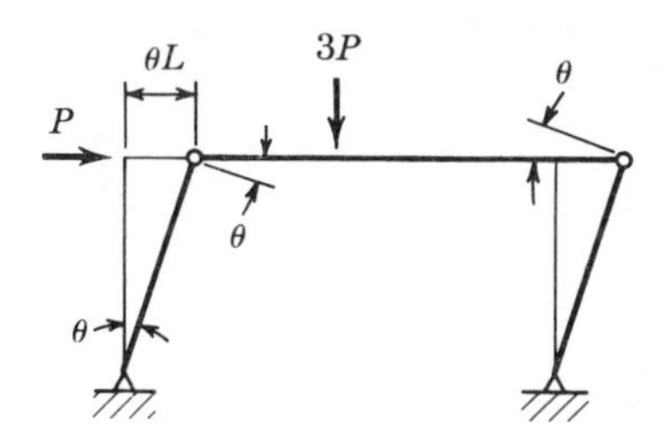

External work = internal work

$$P\theta L = M_P(\theta + \theta)$$

$$P = \frac{2M_P}{L}$$

(*a*) Panel mechanism

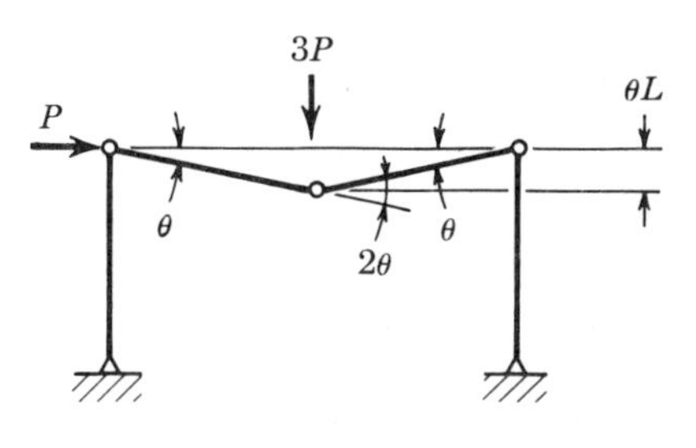

External work = internal work

$$3P\theta L = 4M_P\theta$$

$$P = \frac{4}{3}\frac{M_P}{L}$$

(*b*) Beam mechanism

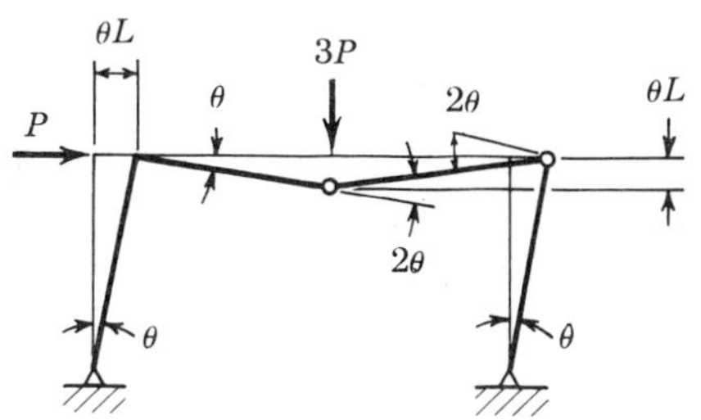

External work = internal work

$$P\theta L + 3P\theta L = M_P(2\theta + 2\theta)$$

$$P = \frac{M_P}{L}$$

(*c*) Combined mechanism

Fig. 7.16

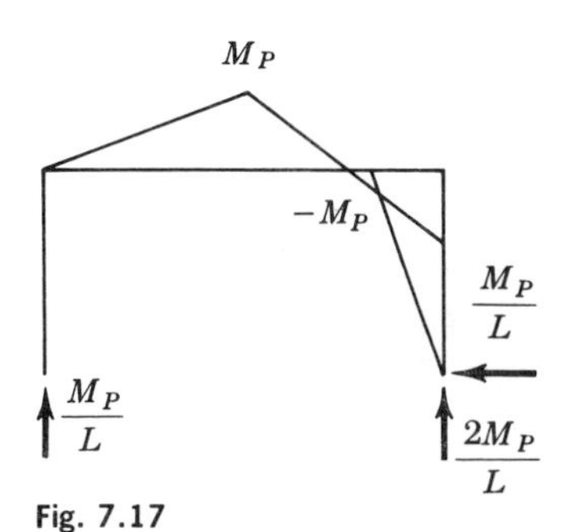

Fig. 7.17

be the correct ones, since the solution is both kinematically and statically admissible. It may be verified at this time that neither solution 1 nor 2 allows construction of such a statically admissible moment diagram.

Example Problem 7.5 Uniform Strength M_P. Neglect effect of axial forces.

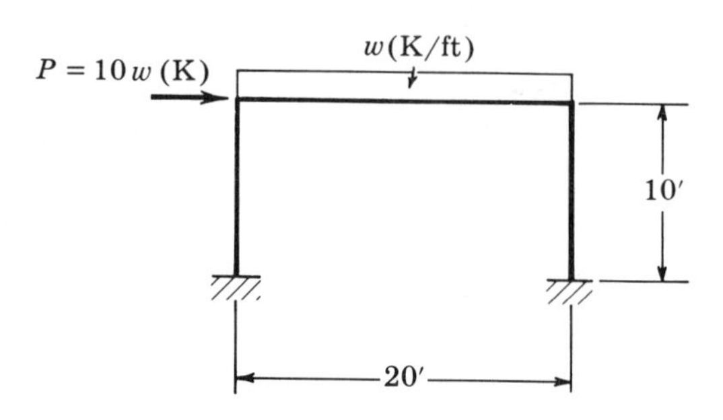

Find an unconservative estimate of the ultimate moment by use of the upper bound theorem. Determine bounds on the amount of error by use of the lower bound theorem.

We try possible collapse mechanisms, using the principle of virtual displacement to satisfy statics. Assuming possible collapse mechanisms, we investigate each in turn.

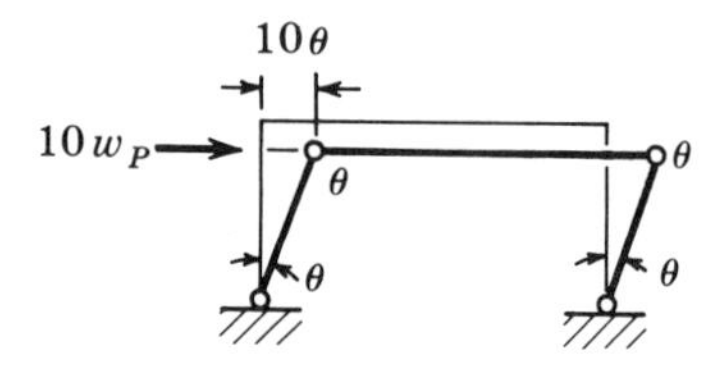

(*a*) Panel mechanism

External work = internal work

$$(10w_P)(10\theta) = M_P 4\theta$$

$$w_{P_{\text{upper}}} = 0.04 M_P, \text{K/ft}$$

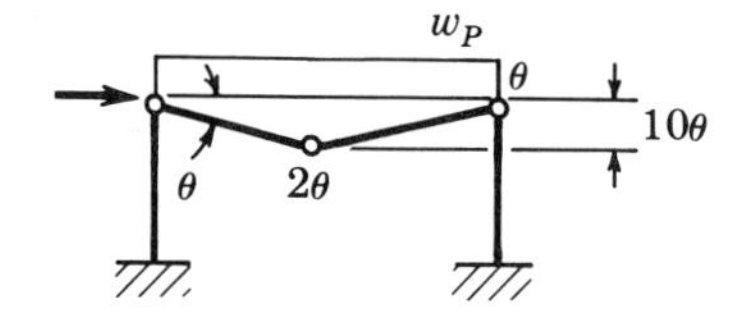

(*b*) Beam mechanism (assume plastic hinge at midspan)

External work = internal work

$$(20w_P)(5\theta) = M_P 4\theta$$

$$w_{P_{\text{upper}}} = 0.04 M_P, \text{K/ft}$$

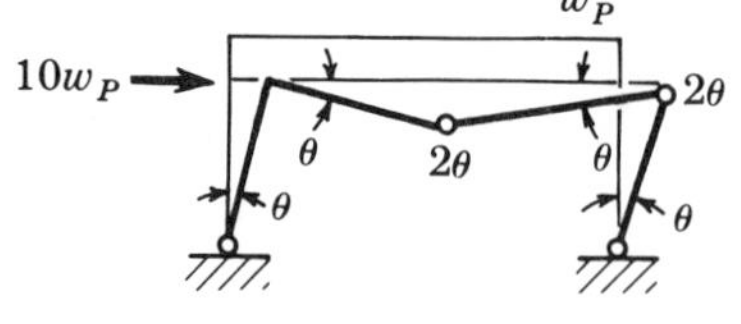

(*c*) Combined mechanism, assume plastic hinge at midspan

$$(10w_P)(10\theta) + (20w_P)(5\theta) = M_P 6\theta$$

or

$$w_{P_{\text{upper}}} = 0.03 M_P, \text{K/ft}$$

Comparing results, we see that mechanism 3 results in the lowest upper bound.
To verify the validity of the result, we draw the moment diagram at collapse according to mechanism 3, using equilibrium conditions of various free bodies.

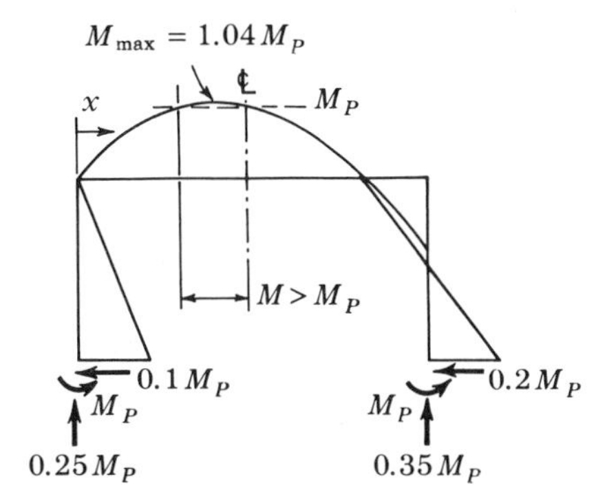

To investigate static admissibility, we determine the maximum moment in the beam (that the static admissibility of the columns is satisfied can be seen by inspection); writing the moment expression, we set

$$M = 0.25M_P x - 0.03M_P \frac{x^2}{2}$$

and, differentiating with respect to x and equating to zero, we determine the point of maximum moment to be of location

$$x = 8.33 \text{ ft}$$

and the magnitude of the largest moment as

$$M_{max} = 1.04M_P$$

Clearly, because of an incorrect assumption regarding the location of one plastic hinge, a statically inadmissible region exists in the beam, in which the member strength is exceeded by 4 percent. A statically admissible moment diagram (and therefore a lower bound) could be obtained by scaling the load, and therefore the moment diagram, down in the ratio 100/104. The resulting lower bound load is

$$w_{P_{lower}} = \tfrac{100}{104} \times 0.03M_P = 0.0289M_P$$

The correct collapse load must therefore lie between $0.289M_P$ and $0.0300M_P$.

Instantaneous center To calculate the rotations of members and displacement of loads resulting from a virtual displacement, the instantaneous-center method is often useful. It will be explained making use of the collapse mechanism of the previously considered rectangular frame, shown again in Fig. 7.18. This mechanism consists of three moving parts or links, ABC, CD, and DE, each one rotating about some point. As before, the virtual displacement will be a rotation of angle θ of the leg DE of the frame. Required is the rotation of the link CD and the travel of the applied vertical load, caused by this virtual movement.

Since link ABC rotates about pin A, the direction of travel of point C will be normal to a radius from C through A. But point C is also a point on link CD, and therefore this link must rotate about a point located somewhere on the extension of this radius CA. Point D, being on link DE, rotates about point E, and its direction of movement is therefore at right angles to the radius connecting D and E. The direction of travel of point D, being also part of link CD, determines another radius through the center of rotation of link CD. This instantaneous center of link CD must therefore lie on the intersection of the two radii and, by proportion, is found to be a distance L above point D.

The angle of rotation of link CD about its instantaneous center is determined by dividing the tangential travel of point D, θL, by the radius to the instantaneous center, to get an angle of rotation of θ. The vertical

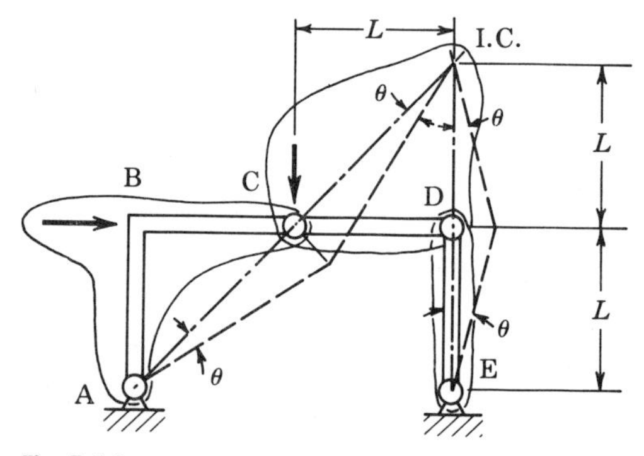

Fig. 7.18

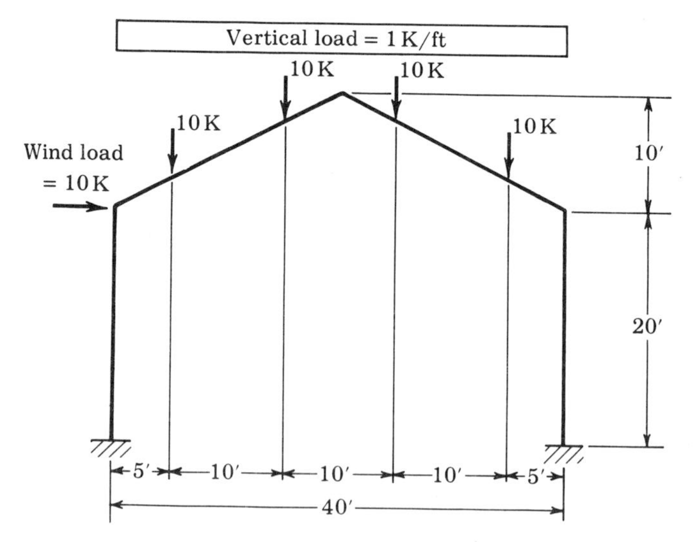

Fig. 7.19

travel of the load at C, traveling with the link CD, is the angle of rotation times the horizontal distance from the load to the instantaneous center, or θL, as obtained before. With command of this technique, a more complicated problem may be attacked.

Example Problem 7.6

Given: Gable frame of dimensions as shown in Fig. 7.19, subject to both vertical and lateral loads; uniform cross-section members.

Required: Ultimate moment capacity of member, if a factor of safety against collapse of 2 is required. Assume structure to be laterally braced.

For ease of calculation, the uniform loads are replaced by equivalent concentrated loads, each one of magnitude equal to one-quarter of the total vertical load and located at the quarter points of each gable member. Since the statically determinate moment diagram for this approximate loading forms an envelope for the corresponding diagram for the uniform load, it can be concluded that this approximation leads to conservative results.

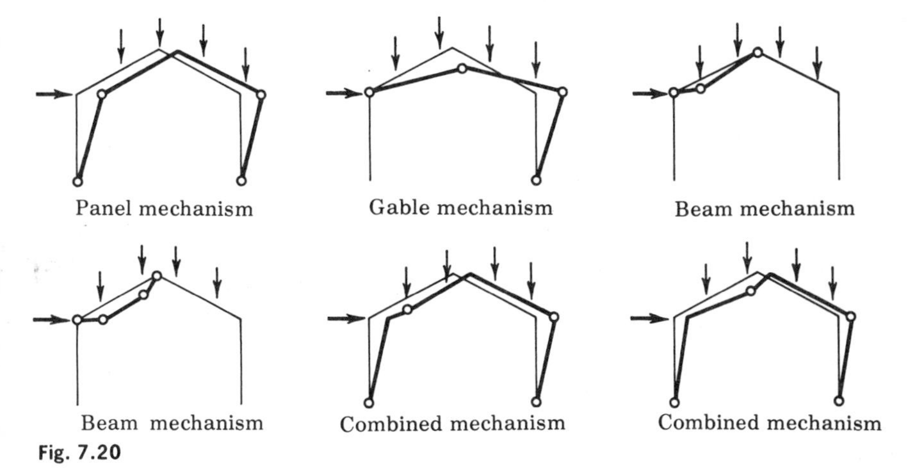

Fig. 7.20

Figure 7.20 shows that a good number of collapse mechanisms is possible. Some of these mechanisms can be eliminated by judgment. So, for instance, it is likely that all the loads will contribute to the external work and therefore to the collapse of the structure. It is seen that only in the case of the combined mechanisms are all applied forces engaged in doing work. Therefore, one of the combined mechanisms will probably turn out to be the critical one.

Accordingly, one of the combined mechanisms is analyzed, making use of the instantaneous-center method to determine relative rotations and displacements of applied forces. It should be borne in mind that the location of the instantaneous center and all other dimensions are determined on the basis of the undeformed frame.

Referring to the Fig. 7.21*a*, the virtual work calculations are as follows:

$$\text{External work} = 10[\tfrac{5}{8}\theta(20\text{ ft} + 5\text{ ft} + 15\text{ ft}) + \tfrac{3}{8}\theta(5\text{ ft} + 15\text{ ft})] = 325\theta$$

$$\text{Internal work} = M_P\theta(\tfrac{5}{8} + 1 + \tfrac{11}{8} + 1) = 4M_P\theta$$

therefore, equating,

$M_P = \frac{325}{4} = 81$ kip-ft

It is therefore seen that for this mechanism the collapse moment is equal to 81 kip-ft.

Rather than investigate other possibilities, the statically determinate moment diagram is drawn for this mechanism in Fig. 7.21*b*. It is seen that at no place is the ultimate moment exceeded. According to the previous discussion, under the required factor of safety of 2, the required ultimate moment for the section is then

$M_{P\,reqd} = 2 \times 81 = 162$ kip-ft

This mechanism is therefore the critical one; the solution corresponding to it is both kinematically and statically admissible, and therefore must furnish the correct conditions for collapse.

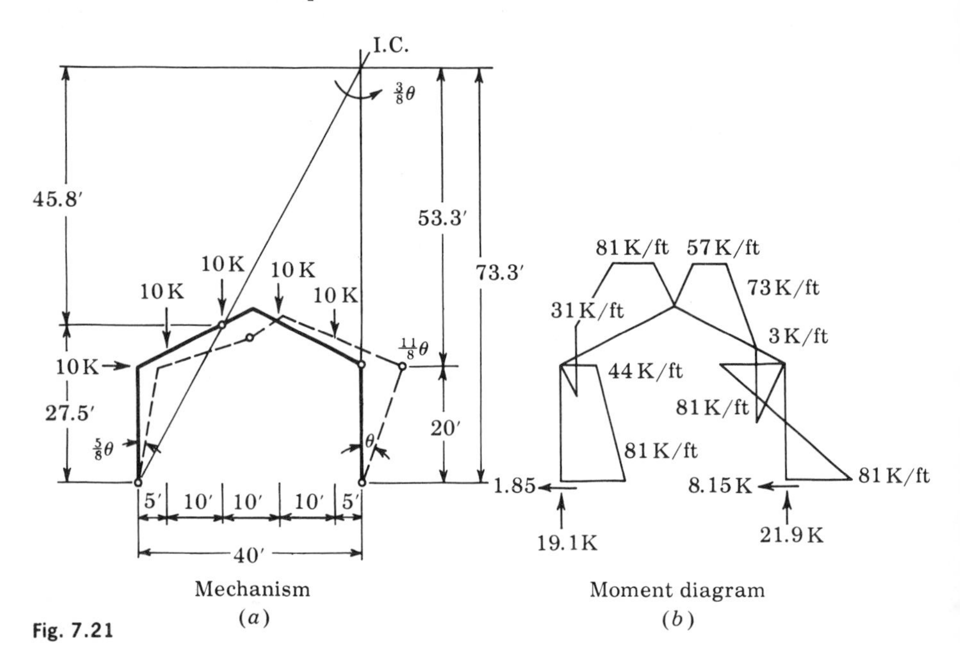

Fig. 7.21

7.6 Graphical Method of Analysis

The graphical method of plastic analysis which has been suggested in Sec. 7.4 for beams can also be extended for use with rigid frames; it is particularly useful in cases of structures with haunched or tapered members and unusual loading conditions.

As explained previously (see Sec. 7.3), the moment diagram at collapse can be drawn by parts, one part resulting from the action of the applied loads on the frame made statically determinate, the other from the effects of the redundant force. The statically determinate moment diagram is drawn first, and the redundant diagram is superimposed of such magnitude that plastic hinges form at sufficient points to ensure formation of a mechanism. By comparing the furnished strength M_P with the total moment (composed of determinate and redundant parts) at any point, the location of plastic hinges can be determined, and the value of ultimate moment is then scaled off the diagram.

For the case of a pin-ended gable frame, Fig. 7.22 illustrates the procedure. The frame of constant section shown in Fig. 7.22*a* and subjected to loads w and T is to be analyzed for the required ultimate moment capacity. In Fig. 7.22*c* the moment diagram for the statically determinate frame of Fig. 7.22*b* is drawn to scale on the developed axis. To this is added the moment due to the redundant reactions R of Fig. 7.22*d*, shown in Fig. 7.22*e*. The geometrical construction indicates that this moment diagram may be

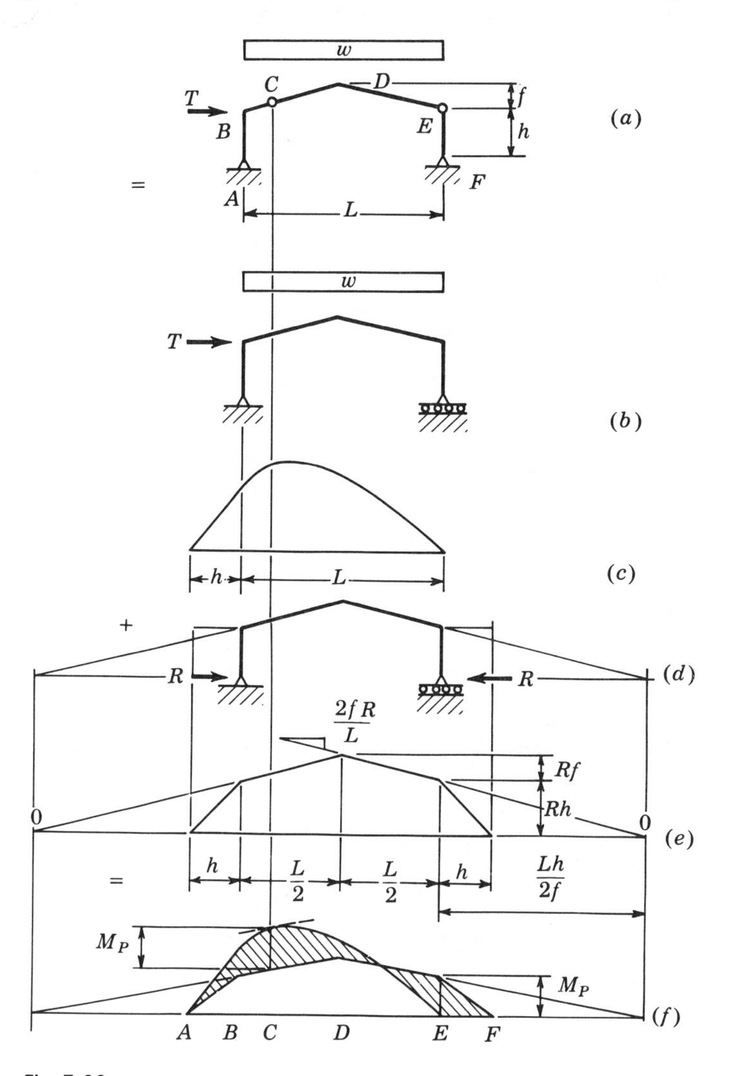

Fig. 7.22

drawn by means of a line passing through a center of rotation O located a distance of $Lh/2f$ from the knee. To draw this moment diagram for varying magnitudes of the redundant R, it is only necessary to rotate this line through O and to connect it from the knee to the zero moment ordinate corresponding to the pinned column base.

The redundant-moment diagram is superimposed upon the determinate diagram as shown in Fig. 7.22f, and the magnitude of the redundant diagram is adjusted as described till the moment at point C becomes equal to the moment at point E. The moment at these points is scaled off to determine the ultimate moment. It should be noted that this mechanism must be the correct one, since all three conditions for collapse outlined in Sec. 7.4 are satisfied.

Example Problem 7.7 Design the gable frame shown in Fig. 7.23a for the loads shown. Steel to be A-7 (σ_{YP} = 33 ksi). Safety factor against collapse to be 2. Effects of axial forces to be neglected.

(a) Design the frame of uniform section.
(b) Investigate the effect of haunching the members at the knees on the weight of the frame. Consider a haunch as shown in Fig. 7.23b. Design haunch, compare weights, and draw conclusions.

Figure 7.23c shows the graphical computation. The frame is made determinate by removing the horizontal reactions, and the statically determinate moment diagram is drawn by parts (the parabola of maximum ordinate $wL^2/8$ = 450 kip-ft resulting from the uniform load, the straight-line diagram from the concentrated load) and superimposed. The scale is established when this moment diagram is drawn.

For part (a) of the design, the redundant part of the moment diagram is drawn as previously described. The straight line segments were rotated about their pivot points H and J till the moments at points B, F, and D are seen to have attained identical maximum values. This redundant diagram is shown in solid line. The critical moment values are scaled off as 215 kip-ft. Note that no kinematically admissible mechanism arises with plastic hinges only at the knees, because a sway mechanism would require moments of opposite signs. In this case, therefore, three hinges are required for collapse, and the mode of collapse may be characterized as a beam mechanism involving spreading of the frame, or *arch mechanism*.

With the safety factor of 2, the required fully plastic moment is 2 × 215 kip-ft = 430 kip-ft, the required plastic section modulus is

$$Z_{reqd} = \frac{430 \times 12}{33} = 156 \text{ in.}^3$$

and this is furnished by a 21 WF 68 (Z_{furn} = 159.8 in.^{3}). The total weight of this frame is computed as

$$(40 \text{ ft} + 44.7 \text{ ft} + 28.3 \text{ ft})68 = 7{,}700 \text{ lb}$$

Turning now to part (b), we conceive the haunches, composed of cover plates, to be strong enough to prevent hinge formation. Plastic hinges must thus form outside the reinforced part, and the graphic analysis is performed on this basis. The dashed redundant-moment diagram, when superimposed upon the previously drawn determinate part, indicates plastic hinges at points G (under the concentrated load) and K (just inside the right haunch). Note that this corresponds to a sway mechanism, in contrast to part (a). Since we did not consider plastic hinges in the columns, we will have to ensure sufficient strength there by suitably extending the cover plates.

We again scale off the critical moments, this time measuring $M_{max} = 170$ kip-ft, leading to a required $M_P = 2 \times 170 = 340$ kip-ft, corresponding to a plastic

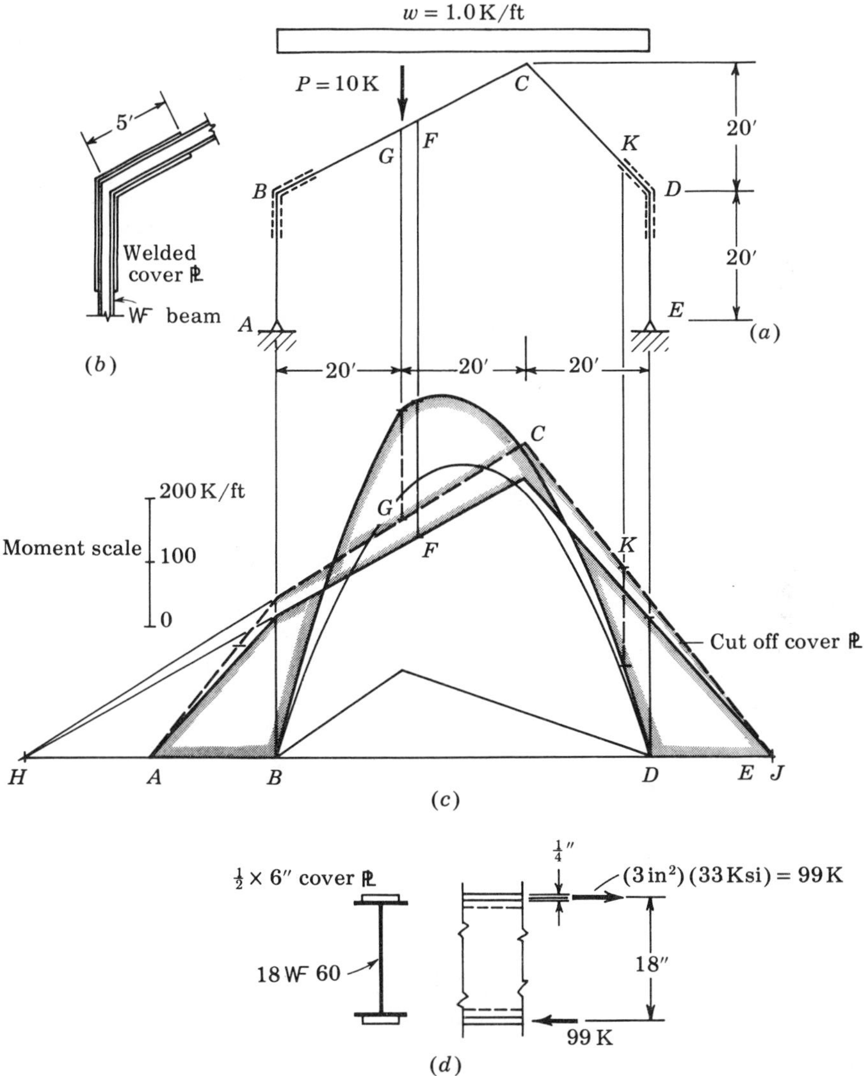

Fig. 7.23

section modulus

$$Z_{reqd} = \frac{340 \times 12}{33} = 123 \text{ in.}^3$$

provided by a 18 WF 60 beam ($Z_{furn} = 122.6$ in.^{3}).

Turning now to the haunches, we scale off the maximum moment at the knees as 245 kip-ft, so that the cover-plated section must have a strength of $2 \times 245 = 490$ kip-ft, of which 340 kip-ft are furnished by the beam. The cover plates must thus resist an additional moment of $490 - 340 = 150$ kip-ft, and, referring to Fig. 7.23*d*, this is provided by $\frac{1}{2} \times 6$ in. cover plates inside and outside. To find the required length of the plates along the column, we determine the cutoff point from the moment diagram as the point where the moment drops below the capacity of the beam alone, at a distance of 6 ft below the knee. This completes the design according to (*b*).

The weight of alternate (*b*) is 113 ft $\times$ 60 lb/ft + 44 ft $\times$ 10.2 lb/ft = 7,250 lb. We see that a weight saving of 450 lb or 6 percent has been achieved by haunching the frame. Further economic studies must decide whether this saving compensates for the added fabrication cost.

7.7 Inelastic Deformations

In the general case of inelastic beams, Eq. (7.7) serves to establish a relationship between the applied moment and the resulting curvature. For the particular case of a rectangular cross-section beam of elastic–perfectly plastic material, it was shown that this relation is given by Eq. (7.16):

$$\phi = \frac{M_{YP}}{EI} \frac{1}{\sqrt{2}\sqrt{\frac{3}{2} - M/M_{YP}}} \tag{7.16}$$

above the yield point of the material. For wide-flange and other beams, corresponding relationships may be found, utilizing identical principles.

Once the relation between moment and curvature is known, the complete deformation of any statically determinate structure may be determined by methods analogous to those utilized in the elastic range.

Recalling that for small deflections $\phi = d^2y/dx^2$, the curvature may be written in terms of the moment at any point, and by successive integration slope and deflection at any point may be determined.

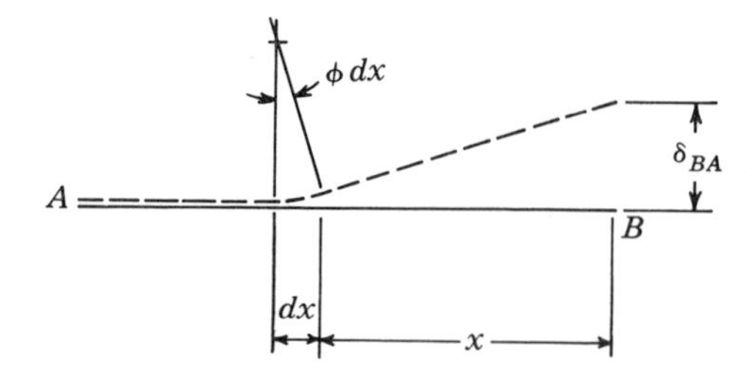

Fig. 7.24

The double-integration method has the same deficiencies in the inelastic as in the elastic range; whenever mathematical discontinuities occur along the structure, such as a concentrated load, a new expression must be set up, and boundary conditions must be met. The use of double integration becomes prohibitive for most practical problems.

Another commonly used approach to the problem of calculating deformation is by consideration of the geometry of the deflected shape of the structure. Figure 7.24 shows a portion of a beam AB, containing a short length dx which is being bent into a curvature ϕ. The effect of this bent length dx on the total angle change $d(\theta_{AB})$ between two tangents at A and B, and on the tangential deviation $d(\delta_{BA})$ of point B on the elastic curve from the tangent at A, is to be investigated. From the geometry of the deflected shape,

$$d(\theta_{AB}) = \phi\, dx$$

and $$d(\delta_{BA}) = \phi x\, dx$$

If now the total angle θ_{AB} resulting from the bending of all parts of the beam between A and B is to be evaluated, the individual effects must be summed up:

$$\theta_{AB} = \int_A^B \phi\, dx \tag{7.19}$$

and, analogously for the tangential deviation,

$$\delta_{BA} = \int_B^A \phi x\, dx \tag{7.20}$$

Deformations may be evaluated using Eqs. (7.19) and (7.20).

Figure 7.25 interprets the physical meaning of the quantities θ_{AB} and δ_{BA}. Referring now to the curvature diagram Fig. 7.25c, which has been

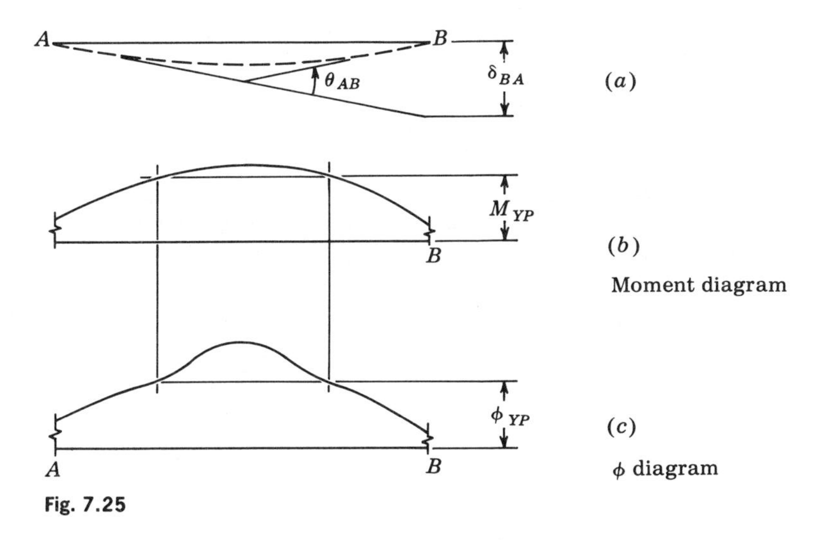

Fig. 7.25

plotted utilizing the moment diagram and the previously found moment-curvature relationship, it can be seen that

$$\text{Area between } A \text{ and } B \text{ of } \phi \text{ diagram} = \int_A^B \phi\, dx = \theta_{AB}$$

and Static moment of area between A and B about point B =

$$\int_A^B \phi x\, dx = \delta_{BA}$$

From these two relations, the *curvature area theorems* may be written:

1. The area under the curvature diagram between two points A and B on a member is equal to the total angle between two tangents drawn to the deflected curve at A and B.
2. The static moment about point B of the area under the curvature diagram between any two points A and B is equal to the deviation of point B on the deflected curve from the tangent to the deflected curve drawn at A.

It is apparent that the well-known moment-area theorems constitute the special case of elastic behavior for which

$$\phi = \frac{M}{EI}$$

Example Problem 7.8

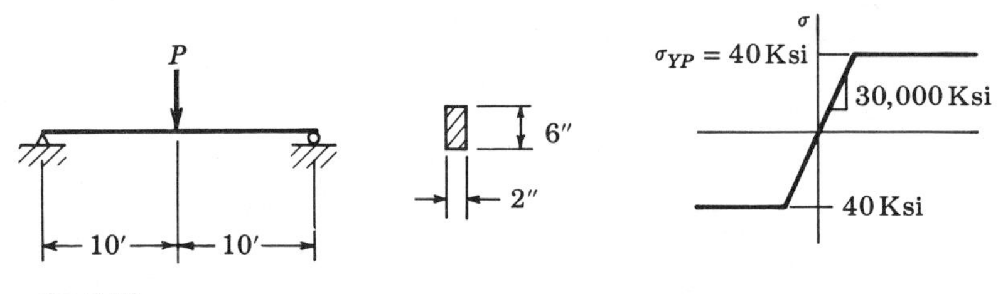

Fig. 7.26

A simply supported rectangular cross-section beam of elastic–perfectly plastic material, as shown, is subjected to a concentrated midspan load. Plot the load (in kips) versus midspan deflection (in inches) of this beam to failure.

We begin by considering the deflected shape of the beam and realize that by symmetry the midspan deflection is equal to the tangential deviation δ_{CB} and can thus be expressed as

$$\Delta_{\max} = \delta_{CB} = \int_0^{L/2} \phi x\, dx$$

where ϕ is the curvature (related to the moment by either elastic or inelastic relations) and x is measured from point C.

To express the curvature, we first write the statically determinate moment as

$$M = \frac{P}{2} x$$

and the moment-curvature relations as

$$\phi = \frac{M}{EI}$$

in the elastic portion of the beam, and

$$\phi = \frac{M_{YP}}{EI}\frac{1}{\sqrt{2}}\frac{1}{\sqrt{\frac{3}{2} - M/M_{YP}}}$$

in the inelastic portion of the beam.

Considering first the deformations under a load $P \leq P_{YP} = 4M_{YP}/L$, such that elastic action prevails throughout, we use elastic theory:

$$\Delta_{max} = \int_0^{L/2} \frac{P}{2EI} x^2\, dx = \frac{1}{48}\frac{PL^3}{EI} \qquad P \leq P_{YP}$$

As soon as the load exceeds P_{YP}, the maximum moment will exceed M_{YP}, and an inelastic zone will develop near midspan. This is indicated in the moment diagram drawn in solid line in Fig. 7.27*b*, the geometry of which shows that the elastic zone

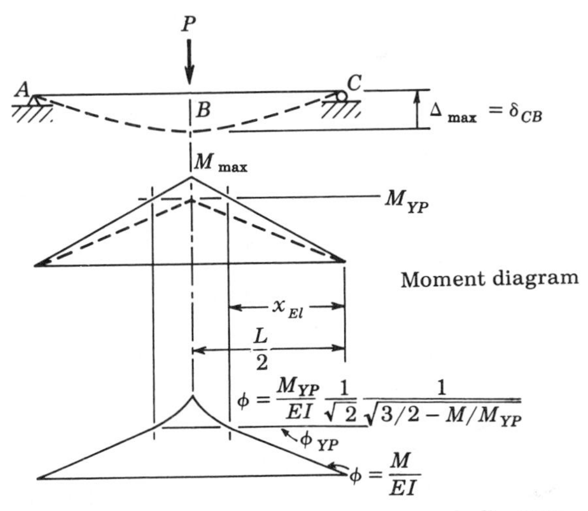

Fig. 7.27

extends from

$$0 < x \leq x_{El} = \frac{M_{YP}}{M_{max}}\frac{L}{2} = \frac{P_{YP}}{P}\frac{L}{2}$$

For this portion of the beam, the elastic moment-curvature relation is appropriate. For the inelastic zone, for

$$\frac{P_{YP}}{P}\frac{L}{2} = x_{El} \leq x \leq \frac{L}{2}$$

the inelastic moment-curvature relation

$$\phi = \frac{P_{YP}L}{4EI}\frac{1}{\sqrt{2}}\frac{1}{\sqrt{\frac{3}{2} - Px/2M_{YP}}}$$

is indicated, so that, for the inelastic loading range

$$\Delta_{\max} = \int_{x=0}^{(P_{YP}/P)(L/2)} \frac{P}{2EI} x^2\,dx + \int_{(P_{YP}/P)(L/2)}^{L/2} \frac{P_{YP}L}{4\sqrt{2}\,EI} \frac{x\,dx}{\sqrt{\frac{3}{2} - 2Px/P_{YP}L}}$$

$$= \frac{P_{YP}L^3}{48EI}\left(\frac{P_{YP}}{P}\right)^2 \left[5 - \sqrt{2}\left(3 + \frac{P}{P_{YP}}\right)\sqrt{\tfrac{3}{2} - P/P_{YP}}\right] \qquad P > P_{YP}$$

Summarizing these results, we plot them in both dimensioned and dimensionless form:

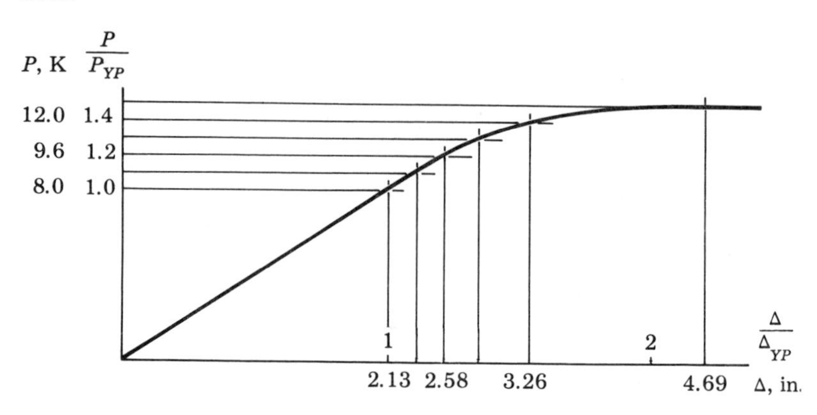

Fig. 7.28

In the case of statically indeterminate structures, the application of these methods becomes extremely laborious; conditions of geometry must be satisfied (as long as the number of plastic hinges is less than the number required to reduce the structure to static determinacy), and the principle of superposition, the mainspring of elastic methods of indeterminate analysis, is invalid owing to the nonlinearity of moment-curvature relations. Methods of trial and error or successive approximations have been applied.[1]

To avoid these difficulties, an approximate method, the so-called "plastic-hinge method," for calculating deflections during the entire range

[1] K. H. Gerstle, Deflections of Structures in the Inelastic Range, *Proc. ASCE, J. Struct. Div.*, July, 1957.

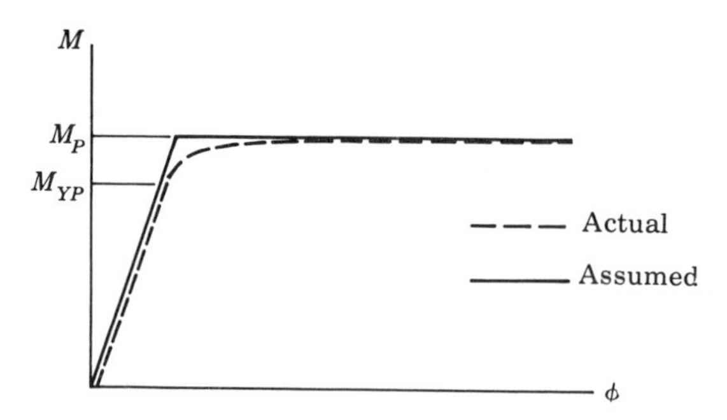

Fig. 7.29

of loads up to collapse has been devised. It is based on an assumed moment-curvature relationship which is shown in Fig. 7.29. According to this assumption, elastic conditions will prevail up to the instant when the applied moment reaches a value of M_P, at which time the member will suddenly form a plastic hinge localized along an infinitely short section of the beam. At other points of the structure elastic relationships hold, and the required deformations can therefore be calculated using conventional elastic methods.

The plastic-hinge method, then, consists of introducing pins at the point or points of the structure at which plastic hinges have been formed, and of analyzing the moments and stiffness of the structure, thus modified, according to elastic principles.

Example Problem 7.9 Let us apply this method to the fixed-ended, uniformly loaded beam shown in Fig. 7.30*a*. The beam will deflect elastically till the end moments reach a value of $M_P = fM_{YP}$. This will occur when the uniform loads reach a value of $w = 12M_P/L^2$. According to elastic analysis, the center deflection will be

$$\Delta_{\text{center}} = \frac{wL^4}{384EI} = \frac{M_P L^2}{32EI}$$

at that instant.

The behavior of the beam is denoted by "stage 1" in the load-deflection curve, Fig. 7.30*b*. At this instant, plastic hinges form at the ends, so that from then on the beam will act as if it were pin-ended. The subsequent deflections resulting from additional loads will therefore be calculated as elastic deflections of a simply supported beam. This behavior, labeled "stage 2," will prevail till the center moment reaches a value of M_P, which, according to previous calculations, will occur when the load reaches a value of $16M_P/L^2$; the additional load during stage 2 is therefore $(16 - 12)\ M_P/L^2 = 4(M_P/L^2)$.

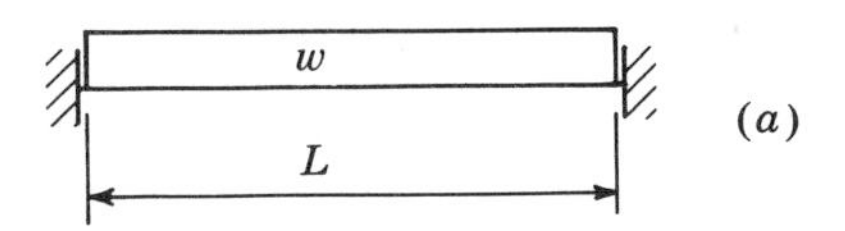

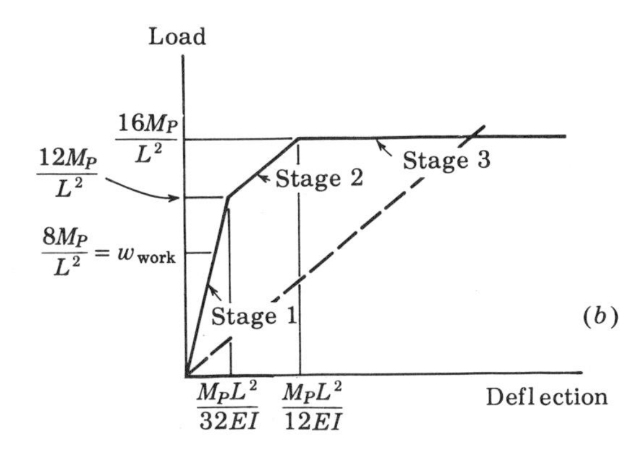

Fig. 7.30

The center deflection of a simply supported beam acting elastically is

$$\Delta = \frac{5wL^4}{384EI}$$

The added deflection during stage 2 is therefore

$$\Delta = \frac{5M_PL^2}{96EI}$$

At the instant when the third plastic hinge forms, the structure becomes a mechanism and will deflect indefinitely under constant load, as shown by the portion of the plot labeled "stage 3."

The complete load-deflection diagram is shown in Fig. 7.30*b*.

The load-deflection diagram of Fig. 7.30*b* gives a reasonably close measure of deflections up to the instant of collapse. The calculations involved consist essentially of two parts: determination of the sequence of hinge formation and the magnitude of load under which this formation occurs, and a series of elastic deflection calculations. For complicated structures, the sequence of hinge formation is not always easily found, and the deflection calculations are laborious. Still faster means of indicating whether deflections will control are therefore desirable.

Assuming that the structure analyzed for deflections in this example was designed with a factor of safety of 2 against collapse, then the working load would have a value of $8M_P/L^2 = \frac{1}{2}w_P$. It is seen, referring to Fig. 7.30*b*, that elastic action still prevails under this load. Indeed, this will often be found to be the case in plastically designed structures. A possibly unconservative estimate of deflections under working loads can thus be obtained by elastic means.

It should be noted that after initial hinge formation the structure becomes more flexible. It is thus possible that it will lose its usefulness owing to excessive deflections under a load smaller than the ultimate. In this case the design safety factor becomes illusory; we can see that a knowledge of deflections under ultimate load is as important as under working load.

The complete load-deflection curve offers one approach to finding the deflection at the instant of collapse; an upper bound to this deflection may be found by assuming the structure to have the stiffness of its statically determinate form just prior to collapse. The dashed line of Fig. 7.30*b* indicates this assumption. It is seen that its value at ultimate load gives a safe measure of the order of magnitude of the deflection at collapse.

7.8 Code Requirements for Plastic Design of Steel Structures

Part 2 of the 1963 AISC Specifications is concerned with the correct application of the principles of limit analysis to the design of steel structures.

No mention is made, or formulas given, for the calculation of plastic moments of sections or collapse loads on structures. It is assumed that the designer is conversant with the basic approach and methods to obtain these quantities. Section 2.1, however, does provide for safety factors on the allowable loads (called *load factors*) which are approximately the same as those which prevail in statically determinate beams designed according to current elastic design provisions. Thus, for continuous beams the load factor against collapse is to be 1.70, while for rigid frames the somewhat more conservative factor of 1.85 is prescribed. This means that the first step in performing a plastic design is to multiply the given design loads by the appropriate factor. Provision is also made for reducing the load factor for combinations of vertical and lateral loads to three-quarters of the value for vertical loads only, analogous to the provisions for elastic design.

An important provision of sec. 2.1 is the limitation of plastic design methods to one- or two-story frames; this is because our understanding of the behavior of columns under high axial loads combined with bending is so incomplete that it seems prudent to await results of further research before extending these practices to tall structures subject to sidesway.

We are also reminded that plastic-design methods are inappropriate to dynamically stressed members such as crane rails.

Section 2.3 is concerned with plastic design of columns. Since this chapter refers only to the strength of members under bending without axial load, a rational approach and explanation of these provisions must await our Chap. 11. It may, however, be noted here in passing that plastic as well as elastic design of compression members is a trial-and-error procedure.

The plastic design of members in which high shear force prevails has not been discussed in these notes because it requires an understanding of the conditions under which a material yields when subjected to combined stresses. Rational attacks on this problem may be found in texts on plasticity;[1] the formula of sec. 2.4 is based on the simple premise that the web of an I beam of cross-sectional area wd is available to carry the shear at a yield stress in shear equal to 0.55 times the normal yield stress, so that the total shear force which can be carried is

$$V_u = 0.55 F_{YP} wd$$

Note that this formula does not consider any interaction between moment and shear force, as a rational formula should. Nevertheless it has been found safe by tests. If insufficient web area is available, the beam should be reinforced with doubler plates or diagonal stiffeners.

Section 2.5 makes provision against web crippling at points of concentrated force application.

Local buckling of outstanding compression flanges is specially critical if plastic compressive stresses are caused by bending; sec. 2.6 establishes limiting width-thickness ratios for both compression flanges and webs of sections used in plastic design; the requirements of this section are more rigorous than those of the corresponding sec. 1.9 for elastic design; cross

[1] Philip G. Hodge, Jr., "Plastic Analysis of Structures," p. 206, McGraw-Hill, 1959.

sections which satisfy sec. 2.6 are called *compact sections;* because of the lessened danger of local buckling, the allowable elastic bending stress for such sections is increased by 10 percent, as was pointed out earlier.

Connection design is considered in sec. 2.7. The concern here is to provide ultimate strength equal to that of the adjoining members, rigidity and rotation capacity to allow full redistribution of moments.

Lateral or torsional buckling is a critical factor especially in the vicinity of plastic hinges, and sec. 2.8 furnishes rules for lateral bracing near such points. Some experts are of the opinion that even these provisions are too liberal: it cannot be said that the problem of lateral buckling of plastic members is solved.

It can at best be said that part 2 of the AISC Specifications provides some sketchy guidelines for plastic design. A good understanding of the principles of plastic behavior of structures is indispensable for a meaningful application and supplementation of these rules.

General Readings

Inelastic beam theory

The basic theory is presented in most of the elementary and advanced-level strength of materials texts listed in Chap. 2. A more specialized, and therefore more thorough, treatment is the following:

Smith, James O., and Omar M. Sidebottom: "Inelastic Behavior of Load-carrying Members," Wiley, 1965. Chapter 3 deals with bending of beams.

Limit analysis and design

AISC: "Plastic Design in Steel," American Institute of Steel Construction, 1959. Simple treatment and design aids for limit design of standard structures.

ASCE: "Commentary on Plastic Design in Steel," Manual of Engineering Practice no. 41, 1961. An excellent summary of the experimental and theoretical bases of current plastic design practice, based on the research done at Lehigh University.

Beedle, Lynn S.: "Plastic Design of Steel Frames," Wiley, 1958. Written at a somewhat lower level than the book by Hodge, this has a more practical orientation.

Beedle et al.: "Structural Steel Design," Ronald, 1964. An up-to-date undergraduate steel design text which emphasizes plastic design. Makes extensive use of the research findings of Lehigh University.

Hodge, Philip G., Jr.: "Plastic Analysis of Structures," McGraw-Hill, 1959. A thorough presentation of limit analysis and design at an intermediate-advanced level, written by one of the foremost plasticians of this time. The first part deals with classical civil engineering structures, the second with continuum structures.

Problems

7.1 A beam of rectangular cross section of width b and depth d is made of material of equal properties in tension and compression whose mechanical properties are given by

$$\sigma = \sigma_{YP}(1 - e^{-k\epsilon})$$

(*a*) Calculate and plot the moment-curvature relationship.

(*b*) Calculate the maximum moment which can be carried by the member, and plot the corresponding strains and stresses.

7.2 A rectangular-cross-section beam of width b and depth d is made of the material of dissimilar properties in tension and compression whose stress-strain curve is shown.

(*a*) Find a moment-curvature relation for the beam and plot.
(*b*) Describe the movement of the neutral axis under increasing load.

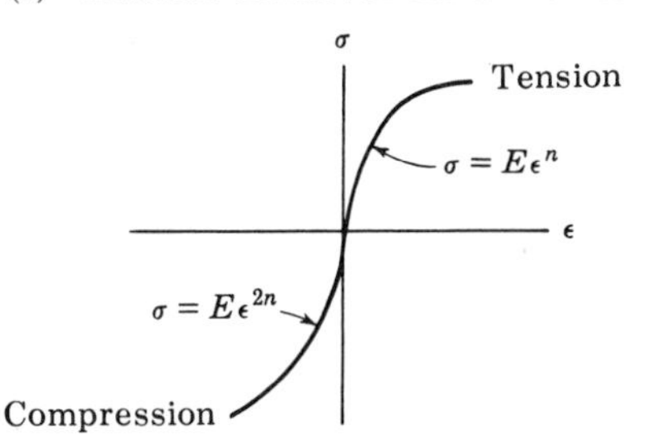

7.3 A beam 6 in. wide by 10 in. deep is constructed of the material whose stress-strain curve is shown.

(*a*) Find the location of the neutral axis during the elastic range.
(*b*) Find the value of the moment at which inelastic action begins. Determine the stresses and strains in the extreme fibers of the beam at this instance.
(*c*) Determine and plot the moment-curvature relation of the beam under further increase of moment. State the limits of validity of this solution.
(*d*) Determine and plot the moment-curvature relation under still further increase of moment.
(*e*) Find the moment at which failure of the beam will occur.

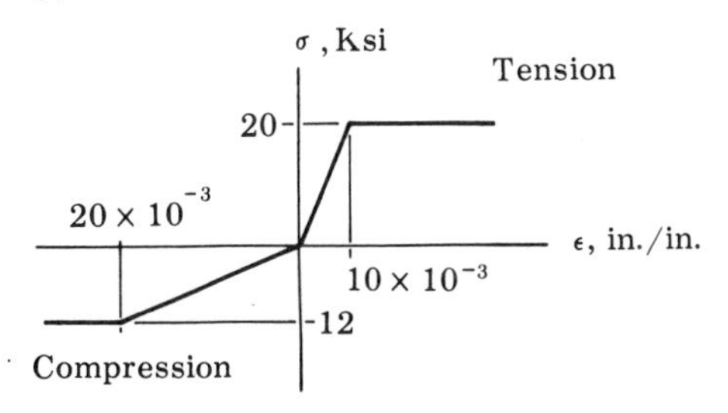

7.4 A material exhibits the properties shown in the stress-strain curve. A beam 6 in. wide by 12 in. deep is fabricated of this material.

(*a*) Calculate the maximum moment which can be carried by this beam.
(*b*) Describe the behavior of the beam if it is bent further. Draw an approximate moment versus curvature curve and find the location of the neutral axis when final separation of the beam occurs.

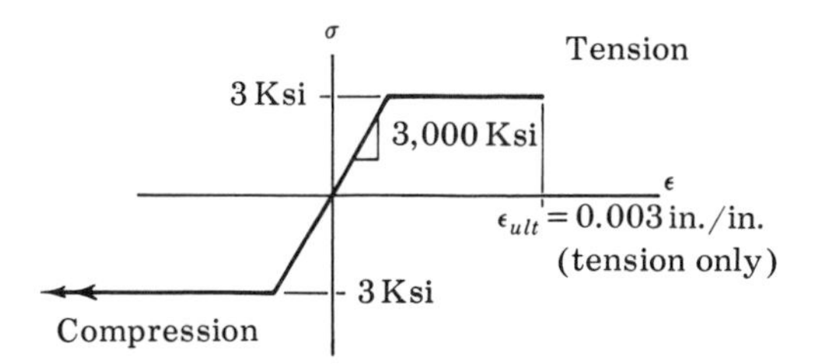

7.5 (*a*) Write an expression relating the applied moment on a beam of the diamond-shaped cross section shown and the resulting curvature. Assume a material of elastic–perfectly plastic behavior with elastic modulus E and yield strength σ_{YP}.

(*b*) Write the expression of part (*a*) in terms of the ratio M/M_{YP} versus the ratio ϕ/ϕ_{YP}. Plot the results and superimpose upon the appropriate curve of Fig. 7.5 in order to check its accuracy.

(*c*) Find the ultimate moment M_P and the shape factor of this section.

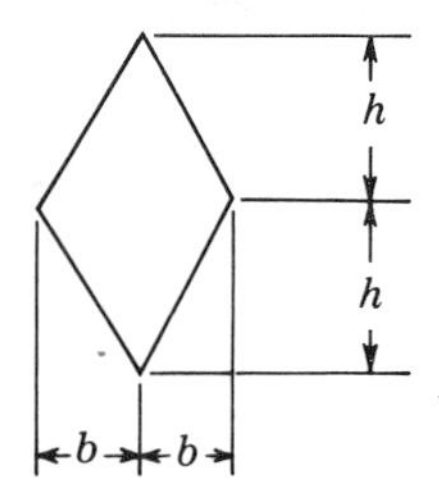

7.6 A beam of triangle cross section is made of elastic–perfectly plastic material of elastic modulus E and yield strength σ_{YP}. Find:

(*a*) The moment M_{YP} at which purely elastic action ceases.
(*b*) The moment M_1 at which yielding of edge AB commences.
(*c*) The maximum moment M_P which may be applied to the section.

Observe the movement of the neutral axis under increasing load.

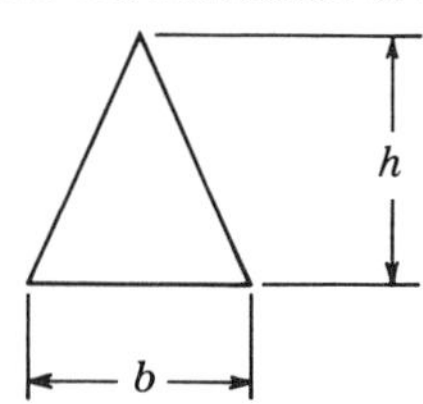

7.7 A rectangular cross-section beam is 4 in. wide, 10 in. deep, and of an elastic–perfectly plastic material of $E = 30{,}000$ ksi, $\sigma_{YP} = 30$ ksi. Calculate the moment necessary to cause a curvature of 0.0004 rad/in.

7.8 The beam of cross section shown is made of perfectly plastic material of $E = 30{,}000$ ksi, $\sigma_{YP} = 30$ ksi. Calculate the moment necessary to cause a curvature of 0.0002 rad/in.

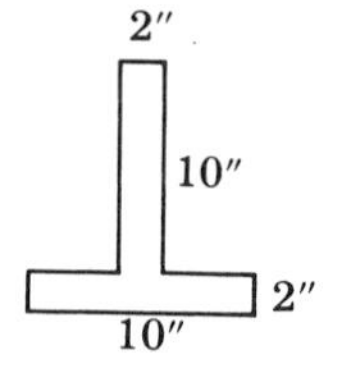

7.9 Determine the ultimate moment, the plastic section modulus, and the shape factor for a 16 WF 36 beam

(*a*) for bending about the strong axis
(*b*) for bending about the weak axis

Take dimensions from the table "Properties for Designing" from the AISC Handbook, and neglect fillets.

7.10 Repeat Prob. 7.9 for a 14 WF 320 beam.

7.11 Repeat Prob. 7.9 for a ST 18 WF 150 T section.

7.12 Repeat Prob. 7.9 for a 18 [58.0 lb channel section.

7.13 Find the ultimate moment and shape factor for the T section shown. Elastic–perfectly plastic material of $\sigma_{YP} = 40$ ksi.

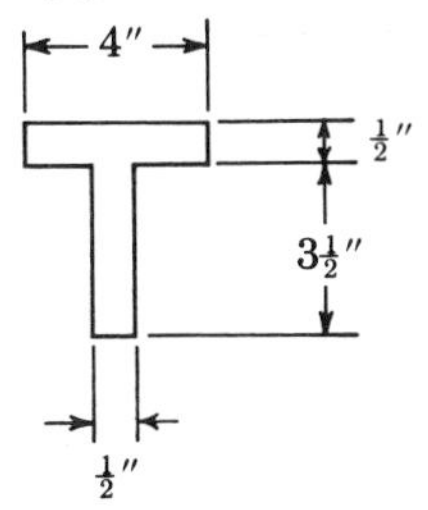

7.14 *Given:* Simply supported beam, $f_{allow} = 0.60 f_{YP}$, shape factor $= f = M_P/M_{YP} = 1.14$, designed according to conventional elastic methods. *Required:* The safety factor against collapse of such a beam.

7.15 Find the allowable value of the midspan loads P, using plastic design methods and a safety factor equal to the one found in Prob. 7.14. $\sigma_{YP} = 33$ ksi.

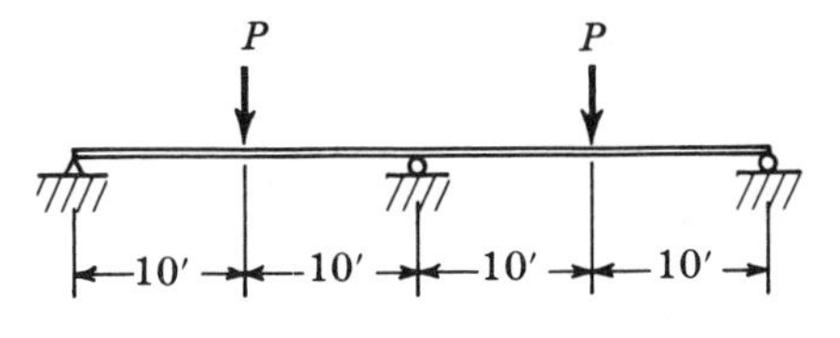

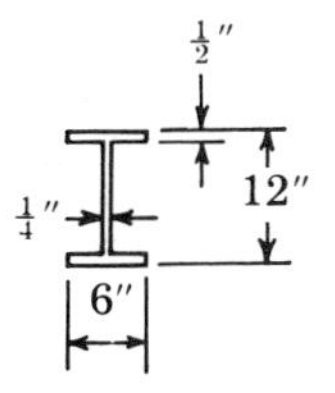

7.16 (*a*) Calculate the load P_p under which collapse will occur.

(*b*) If a safety factor of 1.88 against collapse is used, what will be the allowable load P_w? Will the structure act elastically under this load?

(*c*) Calculate the load which this structure could support according to conventional AISC Specifications.

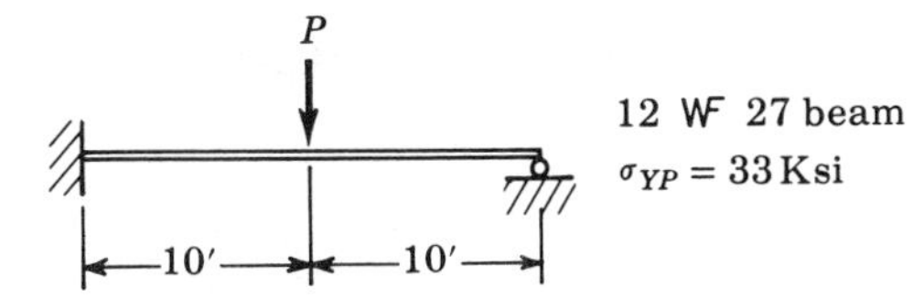

7.17 (*a*) Find the collapse load on the propped cantilever beam shown, of strength M_P. Draw appropriate free bodies and verify your solution by checking the geometry of the moment diagram.

(*b*) Find the value of the collapse load if the beam is a 12 WF 27 and $L = 20$ ft. $\sigma_{YP} = 33$ ksi.

(*c*) Find the safety factor against collapse of the beam of part (*b*) at the instant when the maximum fiber stress in the beam is 20 ksi.

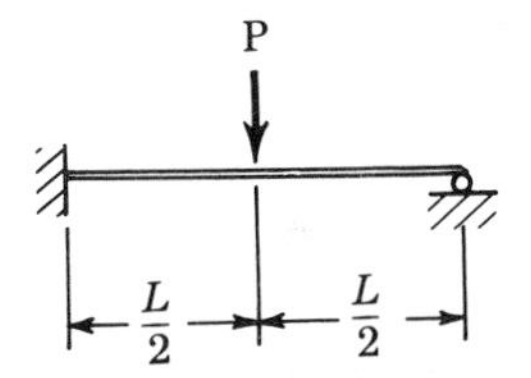

7.18 Strength of all members $= M°$. Neglect effect of axial forces.

(*a*) Find the collapse load w_P on the rigid frame shown.
(*b*) Describe how the collapse load is affected by the ratio k defining the proportions of the frame.

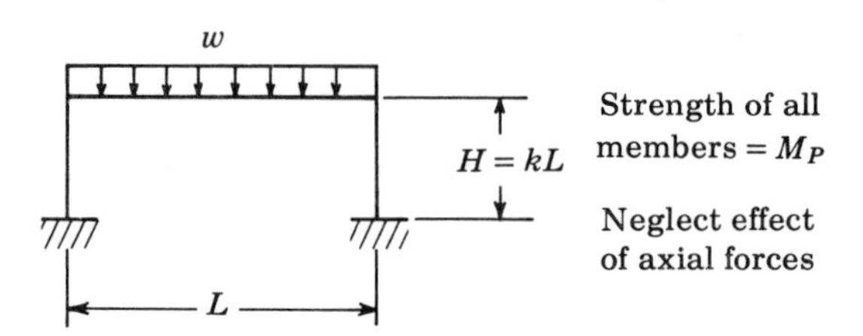

7.19 Find the maximum intensity w_P at collapse of the uniformly varying load on the propped cantilever beam. Constant strength M_P.

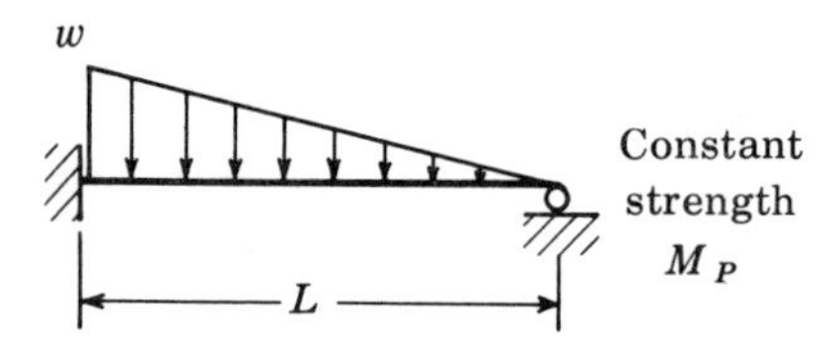

7.20 Strength of beam $= M_P$.

(*a*) Find the uniform collapse load on the three-span continuous beam shown.
(*b*) Describe how the collapse load is affected by the value of k defining the ratio of side span to middle span.
(*c*) Compare results of this problem with those of Prob. 7.18.

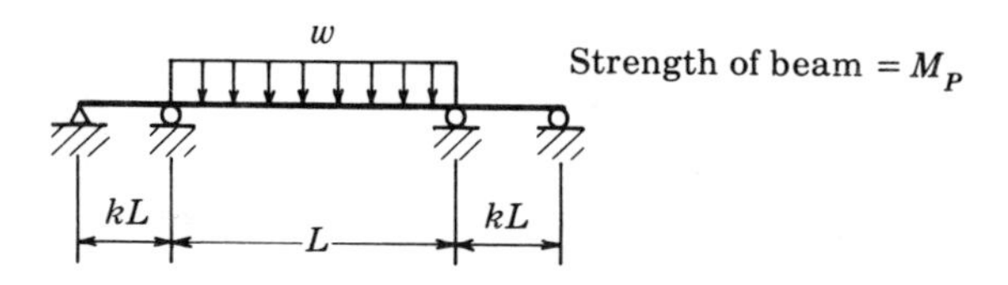

7.21 Constant strength M_P.

(*a*) Calculate the required beam strength M_P to support the ultimate load shown.
(*b*) Select a suitable section to carry the load with a safety factor of 2 against collapse; $\sigma_{YP} = 33$ ksi.

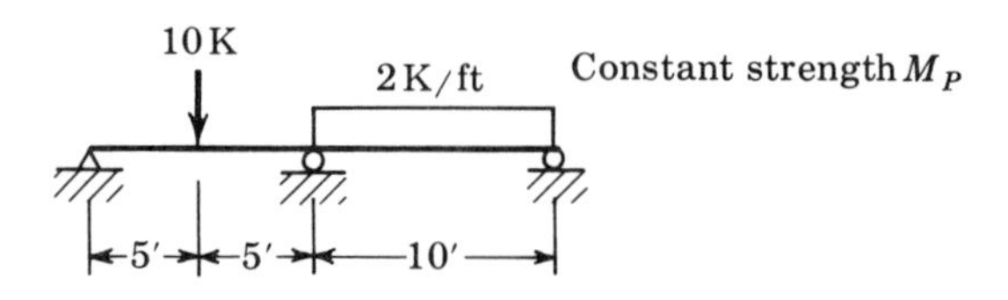

7.22 A rectangular-cross-section beam with uniformly varying depth d is fixed at one end, simply supported at the other. Calculate the collapse load w_P. *Hint:* Plastic hinge in span will occur at point where ratio M/M_P is a maximum.

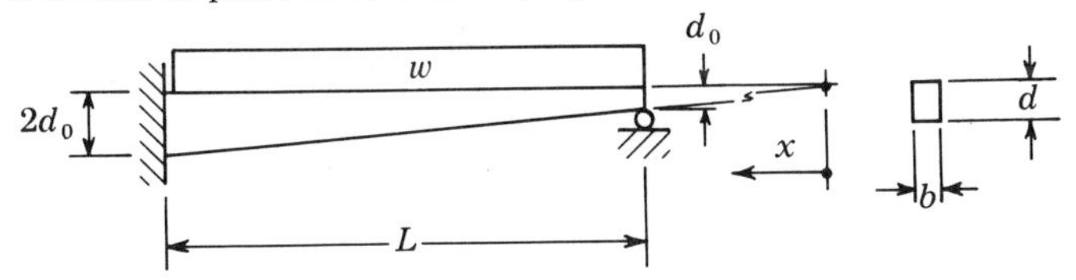

7.23 Check your result for Prob. 7.22 by superimposing the moment diagram due to the load upon a diagram of the moment resistance M_P of the beam and checking the static and kinematic admissibility of your solution.

7.24 Calculate the required ultimate moment capacity required for the three-span continuous beam

(*a*) with full load acting
(*b*) with center span only loaded

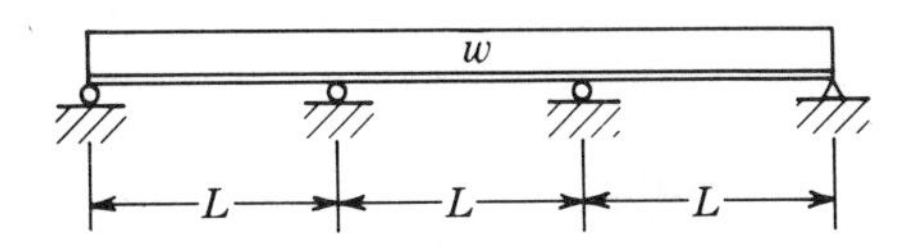

7.25 A tapered beam is fabricated by cutting a 16 W 36 along the skew line shown, reversing one part, and welding longitudinally. A beam made in this fashion is called a *wedge beam.* This member is used as a propped cantilever beam subject to a uniform load; $\sigma_{YP} = 33$ ksi. Calculate the collapse load of this structure in two ways:

(*a*) Analytically, by finding the location of the plastic hinge in the span as the point for which the ratio of applied moment to the moment resistance is a maximum.
(*b*) Graphically, by simultaneously approaching the moment capacities at critical points.

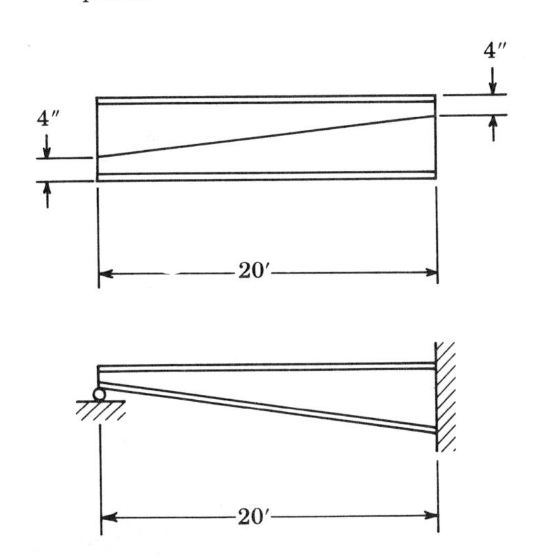

7.26 A 10 W 21 beam is used as a continuous beam to support the loads shown. To increase its capacity, it is to be cover-plated with plates 8 × $\frac{1}{4}$ in. top and bottom over the middle support.

(*a*) Calculate the collapse load P if the cover plates are 5 ft long.
(*b*) Calculate the required length of the cover plates to force formation of a plastic hinge over the middle support, and calculate the collapse load for this case.

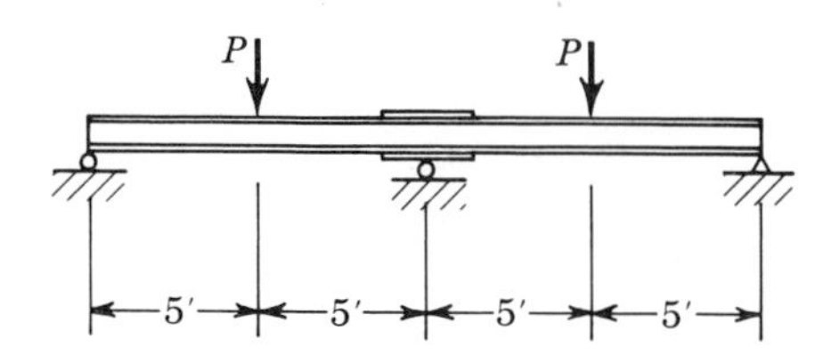

7.27 The propped cantilever beam of constant strength is to be designed.

(*a*) Find the ultimate moment M_P.
(*b*) If $P = 10$ kips, $L = 40$ ft, and a safety factor of 2 is required, determine a suitable beam section of A-7 steel.
(*c*) Demonstrate static and kinematic admissibility of your solution.

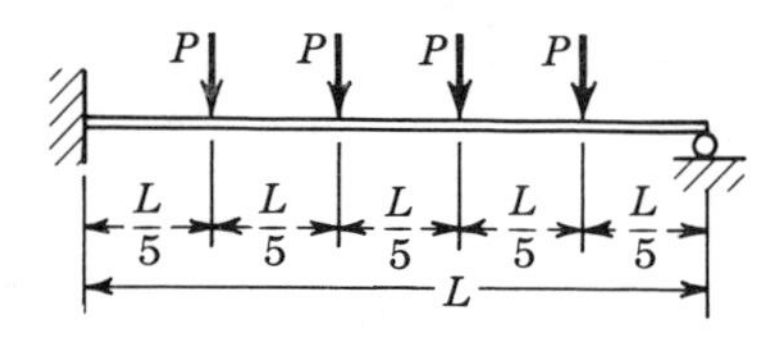

7.28 The frame shown is subject to the working loads shown and is to be designed of the same members throughout. The safety factor against collapse is to be 1.88.

Find the required ultimate moment, and select members of A-36 steel. Verify static and kinematic admissibility of design.

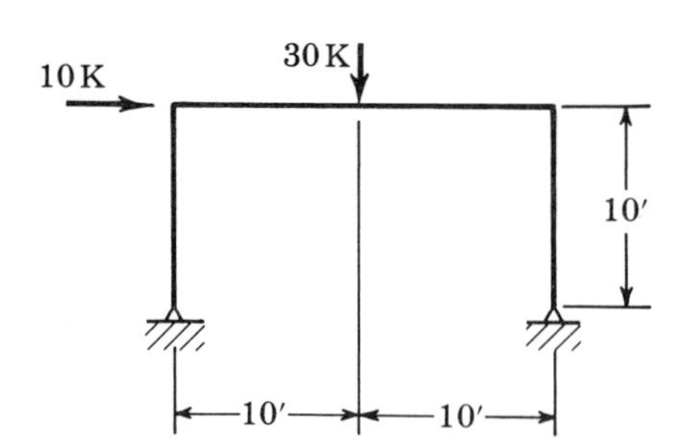

7.29 The Vierendeel girder shown is made of perfectly plastic members of constant strength M_P. Find the collapse load P_P on this structure. Show by use of the upper and lower bound theorems that your solution is correct.

If you are unable to arrive at the exact solution, determine reasonably close upper and lower bounds for the collapse load.

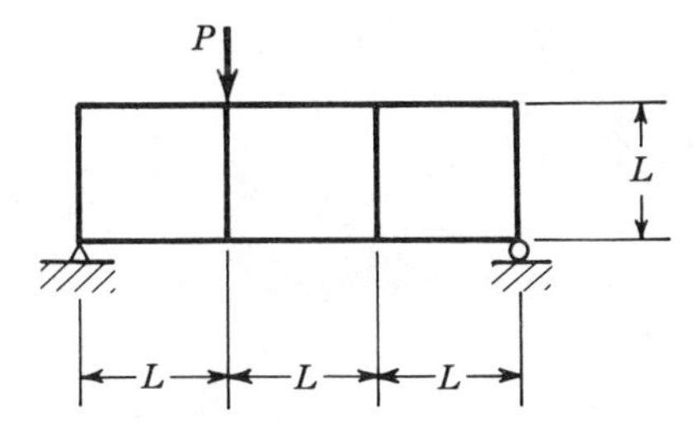

7.30 The structure shown is made of perfectly plastic members of uniform strength M_P.

(*a*) Bracket the correct value of the collapse load w_P by use of the upper and lower bound theorems.
(*b*) Find exact value of collapse load.
(*c*) Compare amount of work of parts (*a*) and (*b*).

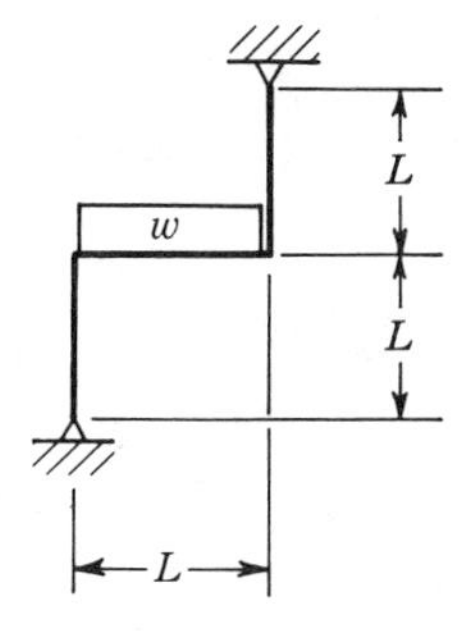

7.31 The structure shown is of perfectly plastic members of uniform strength M_P. Bracket the collapse load by use of the upper and lower bound theorems. Neglect effect of axial forces.

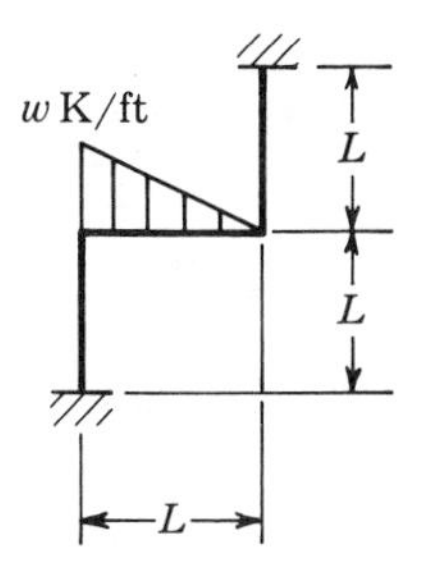

7.32 The rigid frame shown, under the indicated ultimate loads, is to be designed with beams of two times the ultimate capacity of the columns. Find the required member strengths. Demonstrate static and kinematic admissibility of your design. Neglect effect of axial forces.

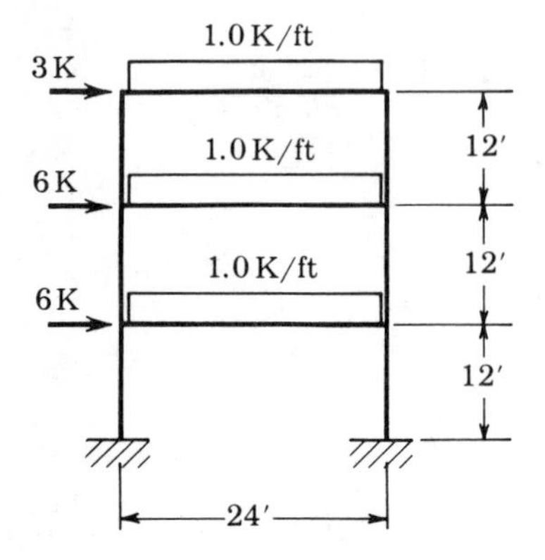

7.33 The rigid frame shown is fabricated of prismatic members of the strengths shown. Calculate the collapse load P of the structure. Draw ultimate-moment diagram to check exactness of solution. Neglect effect of axial forces.

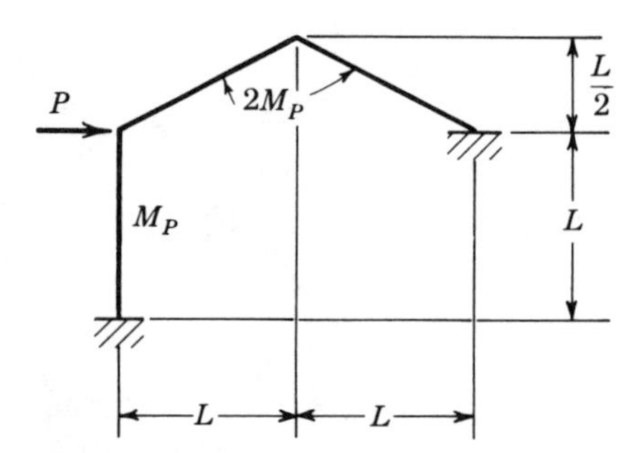

7.34 Find the collapse load w_P for the perfectly plastic structure of strength M_P. Verify static and kinematic admissibility of solution. Neglect effect of axial forces.

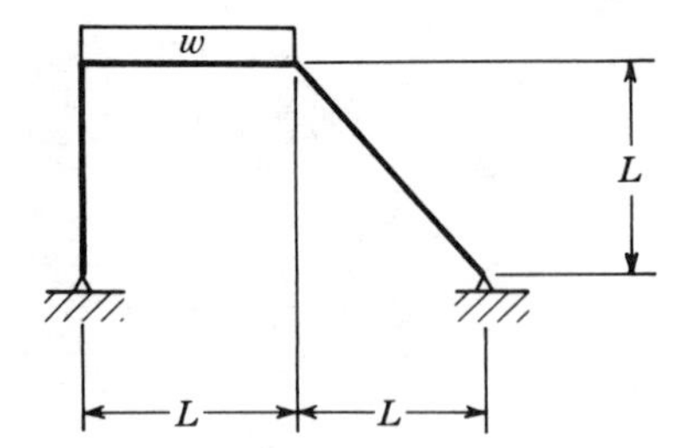

7.35 The rigid frame shown is to be designed to carry a uniform vertical working load of 70 lb/ft² and a uniform lateral wind load of 25 lb/ft². Frame is to be designed of members of constant strength, and load factors according to AISC Specifications are to prevail

(1.85 for vertical load only, 1.40 for combinations of vertical and lateral loads). Select safe and economical members. Neglect effect of axial forces.

Frames spaced at 20-ft centers

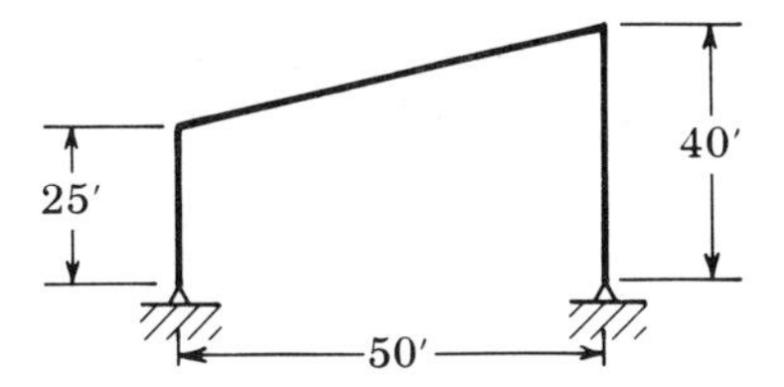

7.36 The gable frame subject to the vertical and wind loads shown is to be designed. Safety factors are to be identical with simple beam factors according to present AISC Code (1.85 for vertical and $\frac{3}{4}$ of this value for wind and vertical loads combined). Assume the structure to be laterally braced. Steel to be A-7; $\sigma_{YP} = 33$ ksi. Select the lightest uniform section beam for this frame.

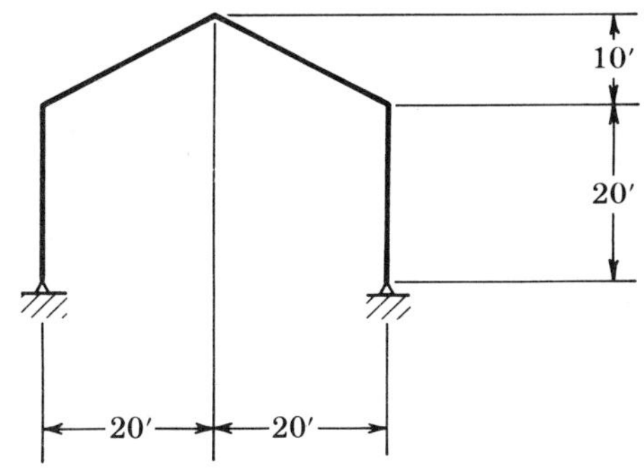

The gable frame of the dimensions shown and of uniform cross section is to be designed.

Loads are: Roof and frame dead loads = 20 lb/ft² } of horizontal
Snow load = 30 lb/ft² } projection
Wind load = 15 lb/ft² of vertical projection

The frames are spaced at 20-ft centers.

7.37 Find the collapse load P_P on the semicircular arch, of $M_P = 1{,}000$ kip-ft. Neglect effect of axial forces.

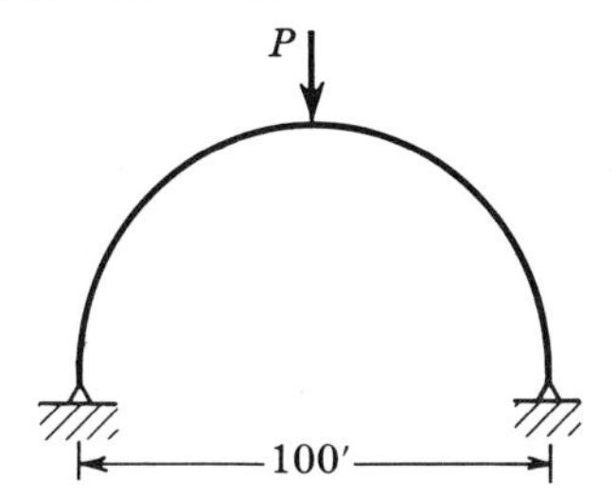

7.38 Repeat Prob. 7.35 by the graphical method.

7.39 Verify static and kinematic admissibility of your design of the frame of Prob. 7.36 by the graphical method.

7.40 Repeat Prob. 7.37 by the graphical method, compare results, and discuss amount of work involved by both methods.

7.41 Redesign the frame of Prob. 7.36 with haunches at the knees. Proportion the haunches to minimize use of material.

7.42 The elastic–perfectly plastic beam of rectangular cross section is subjected to a cranked-in end moment M. Determine and plot the vertical deflection of the free end as a function of the applied moment.

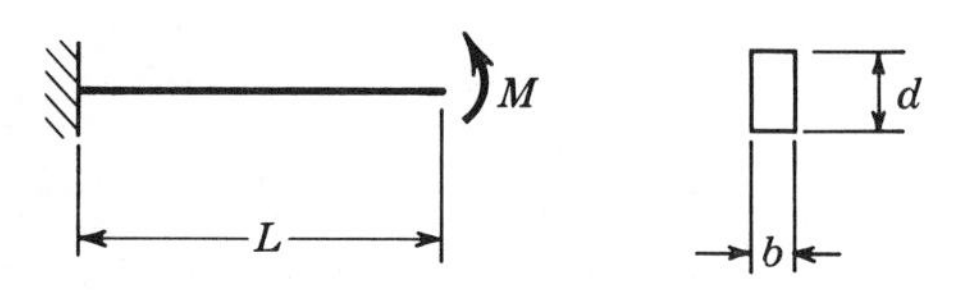

7.43 The beam ABC is of rectangular cross section of width b, depth d, and of a material following the stress-strain relation

$$\sigma = E\epsilon^{\frac{1}{2}}$$

(a) Find the moment-curvature relation for this beam.
(b) Determine and plot the rotation of the beam at point B as a function of the applied load P.

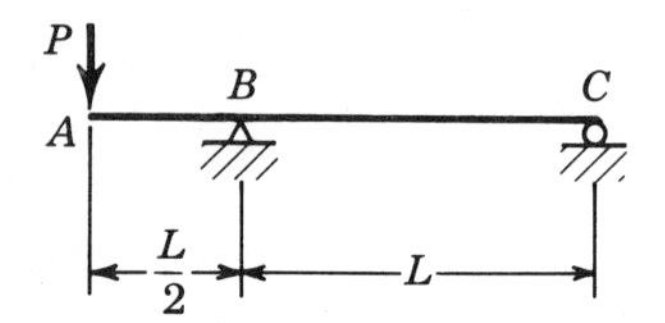

7.44 The cantilever beam of length L and rectangular cross section is made of an elastic–perfectly plastic material. Determine and plot the end deflection of the beam versus the applied load P.

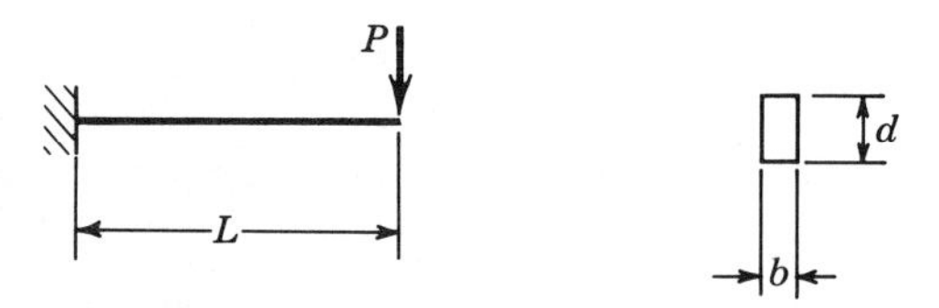

7.45 A load-deflection curve is to be plotted for the elastic–perfectly plastic propped cantilever beam of elastic bending stiffness EI and fully plastic strength M_P. Use plastic-hinge method.

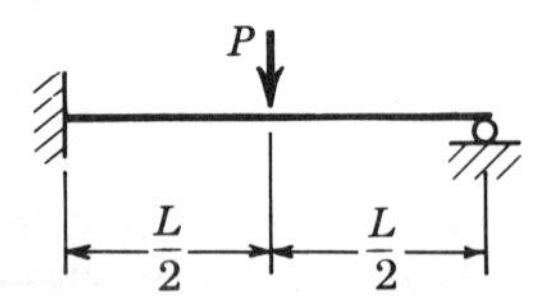

7.46 The continuous two-span beam is a 12 W 27, of A-7 steel. Plot the complete load-deflection curve for one span, using the plastic-hinge method. Note that the point of maximum deflection in the span varies with increasing inelastic load. Clearly state any assumptions you may have made with regard to this.

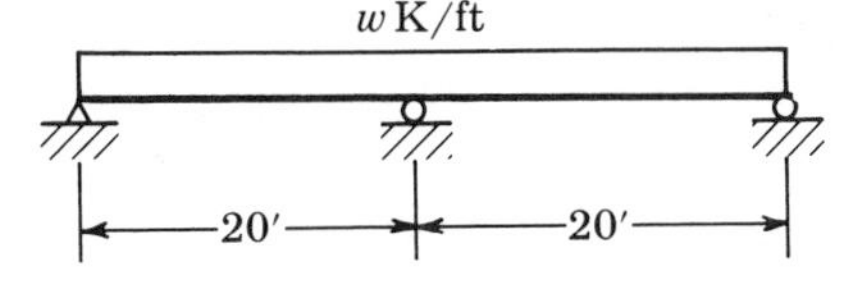

7.47 The midpoint deflection F of the continuous beam is to be determined. The length of the side span stands in the ratio k to that of the center span. Elastic bending stiffness is EI, plastic strength is M_P. Using the plastic-hinge method, find the required midspan deflection as a function of the uniform load w and the span ratio k. Present your results in dimensionless form, plotting wL^2/M_P vertically versus $\Delta EI/M_PL^2$ horizontally, for various values of k. If a maximum allowable deflection is specified, comment on the effect of the value k on the safety factor against excessive deflection.

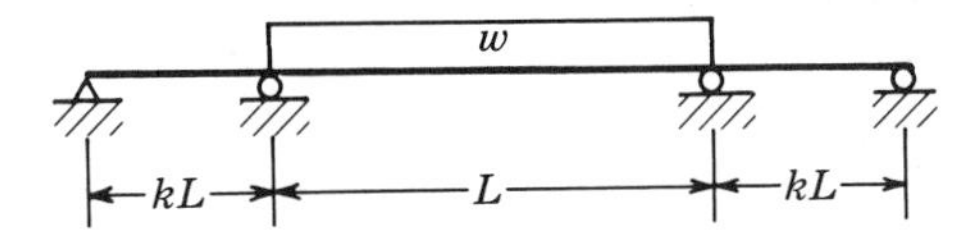

8

Inelastic Theory of Reinforced Concrete Beams

8.1 General Theory

The strength of materials approach will again be called upon for a rational understanding of the inelastic behavior of nonhomogeneous beams, of which reinforced concrete beams constitute an important case. We consider the beam of Fig. 8.1*a*, made of two materials whose mechanical properties are described by the stress-strain curves of Fig. 8.1*d*. We again begin by writing the equilibrium conditions

$$\Sigma F = 0: \qquad \int_{A_1} \sigma_1 \, dA + \int_{A_2} \sigma_2 \, dA = 0 \tag{8.1}$$

and $$\Sigma M = 0: \qquad \int_{A_1} \sigma_1 y \, dA + \int_{A_2} \sigma_2 y \, dA = M \tag{8.2}$$

The strains are again assumed to vary linearly:

$$\epsilon = y\phi \tag{8.3}$$

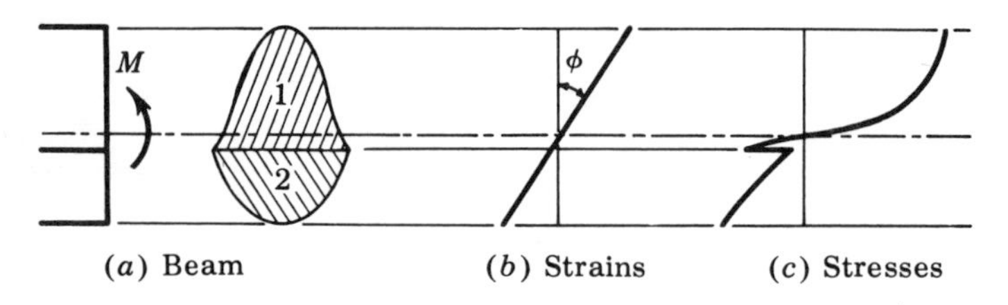

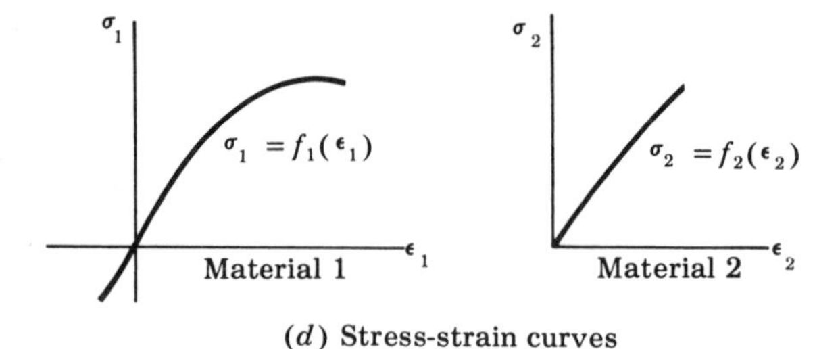

(*d*) Stress-strain curves

Fig. 8.1

and the stress-strain relations are invoked next:

$$\begin{aligned}\sigma_1 &= f_1(\epsilon) = f_1(y\phi) \\ \sigma_2 &= f_2(\epsilon) = f_2(y\phi)\end{aligned} \tag{8.4}$$

and substituting these stresses into the equilibrium equations results in:

$$\Sigma F = 0: \quad \int_{A_1} f_1(y\phi)\, dA + \int_{A_2} f_2(y\phi)\, dA = 0 \tag{8.5}$$

and $$\Sigma M = 0: \quad \int_{A_1} f_1(y\phi) y\, dA + \int_{A_2} f_2(y\phi) y\, dA = 0 \tag{8.6}$$

As in earlier cases of bending of beams, these two equations are to be solved for the location of the neutral axis and for the curvature ϕ corresponding to any given moment M. If the stress-strain relations f_1 and f_2 are nonlinear, then Eqs. (8.5) and (8.6) will also be nonlinear, and a solution may involve considerable labor. Another point to be noted is that, as in other inelastic beams, the location of the neutral axis will in general vary with increasing intensity of applied load.

Rather than talk any further in generalities, we shall apply this theory to the inelastic analysis of reinforced concrete beams, which constitute one case of nonhomogeneous members.

8.2 Inelastic Behavior of Reinforced Concrete Beams

In a previous section we considered the behavior of concrete beams under the assumption that both concrete and steel have elastic stress-strain relations. The actual stress-strain curves, shown by the solid lines in Fig. 8.2, were in that case replaced by the dashed lines. While this procedure

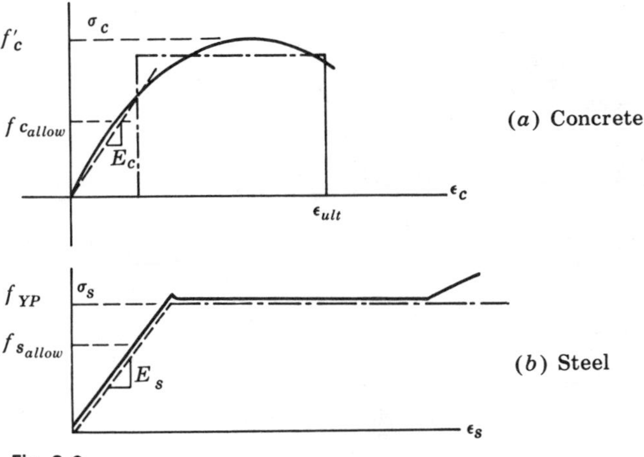

Fig. 8.2

may give good results for cases in which the beam is stressed reasonably low—such as the values of stress marked $f_{c\ allow}$ and $f_{s\ allow}$—it cannot be expected that such a procedure should give valid information about the conditions under which the beam will fail, since this is likely to occur under strains much larger than those for which the elastic approximation is valid.

This means that while the elastic or working-stress theory can give us assurance that a beam will be safe under certain loads, it cannot tell us the actual safety factor against collapse. In the language of limit design which we discussed earlier, an elastic solution is a "statically admissible" solution (since it satisfies statics and provides for fiber stresses which do not exceed the strength of the material), and as such yields a lower bound solution for the collapse load.

If we are agreed that a knowledge of the actual safety factor against failure is necessary for an understanding of the structure, then we must consider stress-strain relations which are realistic for high values of strain. For example, the mechanical behavior of the concrete could be expressed by mathematical functions such as Eqs. (2.15) or (2.16) of Sec. 2.4 or by the rectangular approximation given by Eq. (2.17) and shown by the dash-dot line in Fig. 8.2*a*. For our purposes, we shall for the time being denote the concrete stress-strain relation in compression by $\sigma = f(\epsilon)$. As before, any tensile strength of the concrete is disregarded.

The behavior of the reinforcing steel is assumed elastic–perfectly plastic, as discussed and used earlier, and as shown by the dash-dot line in Fig. 8.2*b*.

An approach which considers the inelastic behavior of the component materials to find the behavior of the member at failure is called an *ultimate-strength theory.*

To analyze a simply reinforced concrete beam in pure bending according to the assumptions just outlined, we begin again with the basic steps, shown in Fig. 8.3.

Subject to the assumption of tensile cracking, the equilibrium equations

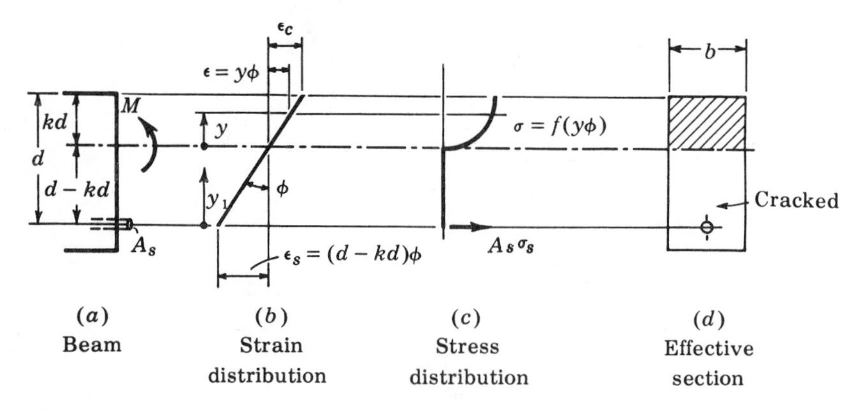

Fig. 8.3

in this case take the form

$$\Sigma F = 0: \qquad A_s\sigma_s - \int_{y=0}^{kd} \sigma \, dA = 0$$

or

$$A_s\sigma_s - \int_{y=0}^{kd} f(y\phi) \, dA = 0 \tag{8.7}$$

and

$$\Sigma M_{N.A.} = 0: \qquad \int_{y=0}^{kd} \sigma y \, dA + A_s\sigma_s(d - kd) = M$$

or

$$\int_{y=0}^{kd} f(y\phi) y \, dA + A_s\sigma_s(d - kd) = M \tag{8.8}$$

Equations (8.7) and (8.8) are generally valid; we shall now distinguish between two cases.

Over-reinforced beams When a beam contains a large amount of tension steel or is subjected to a low moment, then the steel stress and strain will be relatively small, and we shall assume $\epsilon_s < \epsilon_{YP}$, that is, the steel is still acting elastically, so that, using the strain distribution of Fig. 8.3*b*,

$$\sigma_s = E_s\epsilon_s = E_s(d - kd)\phi \tag{8.9}$$

Equations (8.7) and (8.8) then become

$$\Sigma F = 0: \qquad A_sE_s(d - kd)\phi - \int_{y=0}^{kd} f(y\phi) b \, dy = 0 \tag{8.10}$$

and

$$\Sigma M = 0: \qquad \int_{y=0}^{kd} f(y\phi) y \, dA + A_sE_s(d - kd)^2\phi = M \tag{8.11}$$

Having assumed an appropriate concrete stress-strain relation $\sigma = f(\epsilon)$, then the simultaneous solution of Eqs. (8.10) and (8.11) will yield the curvature ϕ and the location of the neutral axis kd corresponding to a given moment M. Once these quantities are known, all strains and therefore stresses can be calculated.

Let us consider the conditions under which such an over-reinforced beam will fail. If the steel remains elastic, then failure must occur when the compressive strain in the extreme concrete fiber reaches its known crushing value ϵ_{ult}. At this stage, the ultimate curvature is

$$\phi_{ult} = \frac{\epsilon_{ult}}{kd} \tag{8.12}$$

and Eq. (8.10) becomes

$$A_sE_s(d - kd)\frac{\epsilon_{ult}}{kd} - \int_{y=0}^{kd} f\left(y\frac{\epsilon_{ult}}{kd}\right) b \, dy = 0 \tag{8.13}$$

from which kd can be found directly.

Equation (8.11) can then be solved for M_{ult}:

$$M_{ult} = \int_{y=0}^{kd} f\left(y\frac{\epsilon_{ult}}{kd}\right) y \, dA + A_sE_s(d - kd)^2\frac{\epsilon_{ult}}{kd} \tag{8.14}$$

or, taking moments about the tension steel rather than about the neutral axis,

$$M_{ult} = \int_{y_1=d-kd}^{d} f\left(y \frac{\epsilon_{ult}}{kd}\right) y_1 \, dA \tag{8.15}$$

where y_1 is the distance from the centroid of the steel. Equation (8.15) is usually easier to apply than Eq. (8.14).

This procedure gives us the ultimate moment which will cause failure in an over-reinforced beam. Because crushing of concrete is a brittle type of failure, it leads to sudden collapse of the beam with little warning.

Under-reinforced beams A beam which is under-reinforced at ultimate strength is one with so little steel that its stress is high enough to cause yielding of the steel. In this case,

$$\sigma_s = f_{YP}$$

and Eq. (8.7) becomes

$$A_s f_{YP} - \int_{y=0}^{kd} f(y\phi) \, dA = 0 \tag{8.16}$$

which again, together with Eq. (8.8), rewritten as

$$\int_{y=0}^{kd} f(y\phi) y \, dA + A_s f_{YP}(d - kd) = M \tag{8.17}$$

can be solved for ϕ and kd corresponding to any given moment.

As the tension steel in an under-reinforced beam yields, it causes the tension cracks in the concrete to open up; the curvature ϕ, and therefore the deflections, increase, and with it the extreme compressive fiber strain ϵ_c. When this reaches ϵ_{ult}, the ultimate moment capacity M_{ult} of the under-reinforced beam will have been reached, and at this time Eqs. (8.16) and (8.17) will become

$$A_s f_{YP} - \int_{y=0}^{kd} f\left(y \frac{\epsilon_{ult}}{kd}\right) b \, dy = 0 \tag{8.18}$$

and $$\int_{y_1=d-kd}^{d} f\left(y \frac{\epsilon_{ult}}{kd}\right) y_1 \, dA = M_{ult} \tag{8.19}$$

Here, Eq. (8.18) may be solved directly for the location of the neutral axis kd, and then M_{ult} can be obtained from Eq. (8.19). In comparing Eq. (8.13) for over-reinforced beams with Eq. (8.18) for under-reinforced beams, we note that the only difference is in the coefficient of the steel area A_s in the first term. This coefficient represents the steel stress which, in the over-reinforced case in which the steel acts elastic, must be related to the strains in the rest of the section, while in the under-reinforced case the steel is at the yield value f_{YP} for any value of strain.

In an under-reinforced beam the ductile yielding of the steel will announce overstress and imminent failure by excessive sag. A ductile failure with previous warning is infinitely preferable to sudden brittle collapse, and under-reinforced beams are therefore preferred in practice. Besides, the relative costs of steel and concrete are such that economical construction usually calls for under-reinforced beams.

It remains to be determined under which conditions a beam will act as an over- or under-reinforced beam, that is, whether concrete will crush or steel yield first under increasing moment. We consider the case under which these events occur simultaneously, in which case we speak of a *balanced beam at ultimate*. In this case, the crushing strain ϵ_{ult} and the yield strain ϵ_{YP}, both known properties of the materials, are related, according to Fig. 8.3*b*, by

$$\frac{\epsilon_{ult}}{(kd)_{bal}} = \frac{\epsilon_{YP}}{d - (kd)_{bal}} \tag{8.20}$$

or $$(kd)_{bal} = \frac{1}{(\epsilon_{YP}/\epsilon_{ult}) + 1} d \tag{8.21}$$

If the distance to the neutral axis kd is larger than this value, then, considering the strain distribution of Fig. 8.3*b*, the compressive concrete strain is relatively larger, and the concrete will crush first; we have an over-reinforced beam. If kd is smaller, then the steel strain is larger, and failure will be initiated by yielding of the steel; we have an under-reinforced beam.

It has also been noted that whether a beam acts over- or under-reinforced depends largely on the amount of reinforcing steel. To find the amount necessary for a balanced beam at ultimate, we introduce the value for $(kd)_{bal}$ found in Eq. (8.21) into Eq. (8.7), and solve for the balanced steel area:

$$A_{s\,bal} = \frac{1}{f_{YP}} \int_{y=0}^{(kd)_{bal}} f(y\phi)\, dA \tag{8.22}$$

In looking over the equations of this section, we note the frequent recurrence of certain integrals which contain the compressive stress-strain relation of the concrete. As soon as such a relation is assumed, these integrals can be evaluated either analytically or by procedures making use of the properties of the stress block. The simplicity or difficulty of these operations depends largely on the nature of the assumed stress-strain relation, as will be seen in the subsequent example. In particular, if we consider the elastic behavior

$$\sigma = f(\epsilon) = E_c\epsilon$$

then we obtain all the equations of Sec. 6.1 pertaining to the working stress analysis of concrete beams.

Example Problem 8.1

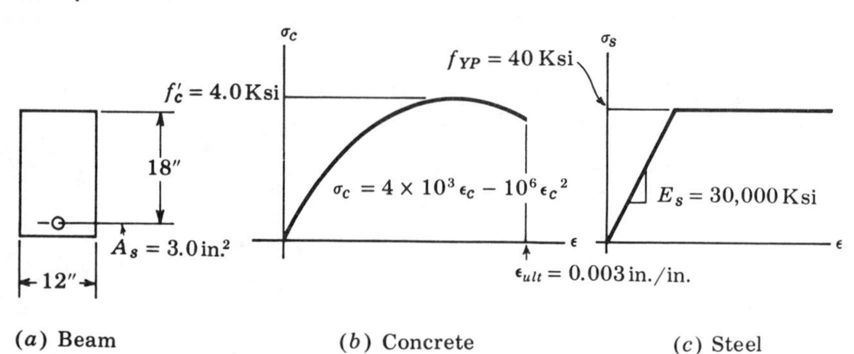

Fig. 8.4

Compute and plot the moment-curvature curve for the reinforced concrete beam shown. Use mechanical properties of materials shown.

We begin by writing the equilibrium equation (8.10), realizing that at low loads the steel will certainly be elastic, and that the stress-strain relation for concrete is that given in Fig. 8.4*b*:

$$\Sigma F = 0: \quad 3.0 \times 30{,}000(18.0 - kd)\phi - 12 \int_0^{kd} (4 \times 10^3 \phi y - 10^6 \phi^2 y^2)\, dy = 0$$

from which, upon integration,

$$10^3 \phi = 6.0 \frac{1}{kd} + 22.5 \frac{1}{(kd)^2} - 405 \frac{1}{(kd)^3}$$

This cubic equation is best solved by assuming values of kd and solving for the corresponding curvature ϕ. A suitable starting value of kd is the one at first load application, when

$$E_c = \left.\frac{d\sigma_c}{d\epsilon}\right|_{\epsilon=0} = 4 \times 10^3 \text{ ksi}$$

and elastic theory can be used. Using transformed section, for instance, we find $kd(\phi \cong 0) = 6.60$ in. Because the concrete stiffness decreases for subsequent load increments, the neutral axis must move down. The numerical results for increasing values of kd are shown in the following table (page 275).

The preceding calculations will be valid till $f_s = 40$ ksi, when the steel becomes plastic and the calculations must be those for an under-reinforced beam. To keep track of the steel stresses, we calculate

$$f_s = E\epsilon_s = E \times (18 - kd)\phi = 30{,}000(18 - kd)\phi$$

and find that the plastification of the steel occurs at $kd = 6.95$ in., $\phi = 0.120 \times 10^{-3}$ rad/in.

Knowing the value of kd for any curvature ϕ, the moment equilibrium equation (8.11) is applied next, and solved for the corresponding moment:

$$\Sigma M = 0: \quad M = 16 \times 10^3 \phi (kd)^3 - 3.0(10^3\phi)^2 (kd)^4 + 90 \times 10^3 \phi (18 - kd)^2$$

The calculations can again be tabulated and the moment plotted against the curvature.

For values of $\phi > 0.120 \times 10^{-3}$ rad/in., $M > 1{,}861$ in.-kips, the steel is plastic, and equilibrium equations for under-reinforced beams apply, as expressed by Eqs. (8.16) and (8.17), which become

$$\Sigma F = 0: \qquad 3.0 \times 40 - 12 \int_0^{kd} (4 \times 10^3 \phi y - 10^6 \phi^2 y^2)\, dy = 0$$

from which

$$10^3 \phi = \frac{3.0}{kd} - \frac{1}{2} \sqrt{\frac{36}{(kd)^2} - \frac{120}{(kd)^3}}$$

and

$$\Sigma M = 0: \qquad M = 16 \times (10^3 \phi)(kd)^3 - 3.0(10^3 \phi)^2 (kd)^4 + 120(18.0 - kd)$$

To evaluate the first of these equations, we take different values of kd, starting with the previously obtained value of 6.95 in., and realizing that after yielding of the steel, the tensile crack in the concrete progresses upward as it widens, pushing up the neutral axis and diminishing the value of kd.

Condensed table of calculations

kd, in.	*$10^3\phi$, rad/in.*	*Moment, kip-in.*	*f_s, ksi*	*ϵ_c, in./in.*	*Remarks*
6.60	0.019	308	6.5	0.000126	Steel elastic
6.70	0.046	748	15.6	0.000308	Steel elastic
6.80	0.080	1,266	26.9	0.000543	Steel elastic
6.90	0.115	1,796	38.4	0.000792	Steel elastic
6.95	0.120	1,861	40.0	0.000834	Steel yields
6.90	0.123	1,878	40.0	0.00085	Steel plastic
6.80	0.127	1,881	40.0	0.00086	Steel plastic
6.50	0.139	1,887	40.0	0.00090	Steel plastic
6.00	0.167	1,908	40.0	0.00100	Steel plastic
5.00	0.253	1,946	40.0	0.00126	Steel plastic
4.00	0.445	1,984	40.0	0.00179	Steel plastic
3.33	0.900	1,981	40.0	0.00300	Concrete crushes

Failure of the beam will occur by crushing of the extreme fiber of the concrete under $\epsilon_{ult} = 0.003$. To keep track of the concrete strain, we compute it as

$$\epsilon_c = \phi(kd)$$

and find that it reaches the ultimate value under $\phi = 0.90 \times 10^{-3}$ rad/in., $kd = 3.33$ in. Note that the neutral axis has moved way up, limiting the compression zone to a narrow strip.

From the values in the above table we plot the moment-curvature curve for the beam shown in Fig. 8.5. It is immediately obvious that the moment-curvature diagram closely resembles the flat-topped type on which the plastic-hinge method for calculating deflections was based, and also that considerable rotation capacity is available in an under-reinforced beam. These results may well lead to the presumption that it may be possible to design reinforced concrete structures by limit-design methods. At this time considerable research is under way to gather further information about this possibility.

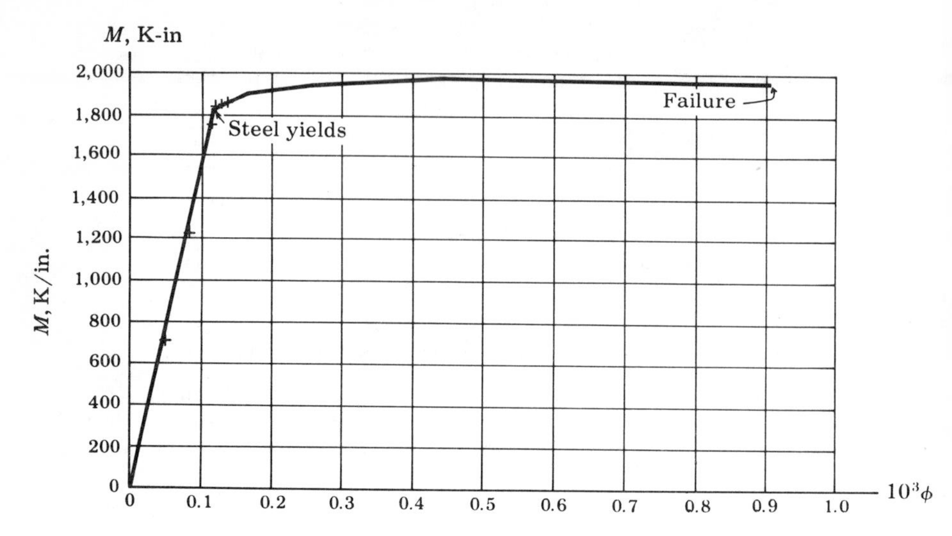

Fig. 8.5

It should be clearly understood that the methods of this section are not those ordinarily followed in structural practice. They have been presented in order to give some insight into the behavior of reinforced concrete beams and a rational plan of attack. The next section will show how the ultimate-moment capacity of such beams can be calculated in a fast, practical manner.

8.3 Ultimate Strength of Singly Reinforced Beams

Whereas in the preceding section a complete description of inelastic behavior has been provided, it suffices in usual structural practice to predict the ultimate moment under which the beam will fail. A simple procedure which gives excellent results makes use of the simplified representation of the concrete stress-strain curve as a rectangular block, as discussed in Sec. 2.4 and shown again in Fig. 8.6a.

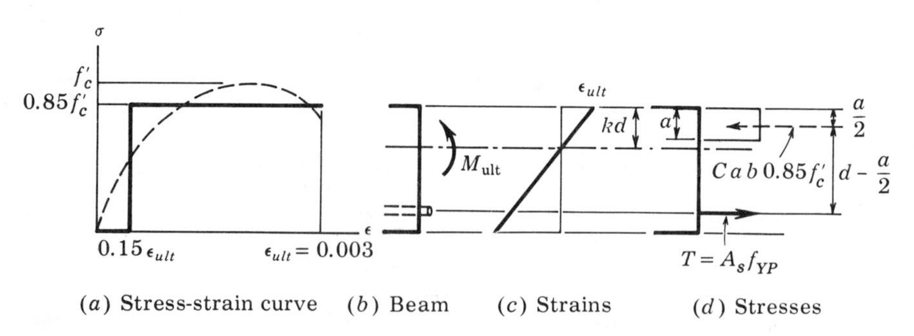

(a) Stress-strain curve (b) Beam (c) Strains (d) Stresses

Fig. 8.6

We consider an under-reinforced beam in which, at failure, the steel has yielded and ultimate collapse is brought about by crushing of the concrete. These conditions are shown in the stress and strain distributions of Fig. 8.6c and d. With the help of these figures, the equilibrium conditions can be written as

$$\Sigma F = 0: \qquad A_s f_{YP} - (ab \times 0.85 f_c') = 0 \tag{8.23}$$

from which

$$a = 0.85\ kd = \frac{A_s}{b} \frac{f_{YP}}{0.85 f_c'} \tag{8.24}$$

and, taking moments about the tension steel,

$$\Sigma M_{tens\ s} = 0: \qquad M_{ult} = (ab\ 0.85 f_c')\left(d - \frac{a}{2}\right)$$

or, taking moments about the compressive resultant, (8.25)

$$\Sigma M_{comp} = 0: \qquad M_{ult} = (A_s f_{YP})\left(d - \frac{a}{2}\right)$$

Substituting the previously obtained value of a into the first moment equation, we obtain the ultimate moment:

$$M_{ult} = [bd^2]\left[f_c' p \frac{f_{YP}}{f_c'}\left(1 - 0.59 p \frac{f_{YP}}{f_c'}\right)\right] \tag{8.26}$$

Note that the first bracketed quantity is a function of the beam dimensions only, while the second depends on the steel ratio and material properties. Thus, for any set of materials, the required quantity bd^2 may be plotted for a given ultimate moment and steel ratio. Such plots are useful for design purposes.

The important expression, Eq. (8.26), for the moment capacity [and its equivalent, the second of Eqs. (8.25)] is found in the ACI Code, chap. 16, eq. 16-1, where the symbol q is used for the reinforcing index $p\ f_{YP}/f_c'$, and where a reduction factor ϕ is applied to the ultimate moment, the latter an added concession to conservative practice.

We again remind ourselves that if we must visualize beam failure at all, we prefer the ductile failure of under-reinforced beams, which was the basis of Eq. (8.26). An added stipulation therefore limits the steel ratio to that which ensures yielding of the steel: according to Fig. 8.6c,

$$(kd)_{bal} = \frac{\epsilon_{ult}}{\epsilon_{YP} + \epsilon_{ult}} d = \frac{E_s \epsilon_{ult}}{f_{YP} + E_s \epsilon_{ult}} d \tag{8.27}$$

Assuming $E_s = 29{,}000$ psi, and $\epsilon_{ult} = 0.003$ in./in., this becomes

$$(kd)_{bal} = \frac{87{,}000}{f_{YP} + 87{,}000} d \tag{8.28}$$

and using this in the force equation (8.23), we obtain the balanced steel ratio

$$p_{bal} = \frac{A_s}{bd} = \frac{1}{f_{YP}bd} 0.85f'_c \times 0.85(kd)_{bal}b$$
$$= \frac{(0.85)^2 f'_c}{f_{YP}} \frac{87{,}000}{f_{YP} + 87{,}000} \quad (8.29)$$

Section 1601 of the ACI Code specifies that the maximum steel ratio shall not exceed 75 percent of this value for balanced design at ultimate.

Example Problem 8.2

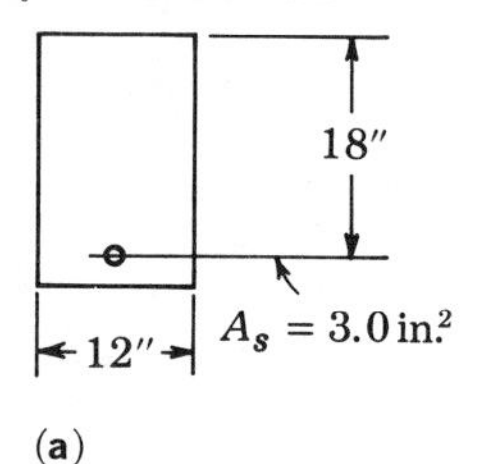

(**a**)

Find the ultimate moment of the beam section shown; $f'_c = 4.0$ ksi, $f_{YP} = 40$ ksi.

$\Sigma F = 0$: $\quad 0.85 \times 4.0 \text{ ksi} \times a \times 12 \text{ in.} - 3.0 \text{ in.}^2 \times 40 \text{ ksi} = 0$
$a = 2.94$ in.

$\Sigma M = 0$: $\quad 120 \text{ kips}\left(18 \text{ in.} - \frac{2.94}{2} \text{ in.}\right) = M_{ult} = \underline{1{,}990 \text{ kip-in.}}$

Note that this beam is identical with that of Example Problem 8.1, where the ultimate moment has been determined more exactly as $M_{ult} = 1{,}984$ kip-in.
(**b**) Find the amount of steel to be placed in the beam of the preceding problem for balanced design at ultimate.

$$A_{s\,bal} = p_{bal}bd = \frac{(0.85)^2 \times 4.0 \text{ ksi}}{40 \text{ ksi}} \frac{87{,}000}{40{,}000 + 87{,}000} \times 12 \text{ in.} \times 18 \text{ in.}$$
$$= \underline{10.7 \text{ in.}^2}$$

We see that the beam of the preceding problem is considerably under-reinforced.
(**c**) Design a beam to resist an ultimate moment of 120 kip-ft; $f'_c = 3.0$ ksi, $f_{YP} = 50$ ksi.
The balanced steel ratio in this case would be

$$p_{bal} = \frac{(0.85)^2 \times 3.0 \text{ ksi}}{50 \text{ ksi}} \frac{87{,}000}{50{,}000 + 87{,}000} = 0.0274$$

To ensure under-reinforced design, we shall provide steel to give only about half this ratio, say 0.014, then,

$$bd^2_{reqd} = \frac{M_{ult}}{f_{YP}p[1 - 0.59p(f_{YP}/f'_c)]}$$
$$= \frac{120 \times 12}{50 \times 0.014[1 - 0.59 \times 0.014(50/3.0)]} = \underline{2{,}390 \text{ in.}^2}$$

This is provided by $\underline{b = 10 \text{ in.}}$, $\underline{d = 15.5 \text{ in.}}$, $\underline{h = 18 \text{ in.}}$ The required amount of steel is then

$$A_s = pbd = 0.014 \times 10 \times 15.5 = \underline{2.17 \text{ in.}^2}$$

8.4 Ultimate Strength of Doubly Reinforced Beams

To determine the ultimate strength of doubly reinforced beams, we again resort to basic principles, drawing the strain distribution at ultimate in Fig. 8.7*b*, characterized by ϵ_{ult} corresponding to crushing of the concrete. The corresponding stresses are drawn next, Fig. 8.7*c*, making again use of the simplified concrete stress block of Fig. 8.7*a*. We proceed with the analysis on the basis that both tension and compression steel will be subject to sufficient strain to cause yielding prior to failure of the beam.

The equilibrium conditions are written using Fig. 8.7*c*:

$$\Sigma F = 0: \qquad C_c + C_s - T = 0$$

or

$$0.85 f'_c ab + f_{YP} A'_s - f_{YP} A_s = 0 \tag{8.30}$$

from which the depth of the stress block is

$$a = \frac{f_{YP}(A_s - A'_s)}{0.85 f'_c b} \tag{8.31}$$

Taking then moments about the tension steel, we find

$$\Sigma M_T = 0: \qquad M_{ult} = C_c\left(d - \frac{a}{2}\right) + C_s(d - d')$$

$$= (A_s - A'_s) f_{YP}\left(d - \frac{a}{2}\right) + A'_s f_{YP}(d - d') \tag{8.32}$$

Note that the first of the two terms in this moment equation denotes the internal couple of the compressive force on the concrete acting with a portion $A_s - A'_s$ of the tension steel, shown in Fig. 8.7*d*, and the second term

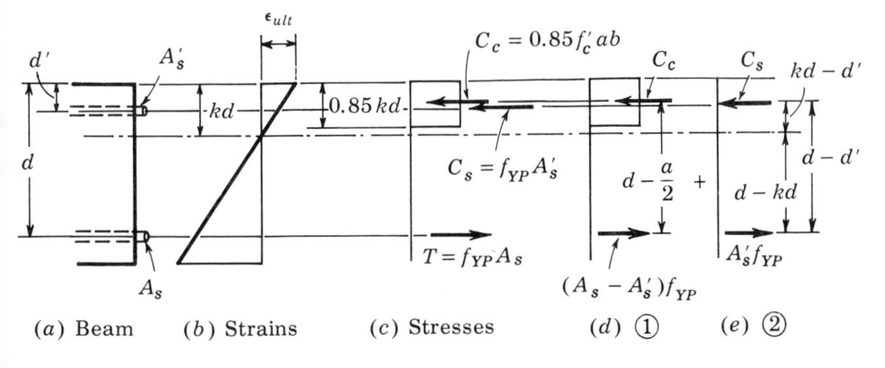

Fig. 8.7

denotes the couple consisting of the compression steel and the rest of the tension steel, shown in Fig. 8.7*e*. The total resisting moment is the sum of these two parts. The value for a is to be inserted here from the force equilibrium condition, Eq. (8.31).

Equation (8.32) appears as eq. 16-3 in sec. 1602 of the ACI Code, modified by the conservative reduction factor ϕ.

It remains to investigate under which conditions the compression steel actually reaches its yield strain, as stipulated. From consideration of strain, Fig. 8.7*b*, we can see that the lower the neutral axis, the larger the ratio of compressive to tensile steel strain. To determine the critical location of the neutral axis, we write

$$\frac{kd_{crit} - d'}{kd_{crit}} = \frac{\epsilon_{YP}}{\epsilon_{ult}} \qquad \text{or} \qquad kd_{crit} = \frac{d'}{1 - (f_{YP}/\epsilon_{ult}E_s)} = \frac{a_{crit}}{0.85}$$

Substituting this critical value for the depth of the stress block into Eq. (8.31), and using the previous values for E_s and ϵ_{ult}, we obtain

$$\frac{A_s - A_s'}{bd} = (p - p')_{crit} = 0.85^2 \frac{f_c'}{f_{YP}} \frac{87{,}000}{87{,}000 - f_{YP}} \frac{d'}{d} \tag{8.33}$$

When the difference between the tensile and compressive steel ratios is smaller than the value of Eq. (8.33), then the compressive steel is so close to the neutral axis that it cannot develop its yield strength. In this case, ACI sec. 1602*c* obliges us either to neglect the compression steel completely in calculating the ultimate moment resistance, or else to calculate the actual stress in the compression steel from its strain and use it in the analysis.

Because even in compressively reinforced beams we must insist that the tension steel yield before the concrete crushes, we combine the reasoning in obtaining the balanced steel ratio of Eq. (8.29) with Fig. 8.7*b* and conclude that tension steel and concrete will fail simultaneously if

$$(p - p')_{bal} = \frac{(0.85)^2 f_c'}{f_{YP}} \frac{87{,}000}{f_{YP} + 87{,}000} \tag{8.34}$$

Section 1602*b* restricts the maximum permissible value of $p - p'$ to 75 percent of this balanced value.

Example Problem 8.3 Calculate the ultimate strength of the doubly reinforced beam; $f_c' = 5{,}000$ psi, $f_{YP} = 40{,}000$ psi.

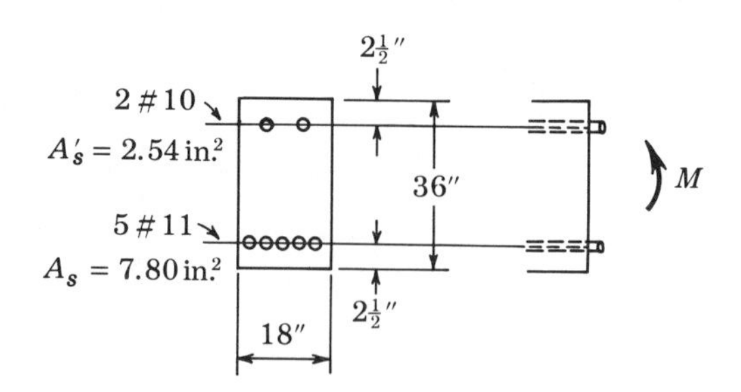

We calculate by basic principles, superimposing two couples:

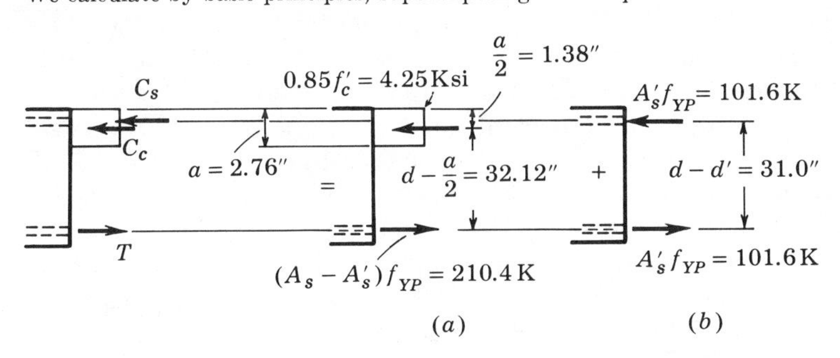

Fig. 8.8

Referring to couple 1, we find the depth of the stress block by summing forces:

$$\Sigma F = 0: \qquad 4.25 \text{ ksi} \times 18 \text{ in.} \times a = 210.4 \text{ kips}; \; a = 2.76 \text{ in.}$$

We next sum the resisting moments 1 and 2, and write

$$\Sigma M = 0: \qquad M_{ult} = \frac{210.4 \times 32.12 + 101.6 \times 31.0}{12} = 827 \text{ kip-ft}$$

The use of Eqs. (8.31) and (8.32) leads to identical results, but it may be observed that basic theory is probably simpler and quicker.

The analysis is completed by verifying that the differential steel ratio $p - p'$ falls within the limits prescribed by Eqs. (8.33) and (8.34):

$$0.85^2 \frac{f'_c}{f_{YP}} \frac{d'}{d} \frac{87{,}000}{87{,}000 - f_{YP}} < p - p' < 0.85^2 \frac{f'_c}{f_{YP}} \frac{87{,}000}{87{,}000 + f_{YP}}$$

$$\text{or} \qquad 0.00124 < 0.0087 < 0.0616$$

This indicates that the strains in both tension and compression steel are above their yield value, as could also be verified by locating the neutral axis and drawing the strain distribution at ultimate.

If we were to delete the compression steel and compute the moment capacity due to the bottom steel only, we would obtain an ultimate moment of 819 kip-ft: the moment gain due to insertion of the compression steel is only 8 kip-ft or about 1 percent. It may be deduced that the addition of compression steel does little to increase the ultimate strength of beams. For this reason, it is little used in ultimate-strength design, and the design of doubly reinforced beams will not be discussed in these notes.

8.5 Code Provisions for Ultimate-strength Theory

The 1963 ACI Code has divided its rational design provisions into two parts: part IV*A*, dealing with working stress design, and part IV*B*, dealing with ultimate-strength procedures. The choice of method is intended to be left to the discretion of the designer. It is the latter of these parts that we wish to peruse in this section in order to see how the previously outlined ultimate-strength theory is specified in order to ensure safe structures.

It is essential that the reader familiarize himself independently with the pertinent code provisions, since only some important ones will be singled out for discussion here.

Section 1502 stipulates that the basic method of analysis of the structure for internal forces and moments is to be elastic; according to previous discussion in conjunction with limit design, this procedure will result in a statically admissible and therefore conservative approximation of conditions near failure of the structure. However, paragraph *d* of this section assumes that a limited amount of redistribution of moments in under-reinforced beams is possible, and allows for a transfer in such ductile beams of up to 10 percent of the moment between positive and negative locations, always making sure, of course, that conditions of statics are satisfied. To ensure that the beam possesses the rotation capacity necessary to achieve this redistribution, the steel ratio in such a beam is held to half of the balanced steel ratio (at ultimate). Example Problem 8.1 indicated that such under-reinforced beams possess considerable ductility.

Section 1503 is a statement of the basic assumptions which we followed in establishing the inelastic theory of concrete beams, including linearity of strains, lack of tensile strength, and ultimate compressive concrete strain. Paragraph *f* rephrases our statement that concrete stress and strain are related in some fashion which, for analytical purposes, we denoted by

$$\sigma = f(\epsilon)$$

Lastly, paragraph *g* allows use of the rectangular stress-strain relation which was discussed in Sec. 8.3, and we see that the formulas obtained in that section are those suggested for design in chap. 16 of the ACI Code.

Sections 1504 to 1506 concern themselves with required safety factors for ultimate-strength design. In order to be able to compensate for possible understrength of the structure, as well as for possible overloading, a mixed approach is used, in that certain safety factors are applied to the strength (note that this is not strictly in accordance with the principles of limit design), as well as load factors to the working loads.

Section 1504 provides for the strength-reduction factor ϕ to be applied to the formulas for ultimate strength of members which have been derived. Different values of ϕ are provided for different types of members in accordance with the certainty with which collapse can be predicted. For flexural members the collapse can be predicted within narrow limits, and accordingly, the coefficient is close to 1; the strength of columns is subject to greater scatter, and a somewhat greater reduction is called for. The coefficient ϕ may be termed a *factor of ignorance.*

Section 1505 is concerned with high-strength reinforcing bars. Because of the reduced ductility of members made with high tensile bars, and because the larger strains associated with the use of higher steel stresses also lead to increased tensile cracking of the concrete surrounding the steel, special precautions are to be taken when using high-tensile bars.

No particular definition is given here of the yield stress f_{YP} of the steel and the compressive strength f'_c of the concrete, except that reference is made in sec. 1500 ("Notation") to ACI sec. 301, where the appropriate ASTM Specifications for the pertinent materials test are cited.

Section 1506, which concerns itself with load factors, splits up the loads into three types:

D Loads the magnitude of which is reasonably well known. Into this category falls above all the dead load of the structure, but also calculable internal forces due to temperature changes, shrinkage of the concrete, and the like. Because we can predict these effects quite well, only a relatively small load factor is required, taken as 1.5 in sec. 1506a1.

L Live loads on structure. Because overloading can easily occur, the L loads will require a larger load factor, given as 1.8, so that the required ultimate capacity of a structure under combination of dead and live load is given by eq. 15-1 of the code as

$$U = 1.5D + 1.8L$$

W,E Wind or earthquake forces of infrequent occurrence and little likelihood of concurrence with critical live load conditions, thus requiring a small load factor, given in sec. 1506a2 and 3.

The way in which load factors may be adjusted to compensate for various probabilities of overloading is an important advantage of the ultimate-load or limit-design theories over the working stress methods, which throw their entire safety factor into the value of the allowable stresses.

Sections 1507 and 1508 are concerned with limiting the deflections and cracking of structures designed according to ultimate-strength methods. The lower the steel ratio and the higher the steel stress, the more severe is the tensile cracking and sag which may be expected in beams.

Chapter 16 presents formulas for the determination of the ultimate capacity of beams, most of which have been derived and discussed in this section.

Chapters 17 and 18 deal with the ultimate strength of members in diagonal tension and in bond respectively. The formulas of these sections are based primarily on experimental results and bear a strong resemblance to those of the corresponding chapters (chaps. 12 and 13) in the part of the code dealing with the working stress method. Shear and bond design for ultimate load is achieved primarily by raising the allowable stress value above those called for in working stress design.

General Readings

All current texts in reinforced concrete design, such as those by Ferguson and by Winter et al. cited in Chap. 6, contain a treatment of the ultimate-strength theory of concrete structures. In addition, the following is a classic discussion:

Rüsch, Hubert: Researches toward a General Flexural Theory for Structural Concrete, *J. ACI*, vol. 57, p. 1, July, 1960. This is a thought-provoking article about the effects of time on the strength of concrete members, based on recent European studies.

The following are useful for intelligent and efficient application of ACI Code provisions to ultimate strength design:

ASCE-ACI Joint Committee: Report on Ultimate Strength Design, *Proc. ASCE, J. Struct. Div.*, vol. 81, October, 1955. Presentation of theoretical and experimental bases which led to present ultimate strength code.

Whitney, Charles S., and Edward Cohen: Guide for Ultimate Strength Design of Reinforced Concrete, *J. ACI*, p. 455, November, 1956. Discussion and design aids.

Problems

8.1 The composite beam is made of two materials whose mechanical properties are shown in the stress-strain curves; tensile and compressive properties are the same.

(*a*) Compute and plot the moment-curvature relation for moments below that causing yielding in material 1. Calculate the moment under which this yielding occurs.

(*b*) Compute and plot the moment-curvature relation for moment above the yield moment calculated in part (*a*). Calculate the ultimate moment which may be applied to the beam.

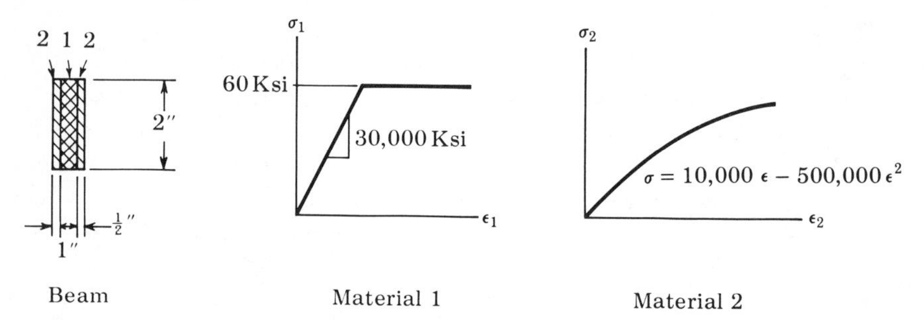

8.2 The composite beam is made of the materials of properties as given in Prob. 1.

(*a*) Compute and plot the moment-curvature relation for moment below that causing yielding in material 1. Also discuss the movement of the neutral axis during this period.

(*b*) Compute and plot the moment-curvature relation for moment above the yield value of part (*a*). Again discuss the movement of the neutral axis, and calculate the ultimate moment on the beam.

Note: The strap of material 1 is thin enough to be considered concentrated at one fiber.

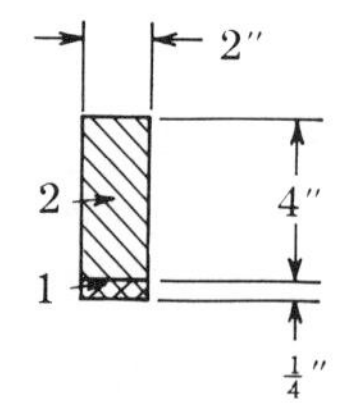

8.3 The reinforced concrete beam shown is made of materials whose stress-strain relations may be represented by the curves shown.

(*a*) Plot a complete moment-curvature curve for the beam up to failure. Carefully distinguish the different stages. Calculate and label the following: elastic limit moment, ultimate moment, and moment at ultimate curvature.
(*b*) Superimpose and compare the moment-curvature curve of part (*a*) with that obtained in the example problem of Sec. 8.2, and draw appropriate conclusions.

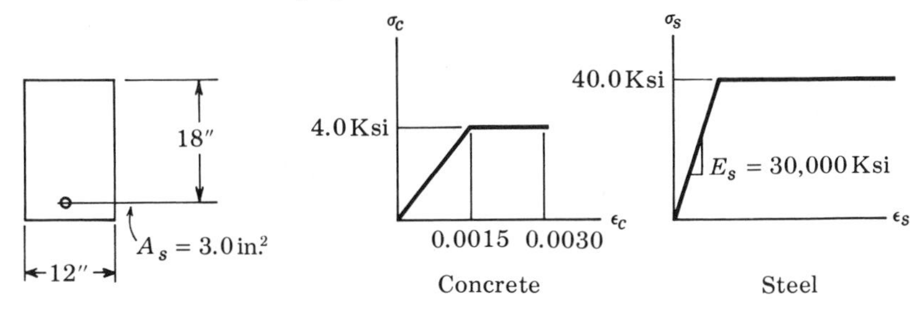

8.4 Calculate the maximum amount of steel under which the beam of Prob. 8.3 will still be under-reinforced at failure (that is, steel yields before concrete crushes in compression).

8.5 If the beam of Prob. 8.3 contains $A_s = 12.0$ in.², calculate and plot the moment-curvature curve. Comment on the type of failure of this beam.

8.6 A rectangular, simply reinforced beam is made of a concrete whose stress-strain curve can be represented as shown. The yield point stress of the steel is f_{YP}.

(*a*) For an under-reinforced beam, calculate the ultimate moment M_{ult} in terms of the stresses f'_c and f_{YP} and the steel ratio $p = A_s/bd$.
(*b*) Find the largest value of p for which this expression is valid.

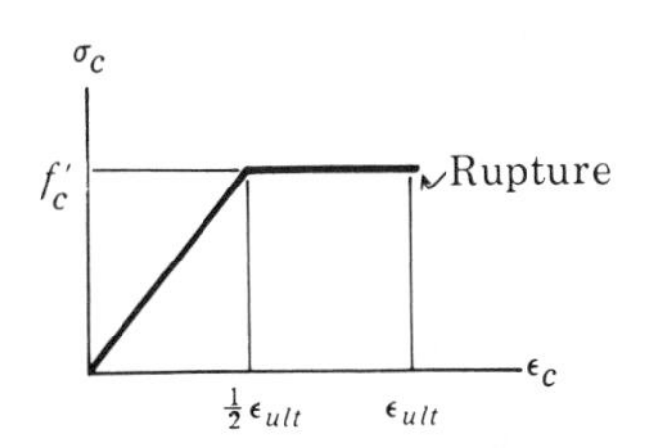

8.7 Calculate the ultimate moment of the beam on the basis of each of the three assumptions for the concrete stress-strain curve shown. Steel is perfectly plastic, $f_{YP} = 40$ ksi. Compare results and draw conclusions regarding the importance of exactness of a concrete stress-strain curve for calculating the ultimate strength of under-reinforced beams.

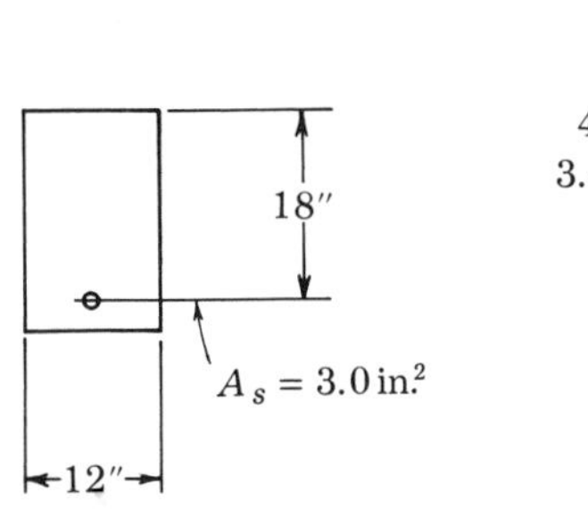

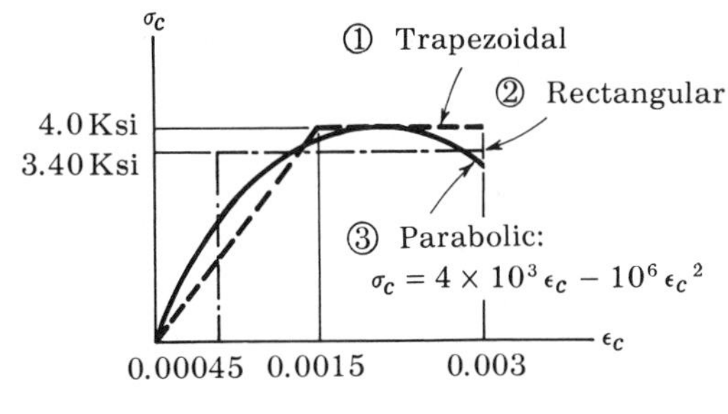

8.8

(*a*) Calculate the ultimate moment on the section, using the Whitney stress block. Make sure that beam is under-reinforced.

(*b*) Find the safety factor for this beam if designed according to the ACI Working Stress Code. (This was done in Prob. 6.2.)

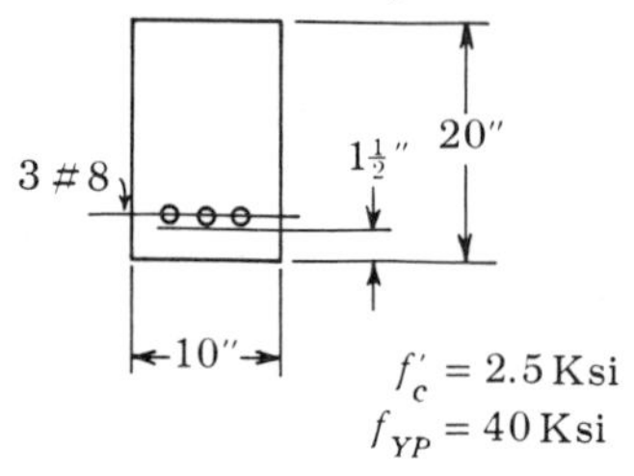

8.9

(*a*) Assuming rectangular stress block, calculate the value P under which the beam will collapse. Do not neglect dead weight of beam. Make sure that beam is under-reinforced.

(*b*) Calculate allowable load P according to the ACI working stress code (see Prob. 6.4).

(*c*) Calculate safety factor of beam of part (*b*) against collapse, first on basis that total load may increase in proportion, second on basis that only the live load may be subject to overloading. Comment on the validity of both assumptions.

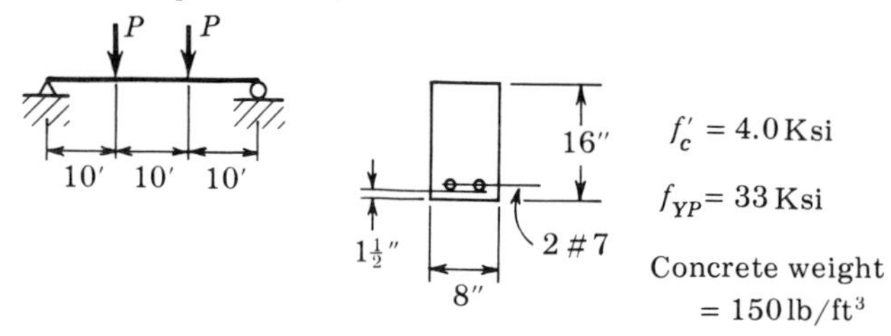

8.10 Calculate the ultimate strength of the one-way slab, if

(*a*) bars are No. 4
(*b*) bars are No. 9

2″
6″
6″ 6″
Typical spacing
$f'_c = 3{,}000$ psi
$f_{YP} = 40{,}000$ psi

8.11 A simply supported beam of 20 ft span carries a uniform load of 1.5 kips/ft (not including beam dead weight). Beam is to be made of gravel concrete (unit weight

150 lb/ft³); f'_c = 3.0 ksi, f_{YP} = 40 ksi. Safety factor against collapse (based on total load) = 2.

(*a*) Assume beam dead weight and design the critical section. Revise as required if beam dead weight was assumed incorrectly. Select steel and draw cross section.

(*b*) Compare the design with the working stress design of Prob. 6.9.

8.12 Repeat the design of Prob. 8.11, this time using a live load factor of 2.0 and a dead load factor of 1.2.

8.13 A cantilever beam of 10 ft length supports a uniform live load of 1.5 kips/ft; f'_c = 3,750 psi, f_{YP} = 33 ksi. Gravel concrete (unit weight = 150 lb/ft³).

(*a*) Using the safety factors of ACI sec. 1504 and the load factors of ACI sec. 1506, design the critical section of the beam.

(*b*) Compare design of part (*a*) with that of Prob. 6.12.

8.14 A fixed-ended beam of span 20 ft is supporting a uniform live load of 1.2 kips/ft; f'_c = 5,000 psi, f_{YP} = 40,000 psi; gravel concrete. Using ACI ultimate-strength design provisions, design the beam; draw cross sections at critical points of the beam.

8.15 (*a*) Calculate the ultimate moment which can be applied to the doubly reinforced beam shown; f'_c = 4,000 psi, f_{YP} = 40 ksi.

(*b*) Compare with ultimate moment on same beam without compression reinforced.

(*c*) Draw conclusions about the effectiveness of compression steel in ultimate-strength design.

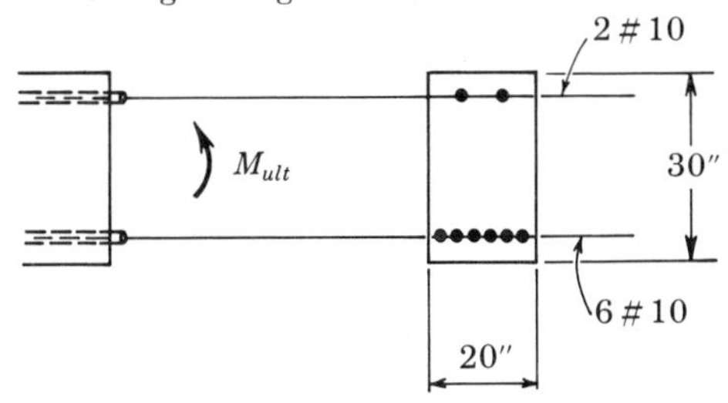

Clear cover = 1½″

8.16

(*a*) Calculate the ultimate strength of the doubly reinforced beam shown. *Hint:* Check compression steel stress at ultimate.

(*b*) Compare the capacity calculated in part (*a*) with the ultimate moment of the same beam without compression reinforcement.

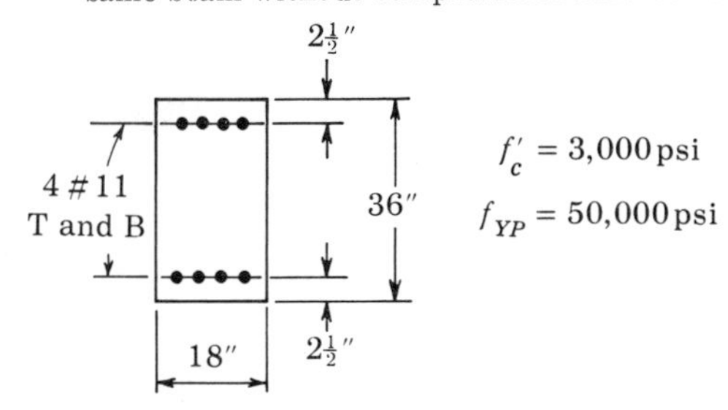

8.17 Using basic principles, calculate the ultimate strength of the T-beam shown.

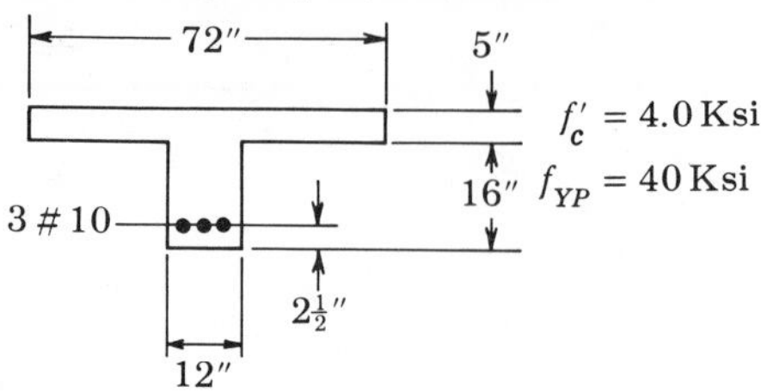

8.18 The typical floor beam of section shown spans 40 ft and is spaced at 12-ft centers. Using basic theory and appropriate sections of the ACI Code, calculate the ultimate moment on the section and the uniform live load on the slab under which collapse of the beam may be expected.

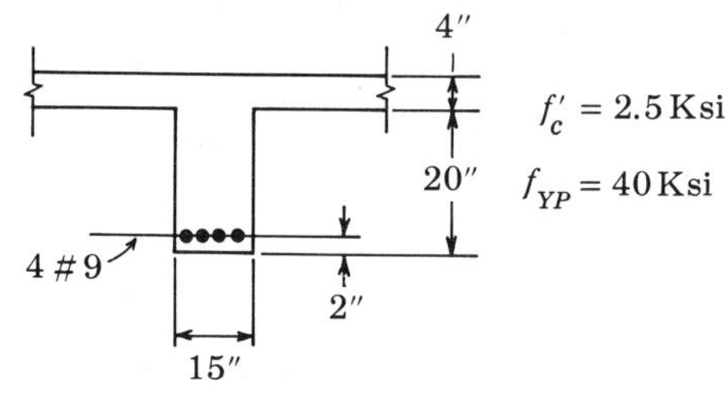

8.19

(*a*) Calculate the maximum amount of steel area for under-reinforced failure.
(*b*) If steel consists of three No. 9 bars, compute the ultimate moment.

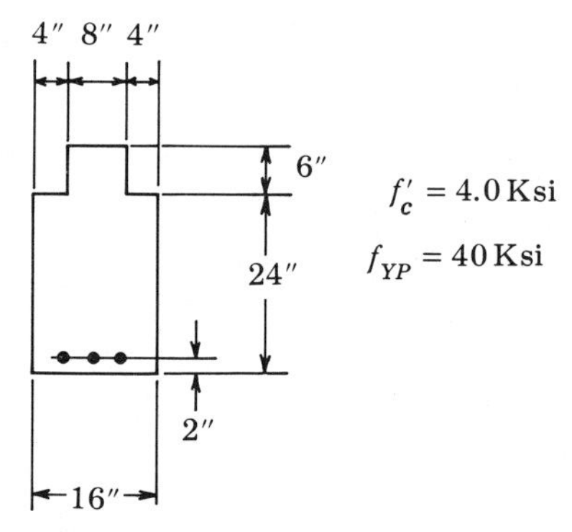

8.20 If the steel in Prob. 8.19 consists of four No. 11 bars, compute the ultimate moment on the section.

part four

Members under Axial Load and Bending

9

Homogeneous Members and Prestressed Concrete

9.1 Stress Analysis for Elastic Members

The four steps of the useful strength of materials approach are again called on to investigate the behavior of members under axial load and bending moment. In structural practice such "beam columns" occur frequently, for instance, as vertical members of rigid frames or as arch ribs, as shown in Fig. 9.1.

We consider a cross section of an elastic, homogeneous member which is subject to force P and moment M, as seen in Fig. 9.2*a*. Equilibrium demands that

$$\Sigma F = 0: \qquad \int_A \sigma \, dA = P \tag{9.1}$$

and $$\Sigma M = 0: \qquad \int_A \sigma y \, dA = M \tag{9.2}$$

where y is measured from the line of action of the load P.

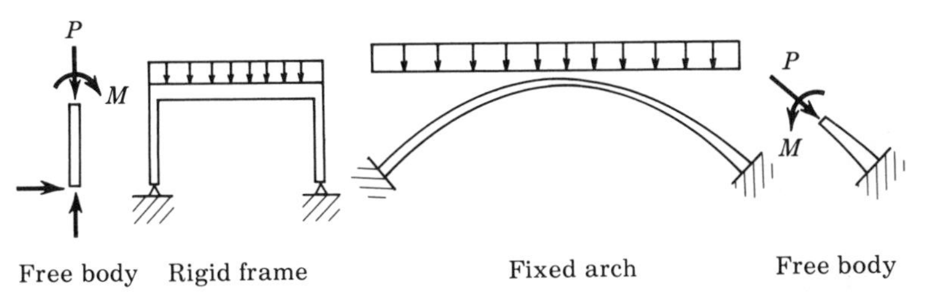

Fig. 9.1

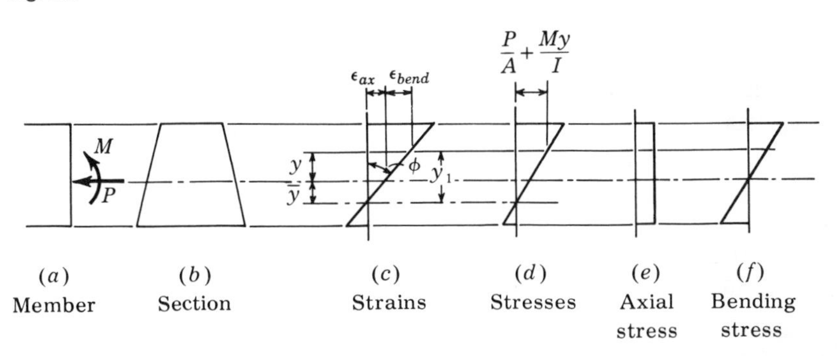

Fig. 9.2

Next we consider the strains, again assuming that plane sections remain plane, as in Fig. 9.2*b*; note that the distance y_1 from the point of zero strain is equal to the previously used distance y plus the distance $\bar{y}$ between the line of action of the force and the neutral axis. Then we write the strain as

$$\begin{aligned} \epsilon &= \phi y_1 = \phi(\bar{y} + y) \\ &= \phi\bar{y} + \phi y = \epsilon_{ax} + \epsilon_{bend} \end{aligned} \tag{9.3}$$

For the elastic material the stress-strain relations are

$$\sigma = E\epsilon = E\phi\bar{y} + E\phi y \tag{9.4}$$

and substituting this into Eqs. (9.1) and (9.2), we get

$$\begin{aligned} \Sigma F = 0: &\quad \int_A E\phi\bar{y}\,dA + \int_A E\phi y\,dA = P \\ &\quad \text{or} \quad E\phi\left(\bar{y}A + \int_A y\,dA\right) = P \\ \text{and } \Sigma M = 0: &\quad \int_A E\phi\bar{y}y\,dA + \int_A E\phi y^2\,dA = M \\ &\quad \text{or} \quad E\phi\left(\bar{y}\int_A y\,dA + \int_A y^2\,dA\right) = M \end{aligned} \tag{9.5}$$

For given P and M, Eqs. (9.5) constitute a set of two simultaneous equations in the unknowns $\bar{y}$ and ϕ.

We can radically simplify this if we apply the resultant force P through the centroid of the section, in which case $\int_A y\,dA = 0$, and Eqs. (9.5) simplify to:

$$\begin{aligned} \Sigma F = 0: &\quad E\phi\bar{y}A = P \quad \text{or} \quad \bar{y} = \frac{P}{AE\phi} \\ \text{and } \Sigma M = 0: &\quad E\phi\int_A y^2\,dA = M \quad \text{or} \quad \phi = \frac{M}{EI} \end{aligned} \tag{9.6}$$

If Eqs. (9.6) are resubstituted into the strain relations (9.3), we find the strains as the sum of the axial strains and the bending strains:

$$\epsilon = \frac{P}{AE} + \frac{My}{EI} \tag{9.7}$$

and if they are substituted into the stress-strain relations (9.4), we find the stresses

$$\sigma = \frac{P}{A} + \frac{My}{I} \tag{9.8}$$

Note that this development is valid only if the resultant load P passes through the centroid of the section; but this can always be accomplished by simple transformation of a couple. The results embodied in Eqs. (9.7) and (9.8) could of course also have been obtained by applying the law of superposition to earlier results obtained for pure axial force and pure bending.

For materials weak in tension, it is sometimes desirable to specify that an eccentric load is to be applied in such a fashion as to prevent any tensile stresses. To analyze this case, we consider Fig. 9.2d and set the extreme fiber stress on the tension side equal to zero:

$$\sigma_{\max} = \frac{P}{A} - \frac{Mc}{I} = \frac{P}{A}\left(1 - \frac{ec}{r^2}\right) = 0$$

where r is the radius of gyration of the section and $Ar^2 = I$.

Solving for the maximum allowable eccentricity of the applied load, we find

$$e_{\max} = \frac{r^2}{c} \tag{9.9}$$

where c is the distance from the centroidal axis to the extreme fiber on the tension side. Note that this maximum eccentricity depends only on the properties of the section. That portion of the cross section bounded by $e_{\max}$ is called the *kern* of the section. An axial force applied within the kern will not cause tension in the section.

9.2 Interaction Curves for Combined Axial Load and Bending

The allowable stress in a section under axial load P_0 and moment M_0 will first be reached in the fiber farthest removed from the centroidal axis, say at a distance c, and is given, according to Eq. (9.8), by

$$\sigma_{allow} = \frac{P_0}{A} + \frac{M_0 c}{I} \qquad \text{or} \qquad \frac{P_0}{\sigma_{allow}A} + \frac{M_0}{\sigma_{allow}I/c} = 1 \tag{9.10}$$

If we now define the allowable pure axial load on the member (neglecting buckling) as

$$P_{allow} = \sigma_{allow}A$$

and the allowable pure bending moment as

$$M_{allow} = \frac{\sigma_{allow}I}{c} \tag{9.11}$$

then Eq. (9.10) can be rewritten as

$$\frac{P_0}{P_{allow}} + \frac{M_0}{M_{allow}} = 1 \tag{9.12}$$

Nondimensional expressions such as Eq. (9.12) are called *interaction expressions;* they can be plotted to give *interaction curves.* Such expressions or curves are useful for design of members under combined loading conditions. Plotting, for instance, Eq. (9.12), we get the linear interaction curve shown in Fig. 9.3, which indicates in dimensionless form the combination of axial load P_0 and moment M_0 which will make the member unsafe.

If we further define

$$\begin{aligned} f_A &= \frac{P_0}{A} = \text{actual stress due to axial force} \\ f_B &= \frac{M_0 c}{I} = \text{actual stress due to moment} \\ F_A &= \frac{P_{allow}}{A} = \text{allowable stress under pure axial force} \\ F_B &= \frac{M_{allow} c}{I} = \text{allowable stress in pure bending} \end{aligned} \tag{9.13}$$

and substitute these into Eq. (9.12), we get the alternate interaction expression

$$\frac{f_A}{F_A} + \frac{f_B}{F_B} = 1 \tag{9.14}$$

In this form, it becomes apparent that this expression may be used even if the allowable stresses F_A and F_B have different values.

Linear interaction equations of the form of Eq. (9.14) are found in the AISC Specifications, sec. 1.6.1, formula 6, and in the ACI Code, eq. 14-9 (in this case for bending about two axes), for the case of members under combined axial load and bending.

We should remark that this discussion has not considered the possibility of buckling, but has been based solely on a criterion of maximum stress. Since the critical buckling stress in a beam column does not vary linearly with loads, such linear equations are inapplicable if the danger of buckling exists. The nonlinear interaction expression eq. 7*a* of the AISC Specifications is an example of this type, and will be discussed in a later section concerned with buckling. Because of the importance of buckling the design of steel beam columns will be covered only in that section.

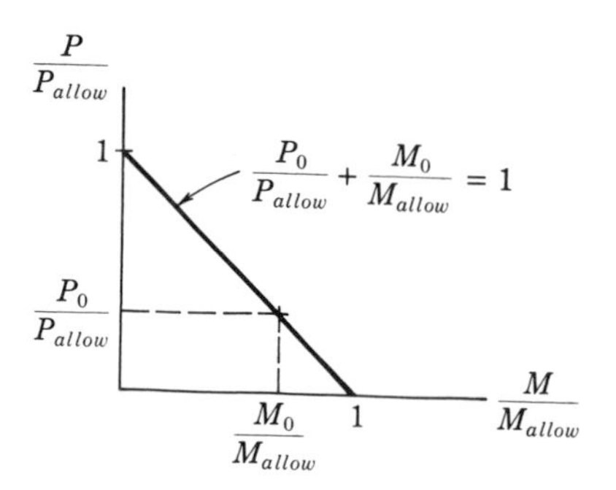

Fig. 9.3

Example Problem 9.1

(**a**) What is maximum compressive axial load which may be applied at the extreme corner of the flange of a 14 WF 142 beam? Neglect buckling. Allowable stresses are 14 ksi in pure compression, 20 ksi in pure bending.

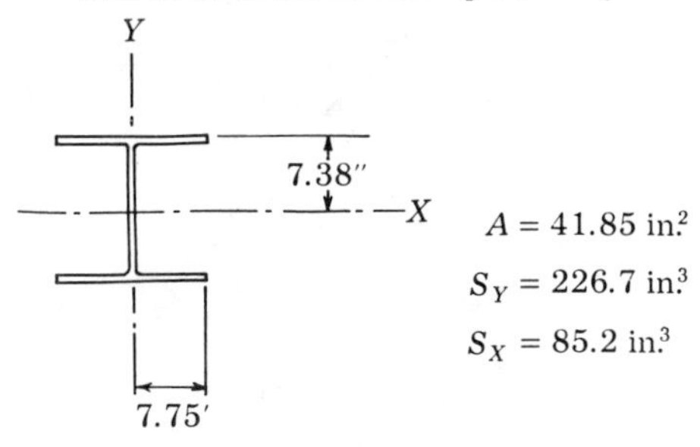

$A = 41.85\ \text{in}^2$

$S_Y = 226.7\ \text{in}^3$

$S_X = 85.2\ \text{in}^3$

We expand Eq. (9.12) for axial load and bending about two axes:

$$\frac{P_0}{P_{allow}} + \frac{M_x}{M_{x\ allow}} + \frac{M_y}{M_{y\ allow}} = 1$$

where

$P_{allow} = 14.0 \times 41.85 = 587$ kips
$M_x = P_0 \times 7.38$ kip-in.
$M_{x\ allow} = 20.0 \times 226.7 = 4{,}530$ kip-in.
$M_y = P_0 \times 7.75$ kip-in.
$M_{y\ allow} = 20.0 \times 85.2 = 1{,}700$ kip-in.

and substituting, we get

$$P_0\left(\frac{1}{587} + \frac{7.38}{4{,}530} + \frac{7.75}{1{,}700}\right) = 1 \qquad \text{or} \qquad P_0 = 126 \text{ kips}$$

The use of an equation similar to Eq. (9.14) would have yielded identical results.

(**b**)

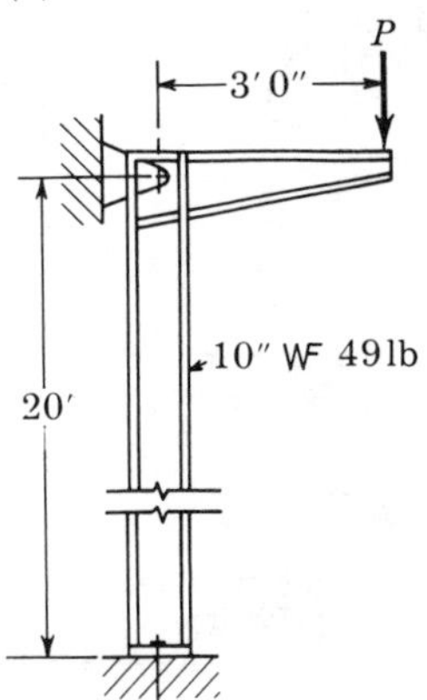

The 10 WF 49 beam column is made of A-36 steel ($F_{YP} = 36$ ksi). Assuming the column is braced in the weak direction, find the allowable load P according to the AISC Specifications.

We start here with the premise that the moment is preponderant, so that f_A/F_A will be less than 0.15. For this case, sec. 1.6.1 of the AISC Specifications assumes that the nonlinear effects due to buckling are sufficiently slight so that the linear Eq. (9.14) may be used.

We compute

$$f_A = \frac{P}{A} = \frac{P}{14.40} = 0.0695P$$

F_A (assuming buckling about strong axis, in which case $L/r = 240/4.35 = 55$) $= 17.90$ ksi [refer to Eq. (3.40)]

$$f_B = \frac{M}{S} = \frac{P \times 36 \text{ in.}}{54.6 \text{ in.}^3} = 0.66P$$

F_B (assuming bracing to prevent lateral buckling) $= 0.6F_{YP} = 21.6$ ksi

Substituting these into Eq. (9.14), we get

$$P\left(\frac{0.0695}{17.90} + \frac{0.66}{21.6}\right) = 1 \qquad \text{or} \qquad P = 29.0 \text{ kips}$$

9.3 Analysis of Prestressed Concrete Beams

An important application of stress analysis for combined bending and axial loads is found in the analysis and design of prestressed concrete beams.

The basic idea underlying prestressing is simple: by the application of initial forces to achieve a favorable stress distribution. In the particular case of concrete beams, in which we must discount that portion of the cross section under tensile strain, we introduce initial compression stresses in order to prevent the occurrence of tensile strains. Several advantages arise from this: because the entire cross section is effective in resisting bending, a more efficient and therefore more economical member results. Because the cracking which accompanies tensile strains is prevented, problems of corrosion of reinforcing and deterioration resulting from severe temperature conditions are avoided. Prestressed concrete may thus be used in exposed locations where the life span of conventional concrete might be limited. Lastly, because diagonal tension is also radically decreased, the use of slender thin-webbed members of I or T section becomes feasible.

The behavior of prestressed beams is best understood by observing the strains and stresses under the applied loads, as shown in Fig. 9.4. Because we anticipate no tension and therefore no cracking of the section, we can

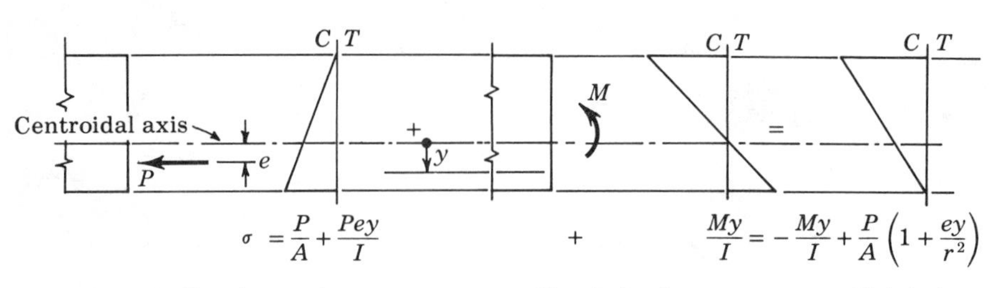

Fig. 9.4

base all our calculations on the entire section, and thus, in contrast to conventional reinforced concrete beams, resort to the simpler elastic theory of homogeneous beams. Because here the strains and stresses are proportional, only stresses are shown in Fig. 9.4. The distance y is measured from the centroid of the section, and compressive stress is here positive.

Figure 9.4 shows the stresses in a prestressed beam in a highly simplified manner. At stage I, the prestressing force is applied, resulting in compressive stresses throughout. The bending stresses caused by the external load are superimposed, resulting in the compressive stress distribution labeled stage II.[1] Several important secondary effects must be included in a more complete analysis, but the very simple basic reasoning is contained in Fig. 9.4.

There are several methods available for prestressing the beam. *Pretensioning* involves stretching wires against end anchorages, and the beam is cast around the tensioned wires. After the concrete has hardened, the wires are cut, thus transferring the compressive force to the beam. Because this transfer depends on the bond between concrete and reinforcing, such beams are called *bonded beams.* Because of this required surface bond, the reinforcing usually consists of small-diameter wires (with high ratio of surface area to cross-sectional area), or of twisted strands which provide surface corrugations. Figures 9.5*a* and *b* show the operation schematically.

Post-tensioning involves prior casting of the beam, which is provided with ducts through which the reinforcing runs freely. After hardening of the concrete, the wires, cables, or rods are tightened against the ends of the beam and secured with end anchorages, thus prestressing the beam. The operation is shown schematically in Fig. 9.5*c* and *d*.

[1] A more precise calculation would consider the fact that any loads applied *after* transfer of the prestressing force would result in a change of strain, and therefore change of stress of the prestressing steel. This fact could be taken into consideration by calculating the stresses resulting from loads on basis of the transformed rather than the concrete section, as is done here. Because the difference in results is slight, this is not done in usual practice, nor in what follows here. It would, however, be reasonably simple to amend Eqs. (9.15) to (9.20) below to account for this effect.

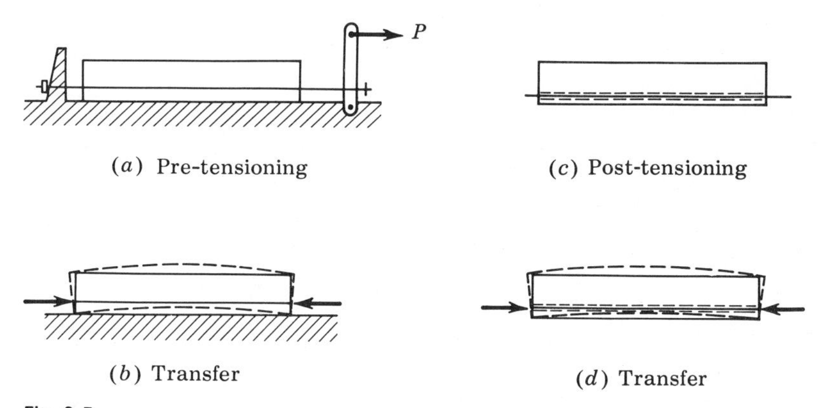

Fig. 9.5

In comparing the advantages and disadvantages of the two systems, we note that pretensioning requires a special casting bed with strong end anchorages, which however may be placed arbitrarily far apart, thus enabling the casting of many beams in one stressing operation. It is therefore well suited to precasting in plants. Post-tensioning, on the other hand, requires no special forms and is therefore preferred for large members which have to be cast and stressed in place. Both methods have their definite spheres of application and are widely used.

No matter which prestressing system is used, the stage I stress and therefore also the strain distribution show negative curvature

$$\phi_I = \frac{\epsilon_{bot} - \epsilon_{top}}{d}$$

where ϵ represents the stage I strains in the top and bottom fibers and d is the beam depth. This will cause the beam to rise off its casting bed as soon as the prestressing force is transferred to the beam, as shown by the dashed lines in Fig. 9.5*b* and *d*. The simple beam moment resulting from the beam dead load will therefore always act simultaneously with the prestressing force, so that we can always include these dead-load stresses in stage I. These additional stage I stresses are of a beneficial nature since they tend to counteract the prestresses and thus enable a larger prestressing force to be applied.

Another important secondary consideration is the inclusion of the effects of compressive creep of the concrete. As a matter of fact, loss of prestress caused by creep prevented use of prestressing for many decades, till the problem was recognized and remedied in the 1920s by Freyssinet.

If a concrete fiber is put under compressive stress, it will creep in the course of time. Let us assume that this creep strain will amount to 0.001 in./in., then this will also allow the embedded steel to reduce its tensile strain by this amount. If the steel had been initially stressed to 30 ksi, then its initial strain was 30/30,000 ksi = 0.001 in./in., and a shortening of this amount will reduce the strain and therefore its stress to zero. In early attempts at prestressing with ordinary steels it was this gradual loss of prestress with time which made prestressing impractical.

If on the other hand, we use high-tensile reinforcing steel and stress it elastically to, say, 150 ksi corresponding to 0.005 in./in. of tensile strain, then a reduction of strain due to shortening of the concrete of 0.001 in./in. will still correspond to a remaining steel strain of $0.005 - 0.001 = 0.004$ in./in., leaving a prestress of $0.004 \times 30{,}000 = 120$ ksi in the steel. In this case then only 20 percent of the prestresses is lost. The higher the elastic steel stress, the less loss will occur because of creep, and use of high tensile steels is mandatory in prestressed concrete.

To account for this loss of prestress (which in practice is often assumed at 20 percent), we simply reduce all stresses due to prestress (but not those due to load) by the percentage loss, as shown by dashed lines in Fig. 9.6. Then, depending on whether stresses in the beam are desired immediately, or after a period of time during which creep relaxation occurs, either the solid or the dashed prestresses are used in the calculations.

A number of other less important causes of loss of prestress are listed in sec. 2607 of the ACI Code, but will not be considered in our elementary discussion.

Considering now the two additional effects which have been mentioned, we redraw our stress distribution more precisely in Fig. 9.6. Here M_B, M_{LL}, and M_{total} denote the moments due to beam dead load, all other loads, and total load respectively, and α is the ratio of prestressing force after creep relaxation to the initial force.

The general formulas for the concrete stress distribution are also shown in Fig. 9.6 underneath the appropriate sketches and may be used to calculate critical stresses. The general requirement is that the extreme fiber concrete stress σ_{ext} be within its allowable values at all times:

$$0 \leq \sigma_{ext} \leq f_{c\ allow}$$

Looking for instance at the stage I stress distribution of Fig. 9.6*d*, we see that for an efficiently designed beam the critical stresses occur before loss of prestress and are in the top fiber, where $y = -c_T$:

$$\sigma_{I\ top} = \frac{P}{A}\left[1 - \left(e - \frac{M_B}{P}\right)\frac{c_T}{r^2}\right] = 0 \tag{9.15}$$

and, in the bottom fiber, where $y = c_B$,

$$\sigma_{I\ bot} = \frac{P}{A}\left[1 + \left(e - \frac{M_B}{P}\right)\frac{c_B}{r^2}\right] = f_{c\ allow} \tag{9.16}$$

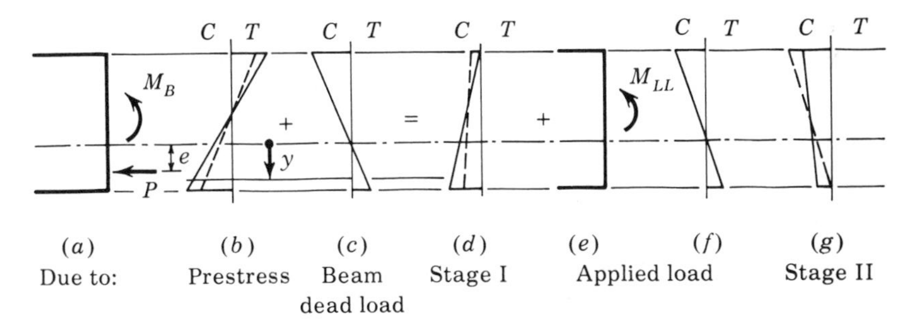

Before loss of prestress:

$$\sigma = \frac{P}{A}\left(1 + \frac{ey}{r^2}\right) - \frac{M_B y}{I}\ \left\{= \frac{P}{A}\left[1 + \left(e - \frac{M_B}{P}\right)\frac{y}{r^2}\right]\right\} - \frac{M_{LL} y}{I} = \frac{P}{A}\left[1 + \left(e - \frac{M_{total}}{P}\right)\frac{y}{r^2}\right]$$

After loss of prestress:

$$\sigma = \frac{P\alpha}{A}\left(1 + \frac{ey}{r^2}\right) - \frac{M_B y}{I}\ \left\{= \frac{P\alpha}{A}\left[1 + \left(e - \frac{M_B}{P\alpha}\right)\frac{y}{r^2}\right]\right\} - \frac{M_{LL} y}{I} = \frac{P\alpha}{A}\left[1 + \left(e - \frac{M_{total}}{P\alpha}\right)\frac{y}{r^2}\right]$$

Fig. 9.6

At stage II, the dashed stress distribution of Fig. 9.6g shows the critical values, after loss of prestress, as

$$\sigma_{II\ top} = \frac{P\alpha}{A}\left[1 - \left(e - \frac{M_{total}}{P\alpha}\right)\frac{c_T}{r^2}\right] = f_{c\ allow} \tag{9.17}$$

and $$\sigma_{II\ bot} = \frac{P\alpha}{A}\left[1 + \left(e - \frac{M_{total}}{P\alpha}\right)\frac{c_B}{r^2}\right] = 0 \tag{9.18}$$

9.4 Design of Prestressed Concrete Beams

A rigorous design method for prestressed concrete beams could be based on a simultaneous solution of the four simultaneous equations, (9.15) to (9.18) for the unknowns depending on the beam section, A, r^2, c_{top}, and c_{bot}, and those related to the prestressing, P and e. Since there are more unknowns than equations, considerable freedom remains. Usually, the cross section is chosen by experience or by trial and error, and Eqs. (9.15) and (9.18) are solved for the magnitude and location of the prestressing force:

$$\left(e - \frac{M_B}{P}\right)\frac{c_T}{r^2} = +1$$

and $$\left(e - \frac{M_{total}}{P\alpha}\right)\frac{c_B}{r^2} = -1$$

from which, solving simultaneously,

$$P = \frac{[(M_{total}/\alpha) - M_B]c_B}{[1 + (c_B/c_T)]r^2} \tag{9.19}$$

and $$e = \frac{r^2}{c_T}\,\frac{(M_{total}/\alpha M_B) + (c_T/c_B)}{(M_{total}/\alpha M_B) - 1} \tag{9.20}$$

Having obtained the stressing force and eccentricity, the maximum compressive stresses can be checked by Eqs. (9.16) and (9.17), or preferably, by use of the stress plots of Fig. 9.6. An example problem will further clarify the procedure.

Example Problem 9 2 The precast T floor beam is simply supported, of span 30 ft. Superimposed load is to be 180 lb/ft² of floor area. Concrete weighs 150 lb/ft³; α = 0.80; $f_{c\ allow}$ = 3.0 ksi, $f_{c\ allow}$ = 150 ksi. Determine magnitude and location of prestressing force at midspan. Check all critical stresses in the beam, and draw conclusions regarding efficiency and appropriateness of design.

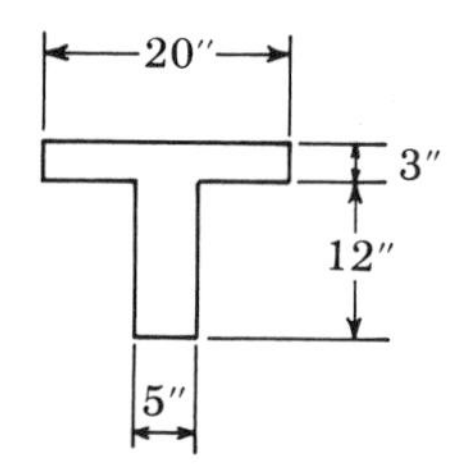

We begin by calculating the cross-sectional properties of the section, and find: $A = 120$ in.², $c_B = 9.75$ in., $c_T = 5.25$ in., $I = 2{,}447$ in.⁴, $r^2 = I/A = 20.3$ in.² The uniform superimposed load is 180 lb/ft² × 1.67 ft = 300 lb/ft, and the beam dead load is 150 × 120/144 = 125 lb/ft, so that the relevant moments become: $M_B = 14.0$ kip-ft, $M_{LL} = 33.75$ kip-ft, $M_{total} = 47.75$ kip-ft. Using Eqs. (9.19) and (9.20), the prestressing should be $P = 92.0$ kips, $e = 5.70$ in.

Note that with an allowable steel stress of 150 ksi, only 0.62 in.² of steel is required, the centroid of which is to be located at 9.75 − 5.70 = 4.05 in. above the bottom of the stem of the T section.

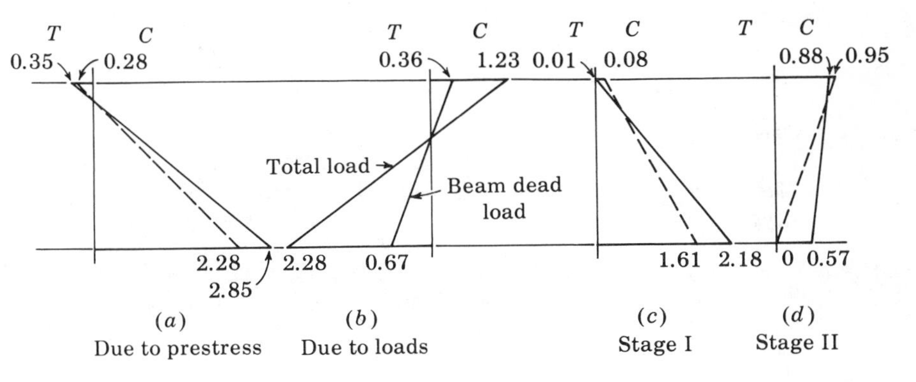

Fig. 9.7

The design is checked by computing and superimposing the stresses resulting from the various effects. This result is shown in Fig. 9.7, and indicates that nowhere is there tension. The maximum compressive stress of value 2.18 ksi prevails at stage I prior to application of live load and prior to creep relaxation. In the course of time this stress should gradually diminish to around 1.61 ksi. Note that with the application of live load the compressive stress diminishes radically.

In thin-webbed members the diagonal tension stress is often critical, as has already been noted in Sec. 6.7. Here again, we utilize the theory of stress transformation, starting with stresses on vertical and horizontal planes at the end of the beam where shear is highest, of value $V = \dfrac{wL}{2} = \dfrac{0.425 \times 30}{2} = 6.38$ kips. The highest shear stress will occur on the neutral axis, of a value (according to the theory of elastic, homogeneous beams)

$$\tau = \frac{V \int y\, dA}{Ib} = \frac{6.38 \times [5 \times (9.75^2/2)]}{2{,}477 \times 5} = 0.128 \text{ ksi}$$

This must be combined with the longitudinal compression stress at the neutral axis due to the prestress, of value

$$\sigma = \frac{P}{A} = \frac{92.0}{120} = 0.767 \text{ ksi} \qquad \text{before creep}$$

or

$$\sigma = \frac{P\alpha}{A} = \frac{73.5}{120} = 0.612 \text{ ksi} \qquad \text{after creep}$$

The latter stress when combined with shear will be critical, and it will be combined with the shear stress in order to find the principal tensile stress by use of Mohr's circle, as shown in Fig. 9.8.

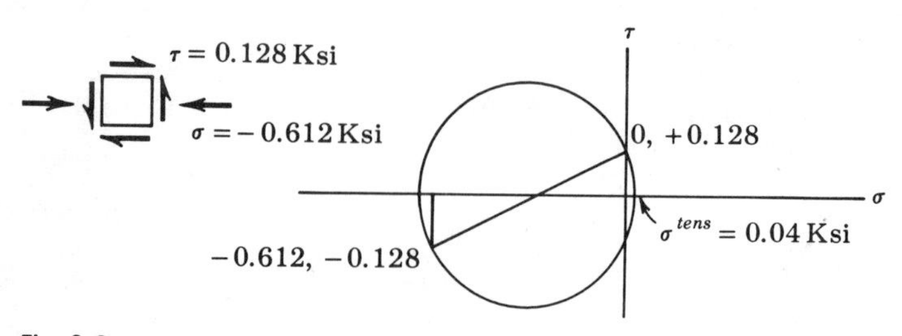

Fig. 9.8

Mohr's circle (or a corresponding analytical transformation) shows that the presence of the compressive prestress serves to decrease the maximum diagonal tension stress radically. The principal tension stress is found to be 0.040 ksi, which is safe.

With a specified allowable stress of 3,000 psi, it must be concluded that while the beam is safe, it is overdesigned and uneconomical. A redesign with somewhat reduced depth should be attempted to bring the maximum compression stress closer to the allowable value.

9.5 Ultimate Strength of Prestressed Concrete Beams

Under sufficiently high moment, tensile strains will develop in the prestressing steel as well as in the surrounding concrete, as shown by the strains of Fig. 9.9*b*. The concrete will then begin to crack on the tension side, and the beam will behave more and more as a conventionally reinforced beam, with stresses as in Fig. 9.9*c*. Stages 1, 2, and 3 of Fig. 9.9 show this progressive change from the uncracked condition under working load to the ultimate stage, when the crack has progressed upward and confined the compression side to a narrow zone subject to high compressive stresses. Failure will occur, as in conventionally under-reinforced beams,

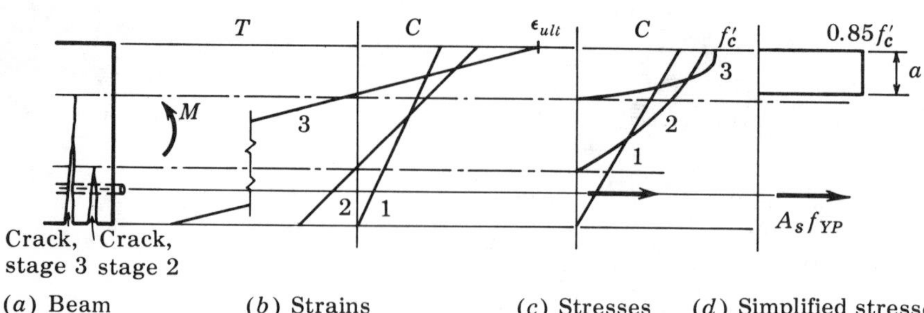

(*a*) Beam (*b*) Strains (*c*) Stresses (*d*) Simplified stresses at ultimate

Fig. 9.9

by compressive crushing of the concrete, and the ultimate moment can be calculated in a simplified manner by applying the equilibrium equations to the stresses of Fig. 9.9*d*. We recognize this as identical to Fig. 8.6*d*, and therefore conclude that the ultimate moment of a prestressed beam is identical with that of the beam reinforced conventionally with the same amount of steel, and can be determined by use of Eq. (8.26).

We see from the strain distribution at ultimate, stage 3 in Fig. 9.9*b*, that, just as in the case of conventional beams, considerable ductility of the steel is required to enable the concrete to reach its ultimate stress distribution of Fig. 9.9*d*. It is important to verify the adequacy of the plastic range of the high-strength prestressing steels used, and in case this is deficient, more elaborate calculations including strain compatibility are in order to ensure the safety of the section.

It is interesting to note that for the case of sufficiently ductile steel the prestress does not affect the ultimate strength of the beam. This is an added verification of the statement of Sec. 3.2 (in connection with initial stresses) that self-equilibrating force systems have no effect on the ultimate strength of ductile structures. Prestress does, however, improve the beam behavior in the elastic range. This means that any savings in the elastically designed beam section due to prestressing will be made at the expense of the safety factor against collapse of the member. Good practice in prestressed concrete design requires a check of the ultimate strength to ensure adequacy of load factor.

9.6 Code Provisions for Prestressed Concrete Design

Chapter 26 of the 1963 ACI Code contains specifications for the design of structural prestressed concrete beams.

Section 2603*a* stipulates that elastic and ultimate strength of beams are to be investigated. 2603*c* calls for rigorous determination of the effects of prestressing in a statically indeterminate structure. Such analyses may be quite involved for sizable building frames.

Section 2604 outlines some of the basic principles of elastic analysis, and secs. 2605 and 2606 give allowable steel and concrete stresses. Note that under some conditions a slight amount of tension stress is permitted in the beam, and also that allowable concrete and steel stresses are held higher at transfer of prestressing force than under full design loads.

Section 2607 calls for consideration of prestress losses due to several causes, without, however, specifying either methods of calculation or numerical coefficients. It also gives factors to calculate the variation of stressing force along the beam due to friction between cables and walls of ducts of post-tensioned beams.

The ultimate strength of prestressed beams is discussed in sec. 2608; formulas 26-4 and 26-5 are results of calculations based on the reasoning of the preceding section, of which the former is similar to formula 16-1 for conventionally reinforced beams. One important difference must be noted: because the high-tensile steel used for prestressing may not possess a pronounced yield point, the steel stress is specified in terms of a value f_{su}, defined as the *calculated steel stress at ultimate load;* guidelines for determination of this stress are given in sec. 2608*a*3. Further discussion of this must be left to texts on prestressed concrete design.

Section 2609 deals with the maximum amount of prestressing steel which can be used to ensure under-reinforcement at ultimate. The rest of the chapter contains a variety of provisions dealing with matters such as shear reinforcing, bond under static and repeated loading, and constructive provisions.

9.7 Inelastic Members—General Theory

We consider a member of a material whose stress-strain curve is shown in Fig. 9.10*a* subjected to a combined loading shown in Fig. 9.10*b*. The approach can again be outlined by applying the four steps of the strength of materials method. Steps 1 to 3, involving statics and geometry, are independent of the stress-strain curve and are therefore identical with Eqs. (9.1) to (9.3):

Statics:

$$\Sigma F = 0: \qquad \int \sigma \, dA = P \tag{9.21}$$

$$\Sigma M_{CA} = 0: \qquad \int \sigma y \, dA = M \tag{9.22}$$

Geometry: Assuming plane sections remaining plane, we set

$$\epsilon = \phi y_1 = \phi(\bar{y} + y) \tag{9.23}$$

The appropriate stress-strain relation $\sigma = f(\epsilon)$ is used to express the stress distribution as

$$\sigma = f[\phi(\bar{y} + y)] \tag{9.24}$$

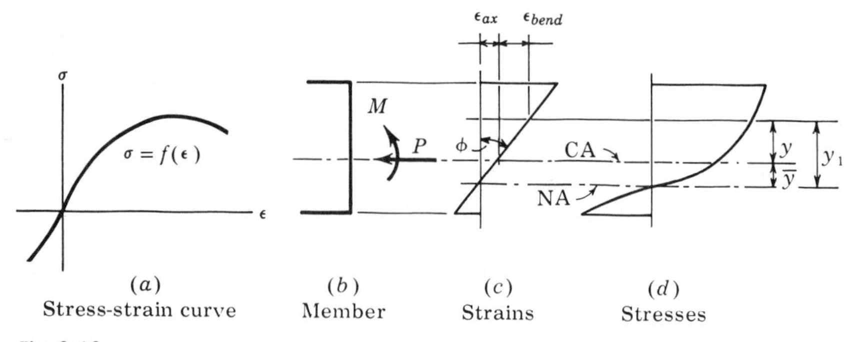

Fig. 9.10

and resubstituting this into the equations of statics, we obtain

$$\Sigma F = 0: \qquad \int_A f(\phi\bar{y} + \phi y)\, dA = P \tag{9.25}$$

$$\Sigma M = 0: \qquad \int_A f(\phi\bar{y} + \phi y)y\, dA = M \tag{9.26}$$

Equations (9.25) and (9.26) must be solved simultaneously for the two unknowns $\bar{y}$, defining the neutral axis, and ϕ, the curvature. If the stress-strain relations are nonlinear, the solution of these equations may be highly involved and laborious. While load-deformation relations of this type may not be often needed in professional practice, they are of great importance in investigating inelastic buckling of beam columns. The calculations of the next section will explain the procedure for a particularly simple case.

9.8 Load-curvature Relations for Elastic-plastic Members[1]

In a later section we shall concern ourselves with the inelastic buckling of beam columns, and for this we shall need the relationship between the applied force and moment and the resulting curvature. To simplify the calculations, we shall consider a member of rectangular cross section of elastic–perfectly plastic material of yield stress σ_{YP}.

The procedure has been outlined in the preceding section; here, because of the piecewise linear stress distribution, the integrations are best carried out numerically.

In the inelastic range, two different types of stress distribution are possible: Case 1, shown in Fig. 9.11*c*, considers that only one side of the member has begun to yield; case 2, shown in Fig. 9.12*c*, assumes that yielding has occurred on both sides of the member. Let us first investigate case 1.

First, applying statics and referring to Fig. 9.11*c*, we can simplify our calculations by breaking up the stresses (1) into two parts—the part labeled

[1] This section may be deleted if it is not intended to cover Sec. 11.5.

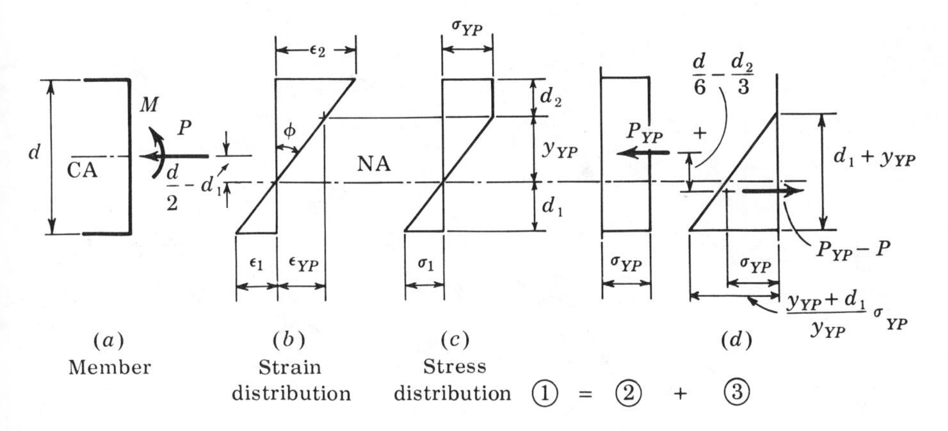

Fig. 9.11

2 corresponding to the axial yield force P_{YP}, and the remaining triangular block, 3. Then force equilibrium can be expressed as

$$\Sigma F = 0: \qquad \text{Force 1} = \text{Force 2} + \text{Force 3} = P$$

therefore

$$P_{YP} - \text{Force 3} = P$$

therefore

$$\text{Force 3} = P_{YP} - P = \frac{1}{2} b \left(\frac{y_{YP} + d_1}{y_{YP}} \sigma_{YP} \right) (y_{YP} + d_1) \tag{9.27}$$

and moment equilibrium, written with respect to the centroidal axis, as

$$\Sigma M_{CA} = 0: \qquad (P_{YP} - P) \left(\frac{d}{6} + \frac{d_2}{3} \right) = M \tag{9.28}$$

Also, from Fig. 9.11*c*,

$$d = y_{YP} + d_1 + d_2 \tag{9.29}$$

Expressions (9.27) to (9.29) contain the three unknown dimensions, y_{YP}, d_1, and d_2; solving them simultaneously for y_{YP} in terms of the ratios P/P_{YP} and M/M_{YP} and the total beam depth d, we get

$$y_{YP} = \frac{9[1 - (P/P_{YP}) - (M/3M_{YP})]^2}{8[1 - (P/P_{YP})]^3} d \tag{9.30}$$

From the strain distribution, Fig. 9.11*b*, we see that

$$\phi = \frac{\epsilon_{YP}}{y_{YP}} = \frac{\sigma_{YP}}{E y_{YP}} = \frac{M_{YP}}{EI} \frac{d}{2y_{YP}} \tag{9.31}$$

and substituting Eq. (9.30) into this, the final expression for the curvature becomes

$$\phi = \frac{M_{YP}}{EI} \frac{4[1 - (P/P_{YP})]^3}{9[1 - (P/P_{YP}) - (M/3M_{YP})]^2} \tag{9.32}$$

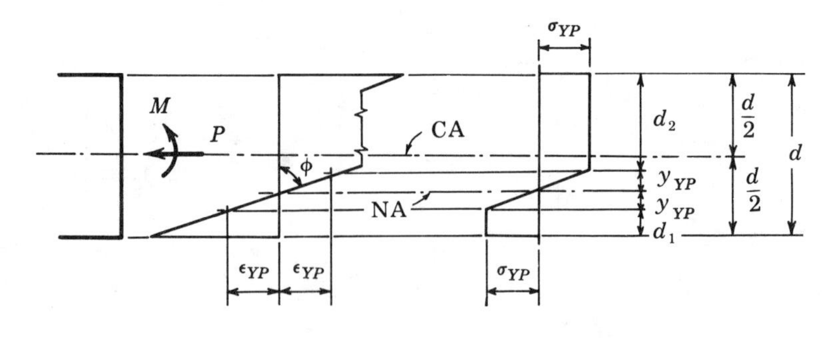

Fig. 9.12

Equation (9.32) gives the curvature for any combination of axial load and moment for case 1. Stresses may be determined by first calculating y_{YP} [Eq. (9.30)] and then d_1 and d_2, using Eqs. (9.27) and (9.29), whereupon the complete stress distribution may be drawn.

Case 2 assumes that yielding has begun on both sides of the rectangular section, resulting in the strains and stresses of Fig. 9.12*b* and *c*.

$$\Sigma F = 0: \qquad -\sigma_{YP} b d_1 + \sigma_{YP} b d_2 = P$$

therefore

$$\frac{P}{\sigma_{YP} b d} = \frac{P}{P_{YP}} = \frac{1}{d}(d_2 - d_1) \tag{9.33}$$

$$\Sigma M_{CA} = 0: \qquad (\sigma_{YP} b d_2)\left(\frac{d}{2} - \frac{d_2}{2}\right) + \left(\frac{\sigma_{YP} b}{2} y_{YP}\right)\left(\frac{4}{3} y_{YP}\right) + (\sigma_{YP} b d_1)\left(\frac{d}{2} - \frac{d_1}{2}\right) = M \tag{9.34}$$

Also $\quad d = d_1 + 2y_{YP} + d_2 \qquad (9.35)$

Solving Eqs. (9.33) to (9.35) simultaneously for d_1, y_{YP}, and d_2, we get

$$y_{YP} = \sqrt{\frac{3}{4}\left[1 - \left(\frac{P}{P_{YP}}\right)^2\right] - \frac{M}{2M_{YP}}}\, d \tag{9.36}$$

The curvature, again given by Eq. (9.31), is obtained by substituting this value of y_{YP}:

$$\phi = \frac{M_{YP}}{EI} \frac{1}{\sqrt{2}\sqrt{\frac{3}{2}\left[1 - \left(\frac{P}{P_{YP}}\right)^2\right] - \frac{M}{M_{YP}}}} \tag{9.37}$$

Stresses may be found using the same procedure outlined for case 1.

It is noted that case 1 will occur under low ratios of applied moment to load, while case 2 will arise under high ratios. To determine the limiting ratio between the two cases, the extreme fiber strain on the less highly strained side of the beam is set equal to ϵ_{YP}, so that the stress distribu-

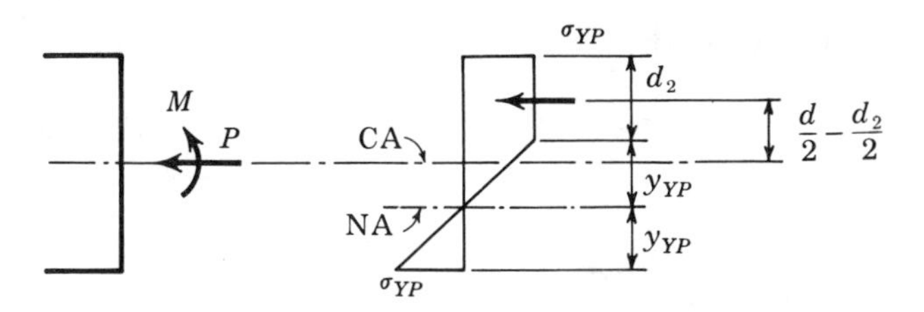

Fig. 9.13

tion of Fig. 9.13 results:

$$\Sigma F = 0: \qquad \sigma_{YP} b d_2 = P \tag{9.38}$$

$$\Sigma M_{CA} = 0: \qquad P\left(\frac{d}{2} - \frac{d_2}{2}\right) + \left(\frac{1}{2}\sigma_{YP} y_{YP} b\right)\left(\frac{4}{3} y_{YP}\right) = M \tag{9.39}$$

$$d = d_2 + 2y_{YP} \tag{9.40}$$

Solving Eqs. (9.38) to (9.40) simultaneously to eliminate d_2 and y_{YP} to obtain the limiting relation between P and M, we get

$$\frac{P}{P_{YP}} = \frac{1}{4} - \frac{1}{2}\sqrt{\frac{9}{4} - 2\frac{M}{M_{YP}}} \tag{9.41}$$

Equations (9.32), (9.37), and (9.41) may be utilized to establish moment-curvature relationships for rectangular members under axial load and bending moment. They will be utilized in a later section to investigate inelastic buckling of beam columns.

9.9 Ultimate Strength of Perfectly Plastic Members

To calculate the carrying capacity of a member of perfectly plastic material subjected to simultaneous axial and bending loads, we utilize the same principles as in previous discussions. At this stage we again disregard the possibility of buckling of the member.

Figure 9.14 shows a rectangular-section member subjected to a moment M and an axial force P. It is required to find the values of M and P under which yielding of the section will occur. Failure of the member will correspond to large strains, so we consider a strain distribution corresponding to infinitely large curvature, as shown in Fig. 9.14*b*, and draw the corresponding rectangular stress distribution of Fig. 9.14*c*. To this we apply the conditions of statics:

$$\Sigma F = 0: \qquad -T + C = P_0$$

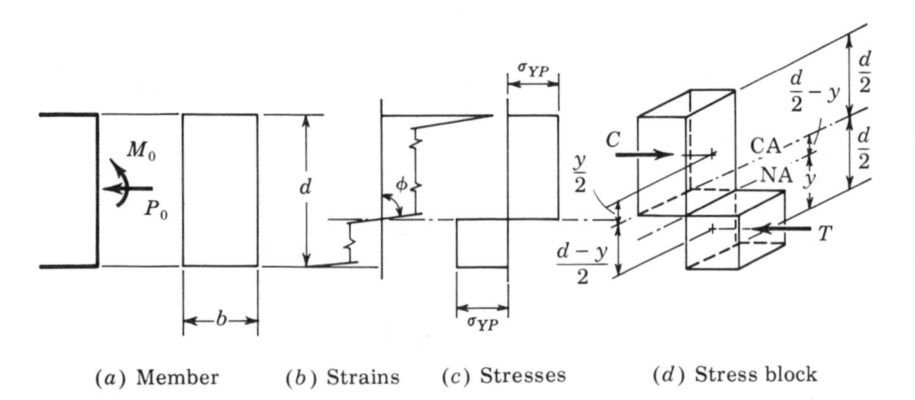

Fig. 9.14

therefore

$$-by\sigma_{YP} + b(d - y)\sigma_{YP} = P_0$$

therefore

$$y = \frac{1}{2}\left(\frac{P_0}{b\sigma_{YP}} + d\right) = \left(1 - \frac{P_0}{P_{YP}}\right)\frac{d}{2} \tag{9.42}$$

where P_{YP} is the axial load which causes complete plastification of the cross section $= \sigma_{YP}A = \sigma_{YP}bd$, and M_0 and P_0 are, respectively, the moment and axial load, applied simultaneously, which cause complete plastification of the section.

Next, setting

$$\Sigma M_{CA} = 0: \qquad C\frac{y}{2} + T\frac{d - y}{2} = M_0$$

or

$$P_0\frac{y}{2} + T\frac{d}{2} = M_0$$

Substituting Eq. (9.42), we obtain

$$P_0\left(1 - \frac{P_0}{P_{YP}}\right)\frac{d}{4} + \frac{1}{2}(P_{YP} - P_0)\frac{d}{2} = M_0$$

Dividing by

$$\frac{bd^2}{4}\sigma_{YP} = \frac{d}{4}P_{YP} = M_P$$

we get the interaction expression

$$\left(\frac{P_0}{P_{YP}}\right)^2 + \frac{M_0}{M_P} = 1 \tag{9.43}$$

Alternately, it is possible to split up the stress block of Fig. 9.14*d* into two parts as shown in Fig. 9.15 and arrive at the result of Eq. (9.43). This expression can be represented by the interaction curve shown in Fig. 9.16, which shows the fraction of the total capacity of the rectangular section used up by the applied axial load and moment respectively. It should be noted that for small axial loads, the ultimate moment capacity M_P is only very

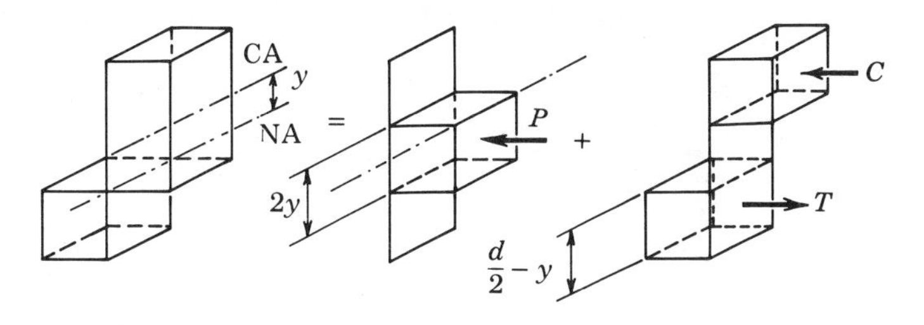

Fig. 9.15

slightly reduced. It must also be recognized that these results may be meaningless for slender members, because of the possibility of axial buckling.

Analogous calculations can be made for members of other cross sections; for wide-flange shapes, two stages must be distinguished:

1. For large ratios of moment to axial force, the neutral axis is in the web, and Fig. 9.17 indicates relations at yielding:

$$\Sigma F_x = 0: \qquad P_0 = \sigma_{YP} 2wy$$

$$\Sigma M_{CA} = 0: \qquad M_0 = \sigma_{YP}\left[\frac{A_F}{2}d_F + w\left(\frac{d_w}{2} - y\right)\left(\frac{d_w}{2} + y\right)\right]$$

$$= \sigma_{YP}\left(\frac{A_F}{2}d_F + \frac{1}{4}wd_w^2 - wy^2\right)$$

But

$$P_{YP} = \sigma_{YP}(A_F + A_w) = \sigma_{YP}A$$

and

$$M_P = \sigma_{YP}\left(\frac{1}{4}A_w d_w + \frac{1}{2}A_F d_F\right) = \frac{P_{YP}}{2}(d - t) - \frac{1}{4}A_w \sigma_{YP} d$$

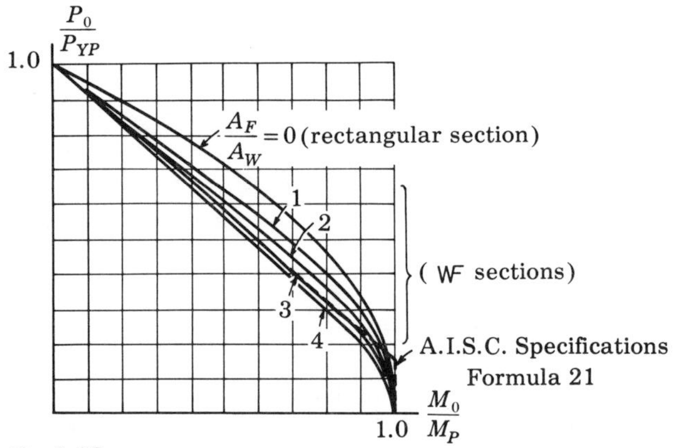

Fig. 9.16

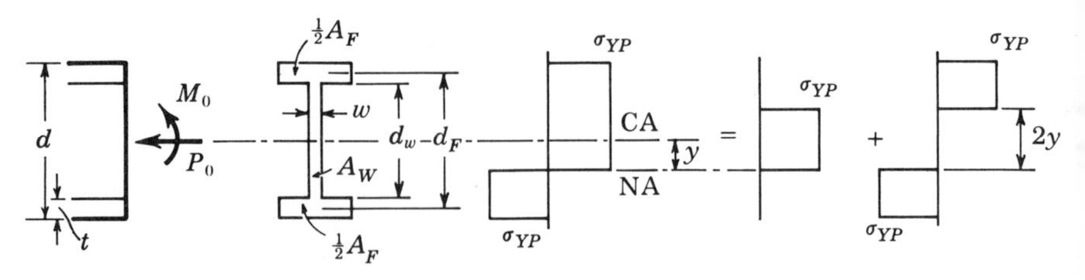

Fig. 9.17

Therefore, solving simultaneously to eliminate y and substituting for the appropriate value of P_{YP} and M_P, we get the interaction expression

$$\left[\frac{1}{\left(\frac{A_w}{A}\right)^2 + 2\frac{A_F}{A}\frac{A_w}{A}\frac{d_F}{d_w}}\right]\left(\frac{P_0}{P_{YP}}\right)^2 + \frac{M_0}{M_P} = 1 \tag{9.44}$$

2. For small ratios of moment to axial force, the neutral axis is in the flange, and the relations of Fig. 9.18 apply:

$$\Sigma F = 0: \qquad P_0 = \sigma_{YP}\left(A_w + \frac{t - \Delta}{t} A_F\right)$$

$$\Sigma M_{CA} = 0: \qquad M_0 = \sigma_{YP}\left(\frac{\Delta}{t} A_F\right)\frac{1}{2}(d - \Delta)$$

Again solving simultaneously and substituting P_{YP} and M_P, we get (neglecting the small quantity Δ in the expression $d - \Delta$):

$$\frac{P_0}{P_{YP}} + \left(\frac{d_F}{d_w} - \frac{A_w}{2A}\right)\frac{M_0}{M_P} = 1 \tag{9.45}$$

The boundary between the two cases will be when the neutral axis is at the junction of web and flange, in which case

$$P_0 = \sigma_{YP} A_w \qquad \text{or} \qquad \frac{P_0}{P_{YP}} = \frac{A_w}{A} \tag{9.46}$$

With this information, it is possible to plot interaction curves for sections with various ratios of d_F/d_w and A_w/A.

For most commercial wide-flange sections, $d_F/d_w \cong 1.05$, and the curves shown on Fig. 9.16 are plotted for this value and various ratios of A_F/A_w.

The ultimate strength of members under combined loading in which buckling plays no role is covered in sec. 2.3 of the AISC Specifications. Case I of this section is intended for "columns bent in double curvature by moments producing plastic hinges at both ends of the columns." Since here the points of maximum moment (the ends) are laterally supported, and midspan corresponds to a point of zero moment, the linear interaction expression, formula 21, is based on the results of this section. It is plotted in dashed line in Fig. 9.16. Because of the ease with which they can be used, such linear interaction curves are often substituted for the more rigorous nonlinear results.

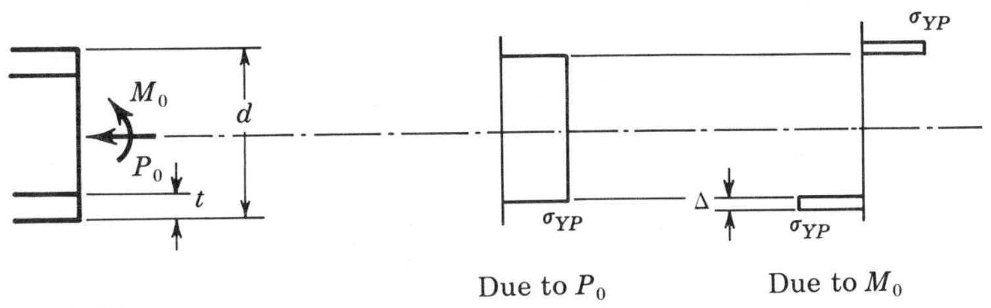

Fig. 9.18

General Readings

Elastic theory is covered in elementary strength of materials texts, interaction curves are used in texts on design, and inelastic behavior is discussed in the tests cited in Chap. 7. The material in Sec. 9.8 follows the outline given in Bleich's book cited in Chap. 3. A standard reference for prestressed concrete is:

Lin, T. Y.: "Design of Prestressed Concrete Structures," 2d ed., Wiley, 1963.

Problems

9.1 Find and plot the stress distribution in the 14 WF 202 beam shown.

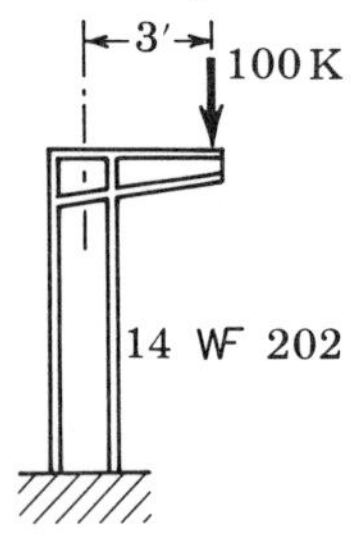

9.2 If the allowable tension stress in the 10 WF 49 is 20 ksi and the allowable compression stress (because of the danger of buckling) is only 16 ksi, what is the maximum allowable eccentricity?

9.3 If the allowable stresses on the 12 WF 27 beam are 20 ksi in tension, 14 ksi in compression, what is the allowable load P?

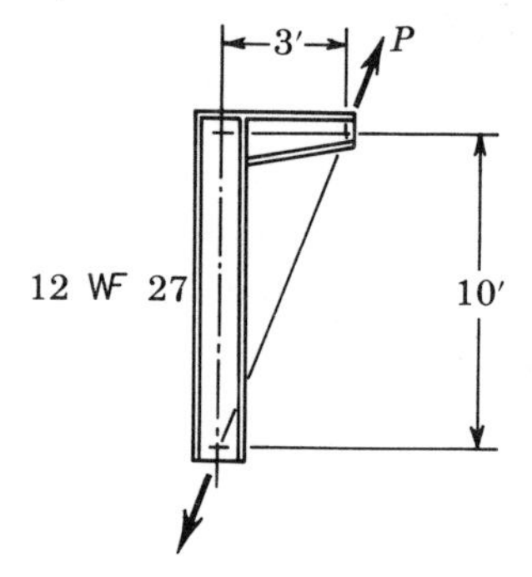

9.4 On the 10 by 10 in. cap plate, outline the area over which an axial force may be applied without causing tensile stresses in the 10 WF 49 member.

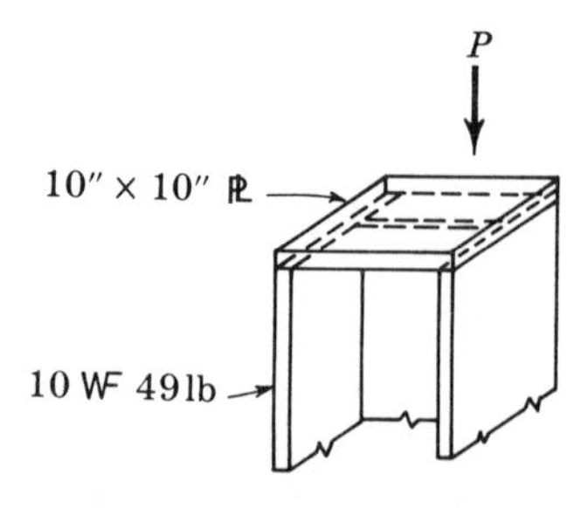

9.5 Define the kern of the unsymmetric I section shown.

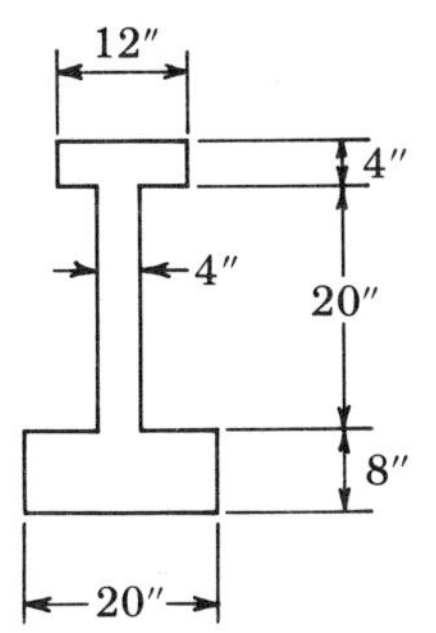

9.6 Assuming linear interpolation of allowable stresses between their limiting values, expand Eq. (9.14) to cover the case of axial force and bending about two axes. Sketch the surface which represents the resulting expression.

9.7 A 16 WF 58 beam has an axial force of 150 kips acting. Allowable stresses are $\sigma_{ax} = 15$ ksi, $\sigma_{bend} = 22$ ksi. Find the additional moment about the strong axis which may safely be applied to this member. Assume linear interaction.

9.8 The unsymmetric I section is subjected to axial load and moment, both of which can suffer a reversal of direction. Allowable stresses are: $\sigma_{tens} = 20$ ksi, $\sigma_{comp,\ ax} = 15$ ksi, $\sigma_{comp,\ bend} = 12$ ksi. Assuming linear interpolation of allowable stresses, draw an interaction curve of the type shown.

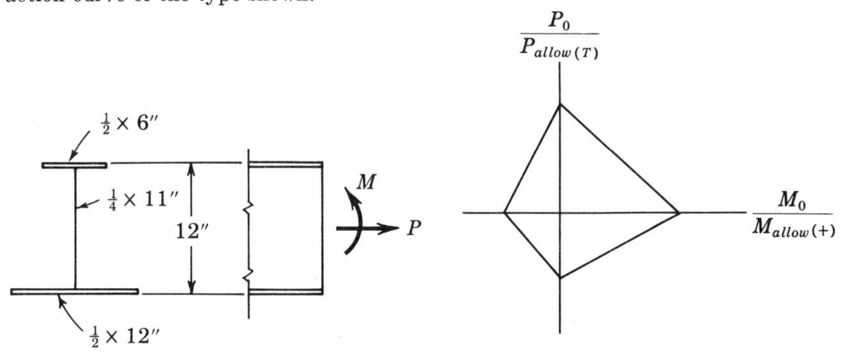

9.9 Allowable stresses: $\sigma_{ax} = 16$ ksi, $\sigma_{bend} = 20$ ksi. Find the allowable load P which may be applied to the frame shown.

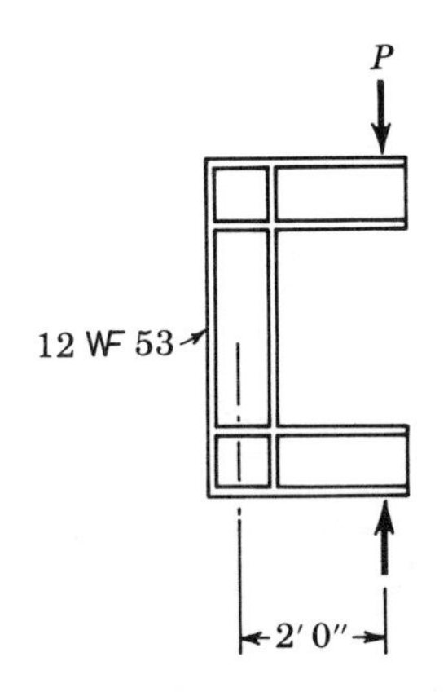

9.10 The semicircular statically determinate arch is formed of a 14 W 142 beam. Allowable stresses are: $\sigma_{ax} = 10$ ksi, $\sigma_{bend} = 16$ ksi. Find the maximum concentrated load which may safely be applied. Assume arch is suitably braced against buckling. *Hint:* Write interaction expression in terms of θ; maximize with respect to θ to find critical section.

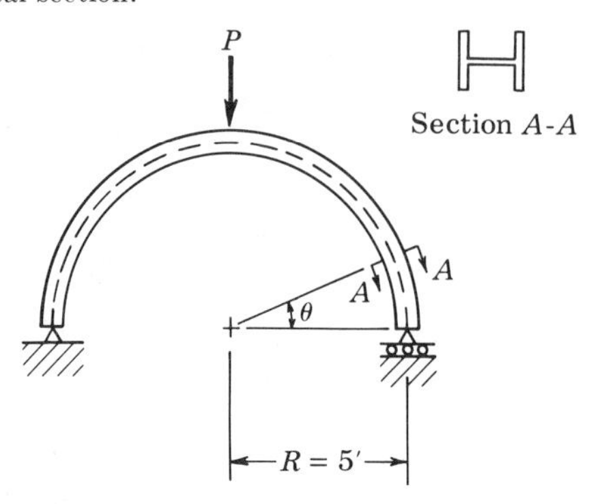

9.11 A rectangular cross-section prestressed beam of width 12 in., depth 24 in. is used to support a uniform superimposed load of 100 lb/ft over a span of 36 ft. Beam is made of lightweight concrete, 100 lb/ft^3. Design the prestressing, assuming a loss of prestress owing to creep of 20 percent. Check all critical bending and diagonal tension stresses in the beam. If the allowable stress in the steel is 150 ksi, and available wire has $\frac{1}{16}$ in. diameter, design steel and draw cross section of member. Comment on the adequacy and efficiency of the section.

9.12 A prestressed one-way slab of thickness $2\frac{1}{2}$ in. is made of gravel concrete to carry a superimposed load of 60 lb/ft^2 over a span of 10 ft. Design the prestressing, assuming a loss of prestress of 25 percent and an allowable steel stress of 150 ksi. Check all critical stresses and comment on the choice of slab thickness.

9.13 $f_{c\ allow} = 2{,}250$ psi, $f_{s\ allow} = 120$ ksi, $\alpha = 0.80$; lightweight concrete, 100 lb/ft^3. The prestressed beam section is used to support a bridge deck over a span of 60 ft. If the weight of the bridge deck plus equivalent uniform live loading is 200 lb/ft^2 of surface, and the girders are spaced 3 ft apart, what will the prestressing have to be? Check all critical stresses and comment on the adequacy of the design.

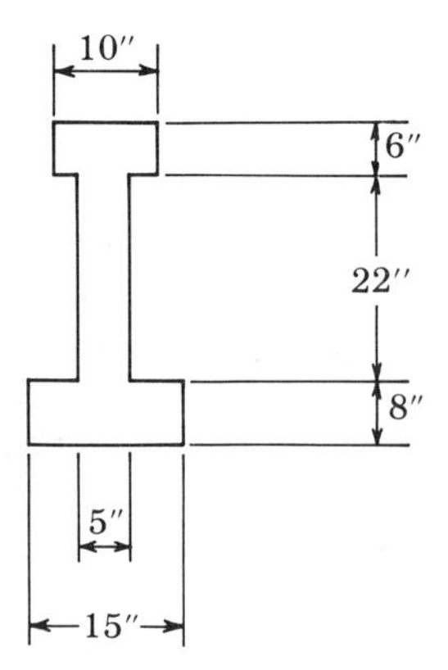

9.14 For the beam of Prob. 9.13, $f_c' = 5{,}000$ psi, and the compressive allowable stresses of the AASHO Specifications govern: Temporary stresses before losses due to creep and shrinkage, $f_{allow} = 0.60 f_c'$; stress at design load after losses have occurred, $f_{allow} =$

$0.40f'_c$. Calculate the maximum allowable beam spacing to support the loads specified in Prob. 9.13.

9.15 A design alternate for the section of Prob. 9.13 suggests use of the section upside down, so that the 8 by 15 in. flange is on top and the 6 by 10 in. flange is on the bottom. Using the allowable stresses of Prob. 9.13, calculate the prestressing and the maximum positive moment which can be resisted by the section. Compare results with those of Prob. 9.13 and decide whether section should be used with big or little flange on top.

9.16 A footbridge, spanning 60 ft as a simple beam, is to be designed of prestressed concrete construction. Width of the walkway is to be 6 ft 0 in.; the live load on the walkway is to be 100 lb/ft² of area. A suggested design envisions use of precast walkway slabs and I-shaped girders, as shown. Concrete weighs 150 lb/ft³ and is of strength f'_c = 5,000 psi. The allowable initial steel stress is 150,000 psi, loss of prestress may be assumed at 25 percent.

Calculate the location and area required of stressing cables. Calculate concrete stresses at critical stages.

Review the design and comment on it. Compare this design with that for the same bridge done in Prob. 6.46 for conventionally reinforced concrete.

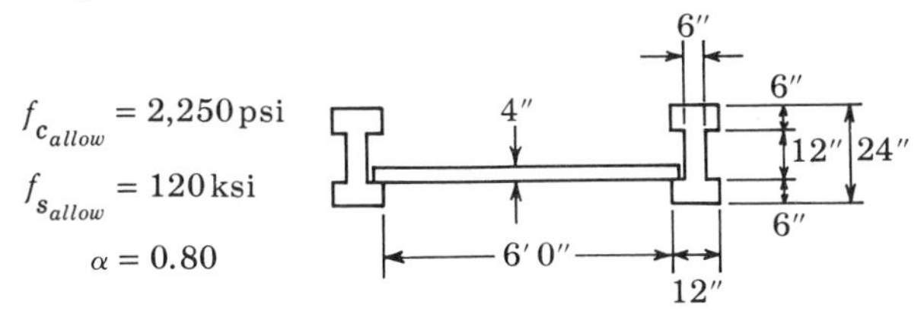

Lightweight concrete, 100 lb/ft³

9.17 Design the walkway slab for the footbridge of Prob. 9.16; reduce the indicated thickness if possible.

9.18 A precast, prestressed floor unit has the dimensions shown. It is made of gravel concrete of f'_c = 5,000 psi, and the allowable stresses of sec. 2605 of the ACI Code govern. If this member is to span across 40 ft and carries a combined superimposed live and dead load of 75 lb/ft², design the prestressing, check the critical stresses, and comment on the adequacy of the design. $f_{s\ allow}$ = 150 ksi; loss of prestress = 20 percent. If available wire is 0.10 in. diameter, dimension the steel and present design sketch of the section.

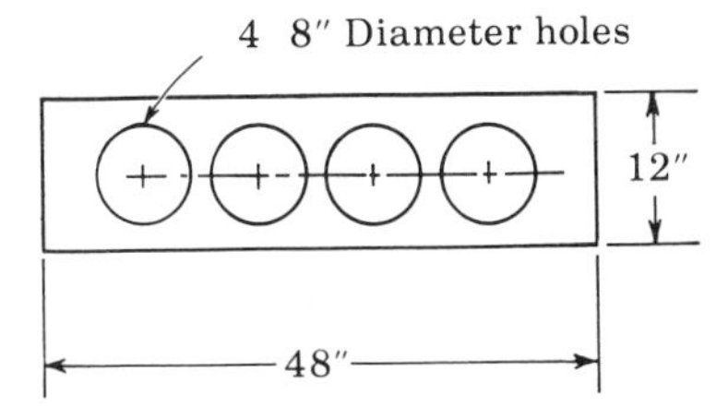

9.19 Compute the interaction curve for ultimate strength under combined axial force P_0 and moment M_0 (applied about the strong axis) for an 8 WF 17 beam. Assume elastic–perfectly plastic material and neglect the possibility of buckling. Superimpose the resulting curve upon Fig. 9.16 and comment on the correlation between the curves.

9.20 Repeat Prob. 9.19, this time considering axial load combined with bending about the weak axis of the 8 WF 17. Again compare with the curves of Fig. 9.16 and comment.

9.21 Discuss how the interaction curves of Fig. 9.16 for ultimate strength under combined load are affected by consideration of initial stresses in the member.

9.22 Calculate the ultimate force P_P, applied as shown. Assume the member is braced against lateral buckling. Use (*a*) basic theory, (*b*) AISC formula 21. Compare results.

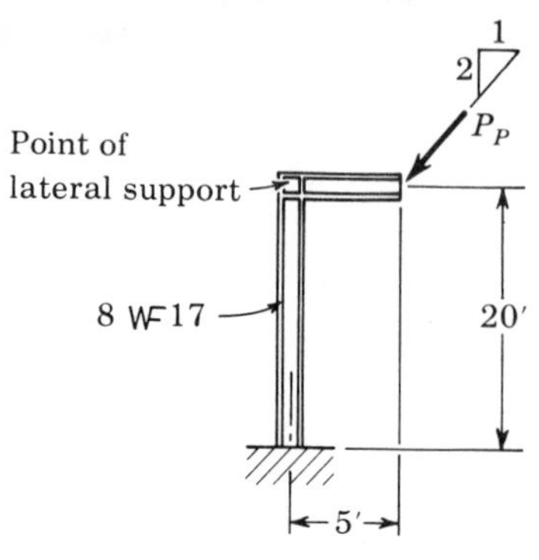

9.23 Calculate the value of load w under which collapse of the rigid frame will occur:

(*a*) by using limit design and interaction curve for critical column
(*b*) by using limit design and AISC Specification for member under combined loading

Assume that lateral buckling is prevented by appropriate bracing.

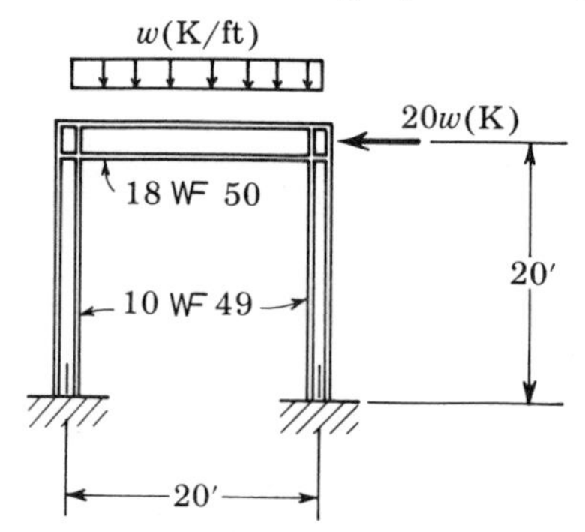

9.24 The closed rigid frame is to resist a diametrically applied load of ultimate value 10 kips. Lateral bracing is provided. Select appropriate WF member to resist the load

(*a*) by interaction theory
(*b*) by AISC Specifications.

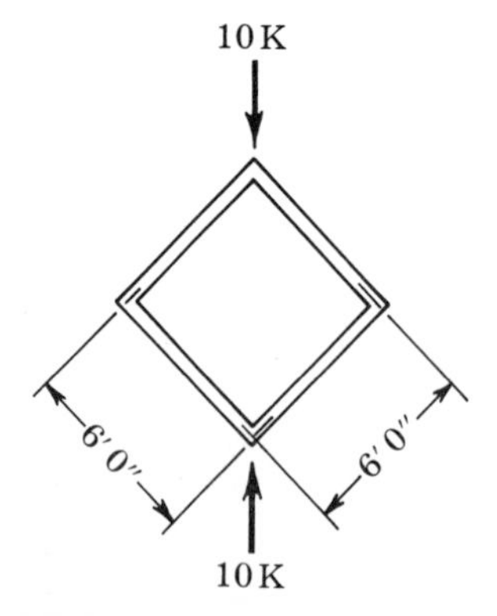

10

Reinforced Concrete Members under Axial Load and Bending

10.1 Working-stress Method

A certain amount of confusion exists with respect to the stress analysis of eccentrically loaded concrete columns; this results in part from the conflicting and inconsistent assumptions which have been used so far in the working-stress theory of reinforced concrete. Considering, for instance, the limiting cases for such members, we remember that, because of the plastic behavior of the concrete, design of axially loaded columns follows inelastic theory, but design of members under pure bending follows elastic theory. A rational and consistent theory which includes as limiting cases an elastic theory on one hand and an inelastic theory on the other does not seem feasible.

It is of course possible, as has been suggested earlier, that an interaction curve can be drawn to interpolate between these limiting cases, and this is one approach which has been followed. But this should be considered a useful rather than a rational procedure, and must be verified by extensive testing.

Another difficulty faced in the design of eccentrically loaded columns is the cracking of the concrete which will occur when the eccentricity exceeds a certain amount corresponding to the kern of the section. We must distinguish between two possibilities—that of an uncracked section when axial load predominates, and that of a cracked section when bending predominates. The next two sections will accordingly be concerned with analyses for these two cases.

A more recent approach to the problem may be symptomatic of a current trend—that is, to perform an ultimate-strength analysis to determine the collapse load of the member, and then to reduce this collapse load to obtain an allowable load comparable with usual safety factors. This is the approach taken in the provisions of the 1963 ACI Code (sec. 1407). It is really fallacious to consider this a working-stress theory; rather, it is typical of the difficulties to be faced during a transition period from working-stress to ultimate-design concepts.

Even though the word "column" is frequently used, buckling considerations usually play no role in reinforced concrete because of the relatively large lateral dimensions of the members. For excessively slender columns, load reduction factors as given in ACI sec. 916 should be used.

10.2 Columns under Low Eccentricity—Uncracked Section

We shall consider here an interaction approach as suggested in the preceding section, valid for cases in which no tensile strain and therefore no cracking of the section occurs.

For the limiting cases, we take those of pure compression and of pure bending of the uncracked section.

For pure axial load, we calculate, according to the inelastic theory of Sec. 3.5 for tied columns,

$$P_{allow} = 0.85A_G(0.25f'_c + f_s p_G) \tag{10.1}$$

For pure bending, we follow the elastic theory of uncracked sections:

$$M_{allow} = \frac{f_{c\,allow}I_T}{kd} \qquad \text{for over-reinforced sections}$$

$$M_{allow} = \frac{1}{n}\frac{f_s I_T}{d - kd} \qquad \text{for under-reinforced sections} \tag{10.2}$$

Note that this allowable moment can never be achieved because the section would crack first, thus voiding the theory (as indicated by the dashed portion of the straight line in Fig. 10.1). It serves only as a limiting value for the interaction line.

We now plot these limiting values in dimensionless form, as shown in Fig. 10.1, and connect them by a linear interaction curve which is supposed to represent allowable values of combined loading. The equation of this curve is, as before, given by

$$\frac{P}{P_{allow}} + \frac{M}{M_{allow}} = 1 \tag{10.3}$$

or, for a given eccentricity, we set $M = Pe$, and solve for P:

$$P = \frac{P_{allow}}{1 + (P_{allow}/M_{allow})e} \tag{10.4}$$

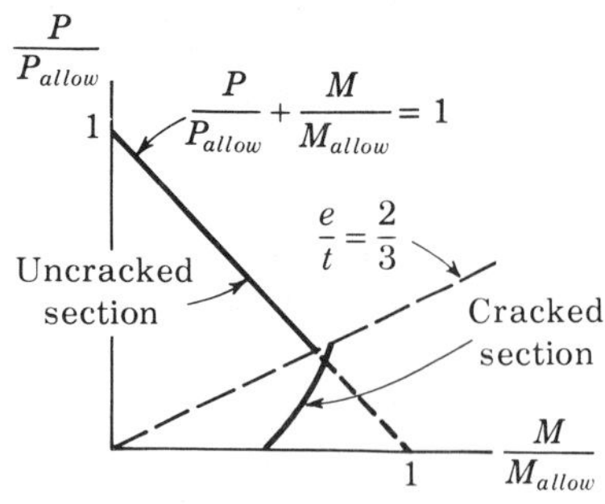

Fig. 10.1

An equivalent form of interaction equation (10.3) is

$$\frac{f_A}{F_A} + \frac{f_B}{F_B} = 1 \tag{10.5}$$

where the actual stresses f_A and f_B and the allowable stresses F_A and F_B must be defined appropriately. Even though this leads to cumbersome expressions, Eq. (10.5) was used in the 1956 ACI Code in preference to the much clearer Eq. (10.3). Here, it is recommended that the analysis be performed by calculating the allowable limiting values P_{allow} and M_{allow} and inserting them into Eq. (10.3) or (10.4).

At this time, we must remember that when the moment becomes excessive, the section cracks, and therefore the results corresponding to predominant bending (and shown by the dashed portion of the interaction curve) cannot be valid.

In the 1956 ACI Code, the limit for uncracked section was somewhat arbitrarily set at an eccentricity

$$e = \tfrac{2}{3}t \tag{10.6}$$

where t is the lateral column dimension in the plane of bending.

For loads at a larger eccentricity than this, a cracked-section theory is called for; this will be discussed in Sec. 10.3

Example Problem 10.1 Analysis of eccentric column, uncracked section. $f_c' = 4.0$ ksi, $f_{yp} = 40$ ksi, $n = 8$; $f_{s\,allow} = 20$ ksi. Find the allowable load which may be applied at an eccentricity of 3 in. from the column center line. Use interaction approach.

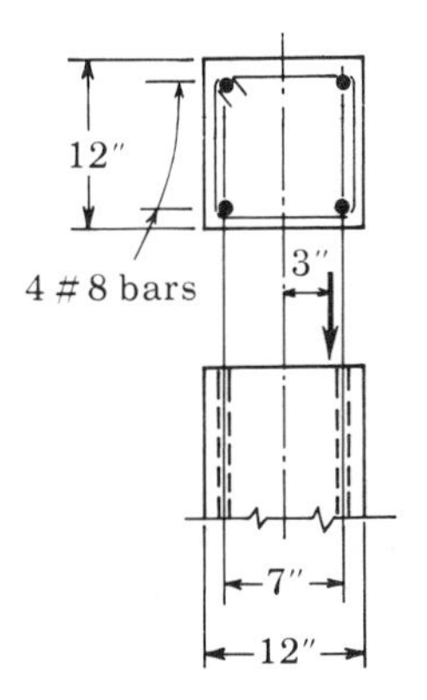

For this low eccentricity, an uncracked section is assumed. For pure axial force,

$$P_{allow} = 0.85 \times 144(1.0 + 20.0 \times 0.0219) = 176 \text{ kips}$$

and for pure bending of the uncracked section (the strains indicate here that the compressive concrete stress is critical), we get, neglecting creep,

$$M_{allow} = \frac{f_{c\,allow} I_T}{c} = \frac{0.45 \times 4.0 \times (1{,}728 + 270)}{6} = 594 \text{ kip-in.} = 50 \text{ kip-ft}$$

Applying Eq. (10.4), we find

$$P = 176 \text{ kips} \times \frac{1}{1 + 0.89} = 93 \text{ kips}$$

The allowable eccentric load is 93 kips. Lateral ties must be provided according to sec. 806 of the 1963 ACI Code.

A possible design procedure would be to assume a column size (this may often be specified anyway by space or architectural requirements), and to solve for the required steel ratio, as outlined in the following:

Example Problem 10.2 A column is to be designed to resist the loads shown. Column dimensions to be 20 by 20 in.; $f'_c = 3.0$ ksi, $f_s = 20$ ksi, $n = 9$, $f_{YP} = 40$ ksi.

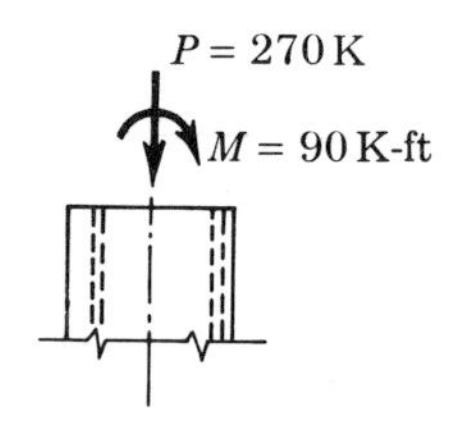

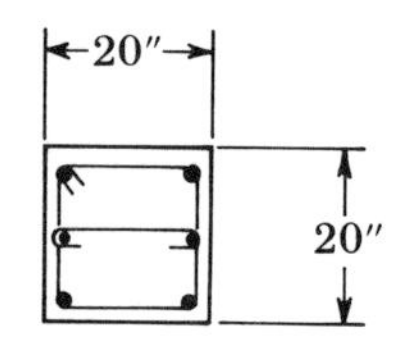

Assuming $2\frac{1}{2}$ in. cover, we can calculate P_{allow} and M_{allow} as functions of the steel area A_s:

$$P_{allow} = 0.85[400 \times (0.25 \times 3.0) + 16.0 \times A_s] = 255 + 13.6A_s$$

$$M_{allow} = \frac{f_c I_T}{c} = \frac{0.45 \times 3.0(\frac{1}{12} \times 20^4 + 8A_s \times 7.5^2)}{10}$$

$$= 1{,}800 + 60.8A_s$$

Substituting these values into Eq. (10.3), we can solve for the required amount of steel A_s:

$$\frac{270 \text{ kips}}{255 + 13.6A_s} + \frac{90 \text{ kip-ft} \times 12}{1{,}800 + 60.8A_s} = 1$$

from which

$$A_{s\,reqd} = 14.52 \text{ in.}^2$$

This amount of steel could be furnished by ten No. 11 bars, five on each side.

An alternate approach would be to design the column by specifying a certain desired steel ratio and solving for the required column dimension. However, this becomes impractical, since a third-degree equation results which must be solved by trial and error. A sounder way is to program the analysis to encompass an entire range of columns and to present the results in graphical or tabular form to be used as design aid.

10.3 Columns under Large Eccentricity—Cracked Section

When the eccentricity of the applied load on a member exceeds the kern dimensions, cracking of the section results. This section will consider the analysis of such members when the concrete is assumed cracked over its entire tension side.

The suggested approach here is an elastic one, entirely paralleling that taken for the analysis of doubly reinforced concrete beams (Sec. 6.9). In that section, it was pointed out that the effects of compressive creep of concrete could be taken into account by increasing the modular ratio of the compression steel, and the same procedure is taken here. In this fashion. the elastic theory is modified for creep effects.

We consider Fig. 10.2 and apply the usual steps of the strength of materials approach; the equilibrium conditions are

$$\Sigma F = 0: \qquad (\tfrac{1}{2}f_c bkd) + A_s'f_s' - A_s f_s = P \tag{10.7}$$

$$\Sigma M_P = 0: \qquad (\tfrac{1}{2}f_c bkd)\left(e - \frac{t}{2} + \frac{kd}{3}\right) + A_s'f_s'\left(e - \frac{t}{2} + d'\right) - A_s f_s\left(e - \frac{t}{2} + d\right) = 0 \tag{10.8}$$

where the moment-equilibrium condition is conveniently taken with respect to the line of action of the eccentric load.

As usual, we assume plane sections remain plane, draw the strain distribution of Fig. 10.2c, and obtain the strain relations

$$\phi = \frac{\epsilon_c}{kd} = \frac{\epsilon_s'}{kd - d'} = \frac{\epsilon_s}{d - kd}$$

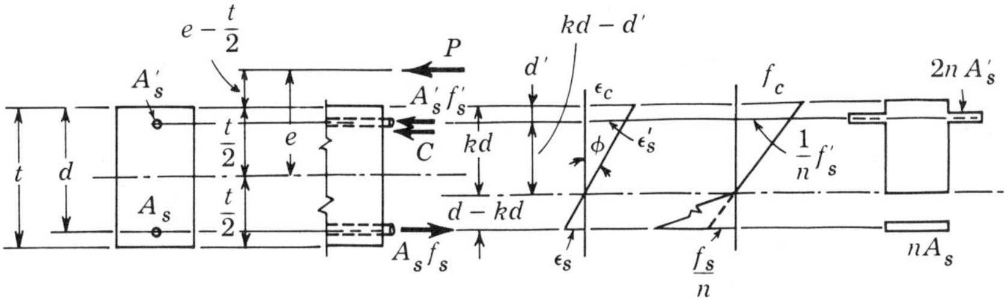

(a) (b) (c) Strains (d) Stresses (e) Transformed section

Fig. 10.2

Introducing the stress-strain relations

$$f_c = E_c\epsilon_c \qquad f'_s = 2nE_c\epsilon'_s \qquad f_s = nE_c\epsilon_s$$

we can express the steel stresses as functions of the concrete stress f_c:

$$f'_s = 2n\frac{kd - d'}{kd}f_c \qquad f_s = n\frac{d - kd}{kd}f_c \tag{10.9}$$

and substituting these into Eqs. (10.7) and (10.8), we obtain

$$\Sigma F = 0: \qquad f_c\left(\tfrac{1}{2}bkd + A'_s n\frac{kd - d'}{kd} - A_s n\frac{d - kd}{kd}\right) = P \tag{10.10}$$

$$\Sigma M_P = 0: \qquad f_c\left[\left(\tfrac{1}{2}bkd\right)\left(e - \frac{t}{2} + \frac{kd}{3}\right) + \left(A'_s 2n\frac{kd - d'}{kd}\right)\left(e - \frac{t}{2} + d'\right) - \left(A_s n\frac{d - kd}{kd}\right)\left(e - \frac{t}{2} + d\right)\right] = 0 \tag{10.11}$$

Equations (10.10) and (10.11) are two simultaneous equations in two unknowns, kd and f_c. Because of the choice of moment center, the cubic Eq. (10.11) contains only the unknown kd which can be found by trial and error. Once kd is known, it can be substituted into Eq. (10.10) to obtain the concrete stress f_c. The other stresses are then proportional to f_c, according to Eqs. (10.9).

The theoretical outline which has been presented, and the accompanying equations, have been cast in very detailed form for the sake of definiteness. In practice, it is preferable to understand the reasoning thoroughly, and, rather than insert values into the rather lengthy Eqs. (10.10) and (10.11), to proceed by calculating first the stresses, then the stress resultants, and insert these in the equilibrium conditions. These are the steps which have been followed in the example problem of this paragraph.

While the outlined procedure sticks to basic principles for the sake of clarity, it may be more convenient in practice to use the transformed-section method; in this, as in the basic method, the cubic equation must be solved.

The procedure which has been described cannot be considered a design method. At best, results from such calculations, suitably nondimensionalized, may be used to draw up design curves; such design plots could take the form of the curve labeled "cracked section" in Fig. 10.1. In practice, establishment of such curves becomes difficult because of the large number of variables which influence the allowable load on the member. It must also be anticipated that, because of the conflicting assumptions between uncracked- and cracked-section theory, a discontinuity arises at the junction between these two cases. This could, contrary to common sense, make the cracked section appear to be stronger than the uncracked section.

At worst, design becomes a trial method in which a member is assumed and the stresses are checked. In any case the question of suitable allowable stresses arises. For concrete, this might vary from $0.18f'_c$ for the limiting case of a tied column under axial load to $0.45f'_c$ for the same member under pure bending. Realizing that cracked sections occur under predominant bending, the 1956 ACI Code permitted the latter value of concrete stress.

Lastly, a criterion to determine cracking of the section is required. According to the 1956 ACI Code, this was simply the critical eccentricity equal to two-thirds of the lateral dimension t. That edition of the ACI Code summarized the outlined procedure as "recognized theory for cracked sections."

Example Problem 10.3 Determine the stresses throughout the section of the eccentrically loaded column, assuming elastic behavior modified for creep. Find the allowable value of the eccentric load P. $f'_c = 4.0$ ksi, $f_{YP} = 40$ ksi, $n = 8$, $f_{s\,allow} = 20$ ksi.

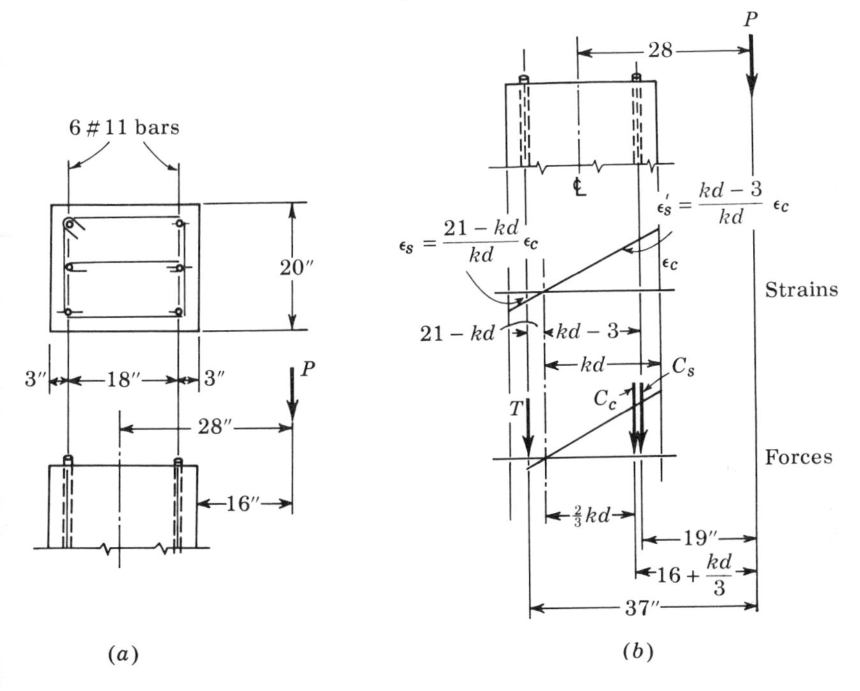

We first plot the strains, assuming the neutral axis at kd; from this we can find the steel stresses as functions of f_c, and write the resultant forces:

$$C_s = 4.68 \times (2 \times 8) \times \frac{kd - 3}{kd} f_c$$

$$= 75.0\left(1 - \frac{3}{kd}\right) f_c$$

$$C_c = \tfrac{1}{2} \times 20 \text{ in.} \times kdf_c$$

$$T = 4.68 \times 8 \times \frac{21 - kd}{kd} f_c$$

$$= 37.5\left(\frac{21}{kd} - 1\right) f_c$$

Next, taking equilibrium conditions, we set

$$\Sigma M_P = 0: \quad \left\{\left[75.0\left(1 - \frac{3}{kd}\right)\right] 19 + 10kd\left(16 + \frac{kd}{3}\right) - \left[37.5\left(\frac{21}{kd} - 1\right)\right] 37\right\} f_c = 0$$

from which we can cancel f_c and solve for the unknown kd in the cubic equation:

$$(kd)^3 + 48.1(kd)^2 + 847kd - 10{,}010 = 0$$
$$kd = 7.9 \text{ in.}$$

Substituting this back to find the forces, we get

$$C_s = 46.5f_c$$
$$C_c = 79.0f_c$$
$$C_s = 62.2f_c$$

and writing now a force equilibrium equation:

$$\Sigma F = 0: \quad C_s + C_c - C_s = P$$

or

$$(46.5 + 79.0 - 62.2)f_c = P$$

From which

$$f_c = \frac{P}{63.3} = \underline{0.0158P}$$

and the other stresses become

$$f_s' = 2 \times 8 \times \frac{4.9}{7.9} \times \frac{P}{63.3} = \underline{0.156P}$$

and

$$f_s = 8 \times \frac{13.1}{7.9} \times \frac{P}{63.3} = \underline{0.209P}$$

Comparing these stresses, we see that under increasing load the tension steel stress will reach its allowable value first, when

$f_s = 0.209P_{allow} = 20$ ksi
from which $\underline{P_{allow} = 96}$ kips.

Example Problem 10.5 will furnish us with a value of the ultimate load on this column, and thus enable us to calculate the safety factor.

10.4 Working-stress Design by Scaling Down Ultimate Loads

Another method which has been proposed in the introduction to this chapter consists of applying a safety factor to ultimate strengths of the member and using the results in working-stress design.

Graphically, this procedure could be shown as in Fig. 10.3, where the interaction curve for the ultimate strength under combined loading is scaled down by the load factor to give another interaction curve to be used in determining allowable loads. This, essentially, is the method followed by the 1963 edition of the ACI Code. The working-stress interaction curve is composed of two straight lines, one representing compression controlling and extending from points 0, N_0 to M_b, N_b, where the first named point corresponds to pure axial load, the latter to balanced design. The second line, corresponding to tension controlling, extends from the point representing balanced design to the point M_0, 0 representing the allowable pure moment.

In the 1963 ACI Code, the former of these curves is represented in terms of stresses by eq. 14-9. The latter curve is described in terms of stress resultants in sec. 1407*c* with all values suitably defined. With reference to the appropriate portion and notation of Fig. 10.3, it can be represented by the equation

$$M_{allow} = M_0 + \frac{M_b - M_0}{N_b} P \tag{10.12}$$

In this as in other portions of this book the order of coverage has been to present elastic or working-stress methods prior to inelastic or ultimate methods, since under increasing loads this would be the order of behavior. However, we realize here that interaction curves for ultimate loads must be available before they can be scaled down to yield working loads; the next three sections should be studied before a set of scaled-down working-stress curves can be rationally derived.

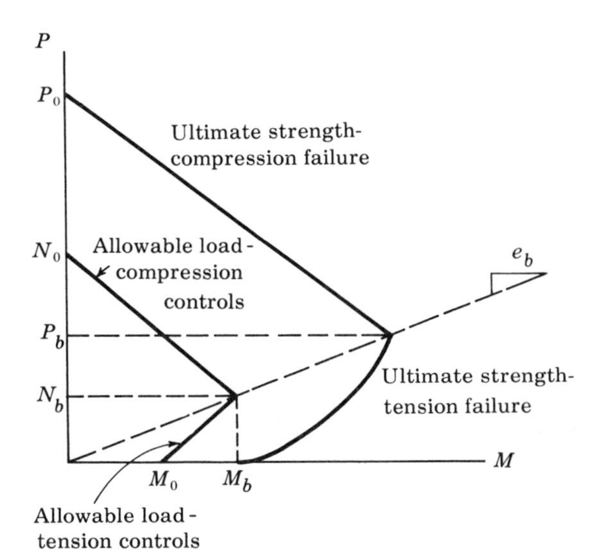

Fig. 10.3

Example Problem 10.4 Following the 1963 ACI Code, find the allowable load P which may be applied at 28 in. from the center line of the column (which is the same as that of Example Prob 10.3);

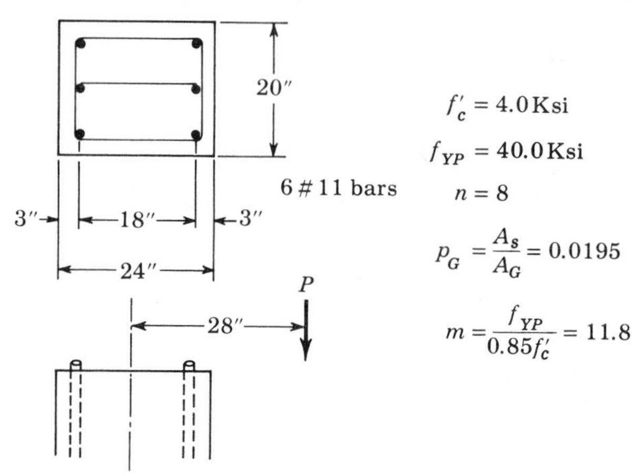

$f'_c = 4.0$ Ksi

$f_{YP} = 40.0$ Ksi

$n = 8$

$p_G = \frac{A_s}{A_G} = 0.0195$

$m = \frac{f_{YP}}{0.85 f'_c} = 11.8$

Balanced eccentricity (ACI eq. 14-7):

$$e_b = (0.67 \times 0.0195 \times 11.8 + 0.17)21 = 6.8 \text{ in.}$$

Since $e = 28 > 6.8$ in., tension failure controls, and the provisions of ACI 1407*c* apply; these can be written by Eq. (10.12):

$$M_{allow} = P_{allow}e = M_0 + \frac{M_b - M_0}{N_b} P_{allow}$$

where

$$M_0 = 0.40 A_s f_{YP}(d - d')$$
$$= 0.40 \times 4.68 \times 40 \times 18 \text{ in.} = 1{,}350 \text{ in.-kips (eq. 14-12)}$$

M_b (from eq. 14-9 and e_b):

$$f_a = \frac{N_b}{480} = 0.00208 N_b \qquad f_b = \frac{6.8 N_b \times 12}{34{,}400} = 0.00237 N_b$$

$$F_a = 0.34(1 + 0.0195 \times 11.8)4.0 = 1.68 \text{ ksi (eq. 14-10)}$$
$$F_b = 0.45 \times 4.0 = 1.8 \text{ ksi}$$

therefore

$$N_b\left(\frac{0.00208}{1.68} + \frac{0.00237}{1.8}\right) = 1 \qquad N_b = 390 \text{ kips}$$

and

$$M_b = N_b e_b = 390 \times 6.8 = 2{,}650 \text{ kip-in.}$$

therefore

$$P_{allow} \times 28 \text{ in.} = 1{,}350 + \frac{2{,}650 - 1{,}350}{390} \times P_{allow}$$

from which

$$P_{allow} = 55 \text{ kips}$$

Note that the approach of the previous section yielded $P_{allow} = 96$ kips, a considerable difference.

10.5 Ultimate-strength Theory

The discussion of working-stress methods for eccentrically loaded concrete columns has brought out some of the inconsistencies with which the designer of such members is faced. To cut through this Gordian knot of conflicting assumptions it seems advisable to resort to the simpler ultimate-strength method.

Here again, we must distinguish between two cases: In members in which bending predominates, failure is likely to occur by yielding of the tension steel. A rather clear-cut ultimate-strength theory can be worked out to predict this failure. In members with predominant axial compression, failure is controlled by compressive crushing of the concrete. Since this crushing is brittle, it does not allow the redistribution of stress which forms the basis of the plastic theory. The exact strength calculation in this case becomes somewhat more difficult and is replaced by a linear interaction curve.

The criterion to distinguish between tension and compression failure involves the strains in steel and concrete and is identical to that used for the balanced design of reinforced concrete beams.

10.6 Failure Controlled by Tension

We shall consider a doubly reinforced concrete member subject to eccentric load at the instant of failure. This failure is initiated by yielding of the tension steel. The strains increase, and total collapse of the member occurs when the ultimate strain of the concrete is reached on the compression side. It is required to determine the value of the failure load, using the strength of materials method and the strain and stress distributions of Fig. 10.5.

Properties of the materials are again shown in Fig. 10.4. Steel is elastic–perfectly plastic, of yield stress f_{YP}; for the stress-strain curve of the concrete we substitute the approximate Whitney stress block as used before.

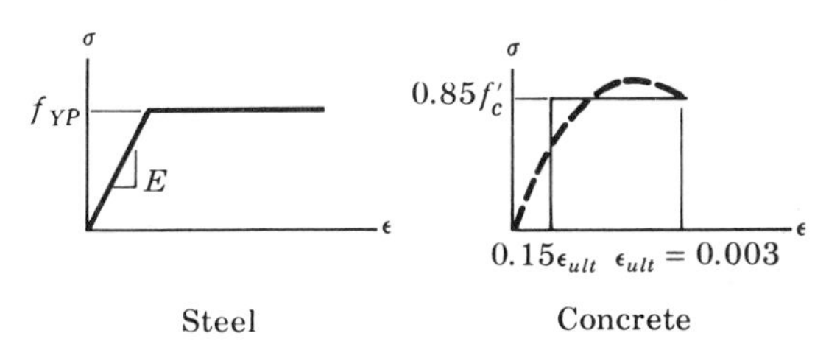

Fig. 10.4

The strains at failure are shown in Fig. 10.5*b*, assuming plane sections remaining plane, and the corresponding stresses are drawn as in Fig. 10.5*c*, using the stress-strain relations. Equilibrium of forces and moments (with moment center on the line of action of the resultant eccentric force) results in

$$\Sigma F = 0: \qquad (0.85f_c'ab) + A_s'f_{YP} - A_sf_{YP} = P_{ult} \tag{10.13}$$

$$\Sigma M_P = 0: \qquad (0.85f_c'ab)\left(e - \frac{t}{2} + \frac{a}{2}\right) + A_s'f_{YP}\left(e - \frac{t}{2} + d'\right) - A_sf_{YP}\left(e - \frac{t}{2} + d\right) = 0 \tag{10.14}$$

This constitutes a set of two equations in two unknowns, the depth of the stress block a and the eccentrically applied ultimate load P_{ult}. (Note that as defined here e is the lever arm from the column center line.) Because of the choice of moment center, Eq. (10.14) is quadratic in only one unknown a, and can be solved directly. Once a is known, it can be inserted in Eq. (10.13) and the value of P_{ult} obtained. The ultimate moment is then given by $P_{ult}e$.

Equations 19-1 and 19-2 of sec. 1902 of the ACI Code are similar to Eqs. (10.13) and (10.14), the only difference being the choice of the centroid of the tension steel as moment center. In that section, the two equations are solved simultaneously for P_{ult}, and the resulting equation is given as eq. 19-4. While equations of such length might be useful for the establishment of design curves such as the one shown in Fig. 10.8 (preferably by use of electronic computer), for usual engineering purposes it may be just as fast, and certainly clearer, to solve the two equilibrium equations as suggested.

For the particular case of symmetrical reinforcement, which occurs very often in practice, Eqs. (10.13) and (10.14) simplify in spectacular fashion. Here $A_s = A_s'$, and the equilibrium equations become

$$\Sigma F = 0: \qquad 0.85f_c'ab = P_{ult}$$

$$\Sigma M = 0: \qquad P_{ult}\left(e - \frac{t}{2} + \frac{a}{2}\right) = A_sf_{YP}(d - d') \tag{10.15}$$

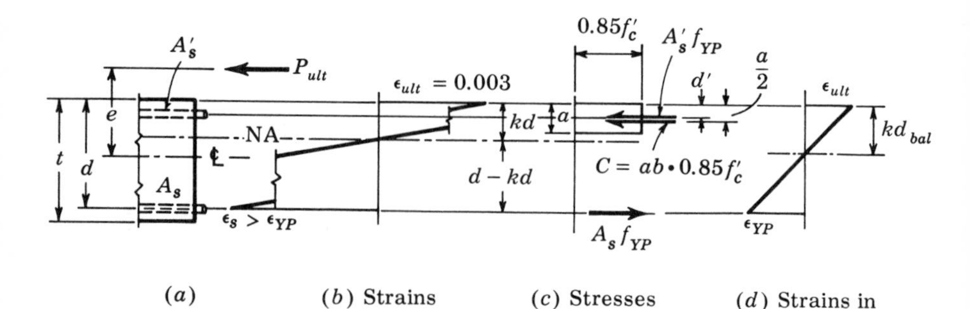

Fig. 10.5

Figure 10.6 shows more clearly the meaning of these equations: since the steel forces balance, the ultimate eccentric force must be equal to the compression force on the concrete, and because of the equality of these forces, moment equilibrium can be established by equating the couple consisting of steel forces to that consisting of concrete and external forces.

It now remains to ascertain the conditions under which failure by yielding of the tension steel takes place, and this is shown by the strains of Fig. 10.5*d*, which indicate simultaneous yielding of the tension steel and crushing of the compression concrete. For this balanced case, we can determine the depth of the compression block a_b by proportion as

$$a_b = 0.85kd_b = 0.85 \frac{\epsilon_{ult}}{\epsilon_{YP} + \epsilon_{ult}} d = \frac{0.85d}{(f_{YP}/90) + 1} \qquad (10.16)$$

The load P_b under which this balanced failure takes place is found by inserting this value for a_b into Eq. (10.13):

$$P_b = 0.72f'_c \frac{1}{(f_{YP}/90) + 1} bd + A'_s f_{YP} - A_s f_{YP} \qquad (10.17)$$

and the corresponding balanced moment can similarly be found by substituting a_b into Eq. (10.14). In the ACI Code, eq. 19-3 gives the corresponding balanced moment obtained from a moment-equilibrium condition.

For a less than the value a_b, or P_{ult} less than P_b, the tension steel will yield first, and the above theory may be used. For loads P_{ult} larger than P_b, failure occurs by compressive crushing of the concrete without prior yielding of tension steel, corroborating our earlier notion that compression failure would occur under predominant axial, tension failure under predominant bending load.

An additional check which should be made (and which is explicitly called for in sec. 1902*a* of the ACI Code) is to ascertain that the compression steel has yielded at the instant of compressive crushing of the concrete, as was assumed in Eqs. (10.13) to (10.17). Figure 10.5*b* indicates that if the compression steel is located too closely to the neutral axis, then the compressive steel strain ϵ'_s may be less than the yield strain ϵ_{YP}. From that

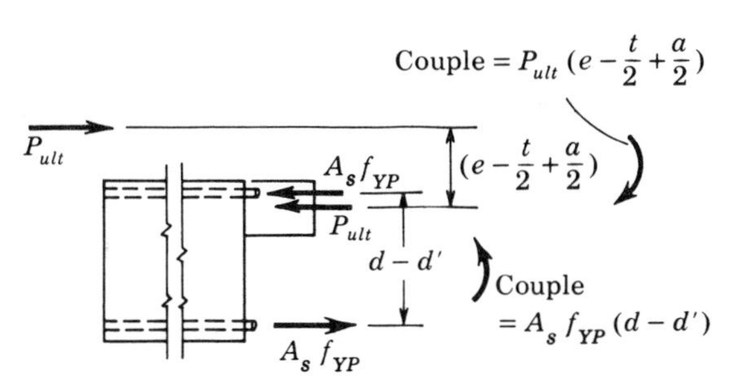

Fig. 10.6

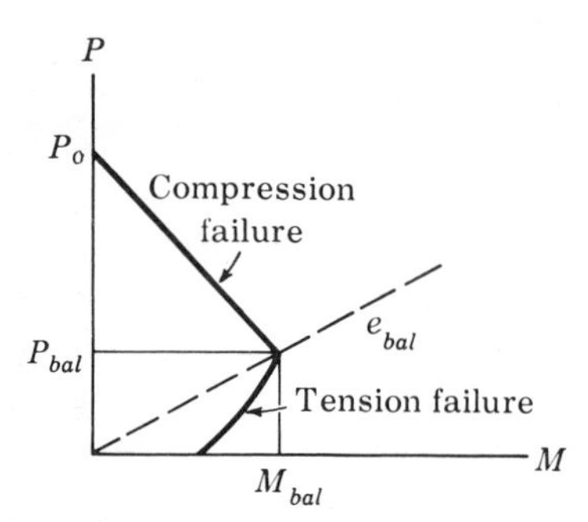

Fig. 10.7

strain distribution we see that this is the case when

$$\epsilon_s' = \frac{kd - d'}{kd}\,\epsilon_{ult} < \epsilon_{YP}$$

or
$$\frac{a}{0.85} = kd < \frac{d'}{1 - (f_{YP}/90)} \tag{10.18}$$

If a, or kd, is smaller than this amount, the actual value of the compressive steel stress should be determined from the strains and used in conjunction with Eqs. (10.13) and (10.14). In this case the analysis becomes considerably more involved, and the analyst may decide to neglect the compression steel entirely in calculating the strength of the member.

The results of the theory of this section can again be plotted in the form of an interaction curve, as shown in Fig. 10.7 (compare this to Fig. 10.1 for working-stress analysis of a similar member).

The applicable portion of this curve is labeled "tension failure." We note that, for this case, a lowering of the compressive ultimate load also results in a lowering of the ultimate moment. This is so because the compressive force will tend to diminish the controlling tension stress in the steel. We should therefore remember that it is possible to fail a member under combined loading by *decreasing* the axial load. For this reason, the critical loading condition on an eccentrically loaded column is not necessarily that which contains the highest axial force.

Example Problem 10.5 Calculate the ultimate eccentric load P_{ult} which may be applied to the column at 28 in. from the column center line. Compare this value to the allowable load on this same column calculated in Example Problems 10.3 and 10.4, and compute the safety factors ($f_c' = 4.0$ ksi, $f_{YP} = 40$ ksi).

We draw the stresses at ultimate, and apply conditions of statics:

$$\Sigma M = 0: \qquad (4.68 \times 40 \text{ ksi}) \times 18 \text{ in.} = P_{ult}\left(16 + \frac{a}{2}\right)$$

$$\Sigma F = 0: \qquad P_{ult} = 0.85 \times 4.0 \times 20 \text{ in.} \times a$$

Solving simultaneously, we find

$$a = 2.8 \text{ in.}$$

$$P_{ult} = 190 \text{ kips}$$

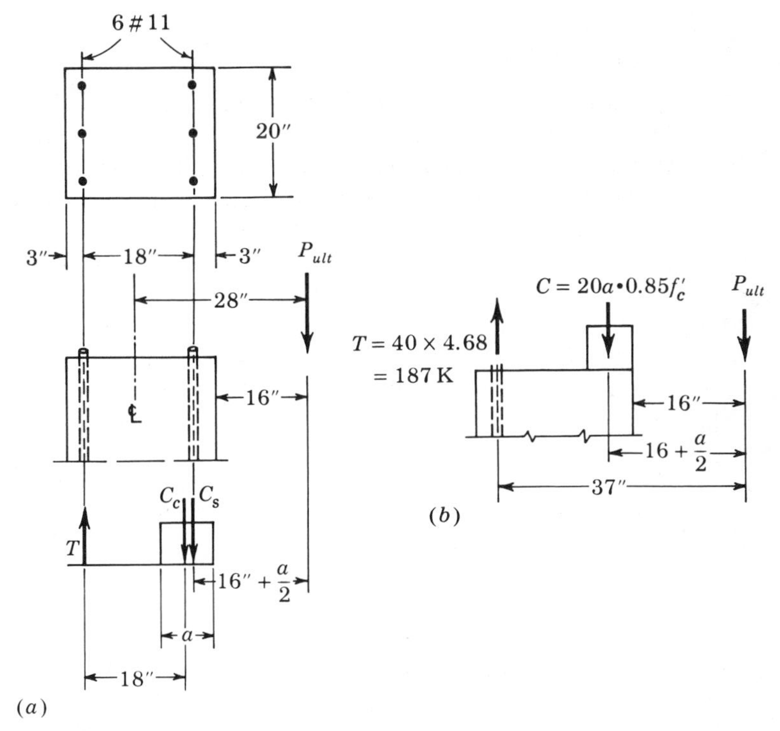

If we look at this result, we notice immediately that, with $kd = a/0.85 = 3.3$ in., the compression steel falls within 0.3 in. of the neutral axis, and its strain is therefore unlikely to reach its yield value, a fact also confirmed by checking the criterion Eq. (10.18). With the compression steel so close to the neutral axis, its stress is likely to be low, and we might perform the analysis neglecting it altogether.

The forces at ultimate are drawn, and the equilibrium conditions are written:

$$\Sigma F = 0: \qquad P_{ult} + (0.85 \times 4.0) \times 20a - 187 \text{ kips} = 0$$

$$\Sigma M_P = 0: \qquad 187 \text{ kips} \times 37 \text{ in.} - 68a\left(16 + \frac{a}{2}\right) = 0$$

from which

$$a = 5.45 \text{ in.}$$

$$P_{ult} = 183 \text{ kips}$$

This is only 4 percent lower than the first value, indicating that compression steel is not particularly effective in raising the ultimate strength. A more rigorous calculation would determine the strain and corresponding actual stress in the compression steel; the corresponding ultimate load should fall between the two calculated values. This calculation is left to the reader (Prob. 10.17).

Based on the more conservative value, the safety factor against collapse of Example Problem 10.3 is

$$\text{S.F.} = \tfrac{183}{96} = \underline{1.91}$$

and that of Example Problem 10.4 is

$$\text{S.F.} = \tfrac{183}{55} = \underline{3.33}$$

10.7 Failure Controlled by Compression

We now turn to the case in which the ultimate load P_{ult} is greater than the balanced value P_b; failure in this case occurs by crushing of the concrete. At that instant, the tension steel will still be elastic, and its stress would have to be calculated by using the strains. Doing this, we can obtain a set of two simultaneous equilibrium conditions similar to Eqs. (10.13) and (10.14), but since the tension steel stress is now a function of the location of the neutral axis, a cubic equation must be solved. In order to avoid this complication, an interaction approach is resorted to.

The assumed interaction curve for compression failure is the straight line marked "compression failure" in Fig. 10.7. It connects the point indicating the ultimate axially applied load P_0 (corresponding to no moment) with the point indicating balanced failure, of coordinates P_b, M_b. The equation of this curve is given by

$$P_{ult} = P_0 - \frac{P_0 - P_b}{M_b} M_{ult} \tag{10.19}$$

or

$$P_{ult} = \frac{P_0}{1 + [(P_0 - P_b)/M_b]e} \tag{10.20}$$

Here, the balanced values P_b and M_b are given by Eq. (10.17) and the accompanying theory, and P_0 is given in the theory of axially loaded columns in Sec. 3.5 as

$$P_0 = 0.85 f_c' A_c + f_{YP} A_s$$

By substituting these values into Eq. (10.19) or (10.20) it is possible to get explicit formulas for the compression case. By plotting these straight lines as shown in Fig. 10.7, we complete the interaction curve covering all types of failure under combined loading. Such curves, drawn in nondimensional form as shown in Fig. 10.8, are used in practice for the design of eccentrically loaded concrete columns.

We might note here that in ultimate-strength design according to the ACI Code (sec. 1901*a*) all columns are to be calculated for a minimum eccentricity to account for accidental eccentricities or faulty assumptions in rigid-frame analysis. This eccentricity is to be taken as $0.05t$ for spirally reinforced and $0.10t$ for tied columns.

Example Problem 10.6 Calculate the ultimate eccentric load P_{ult} which may be applied to the column at 4 in. from the column center line; $f_c' = 4.0$ ksi, $f_{YP} = 40$ ksi. For this low eccentricity, we anticipate compression failure and use the linear interaction expression

$$P_{ult} = \frac{P_0}{1 + [(P_0 - P_b)/M_b]e}$$

where

$$P_0 = 0.85 \times 4.0 \times 470.6 + 40.0 \times 9.36 = 1{,}974 \text{ kips}$$

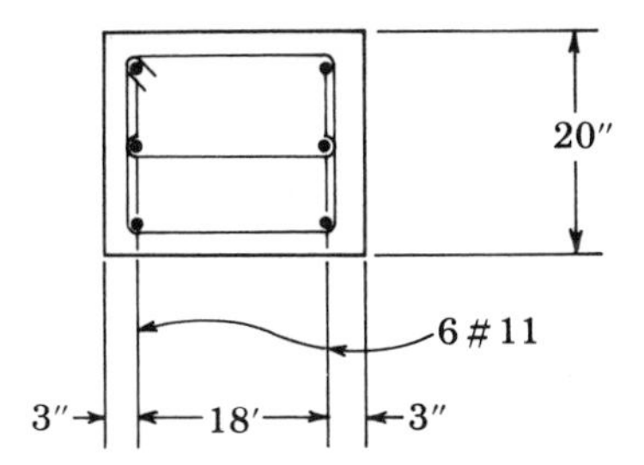

Because of the symmetrical reinforcing we can use Eqs. (10.15) to determine the balanced load and moment values [using a from Eq. (10.16)]:

$$P_b = 0.72 \times 4.0 \times 0.693 \times 20 \times 21 = 838 \text{ kips}$$

and

$$838 \text{ kips}\left(e - 12 \text{ in.} + \frac{0.85 \times 0.693 \times 21}{2}\right) = 4.78 \times 40 \times 18 \text{ in.}$$

from which

$$e_b = 9.93 \text{ in.} \qquad \text{and} \qquad M_b = P_b e_b = 8{,}310 \text{ kip-in.}$$

so that

$$P_{ult} = \frac{1{,}974}{1 + (1{,}136/8{,}310) \times 4.0} = \underline{1{,}278 \text{ kips}}$$

10.8 Design of Reinforced Concrete Columns

We have mentioned earlier that Eqs. (10.13) and (10.14) for tension failure and Eq. (10.19) for compression failure can be plotted in the form of dimensionless design charts, and a typical chart of this type is shown in Fig. 10.8 for the particular case of $d/t = 0.85$. To demonstrate the use of such charts in analysis, we shall apply it in the following example problems.

Example Problem 10.7 Verify the result of Example Problem 10.5 by use of Fig. 10.8. In this case, $d/t = 0.875$, $e/t = 28/24 = 1.17$, and

$$P_T m = \frac{A_s}{bd}\frac{f_{YP}}{0.85f'_c} = \frac{9.36}{20 \times 21}\frac{40}{0.85 \times 4.0} = 0.262$$

From chart,

$$K = \frac{P_{ult}}{btf'_c} = 0.10 \qquad P_{ult} = 0.10 \times 480 \times 4.0 = \underline{192 \text{ kips}}$$

This checks the previous result closely.

We next check Example Problem 10.6, in which $e = 4$ in., consequently $e/t = 4/24 = 0.167$. With all other values the same, we find, from Fig. 10.8,

$$K = \frac{P_{ult}}{btf'_c} = 0.70 \qquad P_{ult} = 0.70 \times 480 \times 4.0 = \underline{1{,}340 \text{ kips}}$$

which falls within 5 percent of the calculated value.

The design of eccentrically loaded columns by use of charts reduces to a rapid trial-and-error process. In designing, the ultimate load and moment have been determined by analysis, and the material properties are given. Several ways of using the charts are then open:

1. We can assume the lateral dimension t and the steel ratio, and solve for the column width b.
2. We can assume overall column dimensions and solve for the steel ratio p_T.

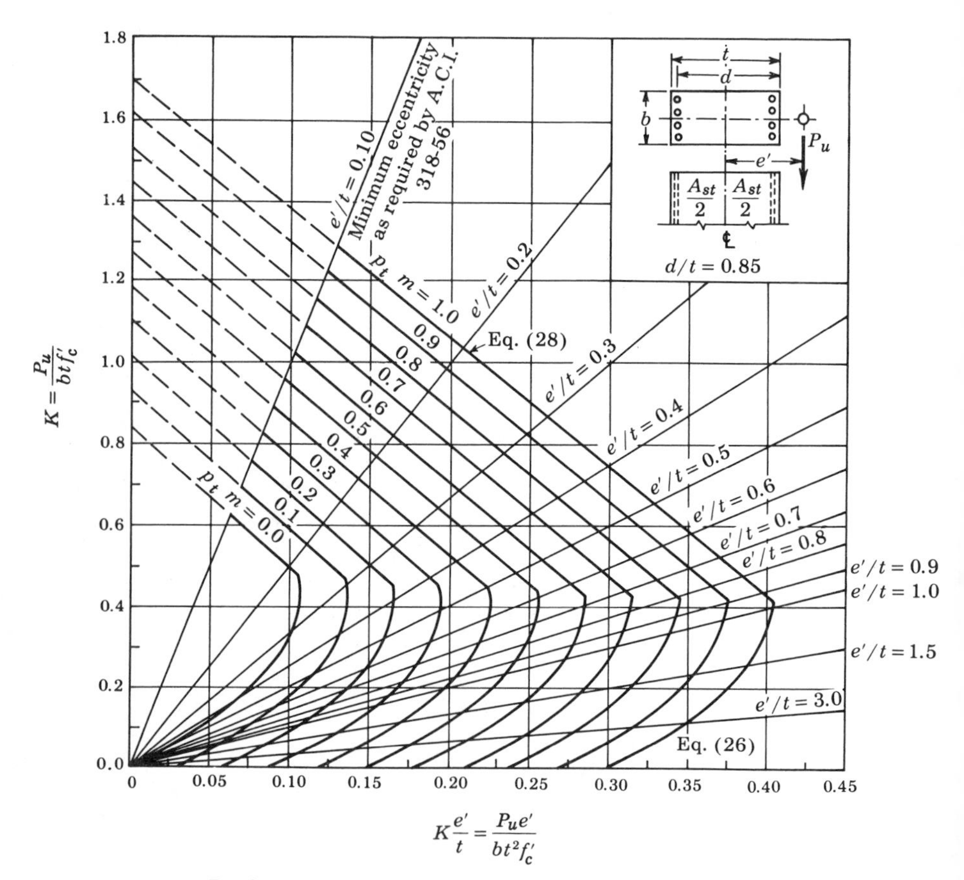

Bending and axial load-$d/t = 0.85$, rectangular sections with symmetrical reinforcement

Fig. 10.8 Ultimate strength of eccentrically loaded column.
(From C.S. Whitney and Edward Cohen, "Guide for Ultimate Strength Design of Reinforced Concrete," Jnl. A.C.I., vol. 28, no. 5, Nov. 1956)

In either case an inadequate assumption may result in a disproportioned member, but the labor involved is so trivial that the work can easily be repeated with an improved assumption.

Example Problem 10.8 Design a square tied column subject to the ultimate loads shown; $f'_c = 4.0$ ksi, $f_{YP} = 40$ ksi.

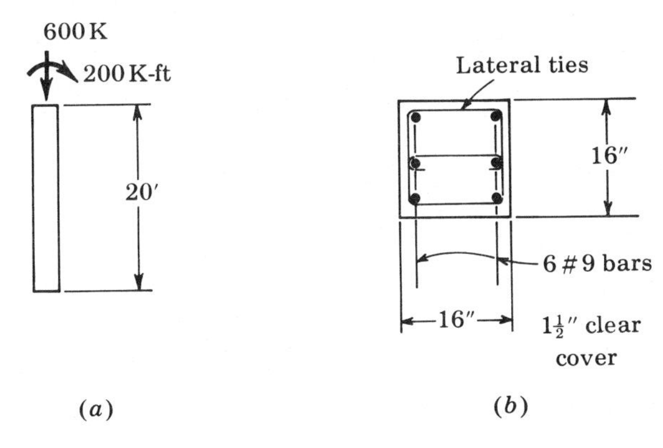

To get an idea of the approximate column size, we could assume that the effect of the applied moment might be about the same as doubling the axial load; if we thus consider a column with average steel ratio of 4 percent to carry a centric load of 1,200 kips, we would get

$$1{,}200 \text{ kips} = A_G(0.85 \times 4.0 + 0.04 \times 40)$$

or

$$A_G = \frac{1{,}200}{5.0} = 240 \text{ in.}^2$$

Try a 16 by 16 in. column; then, anticipating concrete cover to the center of the longitudinal bars to be $2\frac{1}{2}$ in., $d/t = 13.5/16 = 0.85$. With $e = (200 \times 12)/600 = 4$ in., $e/t = 4/16 = 0.25$.

With an ultimate load of 600 kips to be carried,

$$K = \frac{P_{ult}}{btf'_c} = \frac{600}{16 \times 16 \times 4.0} = 0.587$$

we enter the chart, Fig. 10.8, along the line $e/t = 0.25$, and find that, corresponding to a K value of 0.587, we require a value

$$p_T m = p_T \frac{40}{0.85 \times 4.0} = 0.27$$

or

$$p_T = 0.023$$

This steel ratio seems satisfactory, and is provided by $A_s = 5.90$ in.^{2} We place three No. 9 bars in each face ($A_{s\ furn} = 6.00$ in.^{2}); the final section is as shown in the sketch.

General Readings

The references cited in Chap. 6 on working-stress theory and those of Chap. 8 on inelastic theory may be consulted.

Problems

10.1 Find the allowable load which may be applied at 2 in. from the column center line.

(*a*) Follow procedure of Sec. 10.2.
(*b*) Follow procedure of Sec. 10.4.
(*c*) Compare results and discuss.

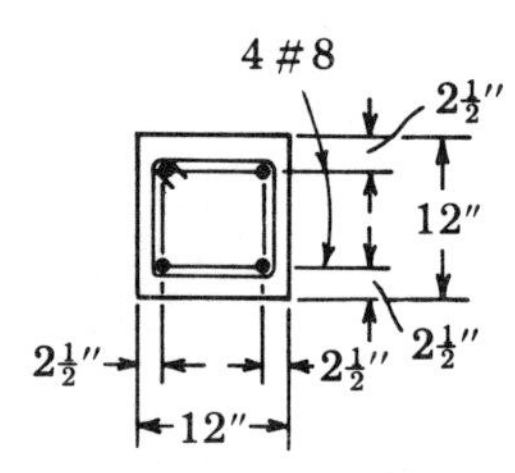

$f'_c = 4.0$ ksi, $n = 8$, $f_{YP} = 40$ ksi, $f_{s\ allow} = 20$ ksi.

10.2 For the column of Prob. 10.1 find the allowable load at 12 in. from the column center line.

(*a*) Follow procedure of Sec. 10.3.
(*b*) Follow procedure of Sec. 10.4.
(*c*) Compare results and discuss.

10.3 Repeat the example problem of Sec. 10.3, but this time use the transformed-section method. Compare the two methods and discuss your preference.

10.4 Where must a compressive load be applied so that the neutral axis falls 6 in. from the center line of the column? Assume cracked section and elastic behavior, with correction for creep.

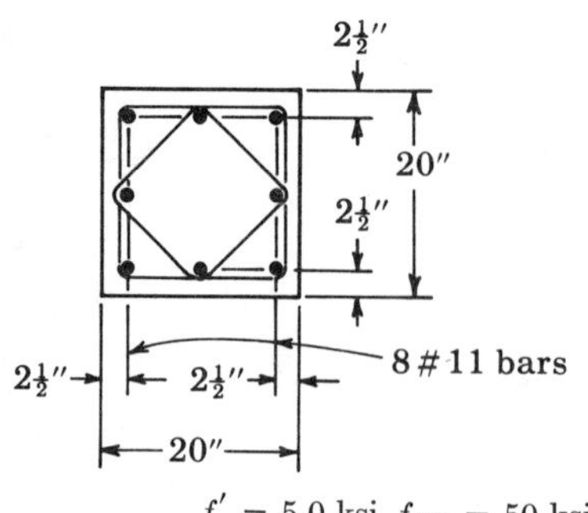

$f'_c = 5.0$ ksi, $f_{YP} = 50$ ksi, $n = 7$, $f_{s\ allow} = 25$ ksi.

10.5 An eccentric load is applied at the corner of the column of Prob. 10.4. Find the allowable value of this load, using an appropriate extension of the theory of Sec. 10.2.

10.6 Outline and discuss a method for analysis for stresses due to a load under large biaxial eccentricities leading to cracked section.

10.7 A reinforced concrete column in a building frame is subject to the loads shown. Design a square tied column. $f_c' = 4.0$ ksi, $f_{YP} = 40$ ksi. Follow ACI working-stress method.

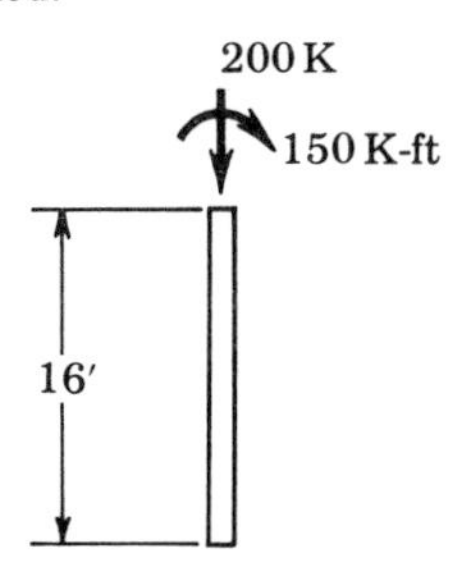

10.8 Design an appropriate rectangular tied column for the loads shown. $f_c' = 5.0$ ksi, $f_{YP} = 40$ ksi. Follow ACI working-stress method.

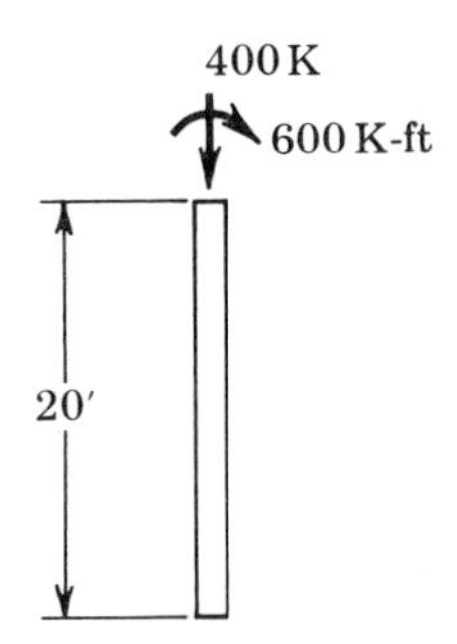

10.9 Calculate the allowable load which may be applied at biaxial eccentricities of 6 in. from each center line. To calculate section properties, consider the steel bars to be replaced by an equivalent steel circle.

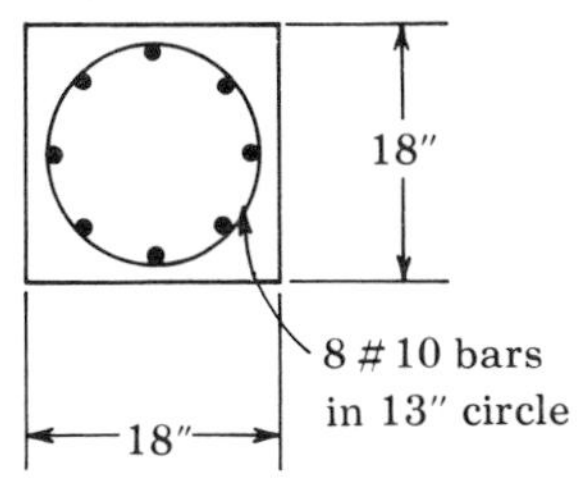

$f_c' = 3.0$ ksi, $f_{YP} = 40$ ksi, $n = 9$.

10.10 A column stack which is part of a building frame is to be designed according to the working-stress provisions of the ACI Code. $f_c' = 4.0$ ksi, $f_{YP} = 40$ ksi. Architectural requirements call for a square column of constant cross section from top to bottom. Section is to be held as small as possible. Design the column stack; present complete design drawing.

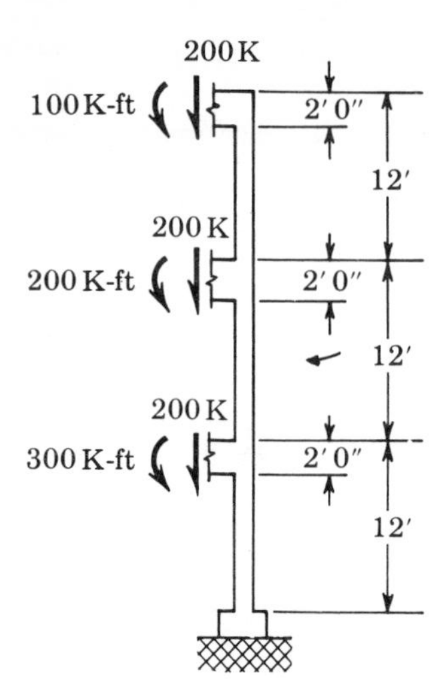

10.11 Verify formulas 19-1 to 19-6 of the 1963 ACI Code; use the tensile-force resultant as moment center for moment-equilibrium equations.

10.12 (*a*) Find the ultimate load which may be applied at 12 in. from the column center line.

(*b*) Find the safety factor of the allowable load on the same column as determined in Prob. 10.2.

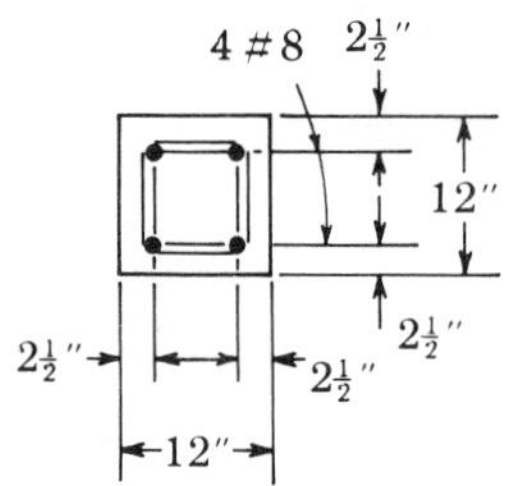

$f'_c = 4.0$ ksi, $f_{YP} = 40$ ksi.

10.13 Find the ultimate load which may be applied at the corner of the column section of Prob. 10.12. Use your judgment in extending the theory for this case, using an interaction approach if considered suitable.

10.14 Outline and discuss an ultimate-load analysis for a column under axial load plus large moments about two axes leading to cracked section and yielding of tension steel.

10.15

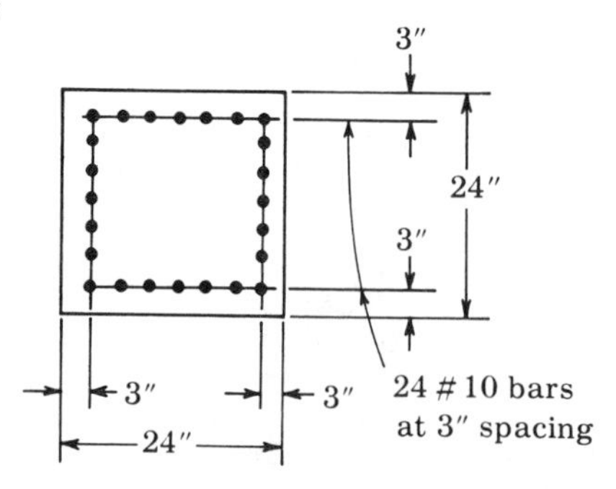

$f'_c = 5.0$ ksi, $f_{YP} = 50$ ksi (lateral ties not shown).

Find the ultimate load which may be applied at 36 in. eccentricity from the center line of the section. Consider what to do about those steel bars which are close to the neutral axis; make suitable engineering assumptions if you feel this is necessary.

10.16 Where must a load be applied to the column of Prob. 10.15 so that at failure the neutral axis is 9 in. from one, 15 in. from the other column face? Assume cracked section. Do this problem two ways:

(*a*) Considering all bars
(*b*) Neglecting those bars which you consider of little effect

Compare amount of work and results, and draw conclusions.

10.17 Rework the example problem of Sec. 10.6, calculating the actual stress in the compression steel. Compare your answer to those obtained earlier, and discuss.

10.18 Using the strength of materials method, set up two equilibrium equations for the ultimate strength of eccentrically loaded columns in which failure is initiated by compressive crushing of the concrete. Use these equations to find the ultimate load which may be applied at 2 in. from the center line of the column of Prob. 10.12.

10.19 Use the interaction approach of the ACI Code discussed in Sec. 10.7 to find the ultimate load which may be applied at 2 in. from the center line of the column of Prob. 10.12. Compare the result with that of Prob. 10.18 and discuss.

10.20 A reinforced concrete column of length 16 ft is to be designed to carry an ultimate load of 400 kips at an eccentricity of 9 in. Follow ultimate-strength provisions of ACI Code. $f'_c = 3.0$ ksi, $f_{YP} = 33$ ksi. Column is to be square tied, of minimum size.

10.21 A square tied reinforced concrete column of length 20 ft is to be designed to carry an ultimate load of 800 kips and an ultimate moment of 1,200 kip-ft; $f'_c = 5.0$ ksi, $f_{YP} = 50$ ksi. Follow ACI provisions for ultimate-strength design.

10.22 The column stack of a building frame is to be designed for the *working loads* shown, using a load factor of 2 against collapse. In design, follow the ultimate-strength provisions of the ACI Code. $f'_c = 4.0$ ksi, $f_{YP} = 40$ ksi. The architect requires a square column of minimum size and of constant section from top to bottom. Design the column stack; compare the design with previous design (Prob. 10.10) according to working-stress method.

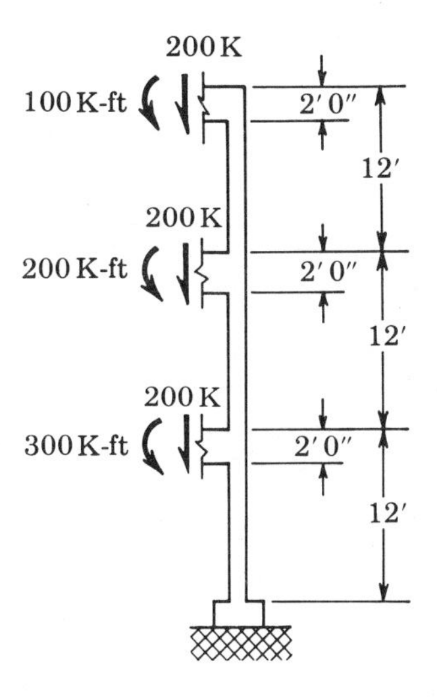

11

Buckling of Beam Columns

11.1 Beam Columns—Introduction

To get a grasp of the buckling behavior of slender beam columns, such as the member under eccentric load which is shown in Fig. 11.1*a*, we shall consider the increase of its lateral deflection Δ with increasing load P. Just as in the case of the axially loaded columns of Secs. 3.6 and 3.7, we must here take into account the effect of the geometry change on the magnitude of moment. Under increasing load P, the deflection Δ will increase, thereby increasing the moment $P(e + \Delta)$, which in turn will lead to a further amplification of the deflection; this results in a nonlinear load-deflection relation as shown by the initial curve of Fig. 11.1*b*.

The bending of the column results in axial and bending stresses. It will be shown that as long as these stresses do not exceed the elastic limit, the load-deflection curve will increase nonlinearly as shown by the dotted line in Fig. 11.1*b* and eventually become asymptotic to the horizontal line corresponding to the Euler load P_e under which the column would buckle if loaded axially. That is, for the elastic column, the axial and eccentric buckling loads are identical.

Actually the stresses will reach their proportional limit long before this, say at a load P_{PL}. After this the decreased stiffness of the fibers will result in increased nonlinearity, till eventually, under a load P_{cr}, a point of instability is reached. Because of the amplification of moment, the column loses all stiffness, and buckling will occur under the ultimate load P_{cr}.

The *elastic* beam-column theory which is outlined in Sec. 11.3 and is the basis of several important code provisions considers that the member

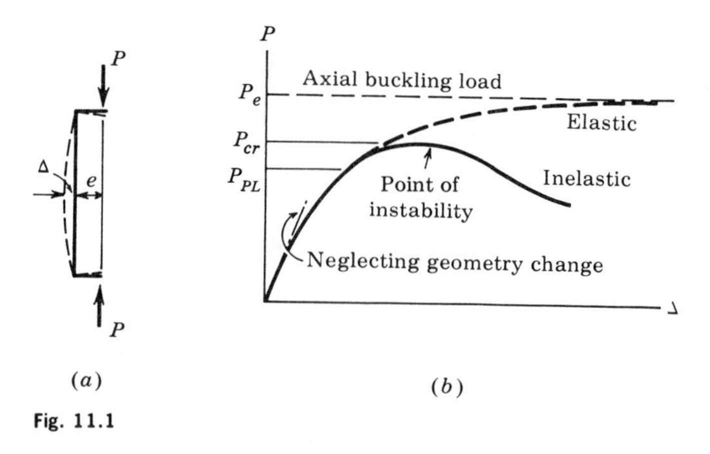

Fig. 11.1

has reached its capacity when the load P_{PL} is reached. In common with other elastic theories of design it considers yielding of any fiber the end of the useful life of the structure. Safe design loads are then determined by applying a suitable safety factor.

On the other hand, *inelastic* beam-column theories such as the one outlined in Sec. 11.5 seek to determine the value of the load P_{cr} under which instability actually sets in. In basic philosophy, this method is a typical limit design approach in which, irrespective of local yielding of fibers, the end of the road is reached only at collapse of the structure.

From a fundamental viewpoint, it is important to note that the beam-column problem, in contrast with the ideal-column problem, is a *response* problem. That means that for a given eccentricity and applied load the response is uniquely determined, and the stress-analysis procedures of Sec. 2.5 (however, including effects of geometry changes), rather than the stability approach of Sec. 2.7, must be used to determine this response.

An additional buckling mode which is particularly dangerous in thin-walled open section is torsional or twisting instability, which results from rotation of the section. Because an analysis of this phenomenon is considered outside the scope of these notes, no further examination of this will be made here, but the interested reader is referred to specialized texts on structural stability.

11.2 Amplification of Moment Due to Initial Eccentricity

To understand more about the amplification of lateral deflection and moment under eccentric loading, let us consider the elastic column of Fig. 11.2 with an initial crookedness y_0, of maximum value e and varying in sinusoidal fashion:

$$y_0 = e \sin \frac{\pi x}{L}$$

We proceed here in a fashion analogous to that used for the Euler column. Calling the additional deflection resulting from bending of the member y,

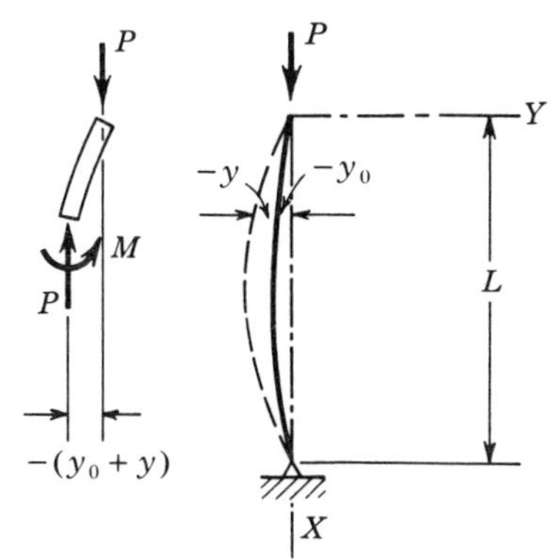

Fig. 11.2

we find the moment at any section

$$M = -P(y_0 + y) = -P\left(e \sin \frac{\pi x}{L} + y\right)$$

and introducing this into the equation of the elastic curve, obtain

$$M = -P\left(e \sin \frac{\pi x}{L} + y\right) = EI \frac{d^2y}{dx^2} \tag{11.1}$$

from which we get the nonhomogeneous differential equation relating the applied load P and the resulting deflection y:

$$\frac{d^2y}{dx^2} + \frac{P}{EI} y = -\frac{P \cdot e}{EI} \sin \frac{\pi x}{L} \tag{11.2}$$

The solution which satisfies this equation and the end conditions is

$$y = B \sin \frac{\pi x}{L}$$

where the coefficient B is found by resubstituting this solution into Eq. (11.2).

$$-\left(\frac{\pi}{L}\right)^2 B \sin \frac{\pi x}{L} + \frac{P}{EI} B \sin \frac{\pi x}{L} = -\frac{P \cdot e}{EI} \sin \frac{\pi x}{L}$$

or $$B = \frac{e}{(\pi^2 EI/L^2)(1/P) - 1} = \frac{e}{(P_e/P) - 1}$$

where $P_e = \pi^2 EI/L^2$ is the Euler load on the perfectly straight column.

Then $$y = \frac{e}{(P_e/P) - 1} \sin \frac{\pi x}{L} \tag{11.3}$$

and the moment

$$M = Pe\left[1 + \frac{1}{(P_e/P) - 1}\right] \sin \frac{\pi x}{L}$$

Pe is the unamplified midspan moment, called M_0, so that the amplified maximum moment is

$$M = M_0\left[1 + \frac{1}{(P_e/P) - 1}\right] = M_0\left[\frac{1}{1 - (P/P_e)}\right] \tag{11.4}$$

The bracketed quantity is the amplifying factor. For the axial load $P = 0$, no amplification occurs. When the load P reaches the Euler load P_e, the amplification is infinite, thereby demonstrating the validity of an earlier statement regarding the identity of axial and eccentric buckling loads for elastic columns.

The useful Eq. (11.4) will be called on later.

11.3 Elastic Beam-column Theory—The Secant Formula

We consider a beam column under eccentric loading as shown in Fig. 11.3, with the larger eccentricity of value e_g, the smaller of value e_s, and the ratio $e_s/e_g = \alpha$.

Our aim is to find the load P_{PL} under which first yielding takes place. The problem is thus not one of determining the buckling load (it has been pointed out earlier that in the elastic case this is identical with the Euler load), but one of stress analysis. This will, according to Fig. 11.1, furnish us with a lower bound on the actual buckling load of the inelastic member, P_{cr}.

We can outline our procedure in the following steps:

1. Find general expression for the moment. Because this depends on the deflection of the beam column, we must do this by introducing the general equation of the elastic curve, similar to the preceding section.
2. Find the maximum value of the moment by differentiation.
3. Find the maximum stress resulting from axial load and moment, and equate this to the yield stress of the material.

The member will deflect an amount y laterally under the applied load P, and the moment at any point can be written by statics as

$$M = P\left\{e_g\left[1 - \frac{x}{L}(1-\alpha)\right] - y\right\} \tag{11.5}$$

and inserting this into the general equation of the elastic curve, we get

$$M = P\left\{e_g\left[1 - \frac{x}{L}(1-\alpha)\right] - y\right\} = EI\frac{d^2y}{dx^2}$$

from which we can write the differential equation relating load and lateral deflection:

$$\frac{d^2y}{dx^2} + \frac{P}{EI}y = \frac{P}{EI}e_g\left(\frac{1-\alpha}{L}x - 1\right) \tag{11.6}$$

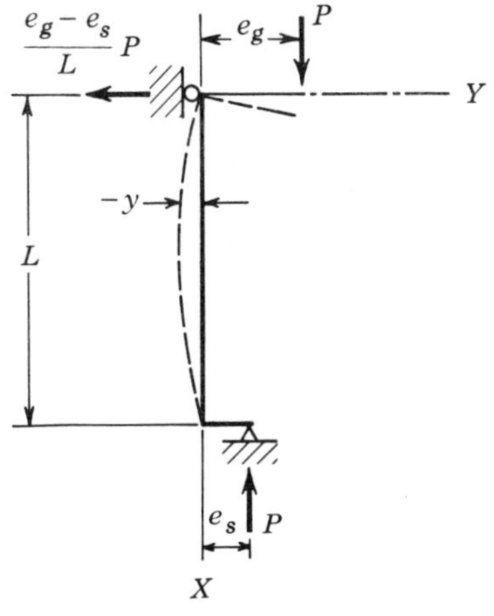

Fig. 11.3

The solution of this nonhomogeneous second-order differential equation consists of the complementary part (setting $P/EI \equiv k^2$)

$$y_c = A \sin kx + B \cos kx$$

and the particular part

$$y_p = e_g \left(\frac{1 - \alpha}{L} x - 1 \right)$$

The boundary conditions are $y_{x=0} = 0$ and $y_{x=L} = 0$, from which we solve for the constants A and B:

$$A = -e_g \frac{\cos kL - \alpha}{\sin kL} \qquad B = e_g$$

so that the total deflection is

$$y = e_g \left(\frac{\alpha - \cos kL}{\sin kL} \sin kx + \cos kx + \frac{1 - \alpha}{L} x - 1 \right)$$

and the moment is, from Eq. (11.5),

$$M = Pe_g \left(\frac{\alpha - \cos kL}{\sin kL} \sin kx + \cos kx \right) \tag{11.7}$$

Since we wish to find the largest stress in the member, we look for the maximum moment by differentiation:

$$\frac{dM}{dx} = Pe_g \left(\frac{\alpha - \cos kL}{\sin kL} k \cos kx - k \sin kx \right) = 0$$

The point of maximum moment is then located at

$$x = \frac{1}{k} \tan^{-1} \frac{\alpha - \cos kL}{\sin kL}$$

and, substituting this back into Eq. (11.7), we find the maximum moment, after some manipulation, as

$$M_{\max} = Pe_g \left(\frac{\sqrt{\alpha^2 - 2\alpha \cos kL + 1}}{\sin kL} \right) \tag{11.8}$$

We are now equipped to determine the maximum fiber stress in the member,

$$\sigma_{\max} = \frac{P}{A} + \frac{M_{\max} c}{I}$$

$$= \frac{P}{A} \left(1 + \frac{e_g c}{r^2} \frac{\sqrt{\alpha^2 - 2\alpha \cos kL + 1}}{\sin kL} \right) \tag{11.9}$$

It is this stress which is to be set equal to f_{YP} to determine the load P_{PL} under which first yielding occurs:

$$\frac{P_{PL}}{A} \left(1 + \frac{e_g c}{r^2} \frac{\sqrt{\alpha^2 - 2\alpha \cos kL + 1}}{\sin kL} \right) = f_{YP}$$

and solving this for the load P_{PL}, we obtain

$$P_{PL} = \frac{f_{YP}A}{1 + (e_g c/r^2)\sqrt{\alpha^2 - 2\alpha \cos kL + 1}\csc kL}$$

By dividing through by A, we can write this in terms of a nominal stress $\sigma_{PL} = P_{PL}/A$ corresponding to first yielding, and by dividing this by a suitable safety factor η we can determine a nominal allowable stress

$$\sigma_{allow} = \frac{\sigma_{PL}}{\eta} = \frac{P_{allow}}{A} = \frac{f_{YP}/\eta}{1 + (e_g c/r^2)\sqrt{\alpha^2 - 2\alpha \cos kL + 1}\csc kL} \tag{11.10}$$

In appendix C of the AASHO Specifications we find formula A, which is almost identical with Eq. (11.10).

$$f_s = \frac{f_{YP}/\eta}{1 + [0.25 + (e_g c/r^2)]B \csc \phi} = \frac{P}{A} \tag{11.11}$$

where B stands for the radical factor and ϕ replaces kL. The term 0.25 which is added to the nondimensional factor $e_g c/r^2$ is to account for unintentional crookedness of the member. A number of measurements on actual steel columns has shown that this is a reasonable average value.

Equations (11.10) and (11.11) are quite difficult to use for design, and their main purpose is to plot column curves such as those in appendix C of the AASHO Specifications and shown in Fig. 11.4. Such column curves, which are similar in form to those for critical stress of axially loaded columns, are reasonably easy to use for design of beam columns by trial and error.

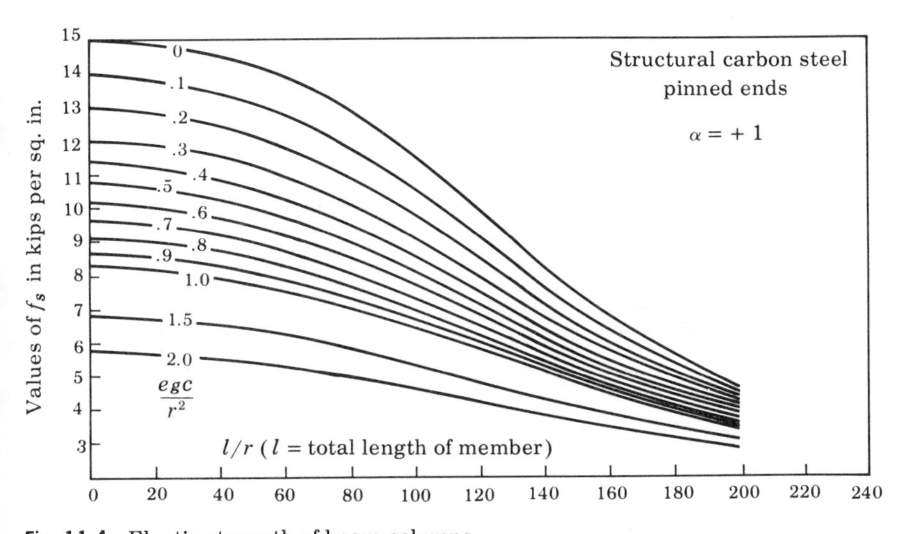

Fig. 11.4 Elastic strength of beam columns. (From "Specifications for Highway Bridges," Am. Assoc. of State Highway Officials, 1961)

Figure 11.3 and the moment expression (11.7) remind us that this theory is valid only for beam columns without end restraint. In practice, such members often occur as parts of rigid frames, so that they possess a certain amount of end fixity which will affect their strength. Such end restraint is usually taken into account by substituting an effective for the actual length, according to the suggestions of Fig. 3.18.

For the particular case in which the eccentricity of the load is a constant value e, we set $\alpha = 1$, and after some slight manipulation reduce Eq. (11.10) to the simpler form

$$\sigma_{allow} = \frac{P_{allow}}{A} = \frac{\sigma_{YP}/\eta}{1 + (ec/r^2)\sec(kL/2)} \tag{11.12}$$

This is called the *secant formula*. In using this well-known expression we should keep in mind that it does not furnish us with a critical buckling stress, but rather ensures that the member stays within the elastic range, thereby preventing attainment of the critical load P_{cr} (of Fig. 11.1) which is associated with inelastic buckling. In a later section we shall consider inelastic buckling of beam columns.

11.4 Beam-column Design by Interaction Formula

A convenient method of beam-column design is to consider the member as a cross between a beam under pure bending and a column under pure axial load, its two limiting cases. In particular, we could assume the linear interaction expression Eq. (9.12), which has already been used in elastic-stress analysis, and is shown again in Fig. 11.5:

$$\frac{P_0}{P_{allow}} + \frac{M_0}{M_{allow}} = 1 \tag{11.13}$$

where P_0 and M_0 are the allowable loads applied jointly, and P_{allow} and M_{allow} are the allowable values applied separately. While this satisfies the limiting cases of pure beam and pure column, there is no reason to consider it theoreti-

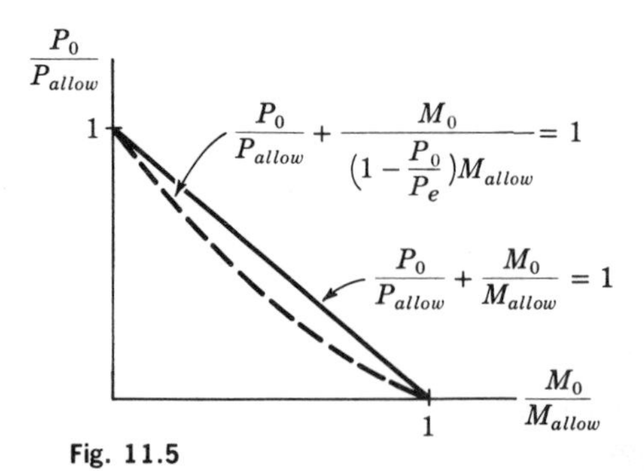

Fig. 11.5

cally valid for intermediate cases. As a matter of fact, it is apparent from the preceding discussion that the effects of deformation must lead to some sort of nonlinear interaction. Because of the amplification of moment caused by lateral deflections [which was not considered in deriving Eq. (11.13)] we can further deduce that the linear interaction formula must give results which are on the unsafe side.

Nevertheless, because of its simplicity Eq. (11.13) has formed the basis for a number of code provisions, such as those of the AISC Specifications prior to 1961. In view of the success of structures built under these provisions, it must be concluded that the unsafe approximation was compensated for by the safety factor provided.

To improve on the linear interaction expression, we substitute for the initial moment of the applied loads the amplified moment according to Eq. (11.4):

$$M = M_0 \frac{1}{1 - (P_0/P_e)}$$

so that the interaction expression becomes

$$\frac{P_0}{P_{allow}} + \frac{M_0}{[1 - (P_0/P_e)]M_{allow}} = 1 \tag{11.14}$$

and this expression is shown in the dashed line in Fig. 11.5.

Casting in terms of stresses similar to Eq. (9.14),

$$\frac{f_A}{F_A} + \frac{f_B}{[1 - (f_A/F_e)]F_B} = 1 \tag{11.15}$$

where f_A and f_B are the axial and bending stresses under the combined loading, F_A and F_B are the allowable stresses for the axially loaded column and the beam, respectively, and F_e is the critical stress according to Euler's column theory.

While the amplification factor for the moment in Eqs. (11.14) and (11.15) has been derived for a column with sinusoidally varying eccentricity, it is nevertheless useful in predicting approximately the column behavior under initial moments of different variations, such as linear ones resulting from eccentrically applied end forces, or those resulting from laterally applied loads on the column.

Equation (11.15) is found in the AISC Specifications, formula 7*a*, sec. 1.6.1, as the basis for design of steel beam columns. To compensate for the idealized conditions for which the amplification factor was computed in Sec. 11.2, the second term in that formula is multiplied by an additional correction factor C_m, whose value is also specified in AISC sec. 1.6.1.

The great usefulness and versatility of interaction formulas or graphs arises from the possibility of including a variety of conditions in at least an approximate manner. So, for instance, if the danger of lateral buckling under applied moment exists, the allowable stress F_B can be reduced according to rigorous theory or according to formulas 4 or 5 of the AISC Specifications. Likewise, effects of torsional buckling under axial load, fatigue, or other secondary effects can be accounted for in similar fashion, drawing upon theoretical or experimental results or upon code provisions.

Interaction formulas also allow consideration of end effects such as partial or complete fixity, or sidesway. This can be taken into account by calculating the allowable axial load P_{allow} or the corresponding allowable axial stress F_A, on basis of an effective buckling length of the member, which can be computed using the factors of Fig. 3.18.

Once an interaction expression or interaction curve has been established, the design of members becomes usually one of trial and error, similar in method to the design of axially loaded columns discussed in Sec. 3.8. A likely member is assumed, either on the basis of judgment or approximate calculation, and checked whether it satisfies the interaction criterion. If it is found wanting, another member is tried.

The establishment of a reasonably efficient design method is based on a suitable tabular form, and will be illustrated by the following example problem.

Example Problem 11.1 A column in a steel building frame is, according to an analysis, subjected to an axial force of 300 kips and a moment of 200 kip-ft at the lower end, 150 kip-ft at the upper end. There is no moment about the other axis, and sidesway may be neglected. Story height is 12 ft. Design the member according to AISC Specifications. Steel is A-36.

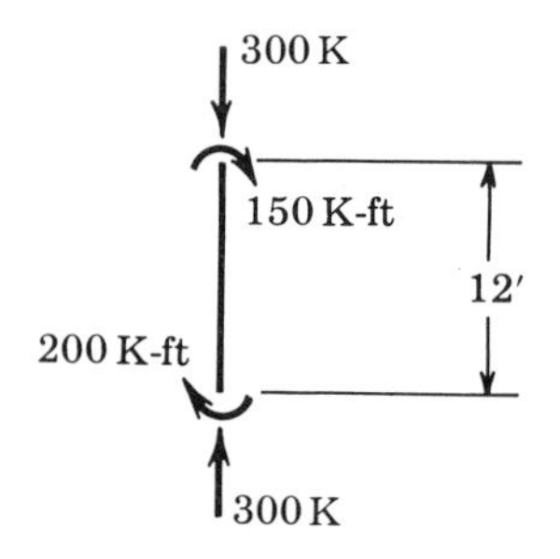

We must first zero in quickly on the range of member size from which to select likely candidates. To do this, we might consider the moment of 200 kip-ft alone, and compute the required section modulus as

$$S_{reqd} = \frac{M}{F_b} = \frac{200 \times 12}{0.66 \times 36} = 101 \text{ in.}^3$$

Assuming that, owing to the presence of axial force, we need twice this value, we look for a member of section modulus about 200 in.^{2} 14 WF shapes are well suited for column members on account of their wide flanges which provide stiffness against weak-axis buckling.

We might, accordingly, consider a 14 WF 127, for which $A = 37.33$ in.^{2}, $S = 202.0$ in.^{3}, $I = 1{,}476.7$ in.^{4}, $r_{max} = 6.29$ in., $r_{min} = 3.76$ in.

On the basis of this we calculate

$$f_B = \frac{200 \times 12}{202.0} = 11.9 \text{ ksi}$$

$$F_B = 0.66 F_{YP} = 23.7 \text{ ksi}$$

$$f_A = \frac{300}{37.33} = 8.05 \text{ ksi}$$

$$(L/r)_{\text{strong axis}} = \frac{12 \times 12}{6.29} = 23$$

$$\text{F.S.} = \frac{5}{3} + \frac{3 \times 23}{8 \times 126.1} - \frac{(23)^3}{8 \times 126.1^3} = 1.73$$

$$\frac{(L/r)^2}{2(C_c)^2} = \frac{23^2}{2 \times 126.1^2} = 0.0166$$

therefore

$$F_A = \frac{\left[1 - \frac{(L/r)^2}{2C_c^2}\right] F_Y}{\text{F.S.}} = \frac{0.9834 \times 36}{1.73} = 20.4 \text{ ksi}$$

$$F'_e = \frac{149{,}000}{(L/r)^2} = 281 \text{ ksi}$$

Inserting these values into Eq. (11.15), we get

$$\frac{8.05}{20.4} + \frac{11.9}{[1 - (8.05/281)]23.7} = 0.39 + 0.52 = 0.91$$

This is considerably less than unity, so that we should try a smaller size.

Note that if tables of F_A and F'_e (as in AISC Handbook, tables 1-33 to 2-50) are available, several lines of the calculations could be eliminated. Having obtained an insight into the nature of the calculations, we shall carry out further trials in tabular form, first trying the next smaller size of 14 WF:

Member	A	S	r_b	L/r_b	f_A	F_A	f_B	F_B	F'_e	$1 - \frac{f_A}{F'_e}$	*Sum*
14 WF 119	34.99	189.4	6.26	23	8.57	20.4	12.7	23.7	281	0.97	0.97
14 WF 111	32.65	176.3	6.23	23	9.19	20.4	13.6	23.7	281	0.97	1.04

We tentatively select the 14 WF 119 member, checking also its weak-axis buckling resistance:

$$L/r = \frac{144}{3.75} = 38.5 \qquad F_A = 19.27 \text{ ksi}$$

With $f_A = 8.57$, and no bending about the weak axis, this is not critical. The column will be made of 14 WF 119.

11.5 Inelastic-buckling Theory

It has been pointed out in Sec. 11.1 that the buckling strength of beam columns depends on their inelastic behavior, and in this section we shall consider an approach to this problem. To simplify the calculations we shall consider a rectangular cross-section member of elastic–perfectly plastic material subjected to an applied load of constant eccentricity. The column ends are without restraint, as shown in Fig. 11.6*a*.

The approach taken here is somewhat similar to that of Sec. 11.3. The first step is to compute the relation between the applied load and the resulting lateral deflection (but this time using inelastic relations, which have already been calculated in Sec. 9.8). This lateral deflection will amplify the moment, so that eventually a load is reached under which the lateral deflection increases faster than bending resistance can be mobilized. This buckling load can be found by maximizing the load with respect to the deflection.

A pin-ended column of rectangular section is shown in Fig. 11.6, subjected to applied load and moment, the former resulting in a moment variation Py, the latter in a moment M_0, so that the total moment is

$$M = Py + M_0 \tag{11.16}$$

Assuming the column to deflect as a sine curve,

$$y = y_m \sin \frac{\pi x}{L}$$

and the curvature at the critical section at the midpoint is

$$\left.\frac{d^2y}{dx^2}\right|_{x=L/2} = \frac{\pi^2}{L^2} y_m \tag{11.17}$$

The curvature, it will be recalled, has already been established in Sec. 9.8 in terms of the applied loading for members of rectangular section of elastic–perfectly plastic material, subjected to axial load and moment. Equations

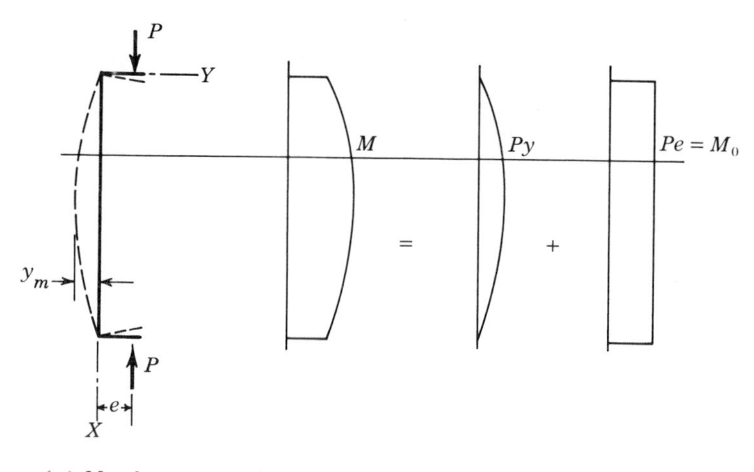

Fig. 11.6

(9.32) and (9.37) give these relationships:

$$\phi = \frac{d^2y}{dx^2} = \frac{M_{YP}}{EI} \frac{4[1 - (P/P_{YP})]^3}{9[1 - (P/P_{YP}) - (M/3M_{YP})]^2} \tag{9.32}$$

for $\dfrac{P}{P_{YP}} \geqq \dfrac{1}{4} - \dfrac{1}{2}\sqrt{\dfrac{9}{4} - 2\dfrac{M}{M_{YP}}}$ (9.41)

and $$\phi = \frac{d^2y}{dx^2} = \frac{M_{YP}}{EI} \frac{1}{\sqrt{2}\sqrt{\frac{3}{2}\left[1 - \left(\frac{P}{P_{YP}}\right)^2\right] - \frac{M}{M_{YP}}}} \tag{9.37}$$

for $\dfrac{P}{P_{YP}} \leqq \dfrac{1}{4} - \dfrac{1}{2}\sqrt{\dfrac{9}{4} - 2\dfrac{M}{M_{YP}}}$ (9.41)

Considering first the case of small eccentricities, we substitute

$$\frac{M_{YP}}{EI}\frac{4}{9}\frac{[1 - (P/P_{YP})]^3}{[1 - (P/P_{YP}) - (M_m/3M_{YP})]^2} = \frac{\pi^2}{L^2} y_m \tag{11.18}$$

where M_m is the midpoint moment of the column. Rewriting Eq. (11.18), we get

$$\frac{M_{YP}}{EI}\frac{4}{9}\left(1 - \frac{P}{P_{YP}}\right)^3 - \frac{\pi^2}{L^2} y_m \left[1 - \frac{P}{P_{YP}} - \frac{P}{3M_{YP}}\left(\frac{M_m}{P}\right)\right]^2 = 0 \tag{11.19}$$

Since the midpoint moment M_m is also given by Eq. (11.16), we can write

$$\frac{M_{YP}}{EI}\frac{L^2}{\pi^2}\frac{4}{9}\left(1 - \frac{P}{P_{YP}}\right)^3 - y_m \left[1 - \frac{P}{P_{YP}} - \left(\frac{P}{3M_{YP}}\right)\left(y_m + \frac{M_0}{P}\right)\right]^2 = 0 \tag{11.20}$$

Equation (11.20) relates the applied axial load P and the resulting midpoint deflection for any eccentricity M_0/P; graphically, it can be represented by the curve of Fig. 11.7, according to which instability occurs under a value of P_{cr}, when

$$\frac{dP}{dy_m} = 0$$

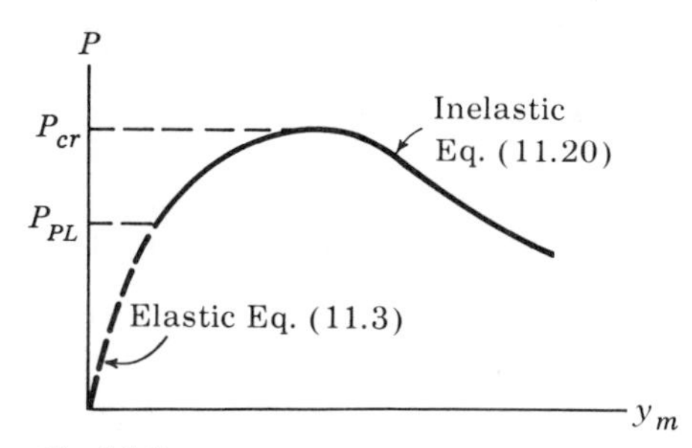

Fig. 11.7

We therefore differentiate Eq. (11.20) and equate to zero:

$$2\left[y_m^{\frac{1}{2}}\left(1 - \frac{P_{cr}}{P_{YP}} - \frac{P_{cr}}{3M_{YP}}\frac{M_0}{P_{cr}}\right) - \frac{P_{cr}}{3M_{YP}}y_m^{\frac{3}{2}}\right]\left[\frac{1}{2}\left(1 - \frac{P_{cr}}{P_{YP}} - \frac{P_{cr}}{3M_{YP}}\frac{M_0}{P_{cr}}\right)y_m^{-\frac{1}{2}} - \frac{3}{2}\frac{P_{cr}}{3M_{YP}}y_m^{\frac{1}{2}}\right] = 0$$

Solving this quadratic equation for y_m, we get

$$y_m = \frac{1}{3}\left(\frac{3M_{YP}}{P_{cr}}\right)\left(1 - \frac{P_{cr}}{P_{YP}} - \frac{P_{cr}}{3M_{YP}}\frac{M_0}{P_{cr}}\right) \tag{11.21}$$

and substituting this expression for the midpoint deflection at the instant of buckling back into Eq. (11.20) and simplifying, we finally get

$$P_{cr} = \frac{\pi^2 EI}{L^2}\left[\frac{1 - (P_{cr}/P_{YP}) - (P_{cr}/3M_{YP})(M_0/P)}{1 - (P_{cr}/P_{YP})}\right]^3 \tag{11.22}$$

or, expressing in terms of the average buckling stress $\sigma_{cr} = P_{cr}/A$ and the eccentricity ratio $e/d = (M_0/P)(1/d)$,

$$\sigma_{cr} = \frac{\pi^2 E}{(L/r)^2}\left[\frac{(\sigma_{YP}/\sigma_{cr}) - 1 - 2(e/d)}{(\sigma_{YP}/\sigma_{cr}) - 1}\right]^3 \tag{11.23}$$

According to the stress distribution underlying Eq. (9.32), these expressions are valid for eccentricity ratios e/d smaller than the one given by Eq. (9.41). By substituting into that equation

$$\frac{\sigma_{cr}}{\sigma_{YP}} = \frac{P}{P_{YP}}$$

and $$\frac{6\sigma_{cr}}{\sigma_{YP}\cdot d}(y_m + e) = \frac{M}{M_{YP}}$$

[where y_m is given by Eq. (11.21)], this can be rewritten in the more useful form

$$\frac{\sigma_{cr}}{\sigma_{YP}} \geq 1 - 2\frac{e}{d} \tag{11.24}$$

Considering next the behavior of the beam column under large eccentricities, we substitute the curvature relation of Eq. (9.37) into Eq. (11.17), resulting in

$$\frac{M_{YP}}{EI}\frac{1}{\sqrt{2}\sqrt{\frac{3}{2}\left[1 - \left(\frac{P}{P_{YP}}\right)^2\right] - (M_m/M_{YP})}} = \frac{\pi^2}{L^2}y_m \tag{11.25}$$

Proceeding in the same way as before, we again express M_m by Eq. (11.16) and find an expression relating the axial load P to the resulting midpoint deflection for any eccentricity M_0/P:

$$\frac{M_{YP}}{EI}\frac{L^2}{\pi^2}\frac{1}{\sqrt{2}} - y_m\sqrt{\frac{3}{2}\left[1 - \left(\frac{P}{P_{YP}}\right)^2\right] - [(M_0 + Py_m)/M_{YP}]} = 0 \tag{11.26}$$

Differentiating with respect to y_m and equating to zero to find y_m due to P_{cr}, we get

$$y_m = \left[1 - \left(\frac{P_{cr}}{P_{YP}}\right)^2\right]\frac{M_{YP}}{P} - \frac{2}{3}\frac{M_0}{P} \tag{11.27}$$

and resubstituting this value into Eq. (11.26), we find the required value for the critical load

$$P_{cr} = \frac{(L^2/\pi^2 EI)^2 M_{YP}^3}{[(M_{YP}/P) - (PM_{YP}/P_{YP}^2) - \frac{2}{3}(M_0/P)]^3} \tag{11.28}$$

or again expressing this in terms of the average buckling stress $\sigma_{cr} = P_{cr}/A$ and the eccentricity ratio $e/d = (M_0/P)(1/d)$,

$$\sigma_{cr} = \frac{[(L/r)^2/\pi^2 E]^2 \sigma_{YP}^3}{[(\sigma_{YP}/\sigma_{cr}) - (\sigma_{cr}/\sigma_{YP}) - 4(e/d)]^3} \tag{11.29}$$

Equations (11.28) and (11.29) are valid for eccentricities larger than the one given by Eq. (11.24).

None of these results is suitable for use in design, but they may be used to plot buckling curves. Equations (11.22) and (11.28) lend themselves to

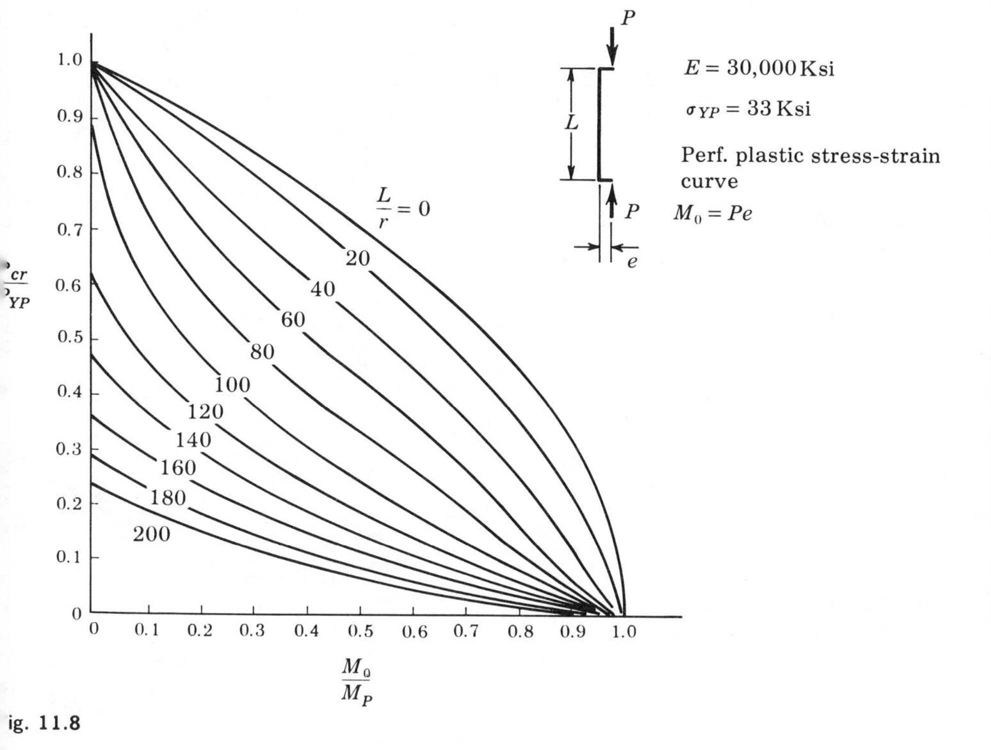

Fig. 11.8

the construction of interaction curves in which the nondimensional values of the applied loads P_{cr}/P_{YP} and M_0/M_{YP} are plotted for a given E and σ_{YP} with the slenderness ratio L/r as a parameter. These curves form an excellent supplement to the interaction curve of Fig. 9.16, which may be considered for $L/r = 0$. Figure 11.8 shows the results plotted in this fashion.

Equations (11.23) and (11.29), on the other hand, lend themselves to representation as column curves in terms of the nominal buckling stress σ_{cr} versus the slenderness ratio L/r, with the eccentricity ratio e/d as a parameter. Here, the buckling curve drawn for axial load in Fig. 3.16b may be added to supply information for the limiting case $e/d = 0$. Figure 11.9 presents these design curves. Figures 11.9 and 11.10 are reasonably easy to use; preference for one or the other form is the personal choice of the engineer.

11.6 Code Provisions and Plastic Design of Steel Beam Columns

The results which have been deduced and presented in Figs. 11.8 and 11.9 are valid for rectangular cross-section members. In practice, steel

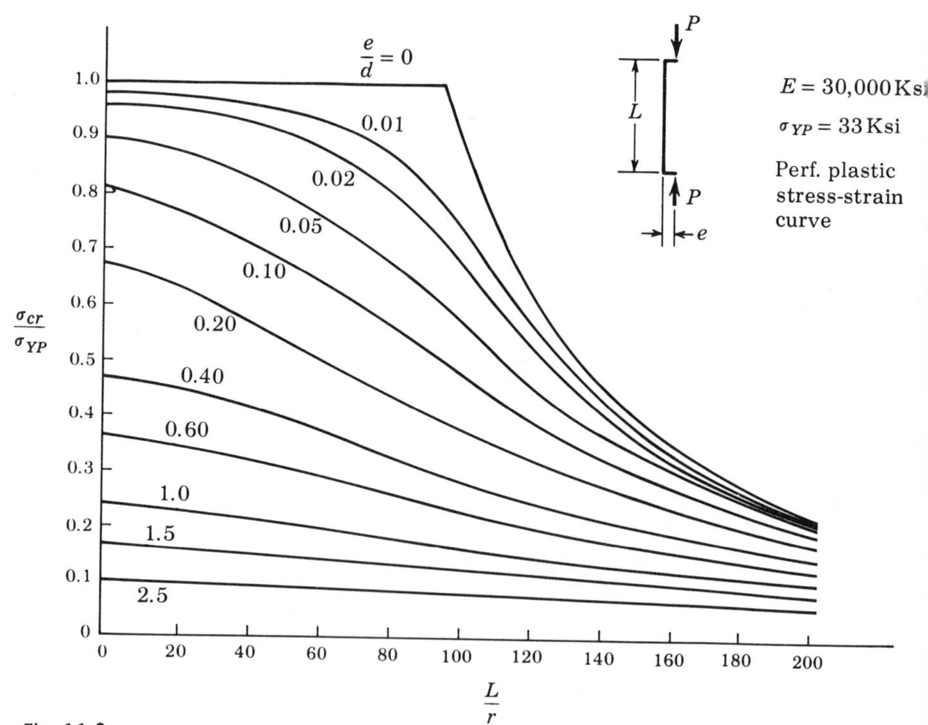

Fig. 11.9

columns are usually of I section, and similar results have been computed for such members.[1] It stands to reason that the carrying capacity of beam columns is influenced by the cross-sectional shape. We have seen in the case of beams that the inelastic strength, indicated by the shape factor, depends on whether most of the material is concentrated adjacent to, or away from, the neutral axis, with the former possessing the larger post-yield strength. According to this reasoning, different buckling curves should be computed for strong- and for weak-axis buckling.

A secondary effect which influences column strength to a considerable extent is the presence of initial stresses in the member, as already discussed in Sec. 3.7. Figure 3.16 of that section indicates that the effect of such stresses is similar to that of eccentricity of the applied load, and it is possible to account for it by increasing the eccentricity in a suitable fashion.

To show how results of inelastic-buckling theory can be incorporated in building codes, we consider sec. 2.3 of the AISC Specifications. Formulas 20 to 24, together with the factors of tables 4-33, 5-33, 4-36, and 5-36, serve for the design of columns under various end conditions. In order to compare our theoretical results with these code provisions, we consider formula 23, for columns bent in single curvature; this is the end condition which most nearly matches the one of the inelastic beam column which was considered theoretically in Sec. 11.5.

Figure 11.10 shows a comparison of theoretical and code results for three representative slenderness ratios. It is seen that in spite of the

[1] ASCE, Commentary on Plastic Design in Steel, p. 87, 1961.

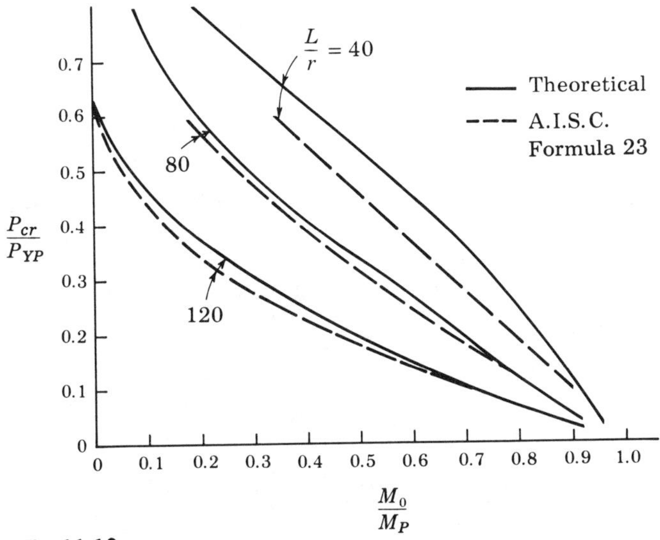

Fig. 11.10

different cross section considered and the neglect of initial stresses in the theory, the results are quite close. The curves representing the code results are not carried above a P_{cr}/P_{YP} ratio of 0.6, since no column may be subjected to an axial force above this value.

It might be noted that use of an interaction representation such as Fig. 11.8 presents considerably less effort for the designer than the use of AISC formula 23. For this reason Fig. 11.11, which plots formula 22 for 36 ksi yield-point steel, has been drawn up and will be used in the example problem of this section.

In common with other column design, the plastic design of steel beam columns must be done by trial and error, using either interaction formulas or the corresponding curves. Section 2.3 of the AISC Specifications provides formulas 20 to 24, which are applicable to columns under different fixity and loading conditions. The maximum axial load which may be applied to the member is specified by formulas 20 and 24 and by the provision that under no conditions may it exceed 60 percent of the yield load P_{YP}. The corresponding reduced plastic-hinge moment M_0 is given by formulas 21 to 23. The first of these, which has been considered earlier, applies to columns at which plastic hinges occur at the supported ends so that no buckling can take place. Failure is thus based on yielding of the section under combined load, a case already discussed in Sec. 9.9.

The linear interaction equation, formula 22, is for columns of double curvature or for pin-ended beam columns which fail in buckling rather than by plastic-hinge formation at the ends, and the nonlinear equation, formula 23, covers the case of single curvature which has already been considered in some detail.

The use of these formulas, and of the corresponding interaction curves, will be demonstrated by means of an example.

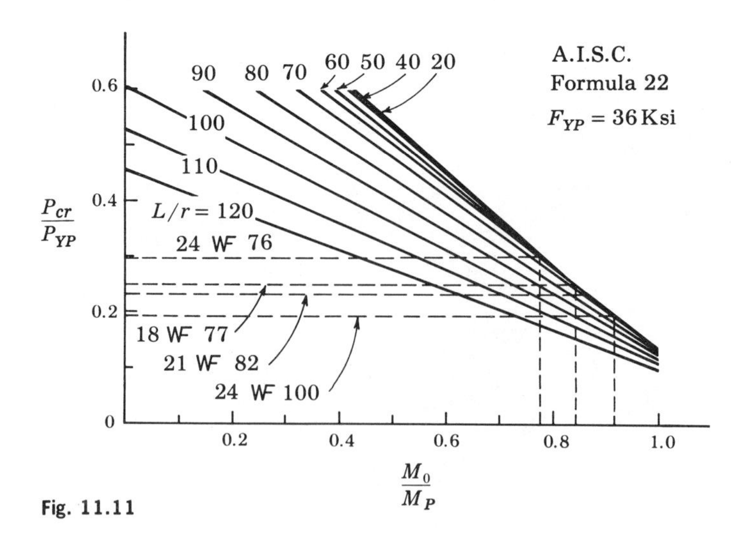

Fig. 11.11

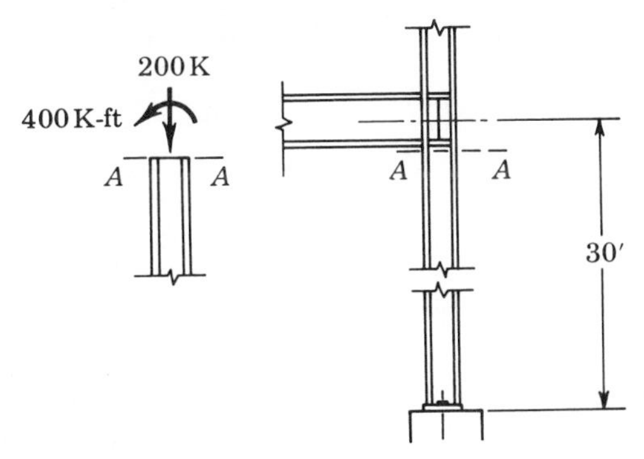

Example Problem 11.2 The exterior rigid-frame column is connected to a welded rigid joint on top, and can be assumed pinned at its foot. The frame is held against sidesway. The column is unbraced over its entire 30-ft length in both directions.

A limit analysis has indicated that at the instant of collapse the column top is subjected to an axial load of 200 kips and a moment of 400 kip-ft. Design the column according to the AISC Specifications, using A-36 steel (F_{YP} = 36 ksi).

Solution: Here again, the first step must be to decide quickly on the approximate range within which to select suitable sections. Let us here assume for the time being that because of the presence of axial force, the plastic hinge will form under, say, half of the fully plastic moment M_P. Then a member is required with $M_P = 2 \times 400 = 800$ kip-ft, and a corresponding plastic section modulus of $(800 \times 12)/36 = 267$ in.³ Entering now the section tables, and looking for members of approximately this value which possess rather wide flanges so as also to provide weak-axis buckling resistance, we consider a 24 WF 100 column.

We now devise a suitable tabular form within which we can check whether AISC eqs. 22 (for strong-axis buckling) and 24 (for weak-axis buckling) are satisfied. To check eq. 22 we can use either the tabulated coefficients or the interaction curve drawn from these coefficients in Fig. 11.11. The following table is suggested:

Member	*A, in.²*	*Z, in.³*	r_{max}, *in.*	r_{min}, *in.*	P_{YP}, *kips*	P/P_{YP}	*L/r, min*	M_P, *kip-ft*	(*eq. 22*) M_0, *kips*	*L/r, max*	(*eq. 24*) (P/P_{YP}) *allow*
24 WF 100	29.43	278.3	10.08	2.63	1,057	0.19	36	833	760		

Note that the 24 WF 100 column will thus form a plastic hinge only when the moment reaches 760 kip-ft. A smaller member is indicated:

24 WF 76	22.37	200.1	9.68	1.85	805	0.248	37	601	467	195	0.229

While the 24 WF 76 still possesses ample strength in the strong direction, it does not satisfy eq. 24, thus showing lack of resistance against weak-axis buckling. We therefore look for a member of somewhat larger weak-axis resistance:

21 WF 82	24.10	191.6	8.53	1.93	869	0.230	42	575	494	187	0.248

While this member satisfies eqs. 22 and 24, it is still not very economical, since the plastic hinge will form only under M_0 = 494 kip-ft. A further search for a section with less weight, but of minimum radius of gyration not less than about 1.9 in., suggests an 18 WF 77:

18 WF 77	22.63	160.5	7.54	1.98	817	0.245	48	482	405	182	0.264

This member of $M_0 = 405$ kip-ft > 400 kip-ft, according to eq. 22, and of $(P/P_{YP})_{allow} = 0.264 > 0.245$, according to eq. 24, seems appropriate, and it is unlikely that a lighter section will work here. An 18 W̄ 77 column will be used.

General Readings

The texts on buckling cited in Chap. 3 contain extensive material on beam columns.
The material of Sec. 11.3 is from the following paper:

Young, D. H.: Rational Design of Steel Columns, *Trans. ASCE*, vol. 62, p. 422, 1936.
This paper has had considerable influence on American specifications.

The material of Sec. 11.5 (which is attributed to Jezek) follows the presentation in Bleich's book cited in Chap. 3. The "Commentary on Plastic Design in Steel" referred to in Chap. 7 gives extensive references to work done on steel beam columns.

Problems

11.1 A 10 W̄ 49 of 40-ft length and 42-ksi steel column is subjected to an axial force. It has an initial crookedness of maximum value e, and sinusoidal distribution. It is required to find the axial load P under which purely elastic action ceases. The effect of initial crookedness is to be investigated by plotting a nondimensional curve of P/P_{cr} versus e/L, under the condition that

(*a*) the eccentricity is in the plane of strong bending
(*b*) the eccentricity is in the plane of weak bending

Compare the results and describe the effects of initial crookedness on the elastic strength of steel columns. Are these effects more pronounced in short or in long columns?

11.2 Repeat Prob. 11.1, this time considering the case of eccentricities of the same magnitude e occurring simultaneously in both planes. Plot the nondimensional elastic limit load P/P_{cr} versus the nondimensional eccentricity e/L, and compare your results to those of Prob. 11.1.

11.3 Repeat Prob. 11.1, this time assuming a uniform eccentricity e rather than a sinusoidal one. Plot results, compare with those of Prob. 11.1, and discuss the importance of the assumed longitudinal distribution of the eccentricity.

11.4 The column is a 12 W̄ 65 of A-36 steel, and bending takes place about the strong axis.

(*a*) Find the load P under which elastic action ceases.
(*b*) Find the allowable load P according to the appropriate provisions of the AISC Specifications.
(*c*) Find the safety factor provided against first yielding by the AISC Specifications.

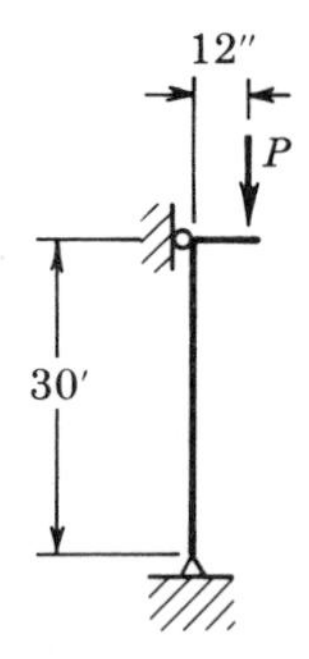

11.5 Repeat Prob. 11.4, this time fixing the lower end of the column against rotation.

11.6 The 12 WF 27 column of A-7 steel (F_{YP} = 33 ksi) is subjected to an axial load P and a uniform lateral load causing bending about the strong axis of the member. The column is braced against weak-axis buckling.

(*a*) Find the load P under which first yielding occurs in the member. Make appropriate assumption if necessary.
(*b*) Find allowable axial load according to AISC Specifications.

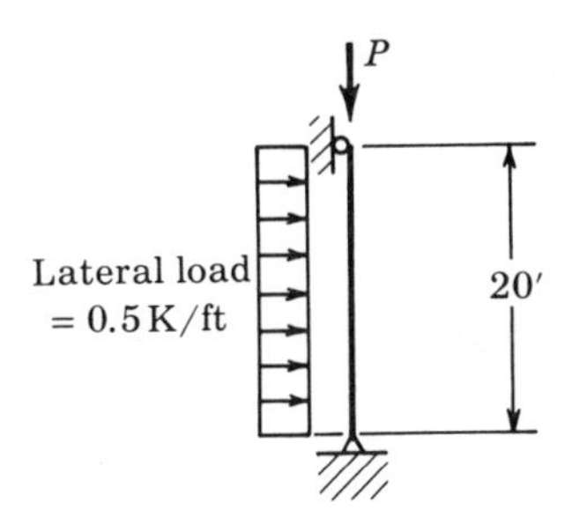

11.7 Select a member to resist the loads shown according to AISC Specifications. Member to be of A-36 steel, and to be braced against buckling in weak direction. No sidesway. Consider both ends of the column rigidly built into the frame.

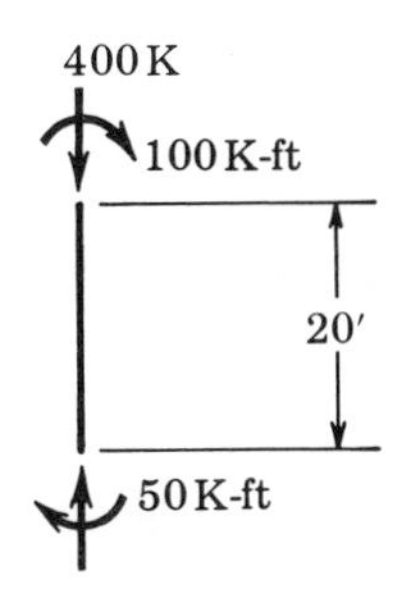

11.8 A pinned truss member of length 16 ft is subjected to an axial load of 200 kips and a uniform transverse load of 1 kip/ft. Design the member of A-36 steel according to the AISC Specifications. Member is unbraced in weak direction.

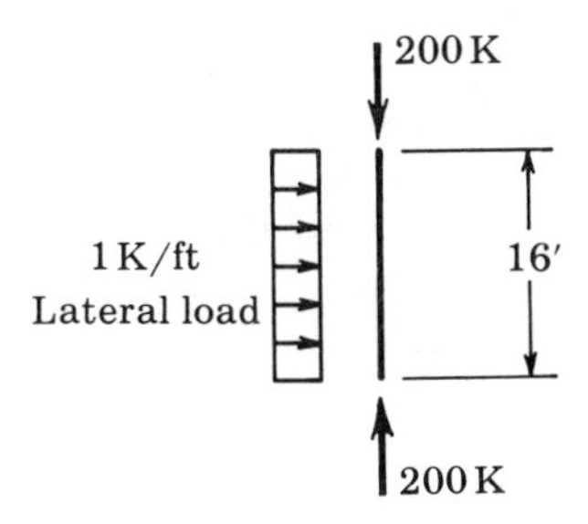

11.9 The column of the rigid frame has been analyzed and found to be subject to an axial force of 100 kips and a moment at the rigid joint of 200 kip-ft. It is unbraced in the weak direction. Select an appropriate member of A-7 steel, following the AISC Specifications. Assume an effective length for the member, considering sway of the frame is possible.

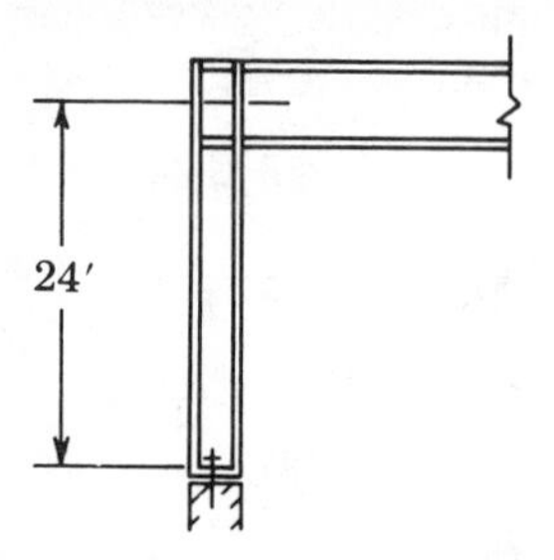

11.10 Draw up a set of interaction curves, similar to those of Fig. 11.9, to satisfy formula 23 of the AISC Specifications for steel of $F_{YP} = 33$ ksi.

11.11 Draw up a set of interaction curves, similar to those of Fig. 11.9, to satisfy formula 23 of the AISC Specifications for steel of $F_{YP} = 36$ ksi.

11.12 Draw up a set of interaction curves, of scale identical to that of Fig. 11.12, to represent formula 22 of the AISC Specifications for steel of $F_{YP} = 33$ ksi. Compare these to Fig. 11.12, and discuss the importance of the value of the yield-point stress in these dimensionless interaction curves.

11.13 The column of Prob. 11.4 is considered again. Calculate the collapse load P_P on the structure according to the AISC Specifications. Compare the result with that of Prob. 11.4, and calculate the safety factor against collapse:

(*a*) if the column is designed according to AISC conventional specifications
(*b*) at the instant that elastic action in the member ceases

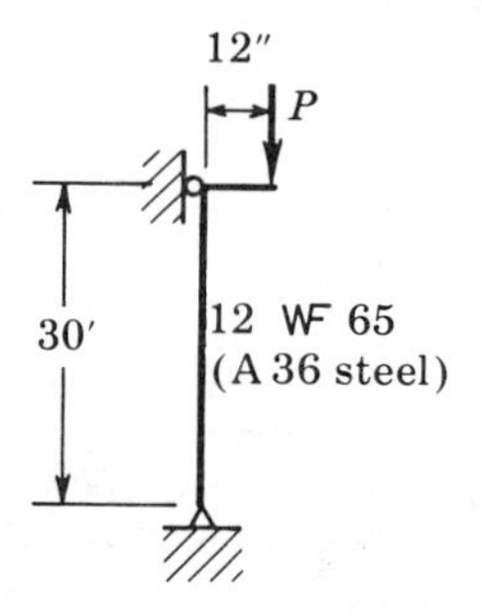

11.14 The column of the previous problem is modified by fixing the foot against rotation. Find the collapse load according to AISC Specifications.

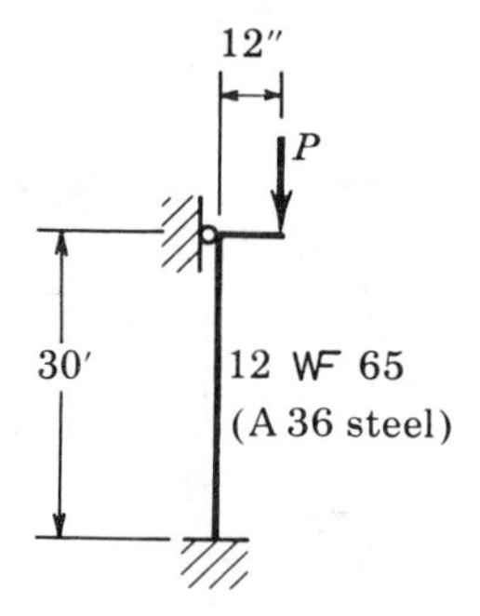

11.15 Find the collapse load P_P on the column:

(*a*) according to AISC Specifications
(*b*) theoretically, assuming rectangular cross section and elastic–perfectly plastic behavior

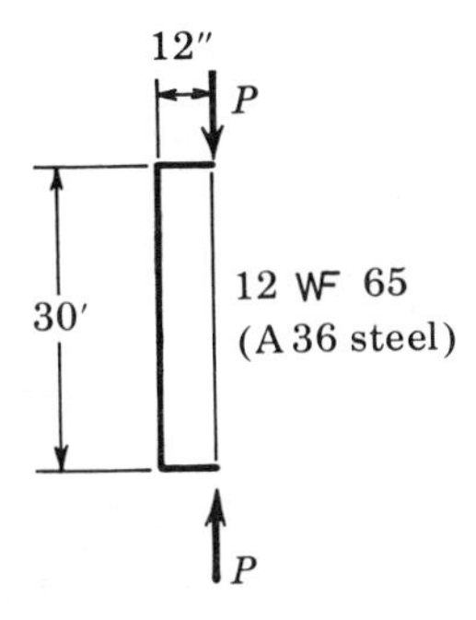

11.16 At the instant of collapse, the column top is subjected to $P = 300$ kips, $M = 200$ kip-ft. The column foot may be considered pinned. The member is unbraced in both directions, and there is no sway of the frame. Design a suitable member of A-7 steel according to the AISC Specifications.

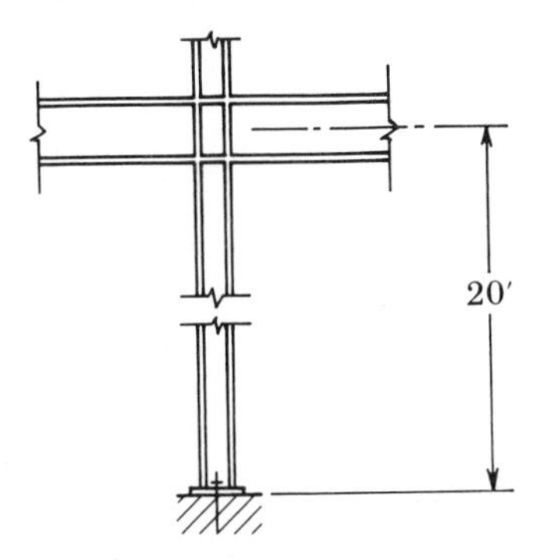

11.17 The rigid frame is subjected to the working loads shown. Apply load factors according to the AISC Specifications, and design the frame of A-7 steel. *Hint:* Select appropriate beam first, then calculate moments on columns at the instant of collapse.

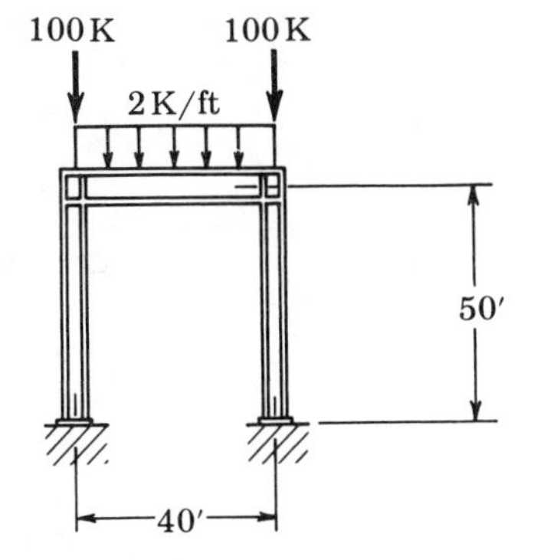

11.18 The member of a rigid frame is subjected to the forces shown at the instant of collapse. Select a member according to the AISC Specifications; use A-36 steel.

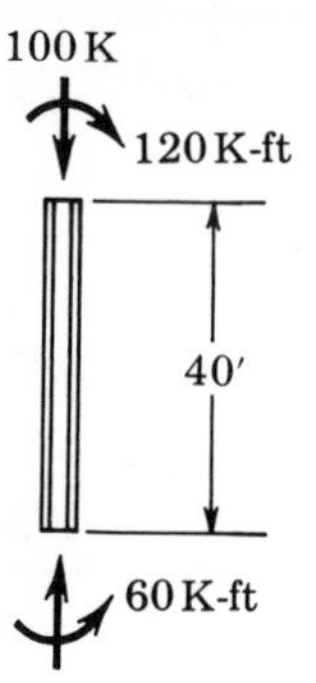

11.19 The two-hinged arch is to be designed according to AISC Specifications of A-36 steel. The load shown is the working load. Weak-axis buckling is prevented by suitable bracing. Select the member, using part 2 of the Specifications.

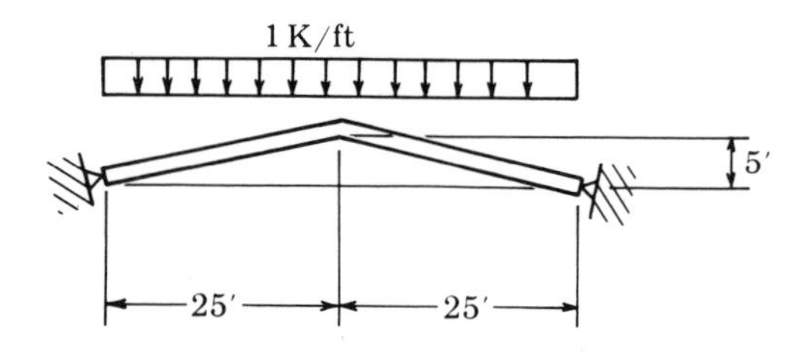

part five

Members in Torsion*

12

Torsion of Elastic and Inelastic Members*

12.1 Introduction

While the torsion of an elastic (or inelastic) prismatic bar of circular cross section is a problem which can be solved by strength of materials methods (see Probs. 2.17, 2.18, and 2.20), the analysis of torsion of noncircular shafts is more involved.

The failure of the earlier approach for analysis of other cross sections can be easily demonstrated. The basic assumption that the strains (and stresses in an elastic member) are tangential and vary linearly with the radial distance from the twisting axis leads to violation of the boundary conditions; consider, for instance, the rectangular cross-section shaft shown in Fig. 12.1. The stated assumptions lead to the presence of shear stresses in the corners which are associated with conjugate shear stresses on free boundaries, an obvious contradiction. The conclusion is that, contrary to the results of this approach, the shear stresses in the corners must be zero.

With this conclusion, it might be possible to assume a stress distribution satisfying the boundary conditions and find values of stress which satisfy overall equilibrium, as was done in Prob. 2.19, but there is of course no assurance that conditions of internal equilibrium and compatibility are satisfied by such a solution, and it must thus be considered as an approximation of questionable worth.

The general problem of torsion is thus established as one which should be solved by the methods of continuum mechanics, as outlined in Sec. 2.6, and here it will be attacked by the inverse method due to St. Venant (1855). According to this approach, a certain likely set of displacements is assumed,

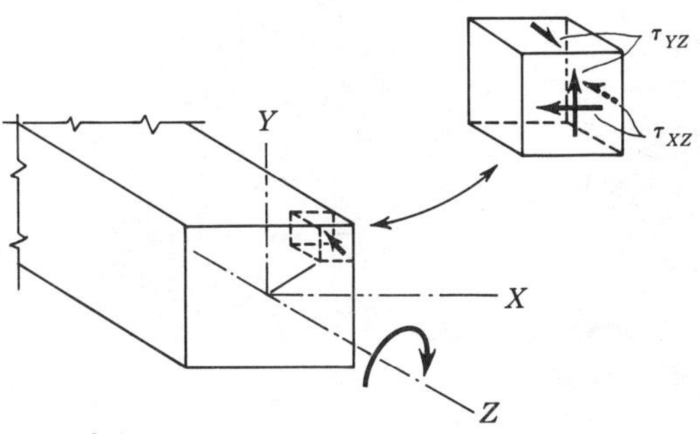

Fig. 12.1

and the equations of Sec. 2.6 are applied to find strains and stresses. Lastly, it must be verified that the assumed set of displacements is able to satisfy all equilibrium, compatibility, and boundary conditions of the problem.

12.2 St. Venant Theory of Torsion

Basic theory We consider a prismatic torsion member as shown in Fig. 12.2, and visualize its deformation under an applied torque T as follows: each cross section will rotate through a small unit angle of twist θ_1 about the twisting or longitudinal Z axis; the total angle of twist θ is assumed to vary linearly along the Z axis, so that

$$\theta = \theta_1 z \tag{12.1}$$

If it is assumed further that each cross section will undergo pure rotation but no in-plane distortion, then the in-plane displacement of any point of coordinates x, y can be described by

$$u = -\theta y = -\theta_1 z y \tag{12.2}$$

$$v = +\theta x = +\theta_1 z x \tag{12.3}$$

We next admit the possibility that warping of the section, that is, deformation parallel to the Z axis, can occur, and that this warping is proportional to the unit angle of twist; thus

$$w = \theta_1 w_1 \tag{12.4}$$

where w_1 is a function of x and y but independent of z.

Starting with these assumed displacements, we find the strains by invoking the strain-displacement equations (2.30a) to (2.31c), and thus

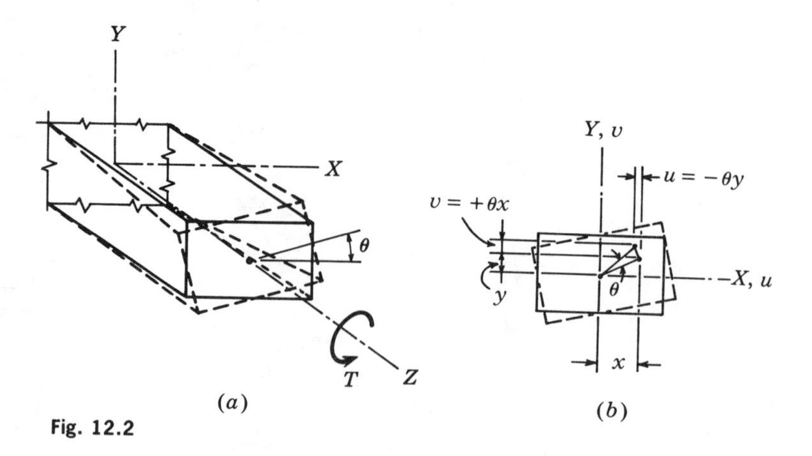

Fig. 12.2

obtain

$$\epsilon_X = \epsilon_Y = \epsilon_Z = \gamma_{XY} = 0 \tag{12.5}$$

$$\gamma_{YZ} = \theta_1\left(x + \frac{\partial w_1}{\partial y}\right) \tag{12.6}$$

$$\gamma_{ZX} = \theta_1\left(-y + \frac{\partial w_1}{\partial x}\right) \tag{12.7}$$

The stresses are found next by use of the elastic stress-strain relations, Eqs. (2.32*a*) to (2.32*f*), as

$$\sigma_{XX} = \sigma_{YY} = \sigma_{ZZ} = \sigma_{XY} = 0 \tag{12.8}$$

$$\sigma_{YZ} = G\theta_1\left(x + \frac{\partial w_1}{\partial y}\right) \tag{12.9}$$

$$\sigma_{ZX} = G\theta_1\left(-y + \frac{\partial w_1}{\partial x}\right) \tag{12.10}$$

These stresses must further be so related that the equilibrium conditions, Eqs. (2.29*a*) to (2.29*c*), are satisfied, so we write

$$0 + 0 + G\theta_1(0 + 0) \equiv 0$$
$$0 + 0 + G\theta_1(0 + 0) \equiv 0$$
$$\frac{\partial^2 w_1}{\partial x^2} + \frac{\partial^2 w_1}{\partial y^2} + G\theta_1(0 + 0) = 0$$

We see that of the three equilibrium conditions, two are satisfied identically, while the third is satisfied if the warping deformations w_1 satisfy the Laplace equation

$$\frac{\partial^2 w_1}{\partial x^2} + \frac{\partial^2 w_1}{\partial y^2} = 0 \tag{12.11}$$

With compatibility satisfied by the assumption of the continuous displacements of Eqs. (12.2) to (12.4), and equilibrium by Eq. (12.11), we next consider the boundary conditions.

We consider a portion of the boundary of the cross section, of length ds, inclined at an arbitrary angle to the X and Y axes, as shown in Fig. 12.3. Because of the stress-free boundary, the resultant of the σ_{ZX} and σ_{ZY} shear stresses on the element of area dA, normal to the boundary, must be zero, or

$$\left[\sigma_{ZX}\left(\frac{dy}{ds}\right) + \sigma_{ZY}\left(\frac{-dx}{ds}\right)\right] dA = 0 \tag{12.12}$$

Lastly, the moment-equilibrium condition about the Z axis demands that, according to Fig. 12.4,

$$\Sigma M_Z = 0: \qquad T = \int(\sigma_{ZX}y - \sigma_{ZY}x)\, dA \tag{12.13}$$

and force equilibrium requires that

$$\Sigma F_X = 0: \qquad \int\sigma_{ZX}\, dA = 0 \tag{12.14}$$

$$\Sigma F_Y = 0: \qquad \int\sigma_{ZY}\, dA = 0 \tag{12.15}$$

It can be shown that conditions (12.14) and (12.15) are identically satisfied by any set of shear stresses which satisfies Eqs. (12.11) and (12.12), so that we must consider only the governing equation (12.11) and the boundary conditions (12.12) and (12.13).

The problem is now completely formulated, but its exact solution is intractable for all but the simplest shapes. For the special case of a circular section, shown in Fig. 12.5, it may be verified that the solution $w_1 = 0$ (no warping), $\sigma_{ZX} = \tau \cos \alpha$, $\sigma_{ZY} = -\tau \sin \alpha$ satisfies all the conditions in Eqs. (12.12) to (12.15), thus proving that the elementary strength of materials solution provides an exact solution of the problem.

Prandtl's stress function To simplify the basic equations of torsion, Prandtl (1903) introduced a new variable called the *torsional stress function* ϕ, defined by

$$\frac{\partial w_1}{\partial x} = \frac{1}{G\theta_1}\frac{\partial \phi}{\partial y} + y \tag{12.16}$$

$$\frac{\partial w_1}{\partial y} = -\frac{1}{G\theta_1}\frac{\partial \phi}{\partial x} - x \tag{12.17}$$

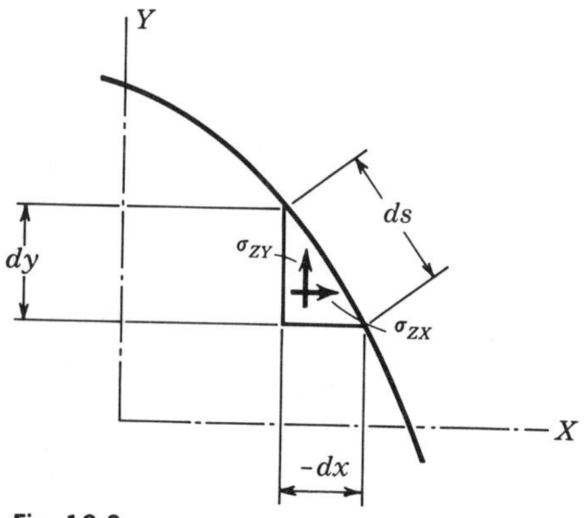

Fig. 12.3

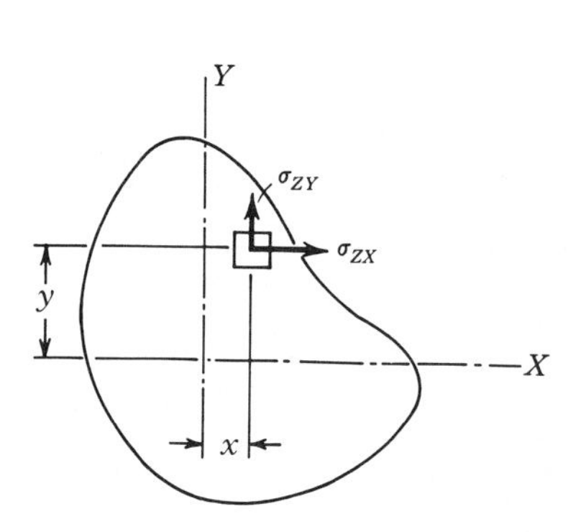

Fig. 12.4

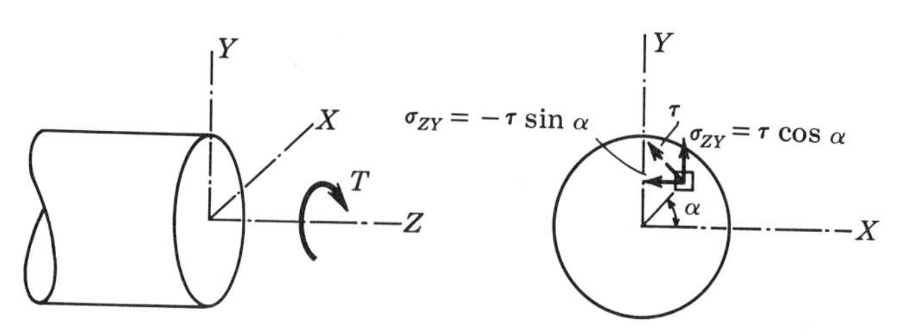

(*a*) Circular torsion member

(*b*) Shear stress

Fig. 12.5

With this definition, we can use Eqs. (12.9) and (12.10) to express the stresses in terms of the stress function as

$$\sigma_{ZX} = +\frac{\partial \phi}{\partial y} \tag{12.18}$$

$$\sigma_{ZY} = -\frac{\partial \phi}{\partial x} \tag{12.19}$$

Thus, the stresses can be obtained by simple differentiation once the stress function ϕ is found. The determination of ϕ therefore corresponds to the solution of the torsion problem.

We next consider the equilibrium equation (12.11) by substituting into it definitions (12.16) and (12.17), to obtain

$$\frac{\partial^2 \phi}{\partial x\, \partial y} - \frac{\partial^2 \phi}{\partial x\, \partial y} \equiv 0 \tag{12.20}$$

The equilibrium conditions are thus identically satisfied for any value of the stress function ϕ, and this provides motivation for adoption of definitions (12.16) and (12.17).

Having checked equilibrium, we ensure compatibility by establishing the condition under which Eqs. (12.16) and (12.17) are satisfied; differentiating (12.16) partially with respect to y, (12.17) with respect to x, and subtracting to eliminate w_1, we get

$$\frac{\partial^2 \phi}{\partial x^2} + \frac{\partial^2 \phi}{\partial y^2} = -2G\theta_1 \tag{12.21}$$

The stress function ϕ must satisfy this governing equation, and also the boundary conditions (12.12) and (12.13); the former of these is written in terms of ϕ by substituting (12.18) and (12.19), yielding

$$\frac{\partial \phi}{\partial y}\frac{dy}{ds} + \frac{\partial \phi}{\partial x}\frac{dx}{ds} = \frac{d\phi}{ds} = 0 \tag{12.22}$$

This means that the variation of ϕ along the boundary is zero, or the stress function ϕ must have a constant value along the boundary of the cross section, which is conveniently assumed zero.

Lastly, by introducing Eqs. (12.18) and (12.19) into Eq. (12.13), applying an integration by parts, and canceling terms, the relation between applied torque and stress function is established as

$$T = 2 \int_{\text{Area}} \phi \, dA \tag{12.23}$$

The stress function is defined by Eqs. (12.21) to (12.23). We can visualize it as a surface extending over the cross section of the bar, as shown in Fig. 12.6. The ordinate of this surface is of magnitude ϕ, its geometry is determined by Eq. (12.21), and Eq. (12.22) demands that its boundary is at a constant elevation, say, $\phi = 0$. Equations (12.18) and (12.19) indicate that at any point of the section the shear stress is equal to the slope of the surface along a line normal to the shear stress. Equation (12.23) states that the applied torque must be equal to two times the volume of the stress function surface.

The introduction of Prandtl's stress function facilitates the solution of some simple problems; considering, for instance, a circular section of radius a, we try, guided by the circular shape of the boundary $r^2 = x^2 + y^2 = a^2$, a stress function of the form

$$\phi = C(x^2 + y^2 - a^2) = C(r^2 - a^2)$$

This function satisfies Eq. (12.21) if $C = -G\theta_1/2$, and also satisfies Eq. (12.22). The torque is calculated by Eq. (12.23):

$$T = 2\left(-\frac{G\theta_1}{2}\right)\int_A (r^2 - a^2)\,dA = G\theta_1\frac{\pi a^4}{2}$$

so that

$$\phi = \frac{T}{\pi a^4}(a^2 - r^2)$$

The tangential shear stress is then $\partial\phi/\partial r$, the rate of change of ϕ normal to the stress direction, or

$$\tau = \frac{T}{\pi a^4}2r = \frac{Tr}{J}$$

and the unit angle of twist is

$$\theta_1 = \frac{T}{G(\pi a^4/2)} = \frac{T}{GJ}$$

as predicted by elementary theory.

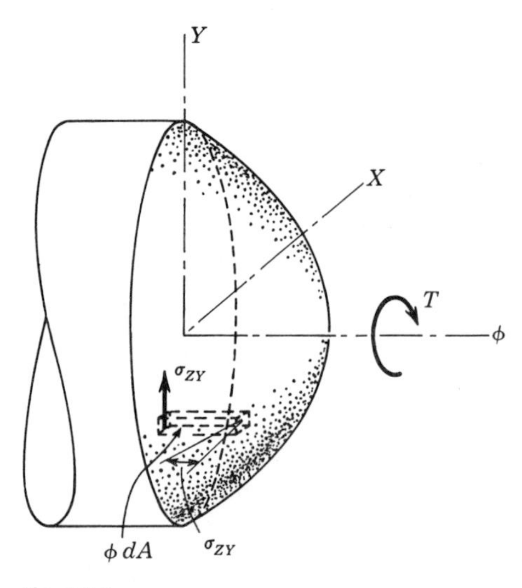

Fig. 12.6

Prandtl's stress function also leads to the useful *membrane analogy*, which is based on the similarity of the governing equation of a shallow membrane (such as a soap film) under uniform pressure with Eq. (12.21). Thus, the experimentally observed behavior of a membrane can be used to predict the behavior of an elastic shaft under torsion. This approach, however, will not be pursued in this text.

12.3 Torsion of Rectangular Cross-section Members and Thin Strips

The important case of elastic torsion members of rectangular cross section, such as the one shown in Fig. 12.7*a*, can be analyzed by solving the partial differential equation (12.21) and the boundary condition (12.22) by assuming a solution of the type

$$\phi = \sum_{n=1,3,5,\ldots} Y_n b_n \cos \frac{n\pi x}{t} \tag{12.24}$$

Here, Y_n is an as yet unknown function of y which represents the variation of the stress function ϕ (shown graphically in Fig. 12.7*b*) parallel to the Y axis. The variation of ϕ parallel to the X axis is represented as the sum of a number of cosine curves, whereby only the odd terms are considered, because those for n even do not satisfy the boundary conditions Eqs. (12.22) at the edges $x = \pm t/2$, which require that $\phi = 0$ there.

The idea is next to express the constant right-hand side of Eq. (12.21) by its cosine series expansion:

$$-2G\theta_1 = -2G\theta_1 \sum_{n=1,3,5,\ldots} \frac{4}{n\pi} (-1)^{(n-1)/2} \cos \frac{n\pi x}{t} \tag{12.25}$$

to substitute the partial derivatives of Eq. (12.24) into the left-hand side, Eq. (12.25) into the right-hand side of (12.21), and then to equate coefficients

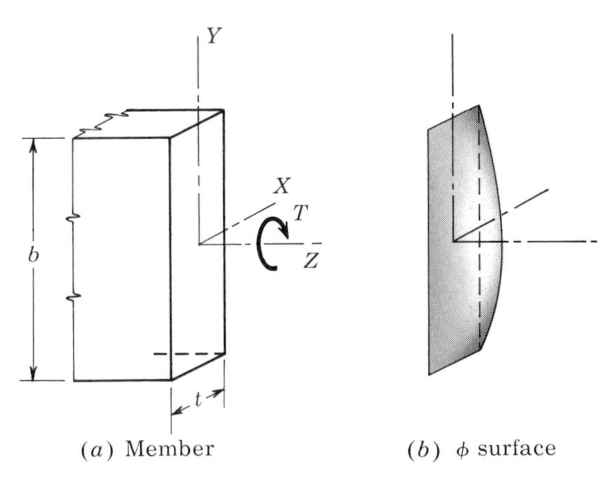

Fig. 12.7

of cosine terms of like order; if we do this for the nth order terms of the series, we get

$$-Y_n\left(\frac{n\pi}{t}\right)^2 b_n \cos\frac{n\pi x}{t} + \frac{\partial^2 Y_n}{\partial y^2} b_n \cos\frac{n\pi x}{t} = 2G\theta_1 \frac{4}{n\pi}(-1)^{(n-1)/2} \cos\frac{n\pi x}{t}$$

from which, canceling $\cos(n\pi x/t)$ and rearranging, we can write the ordinary differential equation for Y_n as

$$\frac{d^2 Y_n}{dx^2} - \left(\frac{n\pi}{t}\right)^2 Y_n = -2G\theta_1 \frac{4}{n\pi b_n}(-1)^{(n-1)/2}$$

This is a second-order, nonhomogeneous equation with constant coefficients whose solution is

$$Y_n = C_1 \operatorname{Sinh}\frac{n\pi y}{t} + C_2 \operatorname{Cosh}\frac{n\pi y}{t} + 8G\theta_1 \frac{t^2}{n^3\pi^3}\frac{1}{b_n}(-1)^{(n-1)/2} \tag{12.26}$$

The constants of integration must be found from the symmetry condition, $dY_n/dy\ (y = 0) = 0$, and the boundary condition, $Y[y = \pm(b/2)] = 0$; using Eq. (12.26) to express these conditions, we obtain

$$C_1 = 0$$

$$C_2 = -8G\theta_1 \frac{t^2}{n^3\pi^3}\frac{1}{b_n}(-1)^{(n-1)/2}\frac{1}{\operatorname{Cosh}(n\pi b/2t)}$$

and Y_n becomes

$$Y_n = 8G\theta_1 \frac{t^2}{n^3\pi^3}\frac{1}{b_n}(-1)^{(n-1)/2}\left[1 - \frac{\operatorname{Cosh}(n\pi y/t)}{\operatorname{Cosh}(n\pi b/2t)}\right] \tag{12.27}$$

and the torsional stress function is found by resubstituting Eq. (12.27) into Eq. (12.24):

$$\phi = 8G\theta_1 \frac{t^2}{\pi^3} \sum_{n=1,3,5,\ldots} \frac{1}{n^3}(-1)^{(n-1)/2}\left[1 - \frac{\operatorname{Cosh}(n\pi y/t)}{\operatorname{Cosh}(n\pi b/2t)}\right]\cos\frac{n\pi x}{t} \tag{12.28}$$

The maximum shear stress acts at the point of, and normal to, the maximum slope of the stress function surface, and Fig. 12.7b shows that this is likely to be at the midpoint of the long edge, that is, $\partial\phi/\partial x$ at $y = 0$, $x = \pm t/2$. Evaluating this slope, we find

$$\tau_{\max} = -\frac{\partial\phi}{\partial x}\left(\frac{t}{2}, 0\right) = 8G\theta_1 \frac{t}{\pi^2} \sum_{n=1,3,5\ldots} \frac{1}{n^2}\left[1 - \frac{1}{\operatorname{Cosh}(n\pi b/2t)}\right] \tag{12.29}$$

The series, Eq. (12.29), converges so rapidly that only a few terms give the stress quite accurately. In particular, if we consider a very oblong thin cross section for which $b \gg t$, then the second term becomes sufficiently small so that it can be neglected compared to unity, in which case

$$\tau_{\max} = 8G\theta_1 \frac{t}{\pi^2} \sum_{n=1,3,5\ldots} \frac{1}{n^2}$$

This series converges to $\pi^2/8$, so that, for a thin strip,

$$\tau_{\max} = G\theta_1 t \tag{12.30}$$

To find the relation between the applied torque T and the resulting twist, we use Eqs. (12.23) and (12.28):

$$\begin{aligned} T &= 16G\theta_1 \frac{t^2}{\pi^3} \sum_{n=1,3,5\ldots} \frac{1}{n^3} \left[\int_{x=-t/2}^{t/2} \cos \frac{n\pi x}{t}\, dx \right] \\ &\qquad \left\{ \int_{y=-b/2}^{b/2} \left[1 - \frac{\operatorname{Cosh}(n\pi y/t)}{\operatorname{Cosh}(n\pi b/2t)} \right] dy \right\} \\ &= 32G\theta_1 \frac{t^2}{\pi^3} \sum_{n=1,3,5\ldots} \frac{1}{n^3} \left(\frac{t}{n\pi} \right) \left[b - \frac{t}{n\pi} \frac{\operatorname{Sinh}(n\pi b/2t)}{\operatorname{Cosh}(n\pi b/2t)} \right] \\ &= 32G\theta_1 \frac{t^3 b}{\pi^4} \sum_{n=1,3,5\ldots} \frac{1}{n^4} \left(1 - \frac{t}{n\pi b} \operatorname{Tanh} \frac{n\pi b}{2t} \right) \end{aligned} \tag{12.31}$$

If we denote a *torsional stiffness factor* K, which depends only on cross-sectional properties, analogous to the polar moment of inertia for circular sections, defined by

$$T = \theta_1 G K \tag{12.32}$$

then, by Eq. (12.31), for rectangular sections,

$$\begin{aligned} K &= 32 \frac{t^3 b}{\pi^4} \sum_{n=1,3,5\ldots} \frac{1}{n^4} \left(1 - \frac{t}{n\pi b} \operatorname{Tanh} \frac{n\pi b}{2t} \right) \\ &= k_1 t^3 b \end{aligned}$$

where k_1 is a numerical factor.

The above series converges sufficiently fast so that only a few terms are necessary to establish the relation between applied torque and the resulting twist. The table on page 374 gives k_1 factors for various ratios of b/t. By combining Eqs. (12.29) and (12.32), it is also possible to write the relation between applied torque and maximum shear stress in the form

$$\frac{T}{\tau_{\max}} = k_2 b t^2 \tag{12.33}$$

where k_2 is again a numerical factor listed in the table for various b/t ratios; note that the quantity $k_2 b t^2$ is analogous to the section modulus in bending of beams.

For the important case of twisting of a thin strip for which $b \gg t$, we can simplify Eq. (12.31). In this case, the second term of Eq. (12.31) can be neglected, and the relation simplifies to

$$T = 32G\theta_1 \frac{t^3 b}{\pi^4} \sum_{n=1,3,5\ldots} \frac{1}{n^4}$$

The series converges to $\pi^4/96$, and

$$T = G\theta_1 \left(\frac{t^3 b}{3}\right) = \theta_1 GK \tag{12.34}$$

For this case, then, the torsional stiffness factor K is $\frac{1}{3}t^3 b$, and the maximum shear stress is related to the applied torque, from Eqs. (12.30) and (12.34), by

$$\tau_{\max} = \frac{Tt}{K} \tag{12.35}$$

Let us compare the stress function surfaces for torsional cross sections of various thin shapes, all of width t and developed overall length b, as shown in Fig. 12.8. If we are willing to overlook stress concentrations, such as at reentrant corners, then we conclude that both volume and maximum slopes of the surfaces are all identical. It follows that the solutions obtained in Eqs. (12.34) and (12.35) are valid not only for thin rectangular sections, but for open thin strip-like cross sections generally.

Torsional constants

$\frac{b}{t}$	$k_1 = \frac{K}{bt^3} = \frac{T}{G\theta_1 bt^3}$	$k_2 = \frac{T}{\tau_{\max} bt^2}$
1.0	0.141	0.208
1.2	0.166	0.219
1.5	0.196	0.231
2.0	0.229	0.246
2.5	0.249	0.258
3.0	0.263	0.267
4.0	0.281	0.282
5.0	0.291	0.291
10.0	0.312	0.312
∞	0.333	0.333

12.4 Inelastic Torsion

We consider a torsion member of arbitrary cross section, of perfectly plastic material of yield stress in shear of value τ_{YP}. The elastic-plastic analysis of this type of problem, that is, the condition under which a portion of the member is still elastic while other portions have already yielded, is a difficult problem in the mathematical theory of plasticity which has so far been solved only partially. Here, we shall consider this stage only in passing, and confine our main efforts to the determination of the ultimate torque, that is, the torque which will produce complete plastification of the section, and therefore failure of the member.

To do this, we marshal our relevant tools of structural mechanics, realizing first that equilibrium and boundary conditions must be satisfied as in the elastic case. We therefore again adopt the concept of the torsional stress function ϕ as defined by Eqs. (12.16) to (12.19). This will ensure satisfaction of internal equilibrium, and, by setting ϕ along the boundary equal to a constant value, say zero, fulfillment of the boundary condition Eq. (12.22). Equation (12.23), relating the applied torque to the stress function, also depends only on equilibrium and is therefore valid.

The only condition which is not valid is Eq. (12.21) which was based on elastic compatibility, which obviously becomes invalid after plastification of the material. Thus, the shape of the stress function surface is not given by the Poisson equation (12.21) and therefore cannot be described by visualizing a smooth surface as in the elastic case.

Rather, we substitute for the elastic relations the "yield condition" which predicts plastification at any point when the maximum shear stress reaches its yield-point value, τ_{YP}:

$$\tau_{\max} = \sqrt{\sigma_{ZX}^2 + \sigma_{ZY}^2} = \tau_{YP} \tag{12.36}$$

Since the maximum shear stress at any point is equal to the maximum rate of change (the "gradient" in the language of vector analysis) of ϕ, that is, the steepest slope of the stress function surface, we conclude that this quantity, according to Eq. (12.36), must have a constant value given by

$$\text{Grad}\ \phi = \sqrt{\left(\frac{\partial \phi}{\partial x}\right)^2 + \left(\frac{\partial \phi}{\partial y}\right)^2} = \tau_{YP} \tag{12.37}$$

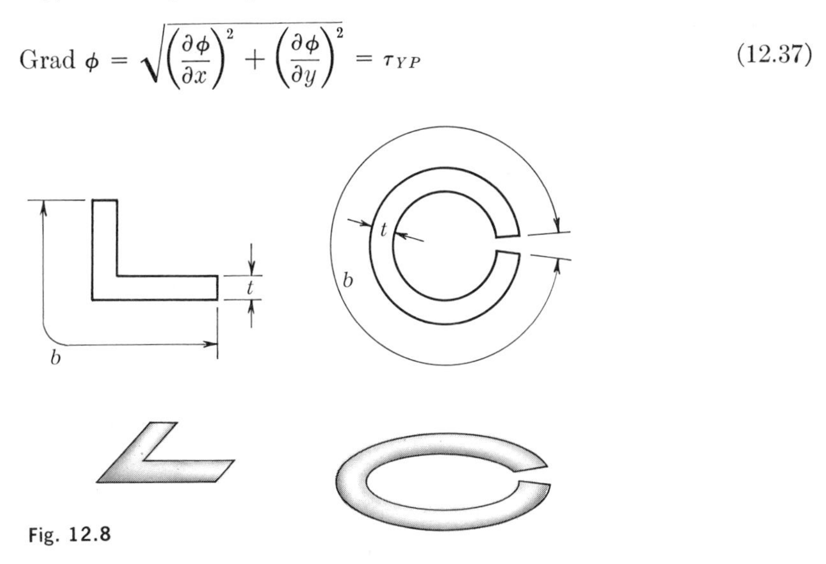

Fig. 12.8

We can now visualize the shape of the ϕ surface over the yielded portion of the cross section. It must consist of a set of planes of slope τ_{YP} which at the same time satisfy boundary condition Eq. (12.22) by having a boundary of constant elevation. This shape is that of a "roof" covering an area of the same plan as the cross section of the member, as shown in Fig. 12.9*b* for rectangular section.

The ultimate torque which produces the complete plastification represented by this ϕ surface is given, according to Eq. (12.23), by two times the volume under this surface. Thus, for the rectangular cross-section shaft of Fig. 12.9 of dimensions b and t $(b > t)$,

$$2 \times \text{volume} = 2 \int_A \phi \, dA = 2 \left[\frac{1}{3} \left(\tau_{YP} \frac{t}{2} \right) t^2 + \frac{1}{2} \left(\tau_{YP} \frac{t}{2} \right) (b - t)t \right]$$

or $$T_{ult} = \frac{\tau_{YP} b t^2}{2} \left(1 - \frac{t}{3b} \right)$$

For a long thin strip in which $b \gg t$,

$$T_{ult} = \frac{\tau_{YP} b t^2}{2}$$

indicating a safety factor against failure at the instant of first yielding [calculating the yield point torque T_{YP} by Eq. (12.35)] of

$$\frac{T_{ult}}{T_{YP}} = \frac{\tau_{YP}(bt^2/2)}{\tau_{YP}(bt^2/3)} = 1.5$$

The intermediate, elastic-plastic stage can be investigated by superimposing the "roof," of slope τ_{YP}, representing the plastic condition, upon

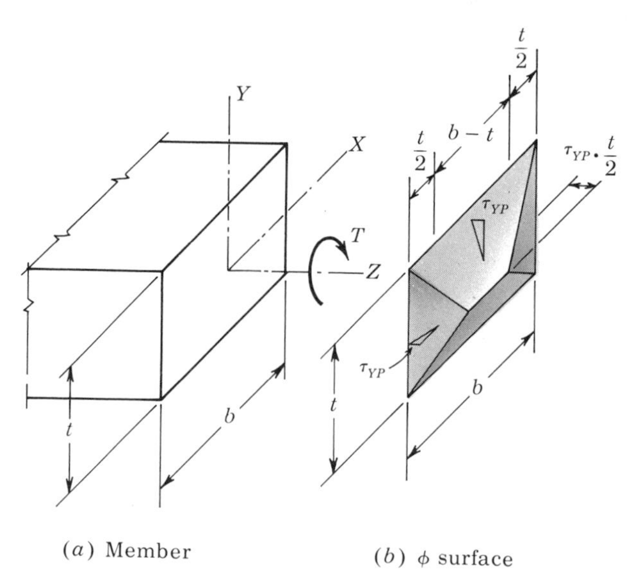

(*a*) Member (*b*) ϕ surface

Fig. 12.9

the elastic ϕ surface. Under increasing load, the ϕ surface will increase in direct proportion to the applied torque; as the slope of the elastic ϕ surface reaches the slope of the roof in some areas, indicating that τ_{max} becomes equal to τ_{YP} over those regions, it is restrained from further distension; thus we can visualize the gradual plastification of the section, till eventually the ϕ surface fills up the roof completely, bringing us to the completely plastic stage represented by the roof. Figure 12.10 shows the rectangular cross-section torsion member at an elastic-plastic stage, indicating elastic and plastic portions of the section.

12.5 Torsion of Closed Thin-walled Sections

We consider a hollow torsion member with thin walls, as shown in Fig. 12.11*a*. In the thin walls, we may reasonably assume that the shear stress τ is uniformly distributed throughout the wall thickness, and is tangential to the wall. In such a member, it is convenient to define the *shear flow* q as the shear stress multiplied by the wall thickness, that is, the shear force in the wall per unit of length:

$$q = \tau t \tag{12.38}$$

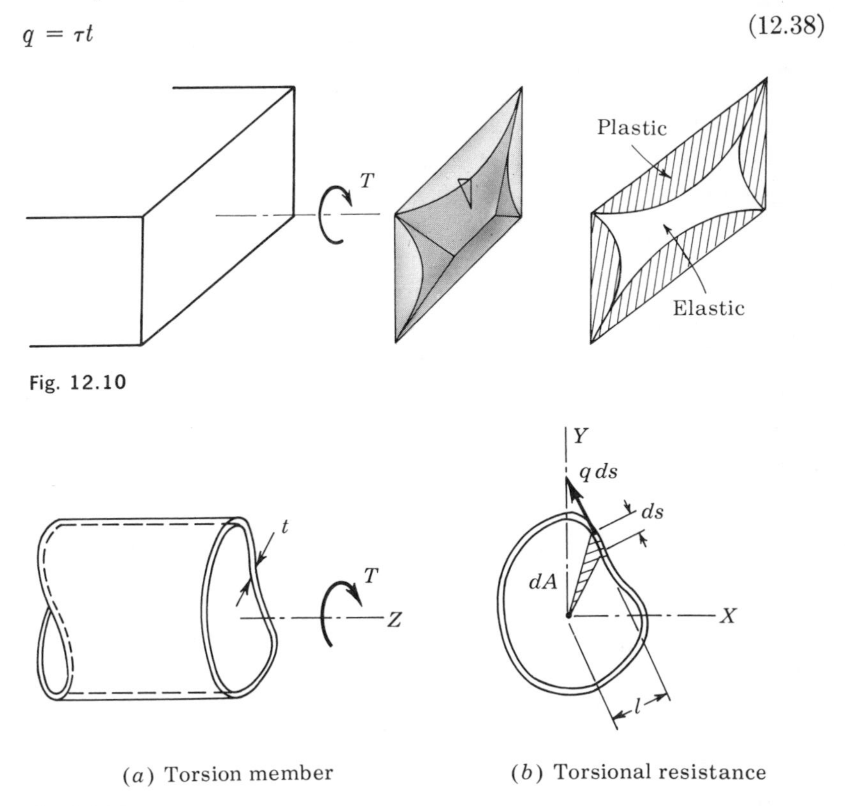

Fig. 12.10

(*a*) Torsion member

(*b*) Torsional resistance

Fig. 12.11

We first prove that in such a section, the shear flow must be constant; consider the element of a variable-thickness wall shown in Fig. 12.12, indicate the shear forces due to torsion, and equilibrate in the Z direction:

$$\Sigma F_Z = 0: \qquad q\,ds - (q + dq)\,ds = 0$$

or $$dq = 0 \tag{12.39}$$

indicating a constant value of q.

Next, we relate the value of shear flow q to the applied torque T by summing applied and resisting moments about the Z axis; the contribution to the moment of the element shown in Fig. 12.11b is $(q\,ds)l$, and summing, we find that

$$\Sigma M_z = 0: \qquad T - \int q\,ds\,l = 0 \tag{12.40}$$

Since q is constant, it may be taken outside the integral, and the remaining integrand can be interpreted as two times the triangular area dA subtended by the line element (shown shaded in Fig. 12.11b); thus

$$l\,ds = 2\,dA \qquad \text{or} \qquad \int l\,ds = 2A$$

so that Eq. (12.40) becomes

$$T = 2Aq \qquad \text{or} \qquad q = \frac{T}{2A} \tag{12.41}$$

where A is the area enclosed by the center line of the wall. With the shear flow found by Eq. (12.41), the shear stress can be determined by definition Eq. (12.38).

The torsional deformation of closed thin-walled members is conveniently calculated using the theorem of virtual work; in Sec. 7.5 one version of this theorem was stated, namely the theorem of virtual displacements, which proved useful in establishing equilibrium equations. Here, another version, the "theorem of virtual forces" is used to calculate deformations. It may be stated as follows:

If a geometrically compatible system is subject to a set of virtual forces, the work done by the external forces is equal to the work done by the internal forces.

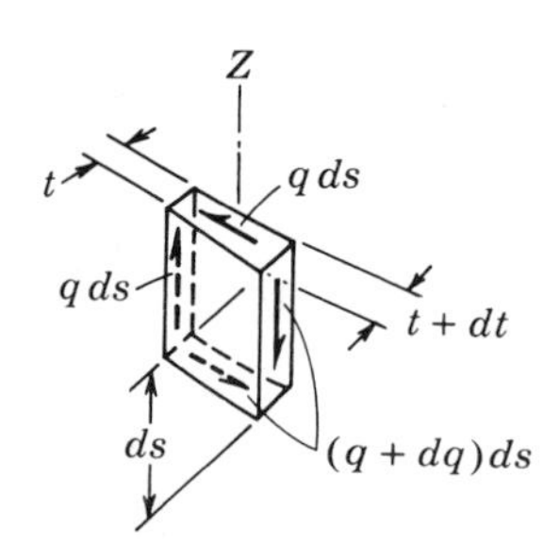

Fig. 12.12

Note that this statement is analogous to the earlier one, but now the displacements are real, and the external and internal forces (which must be in equilibrium with each other) may be assumed arbitrarily.

The real deformation of a typical element of a unit length of the member, as shown in Fig. 12.13, is

$$\frac{\tau}{G}1 = \frac{q}{Gt}$$

and the external deformation is the desired unit angle of twist θ_1. The virtual forces are the external torque T and the internal element shear force $q\,ds$, in equilibrium with T if $q\,ds = T\,ds/2A$, according to Eq. (12.41). The internal virtual work is written by multiplying the virtual element force by the real element deformation, and integrating over the length of the wall; this is equated to the external virtual work:

$$\int \left(\frac{T}{2A}\,ds\right)\left(\frac{q}{Gt}\right) = T\theta_1$$

or $$\theta_1 = \frac{1}{2AG}\int \frac{q\,ds}{t} \tag{12.42}$$

In case q is constant all around, it can be taken out of the integral, and, applying Eq. (12.35), we get

$$\theta_1 = \frac{T}{4A^2G}\int \frac{ds}{t} = \frac{T}{GK} \tag{12.43}$$

where the quantity $K = 4A^2/\int(ds/t)$ is the torsional stiffness factor for the closed thin-walled section. The torsional stiffness is proportional to the square of the enclosed area, which fact can serve the engineer as a guide in designing stiff torsion members.

It is very enlightening to compare the torsional resistance of the open thin-walled sections discussed in Sec. 12.3 with that of the closed thin-walled members of this section. Let us, for instance, compare the shear stresses

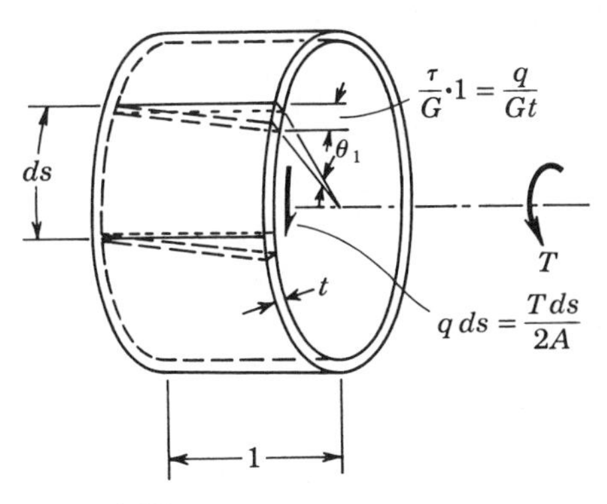

Fig. 12.13

and twisting of two circular sections of radius R, uniform thickness t, as in Fig. 12.14, one slit longitudinally, the other closed, when both are subjected to the same applied torque T.

The maximum shear stress and unit twist in the open section are given by Eqs. (12.35) and (12.34) as

$$\tau_{\max} = \frac{Tc}{K} = \frac{3T}{bt^2} = 0.478\,\frac{T}{Rt^2}$$

$$\theta_1 = \frac{T}{KG} = \frac{3T}{Gbt^3} = 0.478\,\frac{T}{GRt^3}$$

For the closed section,

$$\tau_{\max} = \frac{q}{t} = \frac{T}{2At} = 0.159\,\frac{T}{R^2t}$$

$$\theta_1 = \frac{T}{G4A^2/\int(ds/t)} = 0.159\,\frac{T}{GR^3t}$$

The ratios of the relevant quantities for the open and closed sections are

$$\frac{\tau_{\max,\text{open}}}{\tau_{\max,\text{closed}}} = 3.0\,\frac{R}{t}$$

$$\frac{\theta_{1,\text{open}}}{\theta_{1,\text{closed}}} = 3.0\left(\frac{R}{t}\right)^2$$

Since for thin-walled sections of the type considered here $R \gg t$, we conclude that the closed section has vastly more resistance and stiffness against torsion. This is an important engineering conclusion which should be kept in mind by the designer. Whenever twisting occurs in a structural member, use of closed sections should be considered.

12.6 Torsion of Composite Sections

We consider an elastic torsion member of cross section which can be broken down into elementary shapes, as for instance shown in Fig. 12.15. To analyze this member, we again resort to our elementary tools of mechanics, starting with equilibrium; the total applied torque T is equilibrated by the sum of the individual resisting torques of the individual portions, that is,

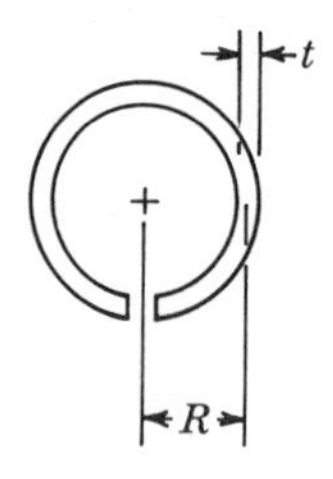

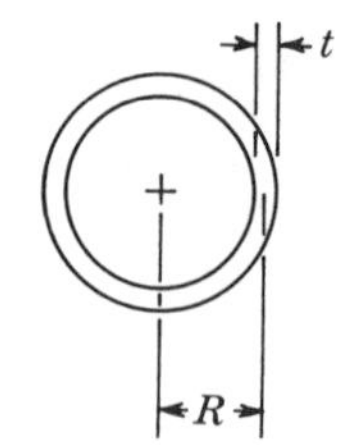

Fig. 12.14

T_i for the ith portion, thus:

$$\Sigma M_z = 0: \qquad T = T_1 + T_2 + \cdots + T_i + \cdots + T_n = \sum_1^n T_i \tag{12.44}$$

Since we have n unknowns and only one equilibrium equation, we must look for geometrical conditions. If the twisted section does not change its shape, then it follows that each individual portion rotates through the same unit angle of twist or

$$\theta_1 = \theta_{1_1} = \theta_{1_2} = \cdots = \theta_{1_i} = \cdots = \theta_{1_n}$$

and relating these angles to the torques by means of the torsional stiffness of each portion,

$$\frac{T}{GK} = \frac{T_1}{GK_1} = \frac{T_2}{GK_2} = \cdots = \frac{T_i}{GK_i} = \cdots \frac{T_n}{GK_n}$$

from which

$$T_i = \frac{K_i}{K} T \tag{12.45}$$

This relation shows that the total torque T applied to the composite section is distributed to the individual parts in direct proportion to their torsional stiffness.

Substituting Eq. (12.45) back into Eq. (12.44), we find that

$$TK = TK_1 + TK_2 + \cdots + TK_i + \cdots + TK_n$$

or $$K = \sum_1^n K_i \tag{12.46}$$

that is, the total torsional stiffness is equal to the sum of the individual stiffnesses of the component parts.

Equation (12.45) enables us to calculate torsional shear stresses; with the torque T_i resisted by the ith component, the shear stresses can be calculated by means appropriate to the portion of the section, as shown in the example problem that follows.

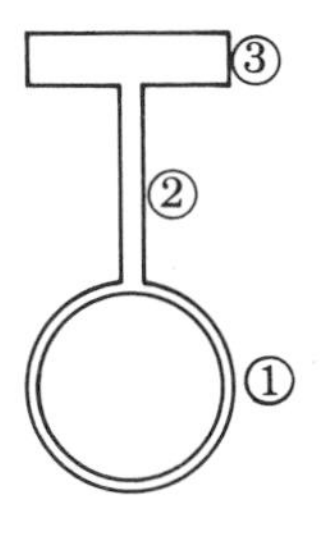

Fig. 12.15

Example Problem 12.1 Calculate the torsional stiffness factor of the section shown. Calculate the maximum shear stress due to an applied torque T.

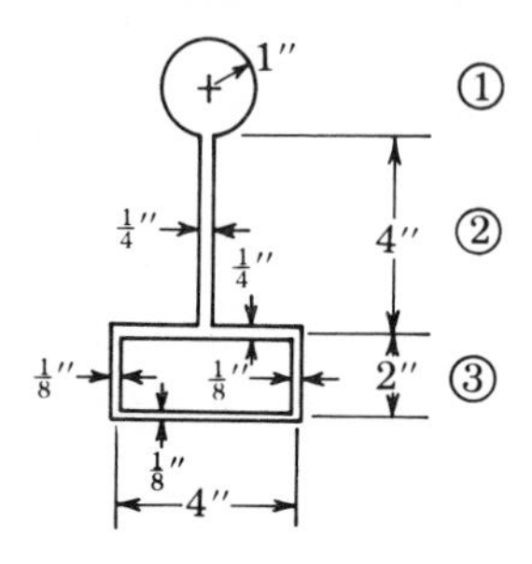

Solution: We separate the section into three components, as shown, and calculate the individual stiffness factors:

$$(1)\qquad K_1 = J = \frac{\pi R^4}{2} = 1.57 \text{ in.}^4$$

$$(2)\qquad K_2 = \tfrac{1}{3}bt^3 = \tfrac{1}{3} \times 4 \text{ in.} \times 0.25^3 = 0.02$$

$$(3)\qquad K_3 = \frac{4A^2}{\int(ds/t)} = \frac{4 \times 8^2}{(4/0.25) + (8/0.125)} = \underline{3.20}$$

$$K = \Sigma K_i = 4.79 \text{ in.}^4$$

The three stiffness factors convey a clear picture of the relative torsional stiffness of different shapes.

To find the torsional shear stress, we first calculate the torque carried by each component:

$$T_1 = \frac{K_1}{K}T = \frac{1.57}{4.79}T = 0.328T$$

$$T_2 = \frac{K_2}{K}T = \frac{0.02}{4.79}T = 0.004T$$

$$T_3 = \frac{K_3}{K}T = \frac{3.20}{4.79}T = \underline{0.668T}$$

$$T = \Sigma T = 1.000T$$

Note that the resisting torques add up to equilibrate the applied torque. We can now calculate the maximum shear stress in each component:

$$\tau_1 = \frac{T_1 R}{J} = \frac{0.328 \times 1}{1.57}T = 0.209T$$

$$\tau_2 = \frac{T_2 \times t}{K_2} = \frac{0.004 \times 0.25}{0.02}T = 0.050T$$

$$\tau_3 = \frac{T_3}{2A \times t} = \frac{0.668}{2 \times 8 \times 0.125}T = 0.334T$$

Because part 3 transmits the major portion of the torque, it has the highest shear stress. But we observe that the ratio of shear stress to torque transmitted is highest in part 2, indicating again the superior performance of closed sections in torsion.

12.7 Torsion of Multicelled Members

A problem which arises frequently in aircraft design and also in the design of box-girder bridges is the analysis of multicelled members in torsion, as shown in Fig. 12.16. Such cross sections are statically indeterminate, which means that along with statics, the compatibility of deformations must be considered, as was already the case in the preceding section.

Before proceeding, we consider the equilibrium of a joint of the section when subjected to shear flows, as in Fig. 12.17. Assuming that the walls of the section adjacent to the joint have shear flows q_1, q_2, and q_3, then by the conjugate shear stress theory there are shear forces of the same magnitude on a unit length of the cut normal planes, acting parallel to the Z axis. Summing these forces, we set

$$\Sigma F_z = 0: \qquad q_1 + q_2 + q_3 = 0$$

which shows that the sum of the shear flows entering a joint must be zero, or in other words, inflow must equal outflow. The analogy with fluid flow explains the term *shear flow*.

Let us now approach the torsional analysis of a multicelled structure as shown in Fig. 12.18. The two cells have areas A_1 and A_2, the wall thicknesses are t_1 to t_3, and the applied torque is T.

We begin by considering each cell by itself, and recall that equilibrium of any wall element demands a constant value of shear flow which we shall call q_1 to q_3 in the three walls. Equilibrium of joint 3 requires that these shear flows are related as discussed; thus, with due regard to sign,

$$q_1 - q_2 - q_3 = 0$$

or $$q_3 = q_1 - q_2$$

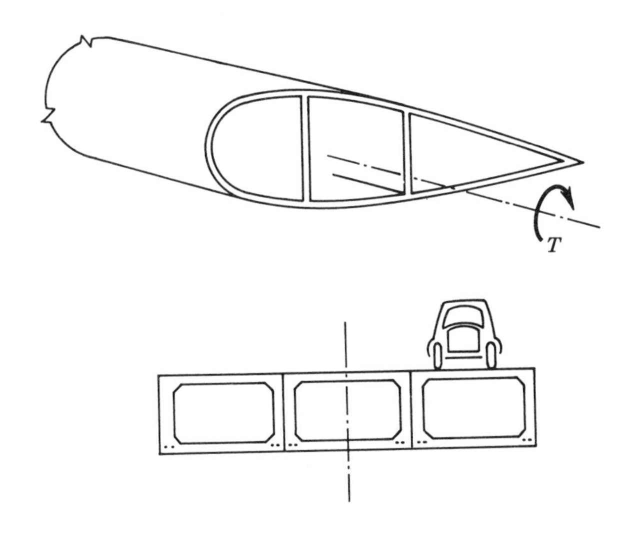

Fig. 12.16

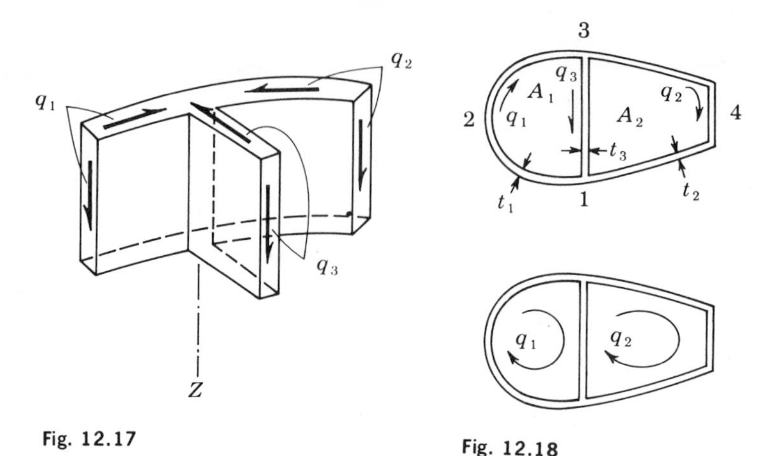

Fig. 12.17

Fig. 12.18

This indicates that the shear flow in the bulkhead 3-1 can be separated into two components, q_1 acting with cell 1, q_2 with cell 2. Now, equating the applied torque to the torsional resistance of the individual cells, we set

$$\Sigma M_z = 0: \qquad T = T_1 + T_2 = 2A_1q_1 + 2A_2q_2$$

With one equation of statics but two unknowns, q_1 and q_2, we resort to the equation of geometry which prescribes that the two cells must twist through the same unit angle; thus, according to Eq. (12.42),

$$\theta_{1_1} = \theta_{1_2}: \qquad \frac{1}{2GA_1}\int \frac{q\,ds}{t} = \frac{1}{2GA_2}\int \frac{q\,ds}{t}$$

or

$$\frac{1}{A_1}\left[\frac{q_1 l_{1-2-3}}{t_1} + \frac{(q_1 - q_2)l_{3-1}}{t_3}\right] = \frac{1}{A_2}\left[\frac{q_2 l_{3-4-1}}{t_2} - \frac{(q_1 - q_2)l_{1-3}}{t_3}\right]$$

With more than two cells, we could write additional joint equilibrium and geometrical conditions, which must then be solved simultaneously with the static condition to determine the unknown shear flows. With known shear flows, the angle of twist can also be found.

Example Problem 12.2 Calculate the unit angle of twist and shear flow in the double-celled section resulting from an applied torque T.

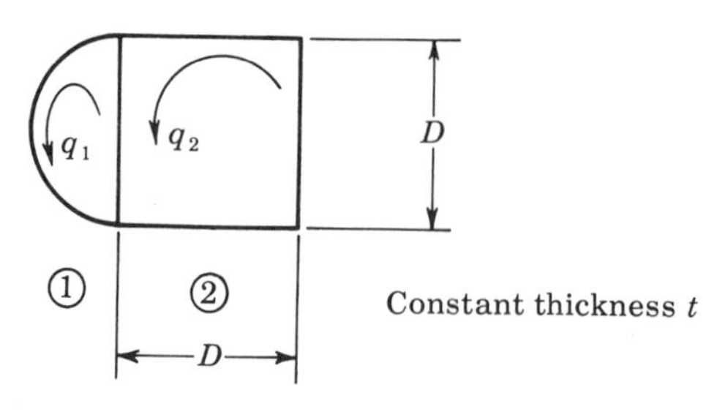

Solution:

Statics: $T = 2A_1q_1 + 2A_2q_2$
$= (0.785q_1 + 2.000q_2)D^2$

Geometry: $\theta_1 = \theta_2$:

$$\frac{1}{G \times 0.785D^2}\left[\frac{1.57D}{t}q_1 + \frac{D}{t}(q_1 - q_2)\right] = \frac{1}{GD^2}\left[\frac{3D}{t}q_2 - \frac{D}{t}(q_1 - q_2)\right]$$

or

$$q_1 = 1.23q_2$$

Resubstituting this into the equilibrium equation, we get

$$q_2 = 0.337\frac{T}{D^2} \qquad q_1 = 0.415\frac{T}{D^2}$$

and, from geometry,

$$\theta_1 = \theta_2 = \theta = 0.933\frac{T}{GD^3t}$$

12.8 Torsion of Plain and Reinforced Concrete Members

The response of plain concrete members to torsion can be predicted conservatively by assuming that the shear stress-strain curve for concrete is perfectly elastic. Under this assumption, the stresses and twist can be predicted by use of the elastic theory according to St. Venant as outlined in Secs. 12.2 and 12.3. The maximum shear stress in a rectangular section is found by Eq. (12.33) as

$$\tau_{\max} = \frac{T}{k_2 \cdot bt^2}$$

where k_2 can be found from the table on page 374 for sections of different proportions.

As in the case of circular sections or those subjected to pure transverse shear (see Sec. 6.7), failure will occur by attainment of a critical diagonal tension stress which for the case of pure shear which prevails here is equal in value to the shear stress, and can be shown by use of Mohr's circle to act along a 45-deg diagonal direction with respect to the member axis. It may thus be expected that the elastic range of the torsional response is abruptly terminated by the onset of a diagonal tension crack under a torsional moment

$$T = f_t' \cdot k_2bt^2 \tag{12.47}$$

where f_t' is the tensile strength of the concrete, which is related to its compressive strength by the approximate relation

$$f_t' = 4\sqrt{f_c'} \qquad \text{in psi}$$

The above analysis of plain concrete sections gives conservative results because any redistribution of shear stresses resulting from plasticity is neglected. Actually, even in a brittle material like concrete, some plastic action may be expected, and, at failure, the stress distribution is probably elastic-plastic, of the type discussed in Sec. 12.4. For specimens of cross sections with reentrant corners, such as I beams of concrete, torsion tests have shown that plastic theory gives better prediction of the ultimate strength than elastic theory.

Once a diagonal tension crack has formed, the torsion section is abruptly destroyed, and immediate brittle failure of the unreinforced member follows. The moment of Eq. (12.47) is therefore the ultimate moment of such sections. The allowable moment might be found by using an allowable rather than an ultimate value for the tensile concrete stress. Such a value is given in ACI sec. 1002 as $v_c = f_{t\,allow} = 1.1\sqrt{f'_c}$ for the case of diagonal tension in unreinforced beam webs, and it might be used here as a suitable value for torsional shear.

To prevent the abrupt failure of the plain concrete section, reinforcement should be provided; the most efficient arrangement of this reinforcing would be along the principal stress trajectories, that is, a 45-deg helix along the lines of principal tensile stress. The resisting torque T_s of such a spiral can be calculated from Fig. 12.19 as

$$\Sigma T = 0: \qquad T_s = 0.707 A_s f_s r \tag{12.48}$$

where r is the radius of the helix.

Forty-five-deg spirals are seldom used as torsional reinforcing because of constructive difficulties, particularly because the possibility of reversal of torque would require two opposing spirals to be used. A more common detail is a combination of longitudinal steel and closed rectangular transverse ties. On account of the diagonal principal stress trajectories, neither longitudinal steel alone nor transverse steel alone can resist torsion, but their combined effect will be to resist separation of the member even after torsional cracking has taken place in the concrete.

We shall consider the behavior of a rectangular-section member reinforced with longitudinal bars and transverse ties under gradually increasing torsion. Till the maximum tensile concrete stress reaches its ultimate value f'_t, strains are small and so are steel stresses. So the initial cracking torque

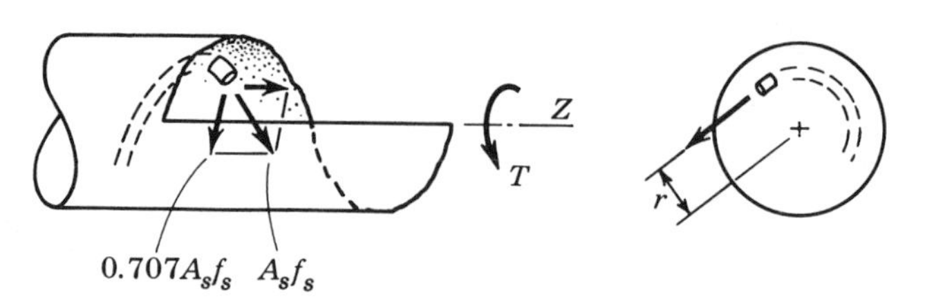

Fig. 12.19

can be calculated by the elastic theory of Sec. 12.3 and Eq. (12.35), neglecting the effect of steel.

After the diagonal concrete cracks open up, starting in the middle of the longer sides of the section, the strain in the steel will increase, and an increasingly larger portion of the total torque will be resisted by the steel. At ultimate torque, spiral cracking has occurred all around the section, and the total torque should be considered logically to be transmitted entirely by the steel; the resisting torque of the steel can be calculated from Fig. 12.20 as

$$\Sigma T = 0: \qquad T_s = \left(A_v \frac{d''}{s} f_v\right) b'' + \left(A_v \frac{b''}{s} f_v\right) d''$$

$$= 2 \frac{Av}{s} f_v b'' d'' \qquad (12.49)$$

where A_v is the area of one transverse bar, b'' and d'' are dimensions (in inches) of the reinforcing cage, and s is the spacing of the ties. The stress f_v in the ties is assumed constant in this analysis, and this appears to be a gross oversimplification. Actually, even after cracking, the steel strains are sufficiently small so as to remain elastic, with stresses in the bars crossing the cracks near corners considerably less than those near mid-depth of the longer sides of the section, where the stress is largest. Based on theory and experiment, an average stress f_v equal to $0.4f_{allow}$ should be inserted into Eq. (12.49) to determine allowable torque, and $0.4f_{YP}$ to determine ultimate torque for torsionally under-reinforced members in which the steel ratio is sufficiently low to enable the steel to yield prior to complete failure of the section.

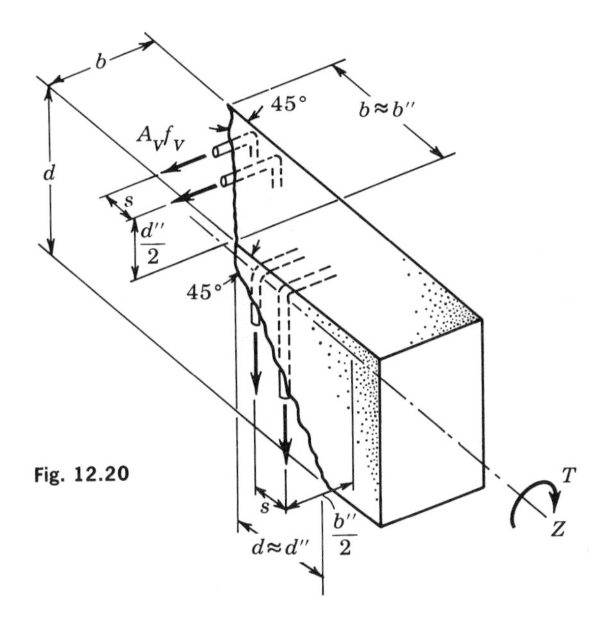

Fig. 12.20

Whether or not this resisting torque of the steel should be added to that of the concrete is an unsolved question, similar to the combination of resisting shears of steel and concrete in a reinforced concrete beam. In view of the fact that the concrete is likely to be cracked at yielding of steel, it would seem logical to calculate the torsional strength as that of the steel alone, when the steel stress is at its yield strength, f_{YP}. In spite of this reasoning, experiments indicate that the actual ultimate moment is best predicted by adding the moment resisted by concrete, given by Eq. (12.47) (or its plastic equivalent), and the moment resisted by the steel, given by Eq. (12.49). Divergences of this type between theory and experiment indicate a lack of understanding of the basic behavior of the reinforced torsion section.

Example Problem 12.3 Design a square torsion section to resist an applied torque of 50 kip-in.; $f'_c = 5.0$ ksi, $f_{s\ allow} = 20$ ksi.

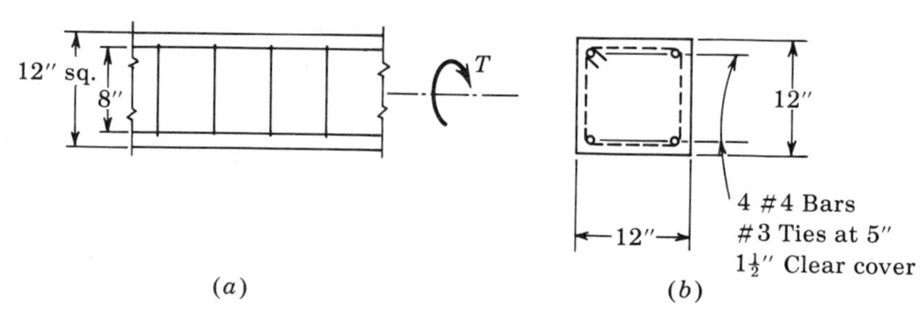

According to the best available information, we shall consider the resisting torque to be the sum of the allowable torque carried by the concrete [Eq. (12.47)], T_c, and that carried by the transverse steel, T_s. Let us try a 12 by 12 in. member, with transverse ties as shown in part *a* of the figure. The allowable diagonal tension stress in the concrete is that given by ACI table 1002*a*, that is

$$v_{c\,allow} = 1.1\sqrt{f'_c} = 78 \text{ psi}$$

Then, using the table on page 374 for $b/t = 1$,

$$T_c = k_2 v_{c\ allow} bt^2 = 0.208 \times 0.078 \times 12^3 = 28.0 \text{ kip-in.}$$

The remaining torque, $T_s = 50 - 28.0 = 22.0$ kip-in., must be carried by the steel, which, assuming No. 3 ties ($A_v = 0.11$ in.²), requires a spacing, according to Eq. (12.49), of

$$s_{reqd} = \frac{2A_v}{T_s} \times 0.4 f_{s\ allow} bd$$

$$= \frac{2 \times 0.11}{22.0} \times 8.0 \times 8^2 = 5.12 \text{ in.}$$

Use No. 3 ties at 5 in.

The 45-deg inclination of the principal stresses requires that an equal amount of longitudinal steel cross the spiral crack; thus,

$$A_{s\ reqd} = \frac{A_v \times 2(b'' + d'')}{s} = \frac{0.11 \text{ in.}^2 \times 32 \text{ in.}}{5 \text{ in.}} = 0.71 \text{ in.}^2$$

Use four No. 4 bars, $A_{s\ furn} = 0.80$ in.^{2}
The final design is as shown in part *b* of the figure above

The ACI Code refers to torsion only in connection with spandrel beams in buildings which must resist the twist caused by the restraining moments of concrete floors framing into them (sec. 921), and only vague provisions are laid down there. In contrast, other codes do spell out torsional provisions in greater detail, and the treatment in this section largely follows some of the approaches used by them.

When concrete members are to be designed to resist torsion primarily, we should keep in mind the conclusions of Sec. 12.5 regarding the superior response to twisting of closed thin-walled sections. Large box-type sections in concrete can be used very advantageously, and for the reinforcing of such members the ideas of this section can be used to advantage. In any case, the diagonal direction of the principal stresses demands equal areas of transverse and longitudinal steel in the section.

12.9 Restrained Torsion

Simple torsion of I-type sections can be handled by the methods of Sec. 12.6. The warping of the section which results can be pictured as shown in Fig. 12.21; this is the deformed shape of the member as predicted by St. Venant's theory of torsion.

Now, if by some means it is possible to prevent the flange from warping, then additional forces are set up which increase the torsional stiffness of the member; the theory which concerns itself with this action is that of *restrained torsion* or *torsion bending*.

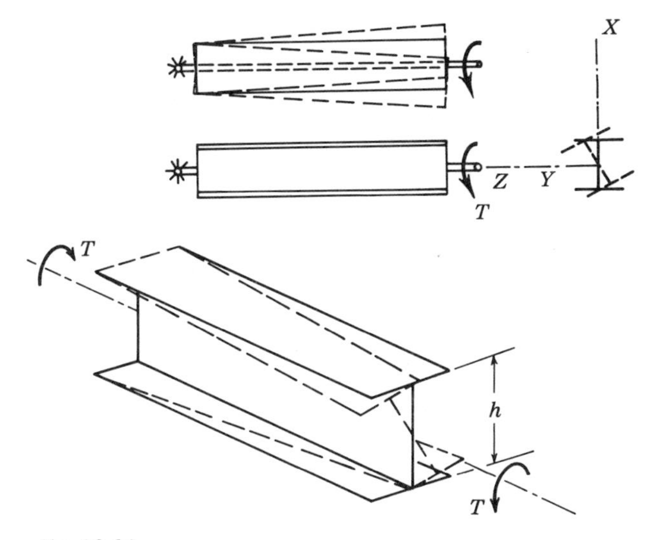

Fig. 12.21

We consider, for instance, the case of a torsion member of I section of length L which is completely fixed at one end, so that all warping at that section is prevented; the other end, to which a torque T is applied, is unrestrained so that it is free to warp. If we consider the deformed shape of this member under twist, we see from Fig. 12.22 that the flanges must bend, thus causing transverse flange shears which are related to their deflections in accordance with ordinary bending theory, that is, using the reference axes and sign convention of Fig. 12.22,

$$Q_F = -\frac{d}{dz}(M_F) = -\frac{d}{dz}\left(EI_F\frac{d^2y}{dz^2}\right) = -EI_F\frac{d^3y}{dz^3} \tag{12.50}$$

Here Q_F, M_F, and I_F are the shear force, bending moment, and moment of inertia of one flange in the YZ plane.

The torsional resisting moment T_F about the Z axis, which is caused by the flange shears, is equal to these forces multiplied by the mean distance between flanges h:

$$T_F = Q_F h = -EI_F h\frac{d^3y}{dz^3} \tag{12.51}$$

This torsional resistance must be added to the resistance resulting from St. Venant torsion T_T, which is proportional to the unit angle of twist

$$\theta_1 = \frac{1}{h/2}\frac{dy}{dz}: \tag{12.52}$$

$$T_T = GK\theta_1 = \frac{2GK}{h}\frac{dy}{dz} \tag{12.53}$$

Equilibrium demands equality of applied and resisting torques:

$$\Sigma M_z = 0: \qquad T = T_F + T_T = -EI_F h\frac{d^3y}{dz^3} + \frac{2GK}{h}\frac{dy}{dz}$$

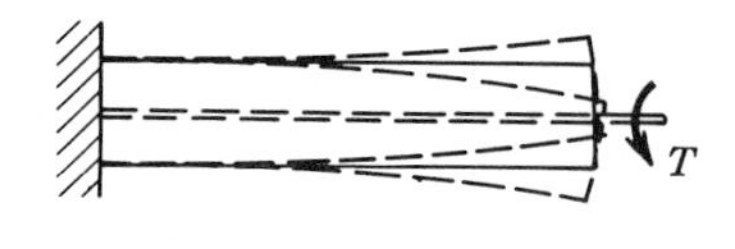

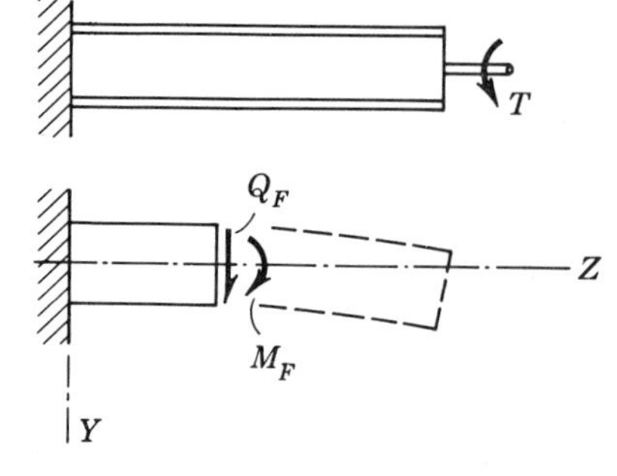

Fig. 12.22

or, setting, for conciseness, $2GK/h^2EI_F \equiv a^2$

$$\frac{d^3y}{dz^3} - a^2\frac{dy}{dz} = -\frac{1}{EI_Fh}T \tag{12.54}$$

With three boundary conditions specified, this third-order, nonhomogeneous, ordinary differential equation with constant coefficients can be solved for the lateral flange deflection y, from which we can determine, in turn, the unit angle of twist θ_1 by Eq. (12.52), the St. Venant torque T_T by Eq. (12.53), the flange bending torque T_F by Eq. (12.51), and the flange shear Q_F and the flange moment M_F by Eq. (12.50). With these stress resultants known, conventional torsion and bending theory is used to find the stresses, whereby we must pay attention to use proper superposition of the shea stresses caused by St. Venant torsion and flange bending.

Let us continue the torsion bending analysis of the beam described earlier. The complementary solution of Eq. (12.54) is

$$y_c = C_1 \operatorname{Sinh} az + C_2 \operatorname{Cosh} az + C_3$$

Since T is constant along the beam, the particular integral is

$$y_p = \frac{1}{EI_Fh}Tz$$

so that the total solution is

$$y = C_1 \operatorname{Sinh} az + C_2 \operatorname{Cosh} az + C_3 + \frac{1}{a^2EI_Fh}Tz \tag{12.55}$$

where the constants of integration are determined by the boundary conditions

$$y(0) = 0 \qquad :C_1\,0 + C_2 1 + C_3 + 0 = 0$$

$$\frac{dy}{dz}(0) = 0 \qquad :C_1\,a\,1 + C_2 0 + 0 + \frac{T}{a^2EI_Fh} = 0$$

$$M_F(L) = EI_F\frac{d^2y}{dz^2}(L) = 0: C_1a^2 \operatorname{Sinh} aL + C_2a^2 \operatorname{Cosh} aL + 0 + 0 = 0$$

These three equations are solved simultaneously to yield

$$C_1 = \frac{1}{a^3}\frac{T}{EI_Fh}$$

$$C_2 = +\frac{1}{a^3}\frac{T}{EI_Fh}\operatorname{Tanh} aL$$

$$C_3 = -\frac{1}{a^3}\frac{T}{EI_Fh}\operatorname{Tanh} aL$$

and the flange deflection is obtained by resubstituting these into Eq. (12.55) as

$$y = [az - \operatorname{Sinh} az + \operatorname{Tanh} aL\,(\operatorname{Cosh} az - 1)]\frac{1}{a^3}\frac{T}{EI_Fh} \tag{12.56}$$

In order to calculate the St. Venant torsion and the flange bending and shear stresses, we need the first, second, and third derivatives of y:

$$\frac{dy}{dz} = (1 - \text{Cosh } az + \text{Tanh } aL \text{ Sinh } az)\frac{1}{a^2}\frac{T}{EI_Fh}$$

$$\frac{d^2y}{dz^2} = (-\text{ Sinh } az + \text{Tanh } aL \text{ Cosh } az)\frac{1}{a}\frac{T}{EI_Fh}$$

$$\frac{d^3y}{dz^3} = (-\text{ Cosh } az + \text{Tanh } aL \text{ Sinh } az)\frac{T}{EI_Fh}$$

For further calculations, let us consider that the torsion member in question is a 10 WF 49 beam, of length $L = 72$ in. For this beam, $I_F = 46.5$ in.^{4}, $K = 1.39$ in.^{4}, the ratio E/G for steel is 2.5, and the mean flange distance is 9.442 in. The section modulus of one flange is 9.3 in.^{3} and its cross-sectional area is 5.58 in.^{2} The torsional stiffness factor of one flange is $\frac{1}{3}bt^3 = 0.58$ in.^{4} With this, we calculate

$$a^2 = \frac{2KG}{h^2EI_F} = \frac{1}{3{,}740} \qquad a = \frac{1}{61.1 \text{ in.}}$$

$$\text{Tanh } aL = \text{Tanh } 1.178 = 0.827$$

$$\frac{1}{a^2EI_Fh} = 2.84 \times 10^{-4}$$

We first evaluate y by Eq. (12.56) at $z = 72$ in., that is, at the free end, and divide the value by $h/2$ to get the maximum angle of twist:

$$\theta(z = 72 \text{ in.}) = \frac{2}{h}\,y(z = 72 \text{ in.}) = 1.27 \times 10^{-3}T \text{ rad}$$

It is instructive to compare this with the total angle of twist had complete warping been permitted, in which case the angle would have been determined by simple torsion theory as

$$\theta(z = 72 \text{ in.}) = \frac{TL}{KG} = 4.31 \times 10^{-3}T \text{ rad}$$

We see that owing to the warping restraint at one end only the torsional deformation is only 29 percent of what it would have been otherwise. However, we should be aware that the flange bending is essentially an end effect; it becomes less pronounced in longer members. If, for instance, the beam has a length $L = 12$ ft $= 144$ in., then the maximum angle of twist at the free end in restrained torsion is

$$\theta = 5.03 \times 10^{-3}T \text{ rad}$$

whereas in simple torsion the maximum twist is twice that of the shorter member, or

$$\theta = 8.62 \times 10^{-3}T \text{ rad}$$

Here, the warping restraint reduces the torsional deformation to 58 percent of its value in simple torsion. For yet longer members, the effect would gradually become negligible.

We now go back to the shorter member and compute some of the stresses resulting from the restrained torsion. Let us first consider the normal stresses resulting from bending of the flanges. The maximum moment arises at the fixed end $x = 0$, where

$$M_F = EI_F \frac{d^2y}{dz^2} \, (x = 0) = 5.37T \text{ kip-in.}$$

and because of this the flange bending stress in the extreme flange fiber is

$$\sigma_F = \frac{M_F}{S_F} = \frac{5.37T}{9.3} = 0.577T \text{ ksi}$$

The maximum flange shear is also at $x = 0$:

$$Q_F = -EI_F \frac{d^3y}{dz^3} \, (z = 0) = 10.6 \times 10^{-2}T \text{ kips}$$

and the resulting flange shear stress occurs in the middle of the rectangular section, of value

$$\tau_F = \frac{3}{2}\frac{Q_F}{A_F} = 2.86 \times 10^{-2}T \text{ ksi}$$

The St. Venant part of the torsion at the fixed end is zero because warping is prevented there, and consequently

$$\frac{dy}{dz} = 0$$

At the free end $x = L$, the flange bending moment is zero, while the torsion is resisted by a combination of St. Venant and restrained torsion. The former is found from Eq. (12.53) as

$$T_T = \frac{2GK}{h}\frac{dy}{dz} \, (z = 72 \text{ in.}) = 0.44T \text{ kip-in.}$$

and the latter from Eq. (12.51) as

$$T_F = -EI_F h \frac{d^3y}{dz^3} \, (z = 72 \text{ in.}) = 0.56T \text{ kip-in.}$$

We see that the sum of the two resistances equilibrates the applied torque T.

At the free end, the flange shear stresses due to the two kinds of torsional resistance combine; the maximum shear stress due to the St. Venant part occurs at the center of the long side of the flange, and has the value

$$\tau_T = \left(\frac{K_F}{K_{total}} 0.44T\right) \frac{t}{\frac{1}{3}bt^3}$$
$$= \left(\frac{0.58}{1.39} \times 0.44T\right) \frac{0.558}{0.58} = 0.177T \text{ ksi}$$

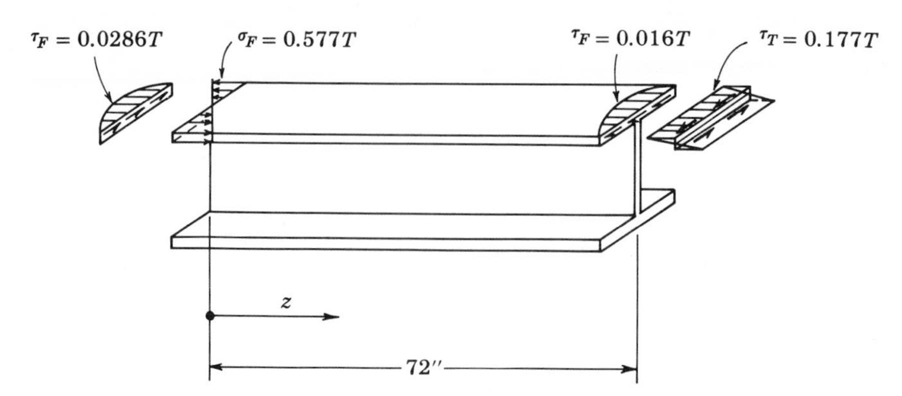

Fig. 12.23

The shear stress due to flange bending has its maximum at the same point, and is calculated as before:

$$\tau_F = \frac{3}{2}\frac{Q_F}{A_F} = \frac{3}{2}\frac{T_F}{h}\frac{1}{A_F} = 0.016T \text{ ksi}$$

Since these shear stresses are concurrent, they can be added to give the critical shear stress as

$$\tau_{\max} = \tau_T + \tau_F = (0.177 + 0.016)T = 0.193T \text{ ksi}$$

The calculated stresses are shown in Fig. 12.23.

We note that the St. Venant torsional shear stress is predominant. Had the member been unrestrained against warping, then all of the torque would have been resisted by St. Venant torsion, with a resulting maximum torsional shear stress of

$$\tau_{\max} = \frac{1.00}{0.44} \times 0.177T = 0.402T \text{ ksi}$$

It is apparent that both shear stress and deformation are considerably reduced by providing suitable torsional support conditions.

General Readings

A simple but complete derivation of the equations of elastic torsion is contained in the book by Murphy cited in Chap. 2. A more thorough presentation is found in chap. 11 of Timoshenko and Goodier, also referred to in Chap. 2. The material of Sec. 12.3 is from sec. 95 of that book.

For the use of the membrane analogy for the solution of torsion problems, see Murphy's book, or the following:

Timoshenko, S. P.: "Strength of Materials," 3d ed., part II, Van Nostrand, 1956. This book contains a wealth of information on this and other topics, including inelastic torsion, torsion of thin-walled and composite sections, and restrained torsion.

Torsion of reinforced concrete sections seems to be bypassed in most texts on reinforced concrete. One exception is the book by Winter et al., which covers this topic lightly in appendix C. A review of torsion provisions of foreign codes is contained in

Fisher, Gordon P., and Paul Zia: "Review of Code Requirements for Torsion Design," *J. ACI*, vol. 61, p. 1, January, 1964.

Problems

12.1 Prove that Eqs. (12.14) and (12.15) are identically satisfied by a set of shear stresses which satisfies Eqs. (12.11) and (12.12).

12.2 By carrying out the indicated steps, verify Eq. (12.23).

12.3 Consider an elastic torsion member of elliptical cross section of boundary $(x/a)^2 + (y/b)^2 = 1$. Verify that all necessary conditions for this member are satisfied by the stress function

$$\phi = C\left[\left(\frac{x}{a}\right)^2 + \left(\frac{y}{b}\right)^2 - 1\right]$$

and find the value of the factor C in terms of the unit angle of twist θ_1. By suitable integration, also find C and the maximum shear stress as function of the applied torque T.

12.4 Evaluate the torsional stiffness factor for a square section. Take sufficient terms of the series to obtain an answer within 10 percent, and compare your result with that in the table of torsional constants, page 374. Discuss the rapidity of convergence of the series.

12.5 Calculate the maximum shear stress in an elastic torsion member of square cross section subjected to an applied torque T. Take three terms of the series and compare result with that given in the table of torsional constants. Draw conclusions regarding the convergence of the series.

12.6 From Eq. (12.28) obtain a general expression for the τ_{yz} component of shear stress, valid over the entire cross section. Represent the variation of this shear stress in an appropriate graphical fashion for the special case of a square section. Note that because of symmetry only one quadrant of the section must be considered.

12.7 Using the results of Prob. 12.6 and symmetry, also obtain the τ_{xz} shear stress components, and combine τ_{xz} and τ_{yz} components to obtain magnitude and direction of the resultant shear stresses over one quadrant of a square torsion section. Plot the directions of the resultant shear stresses, and get a feel for the way in which such a section resists an applied torque.

12.8 Find the torsional stiffness factor for a $3 \times 3 \times \frac{1}{4}$ in. angle section. Calculate the maximum shear stress and the unit angle of twist due to an applied torque T. Comment on possible effects of stress concentration.

12.9 The member of cross section shown is built up of corrugated steel sheet, 0.1 in. thick. The top and bottom sheets were 36 in. long before corrugation, the side sheet

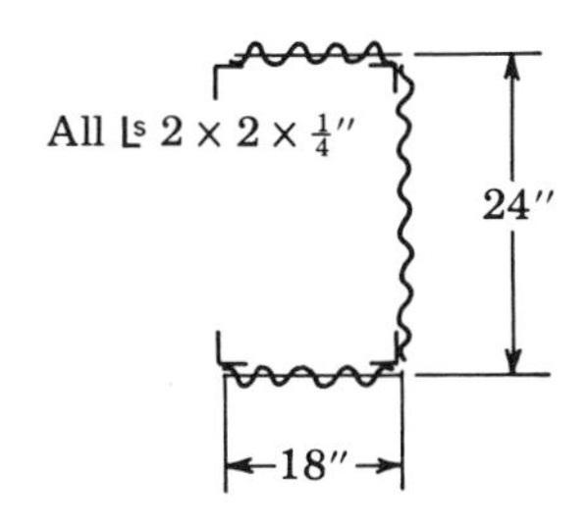

48 in. long. If the allowable maximum shear stress is 10 ksi, find the allowable torque. Find the required strength of the weld in kips/in.

12.10 Find the torsional stiffness factor of a 12 WF 27 wide-flange shape. Neglect the effect of fillets in your calculations, but comment on their probable effect.

12.11 Calculate the ultimate torque on the perfectly plastic section shown. Yield stress in shear is τ_{YP}.

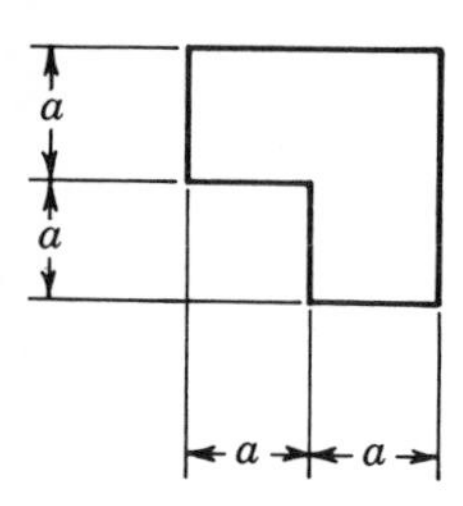

12.12 A circular torsion section, of radius R, is made of elastic-plastic material of elastic shear stiffness G, yield shear stress τ_{YP}. Under a certain torque T, the elastic-plastic boundary is at a radius r. For this condition, sketch the ϕ surface and describe it mathematically. Find the value of the applied torque T and the unit angle of twist θ_1 as a function of r, and plot T versus θ_1 for all values of T up to failure.

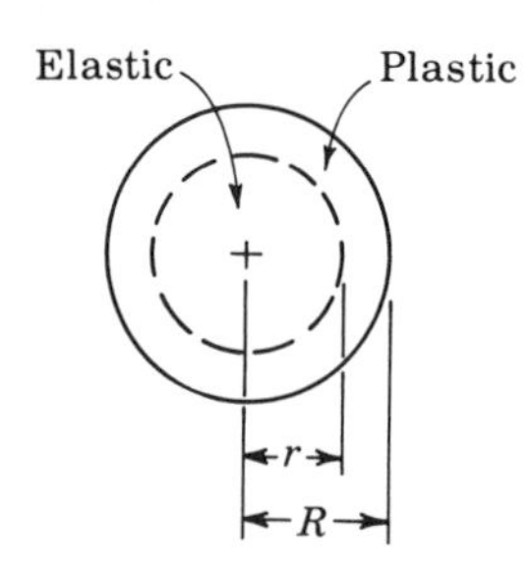

12.13 Repeat Prob. 12.12, this time using only an elementary strength of materials approach. Compare results and draw conclusions regarding the appropriateness of the elementary approach for elastic-plastic torsion of circular sections.

12.14 (*a*) Calculate the maximum shear stress resulting from a torque of 1,000 in.-lb.
(*b*) Calculate the unit angle of twist resulting from 1,000 in.-lb.

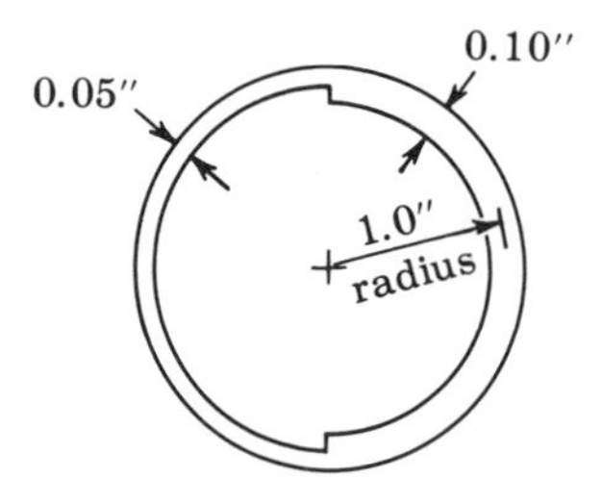

12.15 Consider the corrugated section of Prob. 12.9, and calculate the torsional strength and stiffness of the section if it is closed off by a second side sheet. Compare strength and stiffness of closed and open sections, and draw engineering conclusions.

12.16 Compare the torsional rigidity and strength of the two precast concrete sections, assuming elastic behavior. Draw conclusions regarding shape of section when torsion is involved.

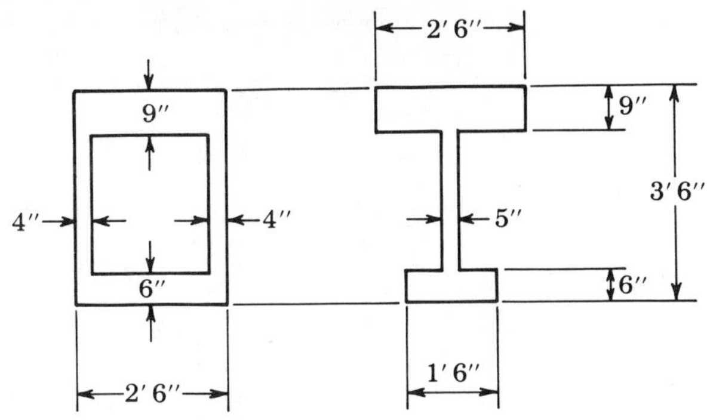

12.17 A torsion member is to be fabricated of two 12 ⊏ 20.7. Compare torsional strength and rigidity of the two alternate arrangements shown. If the maximum torsional shear stress is 10 ksi, calculate the allowable torque on both sections. Calculate the required size of the continuous longitudinal weld for the box section.

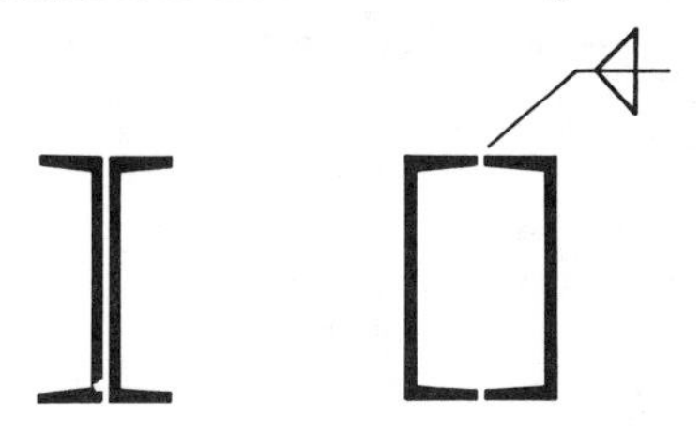

12.18 Find the relation between the applied torque and resulting angle change for the torsion section shown. Modulus of rigidity is G.

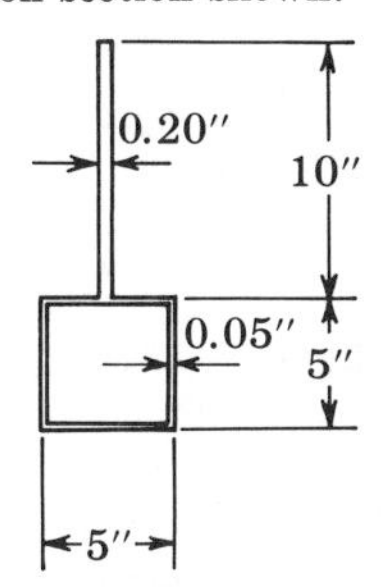

12.19 Find the maximum torque which may be applied to the section shown if the maximum shear stress may not exceed 10 ksi and if the unit angle of twist may not exceed 0.6×10^{-3}. $G = 12 \times 10^3$ ksi.

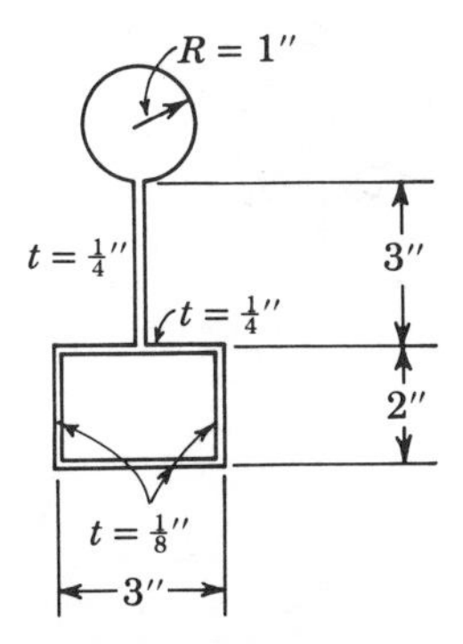

12.20 What is the maximum shear stress in the cross section shown resulting from a twist T?

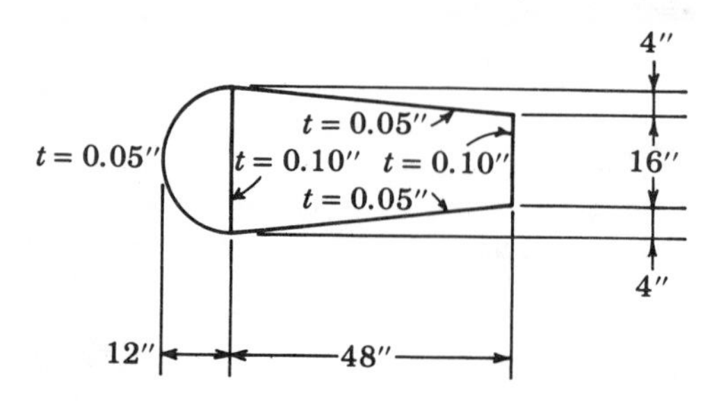

12.21 A deck for a long-span bridge which may be subject to torsion is to be designed. Two alternate designs are contemplated, one consisting of box sections, the other of precast beams carrying a slab.

With the indicated dimensions, compare the torsional stiffness of the two alternates, as well as the torsional strength, assuming elastic behavior. Draw conclusions regarding the sections.

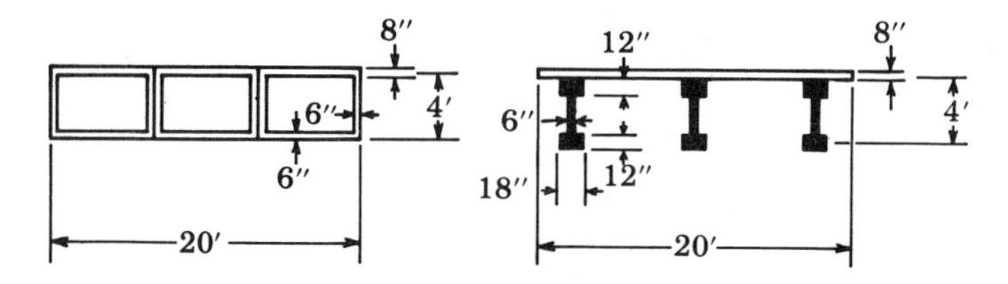

12.22 Design a 12 by 24 in. rectangular concrete member to resist a torque of 200 in.-kips, $f'_c = 3.0$ ksi, $f_{YP} = 40$ ksi. Provide steel as required, and furnish drawing.

12.23 According to your best information, calculate the safety factor against failure of the section designed in Prob. 12.22.

12.24 Design an appropriate box section reinforced concrete member to resist an applied torque of 1,000 kip-ft; $f'_c = 5.0$ ksi, $f_{YP} = 40$ ksi. Select section and reinforcing, and make design drawing.

12.25 A heat-exchanger vessel in a refinery rests on a concrete foundation shown. To remove the coils of the vessel, a force of 20 kips may have to be applied as shown. Design the foundation beams to resist the torque resulting from this pulling force only. $f'_c = 4.0$ ksi; $f_s = 20$ ksi.

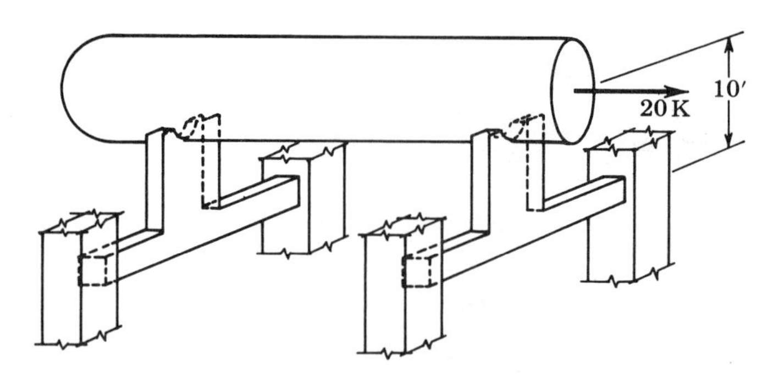

12.26 The vessel of Prob. 12.25 weighs 40 kips full, 18 kips empty. Design the foundation beams to resist any critical combination of bending and torque resulting from gravity and pulling loads. Clearly outline your reasoning and procedure.

12.27 A square-cross-section plain reinforced concrete member is subjected simultaneously to a torque T and an axial prestressing force P.

Write an expression for the principal stresses and make predictions on possible crack formation. Discuss the effect of axial prestressing on plain concrete members.

12.28 Find the values of the constants of integration for the torsion-bending equation for an I beam with two fixed ends, one of which is rotating with respect to the other through an angle θ about its Z axis. Write an equation for the deformed shape of the flange of the beam.

12.29 Write the equation of the deflected shape of the flange for a cantilever beam (one end fixed against torsion and bending, the other free) under a uniform load w applied at an eccentricity e along the full length of the member.

12.30 The 8 WF 31 beam is welded to rigid supports at both ends, so that the ends are perfectly restrained against any motion. It supports the eccentric load as shown. *Required:* To find the locations and values of the critical tensile, compressive, and shear stresses.

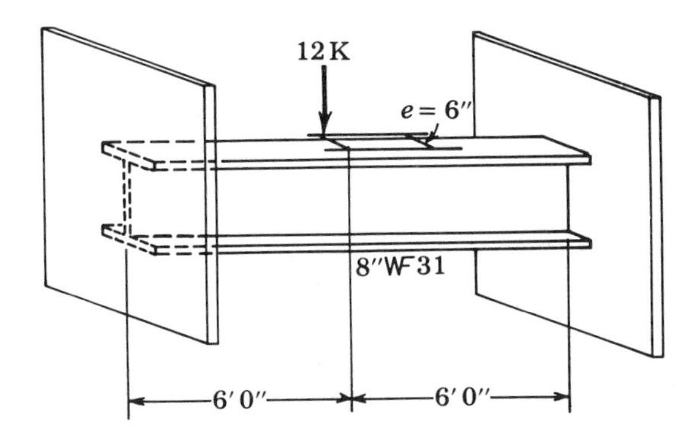

12.31 A 16 WF 36 beam is used as a cantilever as shown. A load of 10 kips is supported at an eccentricity of 3 in. from the free end.

(*a*) Find the largest normal stress in the flanges of the beam.
(*b*) Find the largest shear stress in the web of the beam.
(*c*) Find the total angle of twist of the free end.

Indicate the points where the stresses found in (*a*) and (*b*) act.

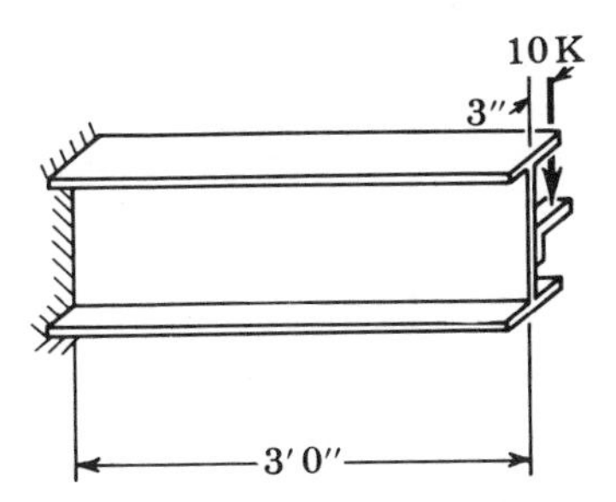

12.32 A 10 WF 49 beam, 6 ft long, is used as a shaft to resist an applied torque T. The far end of the shaft is torsionally fixed. Two alternatives are contemplated for the near end:

(1) the end is torsionally free
(2) the end is boxed in by means of welded plates so as to prevent the flanges from warping with respect to each other.

(*a*) For each case, calculate the angle of twist and the maximum flange stresses. Draw engineering conclusions regarding the effectiveness of the torsional end restraint.
(*b*) Discuss the design of the box at the near end to prevent flange warping.

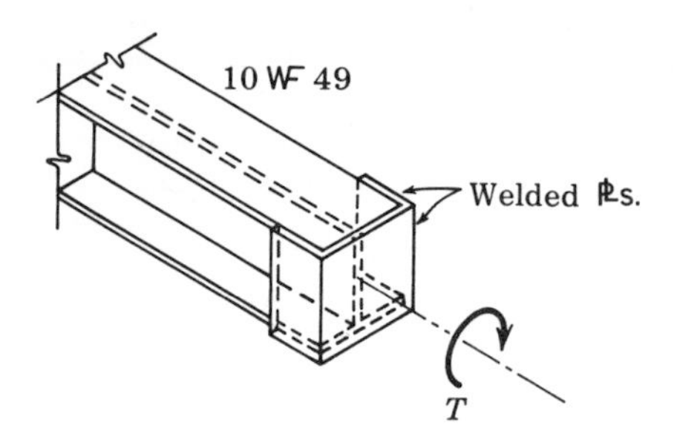

12.33 An elastic slab is subjected to a torque T, as shown. Assuming that a transverse section of the slab remains straight as the slab twists, derive a differential equation for the angle of twist at any section. *Reference:* Torsional Rigidity of Rectangular Slabs, *J. ACI*, page 241, vol. 25, no. 3, November, 1953.

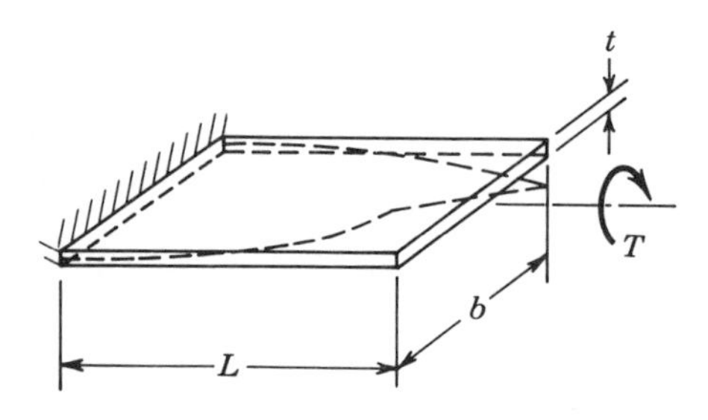

Notation

A	cross-sectional area of member, in.^{2}
A_G	gross area of concrete member, in.^{2}
A_c	cross-sectional area of effective section of concrete, in.^{2}
A_s	tensile steel area in concrete member, in.^{2}
A_s'	compressive steel area in concrete member, in.^{2}
A_T	area of transformed section, in.^{2}
a	$= f_{s\ allow}\, j_{bal}$ = balanced design constant, ksi; depth of compression block in ultimate concrete design, in.
b	width of member cross section, in.
C	internal resultant compressive force, kips
C_c	L/r ratio corresponding to proportional limit stress
c	distance from extreme fiber to neutral axis, in.
D	diameter of reinforcing bar, in.
d	depth of member cross section; effective depth of concrete member, in.
d'	distance from extreme fiber to compression steel, in.
E	modulus of elasticity, ksi
E_T	tangent modulus of stress-strain curve, ksi
e	$= M/P$ = eccentricity, in.; distance to shear center, in.
F	force, kips; allowable stress, ksi
F_A	allowable axial stress, ksi
F_B	allowable bending stress, ksi
F_{YP}	yield stress in steel, ksi
f	stress, ksi; shape factor for elastic plastic beam $= M_p/M_{YP}$
k	effective column length coefficient; ratio of distance to neutral axis to beam depth; stiffness factor
L	length of member, ft; anchorage length of reinforcing bar, in.
M	moment, kip-ft or kip-in.
M_p	fully plastic moment of beam, kip-ft or kip-in.
M_{YP}	moment corresponding to first yielding of extreme fiber, kip-ft
n	modular ratio of nonhomogeneous members; material constant
P	axial force on member, kips
P_{cr}	buckling load on column, kips
P_E	Euler buckling load on ideal column, kips
P_p	fully plastic, or collapse load on structure, kips
p	$= A_s/bd$ = steel ratio
p'	$= A_s'/bd$ = compressive steel ratio
p_G	$= A_s/A_G$ = steel ratio of concrete column
p_s	volume of spiral steel/volume of concrete = spiral steel ratio in concrete column
r	radius of gyration of section, in.

S	$= I/c =$ section modulus, in.^{3}
s	spacing (of bolts, joists, stirrups, etc.), in.
T	torsional moment, kip-in; internal resultant tensile force, kips
t	member thickness, in.
U	virtual work, kip-in.
u, v, w	coordinates along principal axes; displacements along X, Y, Z axes, in.
u	bond stress, ksi
V	shear force in beam, kips
w	distributed load on beam, kips/ft; unit weight of concrete, lb/ft^3; web thickness of I beam, in.
x, y, z	coordinates, in.
y	lever arm from neutral axis in beam; transverse deflection of beam, in.
Z	plastic section modulus, in.^{3}
γ	shear strain
Δ	total deformation or deflection, in.
δ	tangential deviation of deflected beam axis, in.
ϵ	normal strain
ϵ_c	extreme fiber strain in beam
ϵ_{ult}	compressive crushing strain of concrete
ϵ_{YP}	yield strain
θ	slope of deflected beam; rotation of plastic hinge; total angle change between two beam sections
Σ_0	perimeter of reinforcing bars, in.
σ	normal stress, ksi
σ_{allow}	allowable stress, ksi
σ_{cr}	$= P_{cr}/A =$ nominal buckling stress in column, ksi
σ_c	concrete stress, ksi
σ_{in}	initial or locked-in stress in member, ksi
σ_{PL}	proportional limit stress, ksi
σ_s	steel stress, ksi
τ	shear stress, ksi
ϕ	curvature (rate of change of slope of deformed member), 1/in.; strength reduction factor in ultimate strength design; stress function

Index